THE MOLECULAR BASIS OF CELL CYCLE AND GROWTH CONTROL

Edited By

GARY S. STEIN
Department of Cell Biology and Cancer Center
University of Massachusetts Medical Center
Worcester, Massachusetts

RENATO BASERGA
Kimmel Cancer Center
Thomas Jefferson University
Philadelphia, Pennsylvania

ANTONIO GIORDANO
Department of Pathology
Sbarro Institute for Cancer Research and Molecular Medicine
Thomas Jefferson University
Philadelphia, Pennsylvania

DAVID T. DENHARDT
Nelson Biological Laboratories
Rutgers University
Piscataway, New Jersey

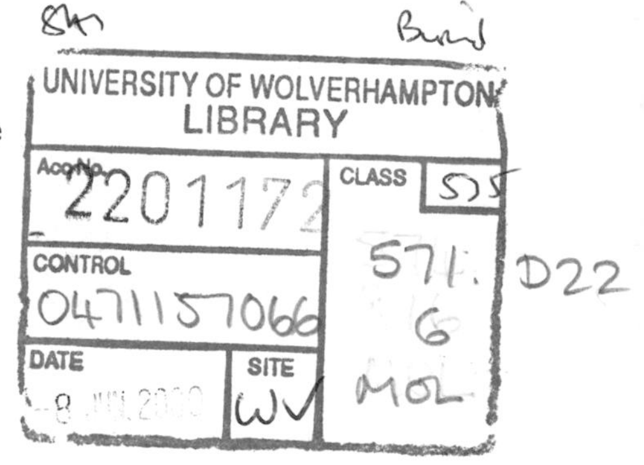

A JOHN WILEY & SONS, INC., PUBLICATION

New York · Chichester · Brisbane · Toronto · Singapore · Weinheim

This book is printed on acid-free paper. ⊗

Published simultaneously in Canada.

Library of Congress Cataloging-in-Publication Data:

The molecular basis of cell cycle and growth control / edited by Gary
S. Stein . . . [et al.].
p. cm.
Includes index.
ISBN 0-471-15706-6 (cloth)
1. Cell cycle. 2. Cellular control mechanisms. 3. Cellular
signal transduction. 4. Cell differentiation—Molecular aspects.
I. Stein, Gary S.
QH604.M6 1998
571.8′4—dc21 98-27547

Printed in the United States of America.

10 9 8 7 6 5 4 3 2 1

CONTENTS

PREFACE

During the past several years, significant advances have been made in understanding control of proliferation at the cellular, biochemical, and molecular levels. As the complexity and interdependency of parameters mediating growth control become increasingly apparent, the demarcations between regulated and regulatory components of mechanisms that promote or inhibit proliferation are eroding. There is recognition of both cause and effect relationships between the activities of factors that modulate the cell division cycle reflecting multidirectional signaling between segments of regulatory cascades that are selectively operative in specific cells and tissues. The necessity of accommodating the integration of positive and negative growth regulatory signals is now appreciated in a broad spectrum of biological contexts. These include but are not restricted to (1) repeated traverse of the cell cycle for cleavage divisions during the initial stages of embryogenesis and continued renewal of stem cell populations, (2) stimulation of quiescent cells to proliferate for tissue remodeling and wound healing, and (3) exit from the cell cycle with the option to subsequently proliferate or terminally differentiate. Equally important is an appreciation of the cell cycle regulatory mechanisms that have been compromised in transformed and tumor cells and in nonmalignant disorders where abnormalities in cell cycle and/or growth control are operative. These insights into regulatory mechanisms operative in cell cycle and growth control are rapidly being translated to clinical diagnosis and therapeutics.

Proliferation is explored in this volume from the perspectives of conceptual and experimental approaches to fundamental principles as well as biomedical applications. The biological problem of cell cycle control is introduced within the historical context of contributions from a broad spectrum of model systems that have provided the foundation for our current understanding of growth control in mammalian cells. The identification of checkpoints and restriction points as well as the factors responsible for proliferation competency and cell cycle progression are presented. The chapters focus on DNA replication and S phase; mitosis and meiosis; regulation of gene expression; growth factors and growth factor receptor–mediated pathways; signal transduction and integration of regulatory information; differentiation; development and programmed cell death; and cellular senescence and immortalization. The concluding chapter addresses cancer and other disorders that are linked to perturbations in control of proliferation. Each chapter was developed with a balance between presentation of biological parameters and gene regulatory mechanisms that mediate growth control and cell cycle progression. A concerted effort has been made to provide a comprehensive bibliography and illustrations that assimilate components of the increasing complexity associated with proliferation.

THE EDITORS

CONTRIBUTORS

RENATO BASERGA, Kimmel Cancer Institute, Thomas Jefferson University, 233 South 10th Street, 624 Bluemle Life Sciences Building, Philadelphia, PA 19107-5541

BRUNO CALABRETTA, Thomas Jefferson University, Department of Microbiology and Immunology, Kimmel Cancer Center, Bluemle Life Sciences Building, Room 630, Philadelphia, PA 19107

JUDITH CAMPISI, Berkeley National Laboratory, 1 Cyclotron Road, Mailstop 70A-1118, Berkeley, CA 94720

M. CRISTINA CARDOSO, Franz Volhard Clinic, Max Delbrueck Center for Molecular Medicine, Humboldt University, Wiltbergstr. 50, 13122 Berlin, Germany

DAVID T. DENHARDT, Department of Cell Biology and Neuroscience, Nelson Biological Laboratories, Rutgers University, Piscataway, NJ 08854

BARUCH FRENKEL, Department of Orthopedics & Institute for Genetic Medicine, University of Southern California School of Medicine, 2250 Alcazar Street, Los Angeles, CA 90033

ANTONIO GIORDANO, Department of Pathology, Sbarro Institute for Cancer Research and Molecular Medicine, Jefferson Medical College, Thomas Jefferson University, Philadelphia, PA 19107

EDWARD H. HINCHCLIFFE, Department of Cell Biology, University of Massachusetts Medical Center, 55 Lake Avenue North, Worcester, MA 01655

DENNET HUSHKA, 1708 William & Mary Common, Hillsboro, NJ 08876

HEINRICH LEONHARDT, Franz Volhard Clinic, Max Delbrueck Center for Molecular Medicine, Humboldt University, Wiltbergstr. 50, 13122 Berlin, Germany

M. LEVRERO, Istituto di Medicina Interna, Universita degli Studi di Cagliari, Cagliari, Italy

JANE B. LIAN, Department of Cell Biology and Cancer Center, University of Massachusetts Medical Center, 55 Lake Avenue North, Worcester, MA 01655

T.K. MACLACHLAN, Department of Pathology, Sbarro Institute for Cancer Research and Molecular Medicine, Jefferson Medical College, Thomas Jefferson University, Philadelphia, PA 19107

P.L. PURI, Fondazione Andrea Cesalpino and Istituto I Clinica Medica, Policlinico Umberto I, Universita degli Studi di Roma "La Sapienza," Rome, Italy

G. Prem Veer Reddy, Senior Staff Scientist, Henry Ford Health System, 1 Ford Place, 4D, Detroit, MI 48202

Conly L. Rieder, Department of Biomedical Sciences, State University of New York, Albany, NY 12222

Tomasz Skorski, Thomas Jefferson University, Department of Microbiology and Immunology, Kimmel Cancer Center, Bluemle Life Sciences Building, Room 630, Philadelphia, PA 19107

Greenfield Sluder, Department of Cell Biology, University of Massachusetts Medical Center, Shrewsbury Campus, 222 Maple Avenue, Shrewsbury, MA 01545

Gary S. Stein, Department of Cell Biology and Cancer Center, University of Massachusetts Medical Center, 55 Lake Avenue North, Worcester, MA 01655

Janet L. Stein, Department of Cell Biology and Cancer Center, University of Massachusetts Medical Center, 55 Lake Avenue North, Worcester, MA 01655

André J. van Wijnen, Department of Cell Biology and Cancer Center, University of Massachusetts Medical Center, 55 Lake Avenue North, Worcester, MA 01655

CHAPTER 1

INTRODUCTION TO THE CELL CYCLE

RENATO BASERGA
Kimmel Cancer Institute, Thomas Jefferson University, Philadelphia, PA 19107-5541

HISTORY OF THE CELL CYCLE

The First Days of Creation

Like Sleeping Beauty, the study of cell division slept for almost a hundred years after mitosis was first described in 1875. Despite countless cytological descriptions of mitosis in its classic four stages (prophase, metaphase, anaphase, telophase), very little progress was made on its biochemical basis, and interphase (the status of cells when not in mitosis) simply remained terra incognita. The measure of our ignorance is best illustrated by an example. It is now well established that colchicine (or Colcemid) blocks cells in mitosis; it is, in fact, used to synchronize cells in mitosis so that we can collect almost pure populations of mitotic cells. But back in the 1930s, scientists, having observed that the number of mitoses increased sharply in tissues of animals given injections of colchicine, came to the conclusion that colchicine was . . . mitogenic, a not unreasonable conclusion given the circumstances. So, let us follow, briefly, how we have come from such a naive view of cell division to the sophisticated cell cycle of the molecular biologist.

The first breakthrough, indeed, the very concept of a cell cycle, came from Alma Howard and Steve Pelc in 1951. Using ^{32}P to label the roots of *Vicia faba* seedlings and autoradiography, Howard and Pelc (1951) concluded that DNA was synthesized during a discrete period of the interphase, that the interval between mitosis and DNA synthesis was a long one, and that the interval between DNA synthesis and mitosis was a short one. Their paper succinctly layed the basis for the four phases of the cell cycle that we all know: G_1, between mitosis and S phase (DNA synthesis); S phase; G_2, between S phase and mitosis; and mitosis itself. The almost contemporary demonstration

The Molecular Basis of Cell Cycle and Growth Control, Edited by G.S. Stein, R. Baserga, A. Giordano, and D.T. Denhardt
ISBN 0-471-15706-6, pages 1–14.

that DNA is the genetic material of cells made their observation even more important: The cell, before dividing, replicated its genetic material. The method used by Howard and Pelc to label cells was, so to speak, primitive, requiring digestion with RNase to distinguish DNA from RNA synthesis, and made even more difficult by the use of ^{32}P, a high energy emitter that gave only an approximate intracellular localization of the radioactivity.

The second breakthrough came with the introduction of high-resolution autoradiography with tritiated thymidine (^{3}H-Tdr). The advantage of ^{3}H-Tdr was twofold: Tritium is a weak emitter and its label can be localized precisely not only in cells but also in cell compartments, and Tdr is a specific precursor of DNA. The original paper by Hughes et al. (1958) merits some discussion. For many years, histologists had observed that, in the lining epithelium of the small intestine of rodents, all mitoses occurred at the bottom, in an area called the *crypts,* while no mitoses were found in the lining epithelium of the villi, intestinal folds that project into the intestinal lumen. Imagine yourself again in the place of an investigator of the 1930s. The number of epithelial cells in the crypts of the adult rodent is constant, so what happens to the siblings of a mitotic cell? Does one of the siblings die in the crypt? Or does it migrate along the villi? The use of autoradiography with ^{3}H-Tdr solved the problem because DNA, once synthesized, remains stable until the cell dies and because Tdr in the living animal, if not incorporated into DNA, is broken down to nonutilizable products in 45 minutes. Only crypt cells were labeled within 30 minutes after an injection of ^{3}H-Tdr, indicating that the epithelial cells of the villi did not synthesize DNA. But, when the animals were sacrificed 24 hours after the single injection of ^{3}H-Tdr, labeled epithelial cells were now visible along the villi, where, with time, they progressed slowly to the tips of the villi before being sloughed off into the intestinal lumen, roughly 48 hours after the initial labeling. It showed that the dividing cells originally labeled in the crypts moved out of the crypts and replaced the lining epithelium of the villi, pushing the older cells upward toward the lumen of the intestine. The similarity of cellular progression in the mucosa of the small intestine to academic or corporate careers is striking: One slowly glides to the top, against much resistance, and, when the summit is finally reached, one is simply discarded into oblivion. Armed with ^{3}H-Tdr, it was not difficult for subsequent investigators to define the phases of the cell cycle in a variety of cells.

The third breakthrough came when Stanley Cohen (1962) purified from the salivary glands of mice a polypeptide that caused early eruption of the incisor teeth and precocious opening of the eyelids in newborn mice. These activities do not strike us as particularly exciting, but what Stanley Cohen had done was to purify the first growth factor, which he called *epidermal growth factor* (EGF). So, by 1962, we had the cell cycle, we had cells that moved from one phase of the cell cycle to another or even out of the cell cycle, and now Stanley Cohen had found the force that moved the cell cycle: growth factors.

The Growth of the Cell Cycle: Kinetics

It would be impossible to give credit to the many scientists who slowly built upon the concepts outlined above: more refined cell kinetics, populations of cells, mathematical models, more growth factors, biochemistry of the cell cycle, genetics of the cell cycle, molecular biology of the cell cycle. But some highlights should be mentioned. Baserga et al. (1960) were the first to apply cell kinetics to the study of tumor growth, and, surprisingly, they found that the cell cycle of tumor cells was not necessarily shorter than that of normal cells. This implied that other parameters, besides cell cycle length, were at work, parameters that allowed tumors to grow while normal adult tissues did not grow. Mendelsohn (1962) then introduced the concept of growth fraction: Not all cells are cycling; in fact, some are fated to die without dividing again, and others are in a state of inactivity from which they can re-enter the cell cycle if appropriately stimulated. We call the former ones *terminally differentiated cells,* the latter ones G_0 cells. It is now generally accepted that G_0 cells are biochemically different from G_1 cells and are therefore entitled to be separately classified. But a vigorous discussion of whether G_0 was simply a long G_1 or not, enlivened the cell cycle circles for many years, a discussion that produced countless papers but little information.

If you now look at Figure 1.1, it is easy to see that, when it comes to cell division, there are three populations of cells: the ones actively dividing (cycling cells), the ones optionally out of the cell cycle (G_O cells), and the ones that are destined to die without further divisions. The third group was all but neglected until recently, when apoptosis suddenly became the fashionable way for cells to die. But Bresciani et al. (1974) had shown a long time ago that even tumor cells die in large amounts, unfortunately not large enough to cause the tumors to regress. From Figure 1.1, we can draw a generalized rule for any increase in cell number. The three mechanisms that produce an increase in cell number of any given cell population are (1) a shortening of the cell cycle, i.e., the cells divide more frequently; (2) more cells participate in the cell cycle, i.e., a decrease in the G_0 fraction; and (3) a decrease in the rate of cell death.

The next question is whether an increase in cell number is the only mechanism for tissue and organismic growth. A tissue can grow by (1) increasing the number of cells, (2) increasing the size of the cells; or (3) increasing the amount of intercellular substance. Because intercellular substance of a tissue, for example, collagen or bone, is usually a secreted product of the cell, it can be considered as an extracellular extension of the cytoplasm. We can therefore take an increase in intercellular substance as a variation of an increase in cell size and reduce tissue growth to two mechanisms, growth in size and growth in the number of cells. This is true regardless of whether we are dealing with normal or abnormal growth, with the intact animal or with cells in culture. However, although both mechanisms may be operative, an increase in cell number is (with very few exceptions) by far the most important component in either

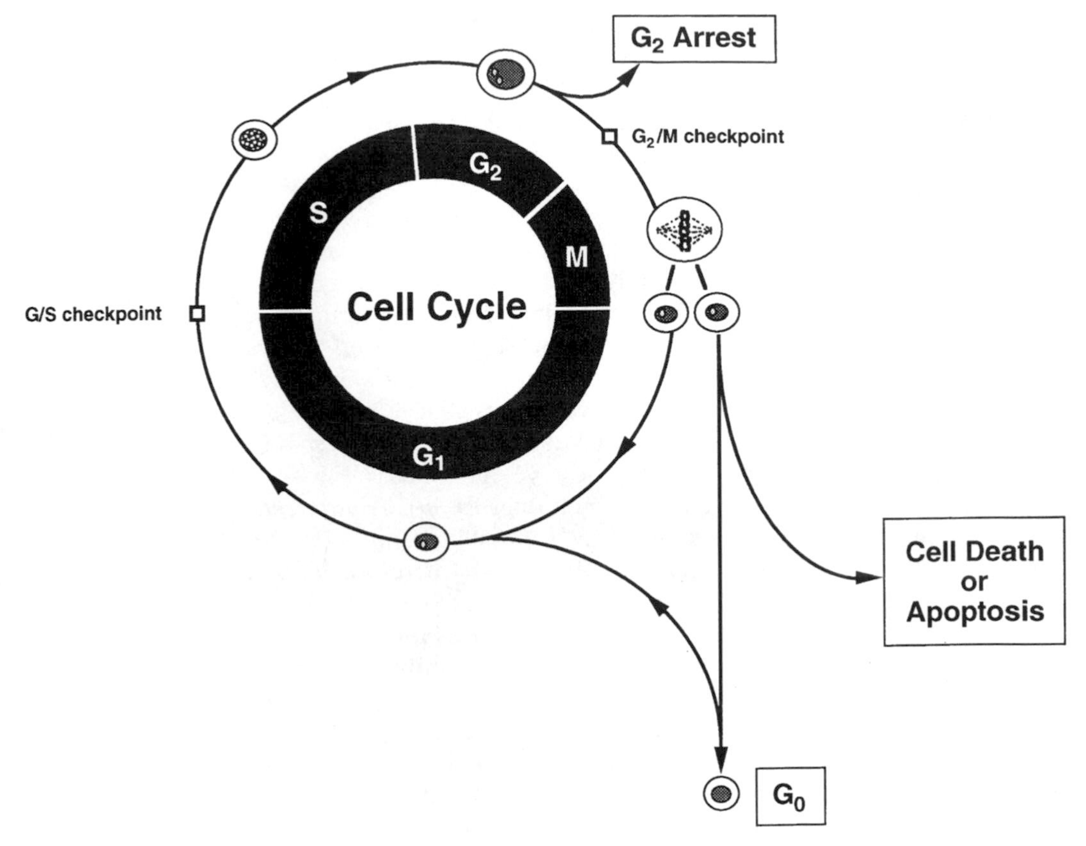

Figure 1.1. Cell Cycle. Note that, as the cell progresses through the cell cycle, there is also an increase in size until, just before mitosis, the size of the cell is twice that of an immediately postmitotic cell. G_2 arrest results in cells with tetraploid amounts of DNA.

normal or abnormal growth (Table 1.1). In abnormal growth, all three mechanisms are operating, but the bottom line is always the same: A tissue will grow only when the number of cells that are produced per unit time exceeds the number of cells that are lost in the same time.

There is another population of cells that deserves to be set apart, a typical example of the fact that some cells are more equal than others. The concept of stem cells originated with the pioneer work of McCulloch and coworkers. Their work identified in the bone marrow a category of pluripotent cells, capable of self-renewal, that could generate from a single cell all the lineages of hemopoietic cells (Becker et al. 1963). It is now accepted that stem cells are present in many other tissues, a reservoir of cells endowed with the ability to regenerate a cell population with several options for differentiation.

From Cell Kinetics to Molecular Biology

Until 1963, the cell cycle was essentially a study in cell kinetics, generating a great amount of useful information on a variety of cells, how long

TABLE 1.1. Postnatal Growth of Rat Liver

Age (Days)	Liver Weight (g)	Protein per Cell (pg)	DNA per Cell (pg)	Cell No. $\times 10^{-6}$
10	0.30	29.3	5.9	168
41	5.7	103.0	11.1	1,060
80	8.1	154	11.4	1,270
182	12.0	155	11.4	1,790

All values are averages of several animals. The protein content is a good measure of cell size. The DNA values at 41 days or after indicate that liver cells undergo polyploidization. Note that whereas cell size stops increasing at day 80 (or before), the number of cells keeps increasing.

Source: Adapted from Baserga (1985).

their cell cycles were, the different phases, the growth fraction, the whole properly spiced with mathematical formulations that looked very clever, but, in reality, were largely descriptive and had little predictive value. It was Irving Lieberman with his collaborators who first suggested that beneath the G_0, G_1, and G_2 phases were a series of biochemical events (Lieberman et al., 1963). A review by Baserga (1968) codified these first experiments and clearly showed that beneath the cell cycle was a whole textbook of biochemistry.

The obvious next steps would be to move to genetics and molecular biology. The genetics of the cell cycle were first studied in the seminal work of Hartwell and collaborators with the yeast *Saccharomyces cerevisiae.* Hartwell (1971), using temperature-sensitive (ts) mutants, showed that cell cycle progression was regulated by the timely expression of specific gene products. Similarly enlightening were the studies of Paul Nurse and coworkers (Beach et al., 1982) with another yeast, *Schizosaccharomyces pombe.* It is regrettable that while the use of ts mutants of the yeast cell cycle generated exceedingly valuable information, ts mutants of mammalian cells were much less informative, with very few exceptions (for a review, see Baserga, 1985). While in yeasts gene expression is cell cycle regulated, it quickly became apparent that in mammalian cells gene products were not only cell cycle regulated but also growth regulated, i.e., they were expressed only in growing cells (for the distinction, see below). Although Cochran et al. (1983) were the first to identify gene products whose expressions was regulated by growth factors, Calabretta et al. (1986) were the first ones to molecularly clone and publish a complete sequence of a mammalian growth-regulated gene (a calcium-binding protein, subsequently called *calcyclin*), i.e., of a gene whose expression was regulated by growth. This first report was followed by many, many others, until now there must be more than 500 genes whose expressions can be said to be growth regulated (see, for instance, the review by Hofbauer and Denhardt, 1991).

The latest developments, regarding cyclins, cyclin-associated proteins, oncogenes, tumor suppressor genes, signal transduction from growth factor receptors, are part and parcel of the present book, and, as such, they do not belong to a historical survey, but to the present.

RULES AND REGULATIONS OF THE CELL CYCLE

The progress of science depends, at least in part, on the fact that scientists define their terms, which makes for economy of language and for understanding across language barriers. Spirits can be heavenly or alcoholic, a musical pitch varies, the soprano prefers it low, the violinist likes it high, but when we say *cell cycle,* all biologists understand what we mean by it, even though the expression may not be perfectly accurate (years ago, a change was proposed to *nuclear cycle;* it really does not matter, as long as we define the term). So, it is helpful if we define terms related to the cell cycle, i.e., if we establish a few rules and regulations, some of which have been consecrated by use, others by common sense.

Mitogenicity

Cells synthesize DNA (i.e., enter S phase) before dividing (mitosis), but cells can synthesize DNA and not divide, something that occurs not too infrequently in both normal and abnormal conditions. Most investigators will say that, for instance, a growth factor is mitogenic because it increases the incorporation of ^{3}H-Tdr into an acid-insoluble fraction of the cell. What the incorporation of ^{3}H-Tdr into an acid-insoluble fraction measures is the incorporation of ^{3}H-Tdr into an acid-insoluble fraction, period. It is very distantly related to mitogenicity and is not even an accurate measurement of how many cells are actually entering S phase. Growth factors increase the permeability of cell membranes (Pardee et al., 1978) and stimulate kinases (e.g., thymidine kinase), which, in turn, change the deoxyribonucleoside triphosphate pool (Walters et al., 1973) so that a measurement of thymidine incorporation into an acid-insoluble fraction can tell us as much about membrane function as about DNA synthesis.

Determining the number of cells entering S phase (by autoradiography, or by FACS analysis, or by bromodeoxyuridine [BUdR] incorporation) is much better, but it still tells us only how many cells enter S phase, not how many cells divide. If we wish to classify a growth factor, or its receptor, as mitogenic, we have to show that it induces mitosis, and by far the best way to show this is by the very old-fashioned method of counting cell numbers. One has to be careful, though, because we are used to a six- to eightfold increase in thymidine incorporation or a 20-fold increase in the number of cells labeled by either ^{3}H-Tdr or BUdR, whereas the increase in cell number may only be one- or twofold. The point is that, when the number of cells labeled, for instance, by BUdR in 24 hours, goes from 1% in unstimulated cells to 95% in stimulated cells, the increase in cell number will only be a doubling. The choice depends on the investigator's taste, whether he or she prefers a large increase in a less accurate value or a modest increase in the correct measurement.

In theory, the number of mitoses ought to be the best indicator of mitogenicity. However, mitoses are fleeting; in most cells they last only 45 minutes, and, unless one looks at precisely the right moment, one can miss them. Furthermore, the duration of mitosis can increase in certain cells, especially in transformed cells (Sisken et al., 1982). Everything else being equal, if the duration of mitosis in cell line A is twice that of cell line B, the number of mitoses in A will also be twice that of B, although the two cell lines may grow at the same rates. In tissues, the number of mitoses per 1,000 cells (the mitotic index) is a reasonable measure of the *proliferating activity* of a cell population, but not of its growth. For instance, in the crypts of the lining epithelium of the small intestine, there are many mitoses. Fortunately for us, the small intestine in the adult individual does not grow, because, for every new cell produced in the crypts, one dies at the tips of the villi. Again, the most reliable method of assessing an increase in cell number is the counting of cells.

The Wind and the Leaves

One of the most charming suggestions I ever read was James Thurber's poetic insistence that it is the leaves of the trees that generate the wind, because every time the leaves move you can feel the breeze coming. Some scientists are like James Thurber: They identify a growth-regulated gene, and they present it as a growth regulatory gene. The two things, of course, are not mutually exclusive, but they are not the same thing, either (for a discussion, see Baserga, 1990). The best illustration is the regulation of expression of the thymidine kinase gene. In growth-regulated cells (like 3T3 cells and human diploid fibroblasts) the expressions of thymidine kinase RNA and protein are not only growth-regulated, but are exquisitely cell cycle regulated (Coppock and Pardee, 1987): They both increase at the G_1/S boundary and decrease when the cell reaches mitosis. Yet thymidine kinase is no growth regulatory gene product: It is not even necessary for cell proliferation; many cells and even some animals grow perfectly normally in the complete absence of thymidine kinase (discussed in Baserga, 1985). There are many genes like that; in fact, one ought to say that if the expression of a gene changes during the cell cycle, or even under different growth conditions, all it tells us is that the expression of that gene is growth regulated or cell cycle regulated. To be growth regulatory, a gene must be capable of altering the growth rate of a cell population.

Distinction Between Cell Cycle–Regulated and Growth-Regulated Genes

This brings us to another important distinction if we wish to understand each other, especially in studies dealing with mammalian cells, and this is the distinction between cell cycle–regulated and growth-regulated genes (Hofbauer and Denhardt, 1991). There are certain genes that are cell cycle regulated (i.e., the proteins are expressed in a cell cycle–dependent manner). For instance, in most mammalian cells, thymidine kinase (Coppock and Pardee, 1987) and histones (Plumb et al., 1983) are expressed only during the S phase of the cell cycle. However, there are many other genes that are simply not expressed in nongrowing cells (G/0 cells) but are differentially expressed when the cells are stimulated to grow. However, once activated, these genes are not cell cycle regulated; in other words, they continue to be expressed as long as the cells are proliferating, without any particular localization to one or the other phase of the cell cycle. Typical of these genes are, for instance, c-myc, vimentin, ornithine decarboxylase, and many others (see review by Hofbauer and Denhardt, 1991). The distinction between these two types of regulated genes is important because it implies diverse regulatory mechanisms.

When Is a Gene a Regulatory Gene of the Cell Cycle?

Thymidine kinase is superfluous: You can delete the gene, and the cells will still grow in a perfectly normal way (see above). But suppose you find another gene that, deleted or inhibited, causes growth arrest; can you call it growth regulatory? This is an important question because there are many gene products in this category, notably the DNA synthesis genes, the cyclins, the cyclin-associated proteins, protooncogenes, and, indeed, many other gene products. Literally, we can say that these genes are growth regulatory because when they are inhibited or suppressed the cells are arrested in their cell cycle progression. But the cells will also stop cycling if you decrease the supply of adenosine triphosphate (ATP), and there are a great number of gene products that behave similarly: When they are inhibited (e.g., by antibodies, or dominant negatives, or antisense oligodeoxynucleotides, or in other ways), the cells stop their progression in the cell cycle. If anything that is required for cell cycle progression is growth regulatory, then we should include ATP and at least a thousand other molecules in this group. Such a broad definition would have scarce informational content (Baserga, 1990).

Perhaps a more rigorous definition of a growth regulatory gene is in order. I have a suggestion. Let us call the genes that are required for cell cycle progression exactly that: *genes required for cell cycle progression.* And let us reserve the term *growth regulatory* only for those genes (a handful) that actually decide whether a cell will enter the cell cycle or not. Let me illustrate this suggestion with an analogy: (1) You can drive a car without windows. It may not be comfortable in winter, but it can be done. That is thymidine kinase. (2) You cannot drive a car without the steering wheel, the four wheels, the engine, the differential, or gasoline. These items are the equivalent of the genes required for cell cycle progression. (3) But you can have all this, and more, and the car, shining in all its splendour, will sit in G_0 in the showroom of the car dealer. Somebody has to turn the ignition key and press the gas pedal to move that car, and that is, for me, a growth regulatory gene. The fact that, under physiological conditions, I think that growth factor receptors activated by their ligands qualify as growth regulatory may reflect my prejudice, or it may not. Inside the cell, the best candidate for growth control is probably the inhibitor of cyclin-dependent kinases, $p27^{kip1}$ (Sherr and Roberts, 1995; Russo et al., 1996).

UNIVERSALITY OF THE CELL CYCLE

The Cell Cycle in Different Cells

I do not know of any *proliferating* eukaryotic cell that does not have a cell cycle. In fact, prokaryotic cells also have a cell cycle, which becomes similar to that of eukaryotic cells when the growth of bacterial cells is slowed down by nutritional manipulations. In this case, bacteria show the presence of a G_2 phase, although the S phase still occupies a preponderant fraction of the cycle (Helmstetter, 1971). Furthermore, from bacteria comes the discovery of thymineless death, which is now used extensively both at the basic science level and for therapeutic interventions. The yeast cells have been a gold mine for our understanding of the cell cycle (see Section 1.1.3.), and the amount of information and insight provided by the use of yeast cells cannot be overemphasized. Their greatest contribution has been the discovery and the elucidation of the role of cyclins and cyclin-associated proteins in cell cycle progression. Because these are the subject of another chapter, I will not discuss them here, but this should not be taken as a disregard for them. On the contrary, I will venture to say that the yeast cell cycle has provided the frame for our understanding of the mammalian cell cycle at the molecular biological level.

Returning to mammalian cells, there are cells that do not have a G_1 phase and others that do not have a G_2 phase, but all dividing cells have three features that are universal: They must grow in size, they must replicate the genetic material, and they must undergo mitosis. Growth in size is self-evident: If the cells were not to double in size before dividing, they would get smaller and smaller at every division and eventually vanish. In yeast, the evidence is substantial that cell size is a determinant for entry into S phase and mitosis (Hartwell, 1978). In mammalian cells, one can occasionally have division without concomitant growth in size, but it is exceptional and hardly normal. The S phase, during which the genetic material is replicated, is obligatory, although the duration may vary from a few minutes in embryo cells (Graham and Morgan, 1966) to several hours in tumor cells (see references in Baserga, 1985). Needless to say, the whole cell cycle in early embryo cells is also very short, something like 14 minutes.

Some cells may grow in size, double their DNA content, and stop in G_2. This happens not unfrequently during normal development, especially in certain organs like the liver and the heart. In this case, the cells become larger and increase their ploidy (i.e., their DNA content), sometimes severalfold over a diploid amount.

But there are more subtle ways in which different cells have different cell cycles, and that is the way, for instance, in which they regulate gene expression. Again, thymidine kinase is a good example of diversity. In growth-regulated fibroblasts (like 3T3 cells or human diploid fibroblasts), thymidine kinase expression is regulated at the transcriptional level: In G_0 cells, there is no protein, no mRNA, and no transcripts (Lipson and Baserga, 1989). In differentiated muscle cells, thymidine kinase expression is regulated at a translational level: Mature mRNA is present, but there is no protein (Gross and Merrill, 1988). Similarly, in HeLa cells, the mRNA levels change very little during the cell cycle, and thymidine kinase expression is regulated at a translational level (Kauffman et al., 1991). The point here is that different cells have different ways of regulating the expression of growth-related and cell cycle–regulated genes. There are many more instances besides thymidine kinase. One only has to think about genes that are expressed only in certain cell types, for instance, genes that are specific for hemopoietic cells or growth factor receptors, also specific for different cell types.

One pitfall that must be carefully avoided is the religious belief that "my *cell* is better than yours." One can certainly learn a lot about the cell cycle, working with yeasts. But if one is interested in how growth factors regulate cell proliferation, yeasts are a poor choice because, in the succinct words of Gabor Miklos and Rubin (1996) "receptor tyrosine kinases appear to be a metazoan invention." It is more reasonable, and probably also more scientific, to accept the fact that each cell type has features that can be exploited to learn molecular mechanisms and that, as long as we do not generalize our findings and pretend to extrapolate them to the universe of cells, each type has its advantages and disadvantages. Indeed, I would like to repeat here what I have been teaching my younger associates for many years, which, unfortunately, is forgotten by too many: What we find in our cells applies only to our cells, and only under the conditions we have used, until otherwise proven.

In Vivo Versus In Vitro

Another debate that recurs constantly involves the validity of the cell cycle of cells in culture vis à vis the reality of the living animal. This debate is an extension of the debate on the universality of the cell cycle, discussed in the preceding section. Investigators working with animals often display a sovereign contempt for those who work with cells in culture. But then, investigators working with yeasts have the same contempt for anyone working with anything else. The solution to this debate is actually very simple and is the same as in the preceding section. If you wish to study the cell cycle of the lining epithelium of the small intestine of mice, there is only one way to study it, and that is in a mouse. But if your concern is about mechanisms, gene expression, growth factors, and so on, then yeasts and other cells in culture are the best choice. One only has to be very careful in avoiding extrapolations. One of the most revealing findings of the past few years is the discrepancy that often occurs when a gene is deleted in the embryo versus its effect on cells in culture. Quite a few genes that seem to regulate cell proliferation in vitro gave no growth phenotype when they were deleted by targeted disruption (knockout mice), or they gave a very limited growth phenotype, limited, for instance, to a tissue or organ.

Undoubtedly, the environment of cells in culture resembles the environment of cells in vivo no more than a zoo resembles an African habitat. The Petri dish is a hostile environment, and when cells are asked to grow in that environment they pull out all stops and start expressing all kinds of genes they do not express in the adult animal. Tolerance is again the watchword. One has to be cautious not to overinterpret the in vitro findings, but, if all experiments were to be done in the living animal, apart from ethical considerations, progress would be much more slow.

The Time Puzzles

Sometimes there is a correlation between time of appearance of a gene product and its function in the cell cycle. For instance, since the time of Lieberman and coworkers (see above), it is known that in growth-stimulated cells at least some of the enzymes involved in DNA synthesis make their appearance at the onset of S phase. But an extrapolation to all gene products is unwarranted. The best-documented example is that of ribonucleotide reductase in the regenerating liver. Ribonucleotide reductase is one of the enzymes required for DNA replication and entry into S phase. However, in the regenerating rat liver, ribonucleotide reductase reaches its peak of activity at 50 hours after partial hepatectomy, when DNA synthesis has essentially ceased (Larsson, 1969). There are other examples (see Baserga, 1985), and one should be cautious in assigning a specific function in cell cycle progression to a gene product simply on the basis that it makes its appearance at the same time as the function under consideration.

Still in the realm of time, one of the most puzzling aspects of the cell cycle is the long time it takes for a quiescent cell (G_0) to enter S phase, close to 10–12 hours, regardless of the stimulus used. If the cell cycle is a succession of enzymatic actions (phosphorylations, dephosphorylations, and so forth), why does it take so long, when blood clotting requires more than a dozen enzymes, and yet blood can clot in 15 seconds? One is tempted to speculate that the cell is counting something that must be accumulated in sufficient amounts to trigger cell cycle progression, a hypothesis that cell cycle devotees have been discussing for several years. The problem is that nobody has a clue as to what the cell may be counting.

Is There Life After Cyclins?

In the past few years, enormous progress has been made on one aspect of the cell cycle that deals with cyclins, cyclin-associated kinases and other cyclin associated proteins. This work, sparked by yeast genetics (see above), has yielded a large amount of very valuable information that is detailed and discussed in another chapter of this book. The importance of this work can hardly be overestimated: The interaction of these various proteins makes for a coherent, fascinating story that devolves almost literally in front of our eyes like a movie. As a veteran of the cell cycle, however, let me sound a word of caution. In nearly 40 years of experience, I have seen many, many things that, at one time or another, were touted as *The* regulator of the cell cycle: amount of ribosomal RNA, number of ribosomes, histone acetylation, histone deacetylation, histone phosphorylation, histone dephosphorylation, cyclic AMP, cyclic GMP, proteases, cell size, nuclear size, DNA polymerase, nonhistone chromosomal proteins, intracellular pH, Na^+ fluxes, ATP, calcium, calmodulin, magnesium, ornithine decarboxylase, p53, c-myc, c-fos, membrane glycoproteins, membrane glycolipids. . . . The list is not intended to disparage the various investigators who proposed the candidates; I have contributed to it myself. The point is that, one by one, each of these regulators have lost some of their glamour and have, so to speak, re-entered their ranks as very valuable components of cell cycle progression, indeed, required for cell cycle progression, but not *The* regulator of the cell cycle.

At present, cell cycle investigators pursue separate lines of investigation, and there are several cell cycles: There is a cyclin cell cycle, a p53 cell cycle, an oncogene cell cycle, a tumor suppressor gene cell cycle, a growth factor cell cycle, and perhaps several others. The challenge for the future is to bring together these separate cell cycles because all of them are valid, and, obviously, all of them are real. The association between the cyclin cell cycle and certain tumor suppressor genes has been a reality for several years (Sherr, 1993), but in other areas the connections are tenuous. Some progress is being made. A relationship between p53 and growth factor receptors, especially the insulin-like growth factor IGF-I receptor, has already been made (Werner et al. 1996), and there have been sporadic reports that the expression of certain cyclins and cyclin-associated proteins is regulated by specific growth factors, the most important being the induction of $p27^{kip1}$ by transforming growth factor-β (Sherr and Roberts, 1995). But the gaps are more numerous than the filled in boxes. Somewhere, somehow, we must relate these various cell cycles to each other in a balanced picture that takes into consideration all the various aspects.

REFERENCES

Baserga R (1968): Biochemistry of the cell cycle: A review. Cell Tissue Kinet 1:167–191.

Baserga R (1985): The Biology of Cell Reproduction. Cambridge, MA: Harvard University Press.

Baserga R (1990): The cell cycle: Myths and realities. Cancer Res 50:6769–6771.

Baserga R, Kisieleski WE, Halvorsen K (1960): A study on the establishment and growth of tumor metastases with tritiated thymidine. Cancer Res 20:910–917.

Beach D, Durkacz B, Nurse P (1982): Functionally homologous cell cycle control genes in budding and fission yeasts. Nature 300:706–709.

Becker AJ, McCulloch EA, Till JE (1963): Cytological demonstration of the clonal nature of spleen colonies derived from transplanted bone marrow cells. Nature 197:452–454.

Bresciani F, Paoluzzi R, BenAssi M, Nervi C, Casale C, Ziparo E (1974): Cell kinetics and growth of squamous cell carcinomas in man. Cancer Res 34:2405–2415.

Calabretta B, Battini R, Kaczmarek L, de Riel JK, Baserga R (1986): Molecular cloning of a cDNA for a growth factor–inducible gene with strong homology to S-100, a calcium-binding protein. J Biol Chem 261:12628–12632.

Cochran BH, Reffel AC, Stiles CD (1983): Molecular cloning of gene sequences regulated by platelet-derived growth factor. Cell 33:939–947.

Cohen S (1962): Isolation of a mouse submaxillary gland protein accelerating incisor eruption and eyelid opening in the newborn animal. J Biol Chem 237:1555–1562.

Coppock DL, Pardee AB (1987): Control of the thymidine kinase mRNA during the cell cycle. Mol Cell Biol 7:2925–2932.

Gabor Miklos GL, Rubin GM (1996) The role of the genome project in determining gene function: Insights from model organisms. Cell 86:521–529.

Graham CF, Morgan RW (1966): Changes in the cell cycle during early amphibian development. Dev Biol 14:439–460.

Gross MK, Merrill GF (1988): Regulation of thymidine kinase protein levels during myogenic withdrawal from the cell cycle is independent of mRNA regulation. Nucleic Acids Res 16:11625–11643.

Hartwell LH (1971): Genetic control of the cell division cycle in yeast. J Mol Biol 59:183–194.

Hartwell LH (1978): Cell division from a genetic perspective. J Cell Biol 77:627–637.

Helmstetter CE (1971): Coordination between chromosome replication and cell division: Cellular response to inhibition of DNA synthesis. In Mihich E (ed): Drugs and Cell Regulation. New York: Academic Press, pp 1–13.

Hofbauer R, Denhardt DT (1991): Cell cycle–regulated and proliferation stimulus–responsive genes. Crit Rev Eukaryot Gene Expression 1:247–300.

Howard A, Pelc SR (1951): Nuclear incorporation of P^{32} as demonstrated by autoradiographs. Exp Cell Res 2:178–187.

Hughes WL, Bond VP, Brecher G, Cronkite EP, Painter RB, Quastler H, Sherman FG (1958): Cellular proliferation in the mouse as revealed by autoradiography with tritiated thymidine. Proc Natl Acad Sci USA 44:476–483.

Kauffman MG, Rose PA, Kelly TJ (1991): Mutations in the thymidine kinase gene that allow expression of the enzyme in quiescent (G_0) cells. Oncogene 6:1427–1435.

Larsson A (1969): Ribonucleotide reductase from regenerating rat liver. Eur J Biochem 11:113–121.

Lieberman I, Abrams R, Ove P (1963): Changes in the metabolism of ribonucleic acid preceding the synthesis of deoxyribonucleic acid in mammalian cells cultured from the animal. J Biol Chem 238:2141–2149.

Lipson KE, Baserga R (1989): Transcriptional activity of the human thymidine kinase gene determined by a method using the polymerase chain reaction and an intron specific probe. Proc Natl Acad Sci USA 86:9774–9777.

Mendelsohn ML (1962): Autoradiographic analysis of cell proliferation in spontaneous breast cancer of C3H mouse. III. The growth fraction. J Natl Cancer Inst 28:1015–1029.

Pardee A, Dubrow R, Hamlin JL, Kletzien RF (1978): Animal cell cycle. Annu Rev Biochem 47:715–750.

Plumb M, Stein J, Stein G (1983): Coordinate regulation of multiple histone mRNAs during the cell cycle in HeLa cells. Nucleic Acids Res 11:2391–2410.

Russo AA, Jeffrey PD, Patten AK, Massague J, Pavletich NP (1996): Crystal structure of the $p27^{kip1}$ cyclin-dependent kinase inhibitor bound to the cyclin A–Cdk2 complex. Nature 382:325–331.

Sherr CJ (1993): Mammalian G1 cyclins. Cell 73:1059–1065.

Sherr CJ, Roberts JM (1995): Inhibition of mammalian G1 cyclin-dependent kinases. Genes Dev 9:1149–1163.

Sisken JE, Bonner SV, Grasch SD (1982): The prolongation of mitotic stages in SV40-transformed vs. nontransformed human fibroblast cells. J Cell Physiol 113:219–223.

Walters RA, Tobey RA, Ratcliff RL (1973): Cell cycle–dependent variations of deoxyribonucleoside triphosphate pools in Chinese hamster cells. Biochim Biophys Acta 319:336–347.

Werner H, Karnieli E, Rauscher FJ III, LeRoith D (1996): Wild type and mutant p53 differentially regulate transcription of insulin-like growth factor I receptor gene. Proc Natl Acad Sci USA 93:8313–8323.

CHAPTER 2

THE INTRINSIC CELL CYCLE: FROM YEAST TO MAMMALS

P.L. PURI, T.K. MACLACHLAN, M. LEVRERO, AND A. GIORDANO

Fondazione Andrea Cesalpino and Istituto I Clinica Medica, Policlinico Umberto I, Università degli Studi di Roma "La Sapienza," Rome, Italy (P.L.P., M.L.); Dept. of Pathology, Sbarro Institute for Cancer Research and Molecular Medicine, Jefferson Medical College, Philadelphia, PA 19107; (T.K.M., A.G.); Istituto di Medicina Interna, Università degli Studi di Cagliari, Cagliari, Italy (M.L.)

INTRODUCTION

The transmission of genetic information from one cell generation to the next requires an accurate replication of the DNA during **S phase** and the faithful segregation of the resultant sister chromatids during **mitosis.** In most eukaryotic cells these two events are normally dependent on each other, and replication of the genome and mitosis occur in alternative oscillating cycles. In addition, a tight control of the number of cell divisions is required for the development and the maintenance of a multicellular organism. Post mitotic growth of several terminally differentiated cells, instead, excludes the progression into the cell cycle and results in other processes such as hypertrophy, cell death regulation, and senescence. Precise coordination of the progression through cell cycle phases is critical not only for normal cell division but also for effective growth arrest under conditions of stress or after DNA damage. As a consequence, a derangement in the cell cycle machinery may contribute to uncontrolled cell growth, which is the principal feature of cancer. In this chapter, the current knowledge of the mechanisms that regulate the cell cycle progression during several biological processes is presented, with a particular emphasis on the function of cyclin/cyclin-dependent kinase complexes. In higher eukaryotes the choice of the cell between proliferation, cell growth arrest, and differentiation is dictated by a complex network of mitogenic and antimitogenic extracellular signals. Following

The Molecular Basis of Cell Cycle and Growth Control, Edited by G.S. Stein, R. Baserga, A. Giordano, and D.T. Denhardt
ISBN 0-471-15706-6, pages 15–79.

mitogenic stimulation, cells progress through the cell cycle and undergo several discrete transitions. Precise coordination of the progression through cell cycle phases is critical not only for normal cell division but also for effective growth arrest under conditions of stress or after DNA damage, and a derangement in the cell cycle machinery may contribute to uncontrolled cell growth, which is the principal feature of cancer.

During progression through the cell cycle, cells undergo several discrete transitions. **Cell cycle transitions** can be defined as *unidirectional changes in which a cell shifts its activity to perform a new set of processes.* How these transitions are coordinated, how DNA replication is initiated, how it is restricted to the S phase, and how replication occurs only once per cell cycle in most eukaryotic cells are current focuses of cell cycle research. Ordering of cell cycle events may be accomplished, according to Hartwell and Weinert (1989) by two alternative mechanisms: (1) the product of an early event is the substrate for a later event (the so-called **substrate–product** relationship) and (2) regulatory mechanisms, termed **checkpoints,** may delay later events until earlier events are complete. Checkpoints are *biochemical pathways that ensure dependence of one process on another process that is otherwise biochemically unrelated.* Although the term *checkpoint* is often used to indicate points in the cell cycle or cell cycle transitions, the usage should be restricted to genetic and/or biochemical pathways that ensure dependency (Nasmyth, 1996). Thus, the DNA damage checkpoint is the mechanism that detects damaged DNA and generates signals that arrest cells in G_1, slows down the S phase, arrests cells in G_2, and induces the transcription of repair genes. To demonstrate the existence of checkpoint controls, Hartwell and Weinert used the G_2 phase arrest that occurs in response to DNA damage in the budding yeast *Saccharomyces cerevisiae* (after irradiation, wild-type yeast cells are delayed in the G_2 phase to allow the repair of DNA breaks) and looked for mutations that "relieved dependence" of mitosis on undamaged DNA. They identified the Rad9 mutation that allowed irradiated cells to enter mitosis with unrepaired chromosomes and predicted the existence of other cell cycle checkpoint genes (i.e., genes that can be mutated to give a checkpoint-deficient phenotype without significantly affecting cell viability).

The **eukaryotic cell cycle** can be divided into four distinct phases that contribute to the correct formation of two fully functional daughter cells. Several cell cycle transitions are dependent on the activity of cyclin-dependent kinases (CDKs), and their inhibition by some checkpoint pathways is needed for cell cycle arrest. (MacLachlan et al, 1995) These enzymes are composed of a serine–threonine kinase subunit, the CDK, and an activating subunit, the cyclin (Morgan, 1995) (Fig. 2.1). Cyclin/CDK complexes are subjected to many levels of regulation: The number of CDK genes varies in yeast and mammals; cyclins, which are absolutely required for kinase activity, contribute to substrate specificity; the activity of the CDK/cyclin complexes is both positively and negatively regulated by phosphorylation and dephosphorylation events; CDKs are further regulated by binding to inhibitors (CKIs) and other proteins such as

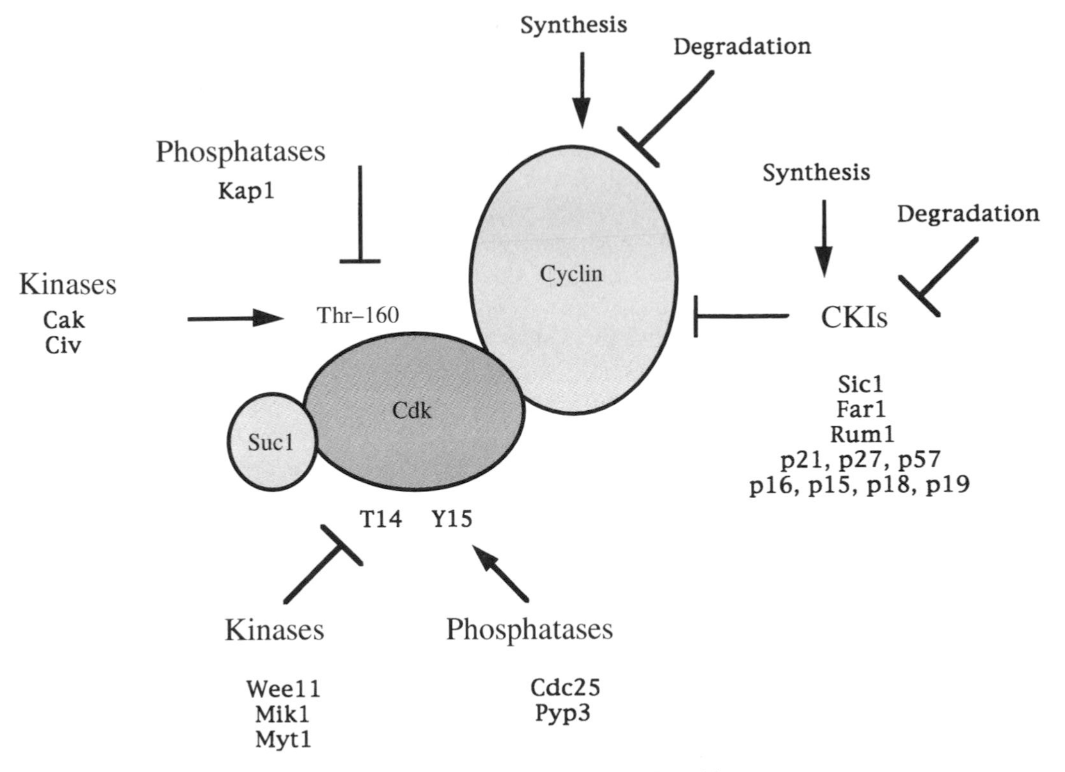

Figure 2.1. Various levels of regulation of cyclin-dependent kinases.

Suc1 (Cks1) that modify their specificity or accessibility to regulators (Patra and Dunphy, 1996); and both cyclins and some CKIs are regulated by synthesis and ubiquitin-mediated proteolysis (Ciechanover, 1994). Finally, each of the proteins that influence CDK activity is potentially affected by signal transduction pathways that regulate transitions along the cell cycle and thus represent targets for checkpoint pathways.

The observation that in fission yeast a single cyclin/CDK complex regulates both replication and mitosis (see below) likely reflects what happened in a more ancient evolutionary condition (Nurse, 1994). The ancestral CDK activated both chromosomal replication and separation possibly by targeting unique protein complex serving at the same time as replication origin and kinetochore. The increase in genome size with evolution and the presence of multiple replication origins and of chromosome condensation before separation made simultaneous S phase and mitosis impossible. The existence of different CDK thresholds for S and M phase substrates likely resulted in the sequential progression through S phase first and then mitosis. Separation of S phase and mitosis required the development of new controls to ensure that S phase can be initiated only when mitosis is completed and that mitosis awaits the completion of the S phase. This was achieved by the development of the prototype cyclin B/CDK complex: S phase was dependent on completion of the previous mitosis and destruction of the mitotic CDK activity; M phase occurred after completion of S phase because ongoing S phase inhibited the rise of CDK to mitotic levels (Murray, 1992). Notably, the ancestral eukaryotic cell cycle did not include much regulation in G_1 phase, which became necessary with the appearance of sexual differentiation (G_1 arrest is a prerequisite for the conjugation of mating partners). The need for a stringent control of the onset of S-phase CDK activity led to the introduction of G_1 cyclins (CLN cyclins in *Sac. cerevisiae* and cig2 in S. pombe) and G_1-specific CDK inhibitors (rum1 in *Schizosaccharomyces pombe*). In multicellular organisms, the ancestral cyclin CDK complex has been replaced by specialized S phase–and M phase–activating CDKs. Only cyclin A, which is structurally related to B-type cyclins, remained multifunctional (as B-type cyclins in yeast). In fact, cyclin A is implicated in the control of the S phase (Girard et al., 1991; Pagano et al., 1992) as well as mitosis (Minshull et al., 1989; Lehner and O'Farrell, 1989) and, at least in *Drosophila,* prevents re-replication (Sauer et al., 1995).

THE YEAST CELL CYCLE

Cyclin/CDK Function and Regulation in the Control of DNA Replication

Although mammalian cells were known to divide long before a cell cycle was proposed, the first evidence of a segmental pathway to cell division arose from studies involving both budding and fission yeasts. Similarly, cyclins were originally identified in sea urchin eggs as proteins rapidly synthesized after fertilization that accumulated during the succeeding cell cycles (Evans et al., 1983), but most of the studies on the cell cycle have then been made in yeast. Checkpoint pathways have indeed been identified through the analysis of cdc (cell division cycle) mutants in yeast. About 20 years ago the fission yeast *Sch. pombe* cdc2 temperature-sensitive mutant was identified as a key regulatory gene that caused growth arrest at the restrictive temperature. Conversely, when the wild-type cdc2 gene product was overexpressed, it accelerated the cell cycle progression, causing the wee phenotype in which cells divide at smaller sizes before the optimal critical mass is reached (Hartwell et al., 1974; Nurse et al., 1976; Nurse and Thuriaux, 1980). Soon after, the homologous protein cdc28 was found in the budding yeast Sac. *cerevisiae* (Beach et al., 1982). Based on sequence and functional similarity, several cdc2-related kinases were subsequently discovered in other eukaryotic organisms (Lee and Nurse, 1987; Draetta et al., 1987). The yeast system has been discovered as of late to have some complicated pathway and feedback systems as well and serves as an amenable model system for this field of research.

In the budding yeast ***Sac. cerevisiae*** the G_1 cyclins CLN1, CLN2, and CLN3 drive cells through G_1 by activating the **cdc28 kinase.** Distinct cyclins promote S phase (**CLB5** and **CLB6**) and mitosis (**CLB1** through **CLB4**). cdc28 is the sole kinase known to be involved in progression through all phases of the cell cycle in *Sac. cerevisiae* (Beach et al., 1982). The functional specificity of the cdc28 kinase is determined by its association with G_1 or G_2 cyclins, and alternation of cell cycle phases is mainly due to mechanisms that ensure one cyclin family to succeed another. In G_1 phase, cdc28 associates with the CLN1, CLN2, and CLN3 cyclins. Deletion of one or two of these genes has no effect on G_1 progression, while loss of all three arrests the cells in G_1. This shows that CLN cyclins at the same time provide redundant functions and are necessary for S phase onset. cdc28/CLN associations are necessary for spindle pole duplication, bud formation, and other processes required for preparation to entry into S phase (Richardson et al., 1989). cdc28 associates with CLN3 early in G_1 and persists throughout the G_1 phase. This complex functions as the activating enzyme for CLN1 and CLN2 transcription, whose association with cdc28 rises in late G_1 (Tyers et al., 1993; Koch and Nasmyth, 1994). The active CLN1 and CLN2/cdc28 complexes act as a positive feedback loop to further stimulate the transcription of CLNs, likely through the phosphorylation of the heterodimeric transcription factor Swi4/Swi6. In G_1, two B-type cyclins, CLB5 and CLB6, also associate with cdc28 (Nasmyth, 1993). Their transcription is concurrent with that of CLN1 and CLN2 in late G_1 (Epstein and Cross, 1992; Schwob and Nasmyth, 1993). CLB5 mutations slow S phase, and deletion of both CLB5 and CLB6 delays progression into S phase of at least 30 minutes. However, because overexpression of CLB5 does not shorten G_1 phase, as CLN cyclins do, it appears that B-type cyclins are not essential for S phase onset.

Molecular cloning of **cdc34,** a gene required for G_1 S transition in the budding yeast, revealed that a ubiquitin conjugation step was required just before the initiation of DNA replication. cdc34 encodes a ubiquitin-conjugating enzyme (Goebl et al., 1988) that participates in the destruction of many proteins involved in the G_1 to S progression, including the G_1 cyclins CLN2 and CLN3 (Deshaies et al., 1995; Yaglom et al., 1995), as well as proteins not directly related to cell cycle control (Kornitzer et al., 1994). However, accumulation of G_1 cyclins does not account for the cell cycle arrest in cdc34 thermosensitive (ts) mutants, and extracts from cdc34ts mutants inhibit S phase CDKs, implying that cdc34 may be required for degradation of a **CDK inhibitor** (CDI). The target for this activity has been found to be the S phase–specific CDI **Sic1** (Mendenhall, 1993; Nugroho and Mendenhall, 1994; Schwob et al., 1994). p40Sic1, whose levels rise during G_1 directly after mitosis, is rapidly degraded before S phase and accumulates in cdc34ts mutants. Because cdc34ts Sic1Δ double mutants initiate DNA replication at the nonpermissive temperature, Sic1 is likely the crucial substrate that blocks the G_1 to S progression in cdc34ts mutants (Schwob et al., 1994). This is further confirmed by the ability of a nondegradable form of Sic1 to block cell division at the G_1/S transition in wild-type cells. Thus, proteolysis via a ubiquitin-dependent mechanism is one control in the onset of S phase. The mechanism responsible for proteolitic regulation of Sic1 has been recently uncovered and involves Sic1 phosphorylation by CDK com-

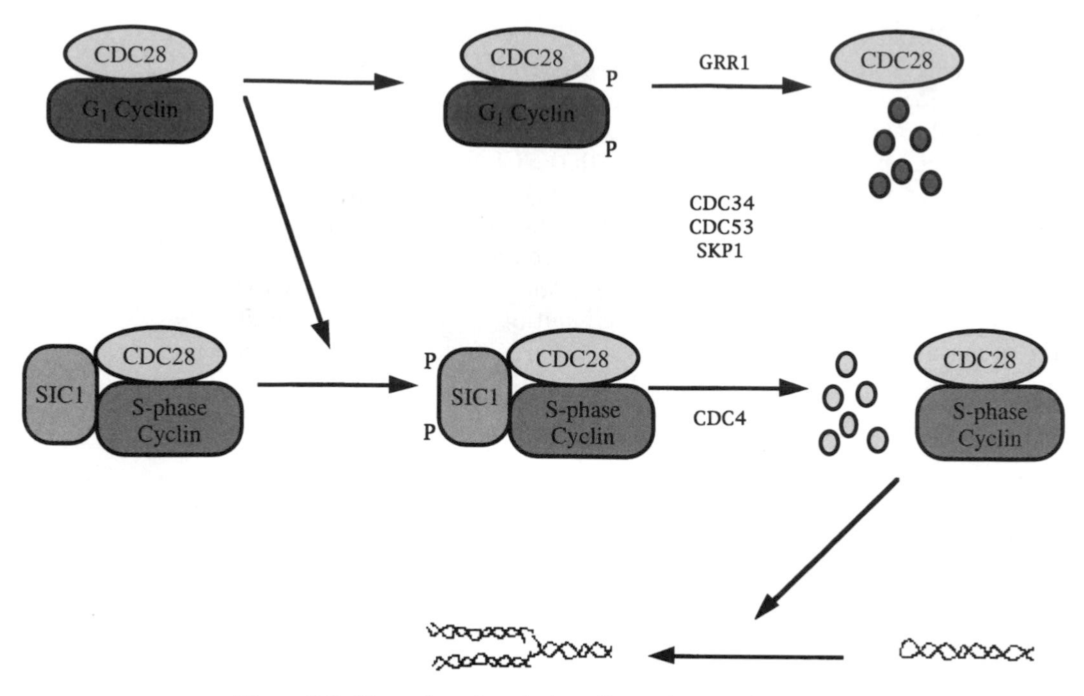

Figure 2.2. How phosphorylation-directed proteolysis affects cell cycle progression in yeast.

plexes activated late in G_1 phase. The phosphorylated Sic1 is then removed from the associated CDK complex by SCFcdc4-mediated ubiquitination and subsequent degradation, thus delivering free CDKs complexes and allowing S phase entry (reviewed by Hoyt 1997). Three additional genes are required for G_1/S transition in budding yeast: **cdc4, cdc53** (Mathias et al., 1996), and **SKP1** (Bai et al., 1996). Termosensitive mutations of any of these genes confer a phenotype similar to that of cdc34ts mutants, and in all cases Sic1 deletion restores the ability to replicate DNA. Although little is known about their biochemical functions, some evidence indicates that cdc53, cdc4, and SKP1 contribute to Sic1 ubiquitination (King et al., 1996). In addition, SKP1 and CDC4 interact with each other (Bai et al., 1996) through a motif present in cdc4 termed the *F box,* named with respect to the human cyclin F, which has been found to be involved in protein ubiquitination and degradation. cdc53 also may interact with other substrates of the cdc34 pathway, such as G_1 cyclins (Willems et al., 1996). Another protein that contains an SKP1-binding F box is **GRR1** (Bai et al., 1996). Mutant GRR1 fails to degrade CLN2 but progresses normally to S phase and destroys Sic1 normally. Stability of the different substrates of the cdc34 pathway is regulated by substrate-specific phosphorylation, which serves as a trigger for cdc34-dependent ubiquitination (King et al., 1996) (Fig. 2.2). Autocatalytic CLN2 and CLN3 phosphorylation by cdc28 is required for their cdc34-dependent ubiquitination (Deshaies et al., 1995; Yaglom et al., 1995), and, indeed, CLN2 and CLN3 phosphorylation mutants are stable in vivo (Yaglom et al., 1995; Lanker et al., 1996). However, because G_1 cyclins in budding yeast turn over rapidly throughout the cell cycle, their abundance, which controls progression through G_1, is mainly regulated by their rate of transcription (Willems et al., 1996). In contrast, Sic1 is stable during early G_1 and prevents the premature activation of S phase CDKs (Schwob et al., 1994). A crucial role of the G_1 cyclin/CDK complexes assembled in late G_1 is the phosphorylation of Sic1, which enables its recognition and ubiquitination by components of the cdc34 pathway (Schneider et al., 1996) (Fig. 2.2). The active cdc28/S phase cyclin complexes are then free to initiate DNA replication by phosphorylating key targets that remain to be identified. Whereas cdc34, cdc53, and SKP1 are required for both G_1 cyclins and Sic1 degradation, it seems that GRR1 and cdc4 may act as substrate-specific components of the G_1/S proteolytic pathway (Bai et al., 1996). It has also been shown that G_1 cyclins inhibit proteolysis of some G_2 cyclins, which enables G_1 cyclins to promote the accumulation of the G_2 cyclins (Amon et al., 1994). In turn, the G_2 cyclins CLB1 through CLB4 inhibit the transcription of the CLN1 and CLN2 genes (Amon et al., 1993; Koch et al., 1996). Because CLN1 and CLN2 proteins are highly unstable, this mechanism results in the rapid loss of CLN1 and CLN2 proteins after the appearance of the CLB proteins. In addition, it has been shown that the G_2 cyclins CLB1 through CLB4 are required for CLN1 and CLN2 proteolysis (Blondel and Mann, 1996), providing an additional mechanism for coupling synthesis of G_2 cyclins with the disappearance of G_1 cyclins. This latter activity of G_2 cyclins involves the Ubc9 ubiquitin-conjugating enzyme (Seufert et al., 1995). In this perspective, the cdc34 pathway may either be redundant with Ubc9 or may be involved in CLN1 and CLN2 turnover by promoting the degradation of Sic1 and liberating active cdc28/CLB complexes. Vertebrate cells express structural and functional homologues of cdc34 (Plon et al., 1993), cdc53 (Kipreos et al., 1996), and SPK1 (Bai et al., 1996; Zhang et al., 1996). The CDI **p27** is degraded in part by a cdc34 pathway in human cells (Pagano et al., 1995), and CDK-dependent initiation of DNA replication in *Xenopus* eggs requires a cdc34 homologue (King et al., 1996).

In *Sch. pombe,* a single CDK encoded by **cdc2** and three B-type cyclins encoded by **cdc13, cig,** and **cig2** govern cell cycle progression (Fisher and Nurse, 1995; Stern and Nurse, 1996). Cig2 is the major partner of cdc2 in G_1 phase (Fisher and Nurse, 1996; Martin-Castellano et al., 1996; Mondesert et al., 1996). Although cig2 levels peak around the G_1/S transition (Mondesert et al., 1996), cig2 is not essential for S phase to take place in that cig2Δ cells are only retarded when entering S phase. Unexpectedly, this mutant cig2 function is compensated by cdc13, previously thought to act exclusively as a mitotic cyclin (Fisher and Nurse, 1996; Mondesert et al., 1996). In the absence of cig2, G_1 phase is extended until the cdc13-associated kinase rises at sufficient levels to bring about DNA replication. Consistent with this model, the onset of the S phase is severely compromised in cig2Δ cdc13Δ double mutants (Fisher and Nurse, 1996; Mondesert et al., 1996). Similarly, although cig1 contributes to cdc2-associated kinase activity, it is not required for the onset of mitosis. In cig1Δ and cig2Δ double mutants, the single oscillation of cdc2/cdc13 protein kinase activity, which peaks in mitosis, is sufficient to ensure not only an orderly onset of S phase and mitosis but also re-replication control, and it avoids the occurrence of more that one S phase per cell cycle. Indeed, cdc13Δ mutants (Hayles et al., 1994) or wild-type cells overexpressing **rum1,** a specific inhibitor of the cdc2/cdc13 complex (Moreno and Nurse, 1994; Correa-Bordes and Nurse, 1995), display an increase of DNA content up to 32–64 chromosomes. Recently, a "quantitative model" has been proposed (Stern and Nurse, 1996) in which different levels of cdc2 activity regulate cell cycle progression. S phase is initiated when protein kinase activity increases from the *very low* "G_1" level to a *moderate* level. Maintenance of the S-phase *moderate* levels would prevent the reinitiation of the S phase, and a further increase to *high* levels initiates mitosis. The model predicts that S phase substrates must be more readily activated by cdc2/cdc13; mitotic substrates and several observations now support this view. The inactivation of the cdc2 kinase activity at the end of the mitosis resets the cell for a new cycle.

Mitosis

Mitosis in all organisms is initiated by the **mitotic CDK** composed of **cyclin B** and **cdc2,** which is historically known as *maturation-promoting factor* (MPF). During mitosis, cdc2 also forms complexes with **cyclin A** (Draetta et al., 1989). Cyclin B accumulates during interphase and rises to a threshold level that culminates in the activation of MPF and entry into mitosis (Murray and Kirschner, 1989; Murray et al., 1989; Solomon et al., 1990). However, cyclin B overexpression does not accelerate mitosis from G_2 in fission yeast (Hagan et al., 1988; Booher et al., 1989), and cyclin A, which is bound to CDK2 in S phase, does not activate cdc2 until the cell has completed DNA replication and is ready to divide. Additional factors, besides cyclin B transcription rate, are involved. cdc2 is subjected to a complex set of post-translational modifications. In early G_2, phosphorylation on Y15 and T14 inhibits activation of the human cdc2 (Draetta and Beach, 1988; Gould and Nurse, 1989), and mutation of these sites causes HeLa cells to rapidly proceed into mitosis (Krek and Nigg, 1991). The **wee1** mutant strain of *Sch. pombe,* phenotypically characterized by a small size due to premature cell division long before completion of G_2 (Nurse and Thuriaux, 1980), lacks of the wee1 serine/tyrosine kinase that catalyzes the phosphorylation of Y15 on cdc2 (Russel and Nurse, 1987; Featherstone and Russell, 1991). wee1 is itself phosphorylated and inactivated by Nim1 and other unidentified kinases, thus promoting mitosis (Coleman et al., 1993; Parker et al., 1993; Wu and Nurse, 1993). The inhibitory phosphorylation of cdc2 by wee1 is counteracted by the **cdc25** phosphatase (Russell and Nurse, 1986; Gauthier et al., 1991; Kumagai and Dumphy, 1991). cdc25 is phosphorylated and activated by cyclin B/cdc2 complexes, thereby defining an amplification pathway (Izumi et al., 1992). Protein phosphatase 1 **(PP1)** inactivates cdc25 by dephosphorylation of the same residue that is modified by cdc2 (Walker et al., 1992).

Full activation of cdc2 also requires the phosphorylation of T161, which is likely to stabilize its association with cyclin A (Atherton-Fessler et al., 1993; Desai et al., 1992, 1995), by the CDK-activating kinase **(CAK)** (also termed *CDK7* due to its association with **cyclin H**) (Fisher and Morgan, 1994; Makela et al., 1994). One of the best-studied targets of cdc2/cyclin B is **histone H1** protein. H1 phosphorylation causes its dissociation from DNA and forces condensation of the chromosome (Ohsumi et al., 1993). In addition, cdc2 phosphorylates **lamins,** the proteoglycans that constitute the nuclear envelope. Phosphorylated lamins do not polymerize, and this causes the nuclear envelope breakdown that is closely associated with the onset of mitosis, and it is necessary for the spatial arrangement of chromosomes in metaphase. The formation of the mitotic spindle is essential for aligning the chromosomes during mitosis. Centrosomes enucleate growing microtubules that will attach to the kinetochore in the centromere of each chromosome. During metaphase this attachment links the chromosome to the metaphase plate. The subsequent anaphase will see separation of the sister cromatids by pulling them apart on the elongated microtubules. Both cyclin A– and cyclin B–containing cdc2 complexes increase microtubule formation from isolated centromeres and favor the whole process (Booher and Beach, 1988; Ohta et al., 1993).

MPF induces its own inactivation by activating the **mitotic cyclin destruction system** (King et al., 1996). Cyclin B degradation is required for the exit from telophase and entry into the subsequent interphase. Degradation of other substrates, namely, anaphase inhibitors, is important for progression from metaphase to anaphase (Fig. 2.3). In early developmental stage embryos, cyclins are continuously synthesized, and it is only the activation of mitotic cyclin destruction machinery that controls cell cycle progression. Mitotic cyclins contain an N-terminal conserved 9 amino acid motif (destruction [or D] box) that is necessary for cyclin ubiquitination and subsequent degradation (Glotzer et al., 1991; Amon et al., 1994; King et al., 1996; Klotzbucher et al., 1996; Zachariae and Nasmyth, 1996) and confers instability during mitosis to otherwise stable proteins (Okazaki et al., 1992; Brandeis and Hunter, 1996; Funabiki et al., 1996). Deletion of the NH2 terminus does not affect the capacity of mitotic cyclins to activate cdc2 and drive the cells into mitosis but provokes the arrest of cell division in telophase (Ghiara et al., 1991; Luca et al., 1991; Holloway et al., 1993; Surana et al., 1993). Regulation of type-A cyclin proteolysis appears to be more complex and involves sequences outside the NH2 terminus (van der Velden and Lohka, 1993; Stewart et al., 1994). In contrast to cdc34 pathway substrates, cyclin B does not require phosphorylation to be targeted to degradation and is the activity of a large E3 complex, know as the *cyclosome* (Sudakin et al., 1995) or the *anaphase-promoting complex* (APC) (King et al., 1995), which is modulated in a cell cycle–dependent

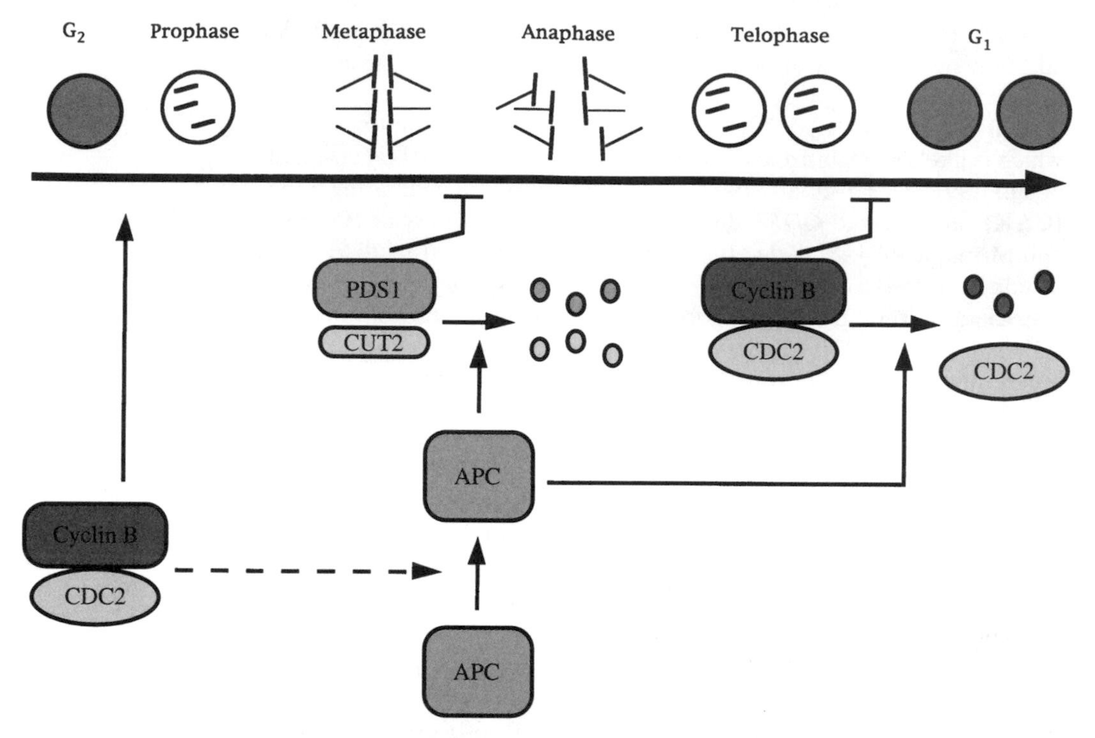

Figure 2.3. Regulation of mitosis and the anaphase-promoting complex.

manner. The ubiquitination machinery of mitotic cyclins utilizes two E2 enzymes, **Ubc4** (King et al., 1995) and **Ubcx/E2 C** (Yu et al., 1996). An additional E2, **Ubc9,** has been implicated in D box–independent proteolysis of mitotic cyclins in budding yeast (Seufert et al., 1995). Known subunits of the APC, which is essential for anaphase progression in multiple organisms, are given in Table 2.1. They were identified by genetic screening in budding yeast as genes required for both anaphase

TABLE 2.1. Subunits of the Anaphase-Promoting Complex

Organism				Required for Anaphase
X. laevis	*Sac. cerevisiae*	*Sch. pombe*	Mammals	
APC1	APC1		TSG24	+
APC2				?
APC3	cdc27	NUC2	cdc27Hs	+
APC4				?
APC5				?
APC6	cdc16	CUT9	cdc16Hs	+
APC7	cdc23			+
APC8				?
	cdc26			+

progression and mitotic cyclin degradation (Irniger et al., 1995) and by reconstitution experiments with purified components of mitotic cyclin ubiquitination in clam and frog egg extracts (King et al., 1995; Sudakin et al., 1995; Hershko et al., 1995). Although the APC functions as an E3 enzyme, it is still unclear how it recognizes its substrates. None of its components share sequence similarity with either **Ubrl** (Bartel et al., 1990) or the **E6AP family** of proteins (Huibregtse et al., 1995), and no ubiquitin thioesters associate with any of the subunits (King et al., 1995). Different from other known E3 enzymes, it does not act as an intermediate carrier of ubiquitin, but favors the contact between ubiquitin-changed E2s and D-box substrates.

The simple model of anaphase and telophase as the result of MPF inactivation has been challenged by a number of observations (i.e., the nondegradable form of the mitotic cyclins arrests cells in telophase rather than metaphase in both *Xenopus* and budding yeast; inhibition of APC activity by different approaches prevents chromosomal segregation) that led to postulation of the existence of noncyclin substrates of the APC that act as inhibitors of the anaphase (King et al., 1996). Two noncyclin proteins that are degraded during anaphase by the APC/**CUT2** in fission yeast (Funabiki et al., 1996) and **PDS1** in budding yeast (Yamamoto et al., 1996) have been recently identified (Fig. 2.3). CUT2 is a nuclear protein that decreases in abundance during anaphase and is not detected in G_1-arrested cells (Funabiki et al., 1996). Deletion of the NH2-terminal D box–like sequences stabilizes the protein, which also accumulates in G_1-arrested cells bearing a mutation in the APC subunit CUT9. Similarly, PDS1, which is required to maintain sister cromatid cohesion in budding yeast, requires an intact D box and functional APC to be degraded in anaphase (King et al., 1996). It remains unclear how CUT2 and PDS1 inhibit anaphase: The current hypothesis is that these proteins function as a chromosomal glue that holds chromosomes together until it is degraded by the APC, thus releasing the chromatids and initiating other anaphase movements. The APC is also likely involved in the proteolysis of other proteins that influence the mitotic spindle.

Activation of the MPF and activation of the mitotic cyclin degradation system occur sequentially. The time lag between these two events allows MPF to remain active to induce mitotic events (nuclear membrane breakdown, chromosome condensation, and chromosome alignment on the metaphase plate), before APC activation extinguishes the mitotic process. Regulation of APC activity, however, is poorly defined (King et al., 1996). Several subunits of the APC become phosphorylated during mitosis in *Xenopus* and *Aspergillus,* and dephosphorylation events inactivate the mitotic MPF in clam and *Xenopus* in vitro (King et al., 1995, 1996). Candidate kinases include MPF itself and cyclin A/cdc2. In budding yeast and mammalian cells, APC remains active during G_1 until the G_1 CDKs are activated (Amon et al., 1994; Brandeis and Hunter, 1996). Thus, G_1 CDK activity has a crucial role in controlling proteolysis during the cell cycle in that it switches off the APC pathway and switches on Sic1 proteolysis.

THE MAMMALIAN CELL CYCLE

Upon mitogenic stimulation, mammalian cells progress from G_0 into an initial phase, called G_1, in which, once a threshold size has been reached and specific proteins are activated, they can enter into S phase. Here cells replicate their genome in a restricted time. The commitment to enter the DNA synthesis occurs at a "restriction point" or "start point" late in the G_1 phase, after which mitogenic signals are no longer required for cells to progress into the cell cycle. After the DNA has been replicated, cells progress through another much shorter growth phase, termed G_2. This phase has several regulatory mechanisms that ensure that the genome has been replicated once and only once. Finally, the cell is ready to divide into two identical cells, which will start a new cell cycle. As in the yeast, the ordered sequence of S phase always preceding M phase is ensured by the timely regulated expression of cyclins. Cyclins are activators of a family of protein serine/threonine kinases, the cyclin-dependent kinases (CDKs), which also share some structural similarities. Complexes of cyclins with CDKs play a central role in the control of the cell cycle by phosphorylating specific substrates, and this is thought to be critical in the regulation of an ordered sequence of events leading to DNA replication and chromosomal segregation. To confer specificity to this mechanism, during evolution, eukaryotic cells have developed a large number of functionally different, cell cycle phase–specific cyclins. Similar to prokaryotes, and also in higher organisms, the new synthesis and the accumulation of a specific cyclin dictate the formation of a cyclin/CDK(s) complex at a distinct point of the cell cycle. The rapid destruction of this cyclin and the synthesis of a new one will ensure the progression into the subsequent phase. Thus, the transient appearance of different cyclin/CDK complexes drives critical cell cycle events such as cell growth (G_1), DNA replication (S), and cell division (G_2/M). The activity of cyclin/CDK complexes is controlled by another family of small proteins, called CDK inhibitors (CDI), that contribute to a spatially and temporally ordered sequence of events along cell cycle progression.

Cyclin/CDK Structure, Function and Regulation

The cyclin family of proteins share homology in a 100 amino acid region, termed the *cyclin box,* through which they bind the CDKs. All the CDKs share a sequence related to the canonical EGVPSTAIRISLLKE motif in domain III, initially found in the first CDKs to be isolated (the fission yeast p34cdc2 and the budding yeast p34cdc28). This region, together with another sequence that includes a threonine residue (Thr-161 in human cdc2, Thr-160 in CDK2) in domain VII, is important for CDK-specific binding to cyclins (Pines, 1996). Many different cyclins with their partner kinases (CDKs) have been cloned and characterized (Table 2.2). Although cyclin/CDK complexes play a central role in regulating cell cycle progression, they are also involved in the control of processes other than the cell cycle. However, here we focus only on the structure and

TABLE 2.2. Description of Known Mammalian Cyclins, CDKs, and CDK Inhibitors

Cell Cycle Protein	Involvement in the Cell Cycle	Function of Protein	Regulation of Activity	Connections to Cancer
Cyclin A	Through S Progression into M	Activate CDK2 Activate cdc2	Transcriptional and ubiquitin degradation	Hepatocellular carcinoma and associated to E1A
Cyclin B1	Progression into M	Activate cdc2	Transcriptional and ubiquitin degradation	?
Cyclin B2	Progression into M	?	Ubiquitin degradation	?
Cyclin C	Through G_1?	Activate CDK8	?	?
Cyclins D1, D2, and D3	Through G_1	Activate CDK4 and CDK6	Transcriptional and PEST-targeted proteolysis	D1 is PRAD1/BCL1 protooncogene
Cyclin E	Progression into S	Activate CDK2	Transcriptional and PEST-targeted proteolysis	Rearrangements and overex. in breast carcinoma
Cyclin F	Through G_2 or Progression into M	Involved in cdc4-like ubiquitination	Transcriptionally activated	?
Cyclin G	Ubiquitous?	DNA repair?	?	Involvement with p53 downstream pathways
Cyclin H	Ubiquitous	Activate CDK7	?	?
cdc2	Progression into M	Phos. histone H1, lamins, etc.	Activated by cyclins A and B	?
CDK2	Progression past START and through S	Associate with and Phos. DNA rep. machinery	Activated by cyclins A and E	?
CDK3	Through G_1 or progression past START	Modulate E2F activity	?	Located near LOH region of BRCA1
CDK4	Through G_1	Phos. pRb tumor suppressor	Activated by cyclins D1, D2, and D3	Gene amplification found in gliomas
CDK5	In G_0?	Phos. tau and neurofilament brain proteins	Activated by p35 subunit	?
CDK6	Through G_1	Phos. pRb tumor suppressor	Activated by cyclins D1, D2, and D3	Non-Hodgkin's lymphoma?
CDK7	Ubiquitous	Activate CDKs and Phos. CTD of RNA pol II	Activity dependent on presence of substrate?	?
CDK8	G_1?	Phosphorylate CTD of RNA pol II	Activated by cyclin C	?

TABLE 2.2. (*Continued*)

Cell Cycle Protein	Involvement in the Cell Cycle	Function of Protein	Regulation of Activity	Connections to Cancer
PITSLRE	G_1?	Apoptosis	?	Localized in region deleted in several cancers
PISSLRE	G_2?	Phos. histones?	?	Localized near LOH in breast and prostate cancer
PITALRE	G_0?	Phos. pRb and transduction of extracellular signals	?	Non-Hodgkin's lymphoma?
p21WAF1	Mostly in G_1 and S	Inhibit CDKs 2, 4, and 6 and PCNA	Downstream of p53 and MyoD transcription factors	Connected to p53 loss in cancer
$p27^{kip1}$	G_1	Inhibit CDK2 upon TGF-β treatment or cell–cell contact	Diluted distribution and ubiquitination	Localized in regions involved in leukemia
$p57^{kip2}$	G_1	Inhibit G_1 kinases	Transcription from a genomically imprinted gene	?
p16INK4	G_1	Inhibit CDKs 4 and 6	Transcribed by pRb- associated transcription factor	Located at multiple tumor suppressor-1 locus
p15INK4B	G_1	Inhibit CDKs 4 and 6 upon TGF-β treatment	Transcribed downstream of TGF-β treatment	Located at multiple tumor suppressor-2 locus
p18INK4C	G_1	Inhibit CDK6 and CDK4	Transcriptional?	?
p19INK4D	G_1	Inhibit CDK6 and CDK4	Transcriptional?	?

regulation of these complexes relative to their function as cell cycle regulators.

Monomeric CDKs are completely inactive in the absence of binding with cyclins. The concentrations of the cyclins vary during the cell cycle. As a result, only when the CDK partner accumulates at a specific point in the cell cycle does a specific cyclin/CDK complex form and can be activated. Crystallographic studies (Jaffrey et al., 1995) have recently unraveled the mechanism for this dependence (Fig. 2.4). Monomeric CDKs are in an inactive form for at least two reasons. First, the ATP is bound to the kinase in a conformation that precludes its nucleophilic attack on the substrate hydroxy group, and therefore the scission of

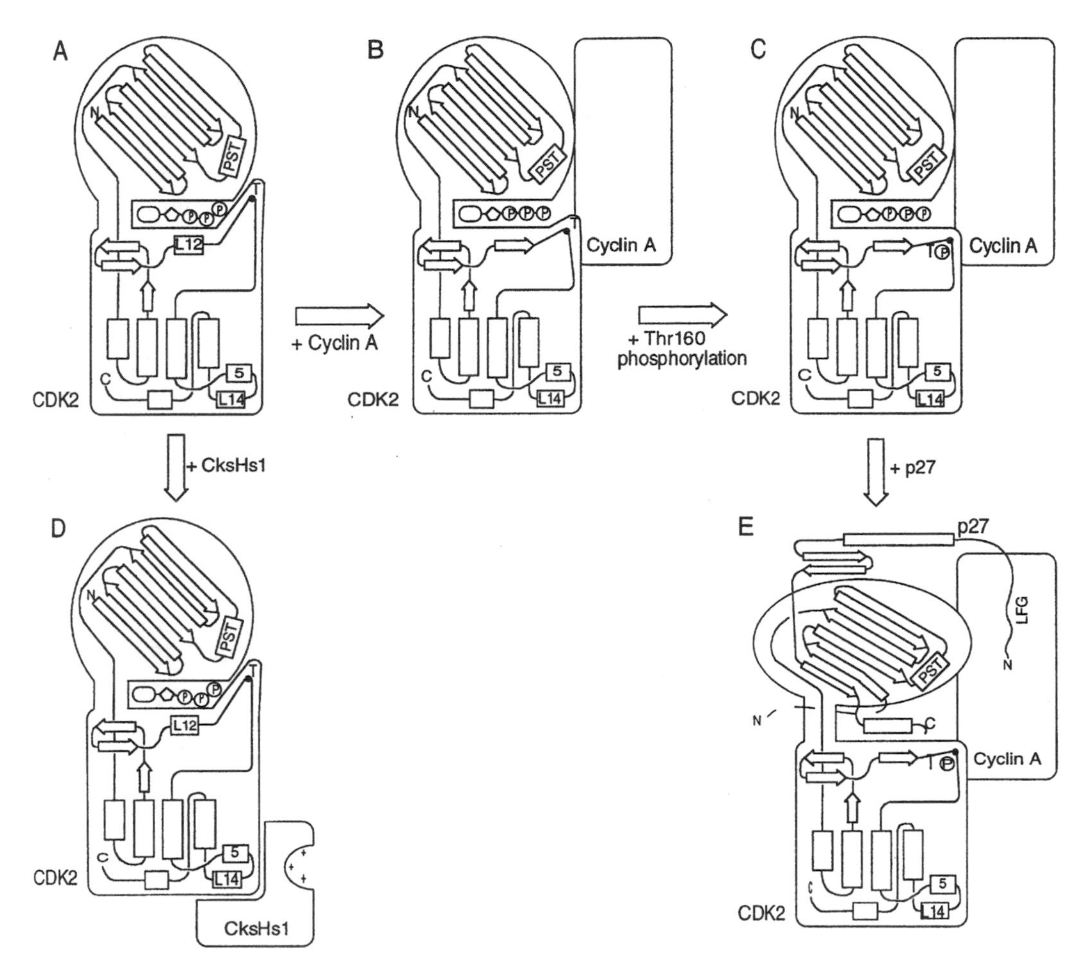

Figure 2.4. Results of the crystal structure of CDK2 **(A)** and how activity is affected by cyclin A binding **(B)**, Thr-161 phosphorylation **(C)**, CksHs1 binding **(D)**, and p27[kip1] binding **(E)**.

the B-y phosphates is inhibited. Second, part of the C terminus of the enzyme—the T-loop domain—blocks the catalytic cleft. Following the interaction of the PSTAIRE region of the CDK with the "cyclin box" of the cyclin, the PSTAIRE helix is reoriented, thus allowing the ATP B-y phosphate bond to became susceptible to the attack from a bound substrate. This is also facilitated by the binding of the partner cyclin to the N-terminal portion of the T loop. This interaction displaces the T loop that partially obscures the substrate binding site. Finally, the interaction between the cyclin and the T loop exposes the threonine Thr-160 in the T loop, which becomes accessible for phosphorylation by CAK, which provides full activation of the cyclin/CDK complex kinase activity and stabilizes the complex itself.

CAK is another cyclin/CDK complex, composed of cyclin H, CDK7, and a third protein of 36 kD that is referred to as MAT1 (menage-à-trois 1) (Devault et al., 1995; Tassan et al., 1995; Fisher et al., 1995).

Most cyclin/CDK complexes are probable substrates for CAK. However, the recent observations that cyclin H/CDK7 is a component of TFIIH (a protein belonging to the basal transcriptional machinery) and that it can phosphorylate the C-terminal domain of the RNA polymerase II (Feaver et al., 1994; Serizawa et al., 1995) (thereby participating in the induction of RNA transcription more than in the activation of cyclin/CDK) raise the question of whether some CDK-specific activating protein exists. Nevertheless, given its role in both transcription and CDK activation, CAK could provide a link between transcription and cell cycle control.

Finally, phosphorylation of two conserved residues (Thr-14 and Tyr-15) within the catalytic cleft of CDKs inhibits their function, while their dephosphorylation represents a mechanism for CDK activation (Lew and Kornbluth, 1996). After exerting its activity, the cyclin/CDK complexes need to be inactivated and replaced by other complexes in the next phase of the cell cycle. A well-established mechanism to regulate cyclin/CDK function during G_2/M phase through phosphorylation has been described in lower organisms and involves two kinases, wee1 and mik1 (Russell and Nurse, 1987; Lundgren et al., 1991). They specifically phosphorylate the CDK subunits on a tyrosine residue (Y15 in the case of cdc2) in the ATP-binding region and inactivate the kinase activity by directly interfering with phosphate transfer to a bound substrate. Phosphorylation of the threonine residue T14 of cdc2 has been observed also (Norbury et al., 1991; Haese et al., 1996), but its function in the regulation of the cell cycle is still unclear. In animal cells, however, the double mutation of Y15 and T14 is required to deregulate cdc2 activation and to accelerate G_2/M progression (Norbury et al., 1991; Krek and Nigg, 1991a,b).

Although it has not been clearly shown whether phosphorylation at these specific sites is important for regulation of other CDKs at other points of the cell cycle, the above-described process provides a simple explanation on how the cell cycle can be either positively or negatively regulated by phosphorylation. As Y15 and T14 phosphorylation plays such an important role in inhibiting the function of cyclin/CDK complexes, the same residues must be dephosphorylated during the other phases of the cell cycle to achieve full CDK kinase activity, and this is accomplished by the cdc25 CDK-activating phosphatases (Russell and Nurse, 1986). In human cells, three CDK phosphatases have been identified, cdc25A, -B, and -C (Galaktionov and Beach, 1991). cdc25A is induced by serum in quiescent cells early in the G_1 phase, cdc25B is expressed at the G_1/S phase transition, and cdc25C is activated in G_2 (Galaktionov et al., 1996). All of these phosphatases are involved in the regulation of cell cycle progression through activation of different CDKs and has been recently implicated as potential oncogenes given their ability to produce colonies in soft agar when overexpressed in primary cells, as discussed later. An example of a direct functional link between CDK phosphorylation and phosphatase activity is provided by the recently proposed mechanism through which transforming growth factor-

β (TGF-β) can induce cell cycle arrest independently from p15 induction. In a human mammary epithelial cell line lacking the CDI p15, TGF-β has been found to increase the level of CDK4 and CDK6 tyrosine phosphorylation, thereby inactivating their function by specifically repressing the expression of the CDK tyrosine phosphatase cdc25A (Iavarone and Massague, 1997). Finally, a new interesting kinase involved in G_2/M phase regulation has come into view. The PISSLRE kinase, named with respect to its homology within the cdc2 PSTAIRE region, does not bind any known cyclins. However, when overexpressed in a variety of cell lines as a kinase inactive mutant (a dominant negative), cells arrest their growth and halt in the G_2/M phase of the cell cycle (Grana et al., 1994; Li et al., 1995). It is unlikely that PISSLRE can carry out its control of the cell cycle by regulating cdc2 activity, as PISSLRE is not able to complement a yeast cdc28 mutation (T.K. MacLachlan, unpublished results). It will be interesting to see if this kinase might define a new mechanism of regulation within the G_2 phase of the cell cycle.

Once activated, cyclin/CDK complexes phosphorylate specific nuclear substrates, and this contributes to the progression in the next phase of the cell cycle. The substrate specificity is likely due to the type of cyclin forming the complex. This fits well with a model in which different cyclins, expressed at specific points of the cell cycle, target the CDKs to different substrates. Although it is now evident that most of the cyclin/CDKs phosphorylate the same basic consensus sequence (K/R)-S/T-P-X-(K/R), their physiological substrates in vivo are only partially known. One of them is the retinoblastoma gene product pRb, which is specifically phosphorylated by a cyclinD1/CDK4 complex at the middle–late G_1/S boundary. This phosphorylation plays a crucial role in the progression of cells toward S phase in response to growth factors, and it is addressed separately.

Mammalian CDK Inhibitors

Besides variations in cyclin abundance and regulation of CDKs by phosphorylation, in mammalian cells additional control is ensured by the action of several cyclin-dependent kinase inhibitory proteins (CDI). CDI inhibit CDK activity just after their function has been exerted, thus providing an efficient mechanism to regulate the sequential activation of specific cyclin/CDK complexes along cell cycle progression. The expression and the functional relevance of the different CDIs have been extensively studied in mammalian cells. CDI are also involved in promoting differentiation, in maintaining the postmitotic state of terminally differentiated cells, and in regulating the cell cycle checkpoint in response to genotoxic stressors.

Two classes of CDI have been defined, based on their sequence similarity and specificity of action. One is the **cip–kip family,** also known as *universal CDI,* that indifferently inhibit all of the G_1 kinases **(CDK2, CDK3, CDK4, CDK6)** and includes **p21, p27** and **p57.** The other is the **INK4 family,** composed of **p15, p16, p18,** and **p19**; it is also called *specific*

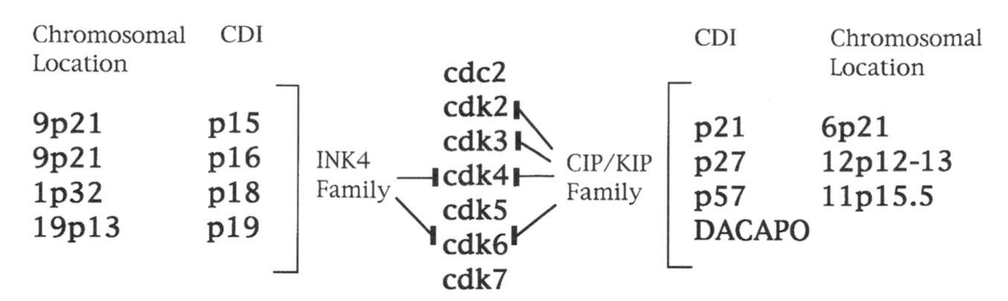

Figure 2.5. The known CDK inhibitors, their chromosomal location, and specificities of inhibition.

CDI, and its inhibitory activity is restricted to **CDK4** and **CDK6** (Fig. 2.5). The first CDKI to be discovered was **p21 (gW A F1, Cip1, Sdi1, mda6).** It is the prototype of universalinhibitors of CDK enzyme activity. p21 is capable of inactivating CDK activity when expressed at high levels and to induce cell cycle arrest. It is noteworthy that p21 has been concurrently cloned using very different approaches. These include a subtraction hybridization for the identification of genes differentially expressed during the induction of differentiation in human melanoma cells (mda6) (Jiang and Fisher, 1993; Jiang et al., 1995) and an expression screening to clone inhibitors of DNA synthesis from senescent fibroblasts (Sdi1) (Noda et al., 1994). However, the approach that better reflected the effective function of p21 during the cell cycle was its identification and cloning through interaction with CDK2. In normal fibroblasts the majority of CDK2 is gathered into quaternary complexes that also contain a regulatory cyclin, the proliferating cell nuclear antigen (PCNA), and the CDK-associated 21 kD protein (Xiong et al., 1992; Zhang et al., 1993). p21 co-precipitates with different CDKs in enzymatically active complexes containing cyclins A, B, D, and E (Harper et al., 1993; Xiong et al., 1993a,b; Zhang et al., 1993). The relative amounts of the different components would determine the kinase activity of these complexes on specific substrates. In more general terms, the stoichiometric ratio between cyclin/CDKs and CDI determines the effective function of the CDK (Zhang et al., 1994). The ability of p21 to inhibit CDK-dependent phosphorylation only when expressed at levels high enough to buffer the cyclin/CDK activity also explains the apparent paradox of p21 presence in most cyclin/CDK complexes in proliferating cells as well as the induction of p21 in both quiescent fibroblasts and in T lymphocytes stimulated to proliferate by mitogens (Firpo et al., 1994; Li et al., 1994a; Noda et al., 1994; Nourse et al., 1994; Sheikh et al., 1994). In these cells, despite the presence of p21, CDK-containing complexes retain their kinase activity. More recently, it has been proposed that p21 might have a role as an adaptor protein that assembles cyclin/CDKs in active complexes (LaBaer et al., 1997). To convert cyclin/CDK-containing complexes from active to inactive, cells are required to further upregulate p21 levels, to downregulate cyclin/CDK expression, or both. An additional mechanism that p21 uses to affect cell cycle progression relies on its interaction with PCNA and the possibility that this interaction might suppress PCNA-mediated activation of the DNA polymerase δ and DNA replication (Zhang et al., 1994; Vaga et al., 1994). This is of particular interest given the ability of p21 to interact with cyclin/CDK complexes and PCNA through two distinct domains, respectively at the N terminus and the C terminus (Chen et al., 1995; Luo et al., 1995). Experiments aimed to discern the functional relevance of these domains in the induction of cell cycle arrest, by using deletion mutants of p21, indicate that both domains are indeed efficient in inhibiting DNA synthesis, although the CDK-binding domain is more potent at blocking cell cycle progression (Luo et al., 1995; Nakanishi et al., 1995).

The tight correlation between p21 levels and functions has led many investigators to study the mechanism regulating its expression levels during the cell cycle. Using a subtractive hybridization approach to identify genes responsive to p53, a protein known to be the effector of the G_1 checkpoint trigger following exposure to DNA-damaging agents, it has been found that the p21 gene is a p53-responsive element (E1 Deiry et al., 1993). Thus, synthesis of p21 is induced by p53 in response to genotoxic stressors and serves to buffer cyclin/CDK activity to block cell cycle progression and allow DNA repair. Cells lacking p53 or p21 and cells in which the p53/p21-dependent growth arrest pathway is impaired are permissive for replication of damaged DNA, thereby increasing the possibility of acquiring chromosomal abnormalities and ultimately to transformation (Brugarolas et al., 1995). When DNA damage cannot be efficiently repaired in p53 +/+ cells, the alternative is programmed cell death as a last defense to eliminate possible chromosomal abnormalities and to prevent cancer development. In this suicide program, despite the evidence of p53 as an important effector in several cell types, the participation of p21 as a critical mediator seems to be unlikely (Ko and Prives, 1996). However, it has been recently revealed that human colorectal cancer cell lines in which endogenous p21 is artificially deleted respond to DNA-damaging agents, such as chemotherapeutic drugs and γ-irradiation, with multiple rounds of DNA synthesis without passing through mitosis instead of cell cycle arrest. These polyploid cells subsequently undergo apoptosis (Waldman et al., 1996). Thus, the function of p21 (and by extension p53) would also be required for normal coupling of the S and M phases of the cell cycle, ensuring that a new round of DNA synthesis is not reinitiated before the cell is passed through a preceding M phase. Expression of p21 is also induced during cell senescence and terminal differentiation in different cell types. In cells undergoing differentiation the induction of p21 has been found to be p53 independent, being attributable to tissue-specific transcription factors that activate the differentiation program (see below).

Another well-known universal CDI is **p27 (Kip1).** This protein was identified following the observation that extracts from either contact-inhibited or TGF-β treated cells do not contain detectable cyclin E–associated kinase activity despite the presence of cyclin E/CDK2 complexes (Koff et al., 1993). This kinase activity was reinducible upon addition of either active or mutant catalytically inactive cyclin D/CDK4 complexes (Polyak et al., 1994a), suggesting that these complexes could buffer a CDK inhibitory component in the extracts. This factor was subsequently purified and identified as a 27,000 Da polypeptide that binds cyclin/CDKs complexes, but not CDKs alone, with the properties of a kinase inhibitory protein (Polyak et al., 1994b). p27 was found to share significant homology with p21 (44%) in an N-terminal 65-amino acid region, a domain necessary for CDK inhibitory activity. The inhibition of CDK-dependent activity by p27 targets all the cyclin/CDK complexes present during G_1/S phase progression, and overexpression of p27 is enough to induce G_1 arrest in many cells independently from the function of the cyclin/CDK substrate Rb (Polyak et al., 1994b; Toyoshima and Hunter, 1994). The mechanism by which p27 inhibits CDK activity has been recently clarified by Pavletich and colleagues (Russo et al., 1996), which solved the structure of the p27/cyclin A/CDK2 ternary complex (Fig. 2.4), showing that the 69 amino acid N-terminal inhibitory domain of p27, which contains the conserved *LFG* sequence motif, anchors to the binding pocket of cyclin A, whereas the C-terminal region enters into the upper lobe of the kinase subunit. Given this kind of interaction, the interference with cyclin/CDK structure and function occurs through multiple approaches. In fact, p27 binding to the β sheet of CDK2 induces conformational changes that alter the shape of the β sheet itself and disrupts the conformation of the catalytic cleft, reducing the affinity of the kinase for ATP. In addition, the C-terminus of p27 binds deep within the active-site cleft to block ATP binding, thus ensuring a further degree of inhibition. Thus, the anchorage of p27 to cyclin A serves as an initial step to form a stable complex, and the interaction with CDK2 with the consequent structural changes in the catalytic cleft are determinant for effective inhibition of cyclin/CDK activity. This model, although restricted to p27 and cyclin A/CDK2, could be a general paradigm for *universal CDKI*–mediated inhibition of cyclin/CDK complex activity.

p27 accumulates in a variety of quiescent cells and disappears in response to mitogens or in concurrence to loss of sensitivity to TGF-β as cells approach the G_1/S boundary. p27 levels inversely correlate with cyclin E/CDK2 activity and represent a critical determinant in the regulation of G_1/S phase progression (Polyak et al., 1994a). As an example, the mitogenic function of interleukin 2 (IL-2) for peripheral blood T cells relies on its ability to induce both the synthesis of cyclin/CDK complexes and the destruction of pre-existing p27 molecules, thereby promoting the formation of active G_1 phase cyclin/CDK complexes and facilitating entry into S phase (Firpo et al., 1994; Nourse et al., 1994). Conversely, the antimitogenic action of the immunosuppressant rapamycin involves the prevention of IL-2–mediated decline in p27 levels. Therefore, p27 accumulates in quiescent cells and declines after mitogenic stimuli, contrary to p21, whose levels rise in response to mitogens (Sherr and Roberts, 1995). This outlines the importance of a tight control exerted by different CDI on the sequentially activated G_1 cyclin/CDK complexes. Although the amount of p27 must rise and fall to coordinate the cell-to-cell contact-induced G_1 arrest and the progression into S phase in response to mitogens, respectively, its transcription is not dramatically affected during cell cycle progression. A role of the ubiquitin–proteosome pathway in regulating p27 levels has also been shown (Pagano et al., 1995), and a regulation of the expression of p27 at the level of translation has been described (Hengst and Reed, 1996). This suggests that p27 levels are controlled by its turnover and that both translational and post-translational regulation are critical to determine p27 abundance and CDK inhibitory activity.

Based on the homology of the N-terminal CDK inhibitory domain to p21 and p27 and on its ability to bind to and inactivate all the G_1/S phase cyclin/CDK complexes, p57 (kip2) is considered another component of the family of universal CDIs (Lee et al., 1995; Matsuoka et al., 1995). Expression of p57 in the developing mouse is tissue restricted. p57 has been found to be especially present in placenta and in postmitotic cell types of developing mouse embryos, where its pattern of expression parallels that of p21 (Lee et al., 1995; Matsuoka et al., 1995). Interestingly, the p57 gene has also been found to be genetically imprinted, being highly methylated from the paternal chromosome contribution. With respect to p57 possibly acting as a necessary cell arrest protein, it should be noted that, due to this imprint, loss of the p57 gene from only the maternal chromosome could result in total loss of p57 expression.

More recently, a gene encoding a *Drosophila* Kip family member, Dacapo (DAP), was identified and characterized (Lane et al., 1996; de Nooij et al., 1996). DAP gene encodes a CDI with structural and functional similarities to the vertebrate kip inhibitors. The similarity between DAP and the vertebrate universal CDI is largely restricted to a few motifs in the N-terminal cyclin/CDK binding domain, that is, the region required for the cell cycle suppressive activity. Furthermore, DAP upregulation during *Drosophila* embryogenesis is required to arrest the epidermal cell proliferation at the correct developmental stage, suggesting a critical function for this CDI in regulating the cell cycle withdrawal of postmitotic cells.

The product of the p16 gene was first identified as a 16 kD protein associated with CDK4 in SV40 DNA tumor virus transformed cells (Serrano et al., 1995). This protein was found to associate to and inhibit the activity of either CDK4 or CDK6, forming a binary complex without the CDK catalytic subunit (cyclin) and therefore lacking kinase activity. A number of observations suggest that p16 might exert its inhibitory action on CDK4 and CDK6 activity by competition for cyclin D binding. Addition of p16 to active cyclin D/CDK4 or CDK6 complexes results in the inhibition of their kinase activity on pRb, a well-characterized CDK4/6 substrate, whereas p16 is without effect on other cyclin/CDK complexes. Furthermore, p16 expression is restricted in late G_1/S phase coincident with dephosphorylation of pRb after previous activation of cyclin D/CDK4/6 in early G_1/S (Tam et al., 1994). Cyclin D–associated CDKs phosphorylate pRb in early G_1/S, releasing transcription factors required for S phase entry (see below). At the same time, p16 transcription would be upregulated possibly by an as yet unknown transcription factor previously repressed by pRb (Y. Li et al. 1994; Parry et al., 1995).

Once p16 levels reach a threshold, it efficiently competes with cyclin D for binding with CDK4/6, thereby interfering with cyclin D/CDK assembly and activity. At this point, cyclin E/CDK2 becomes activated despite the high expression of p16, and cells progress into S phase. Therefore, p16 is an essential component of an autoregulatory loop that coordinates G_1/S phase transition, downregulating CDK4/6 activity once pRb has been inactivated by phosphorylation and allowing the correct activation of distinct cyclin/CDK complexes at the proper time (Elledge et al., 1996). According to this model, overexpression of p16 prevents proliferation in pRb-positive but not in pRb-negative cells. Both the mechanism of inhibition of cyclin/CDK complex activity and the specificity for distinct CDKs define p16 (and other INKs) as CDI unequivocally different from p21, p27, and p57.

Three additional p16-related genes, p15 (INK4b), p18 (INK4c), and p19 (INK4d), have been isolated, all of them encoding for proteins composed of four repeated ankyrin motifs and exerting cyclin D/CDK inhibitory properties (Hannon and Beach, 1994; Guan et al., 1994; Quelle et al., 1995) Overexpression of one of these CDI is sufficient to induce cell cycle arrest, although p19 ability to arrest cell proliferation apparently does not involve direct inhibition of any known cyclin/CDK complexes. p15 was identified as a 15 kD protein upregulated by TGF-β in human epithelial cells. Functionally similar to p16, p15 interacts with and inactivates G_1 CDKs, CDK4, and CDK6 (Hannon and Beach, 1994). Notably, the p16 gene maps, in tandem with the p15 gene, on chromosome 9p21, a region that is known to contain the MTS1 tumor suppressor gene and is mutated in several tumor cell lines. In contrast to p27, p15 is transcriptionally activated by TGF-β and it is through this mechanism that G_1 arrest is induced by TGF-β. In normal proliferating cells, p27 is partitioned to CDKs that are able to bind it, with the majority being sequestered by CDK4 and CDK6. After TGF-β treatment, p15 protein increases, binds, and saturates CDK4 and CDK6, and this results in dissociation of D cyclins as well as in the release of p27 to cyclin E/CDK2. In fact, upon induction by TGF-β, p15 accumulates into the cytoplasm, where interacts with free new synthetized CDK4 or cyclin D/CDK4 complexes. As a consequence, a complex containing cyclin D/CDK4/p15 translocates to the nucleus. Here, the binding of p27 to cyclin D/CDK4 complex is prevented by the steric interference of p15. Thus, the active cyclin D/CDK4/p27 complex of proliferating cells is replaced by an inactive cyclin D/CDK4/p15 complex, with the displaced p27 being free to interact with and to inactivate cyclinE/CDK2 complexes. As a final result, the cyclin/CDK complexes are efficiently buffered, and the G_1/S phase transition is blocked (Reinisdottir and Massague, 1997).

Tumor Promoters and Suppressors: Regulation of G_1/S Phase Progression

As described in the preceeding paragraphs, timely regulated cyclin/CDK expression and function are crucial for normal G_1 phase progression and

S phase entry. During G_1 almost all cell types, before replicating their DNA, must grow and double their size to maintain a constant average size of their daughter cells. Only after this process has been completed do cells begin to duplicate their genome. The time point at which cells commit to enter S phase occurs in mid-to-late G_1 phase and is defined as the restriction (R) point (Pardee, 1989). Before R, G_1 progression is dependent on the presence of mitogens, while after R growth factor stimulation is not longer required for further G_1/S phase progression. Thus, the R point can be defined as a critical cell cycle stage through which cells monitor the correct progression into the next step of the cell cycle. At this time most of the decision about proliferation versus quiescence is made, and the passage through R must be subject to tight regulation. Deregulation may lead to cancerous growth of cells in vivo. The first line of evidence directly linking the cell cycle machinery to oncogenic transformation came from the observation that the adenovirus E1A oncoproteins target the cyclin A/CDK2 complex (Giordano et al., 1989, 1991a,b; Pines and Hunter, 1990; Tsai et al., 1991). E1A proteins are the immediate products translated in host cells after adenovirus infection. In addition to E1A's function to transactivate other viral genes and to compel the replication of both virus and host genomes, E1A oncoproteins also have the capacity to immortalize primary cells and, in cooperation with other oncogenes such as ras or E1B, to transform host cells. The analysis of interactions between E1A and host cellular proteins has provided tangible support for the idea that E1A-mediated transformations are induced by physical interactions with cellular proteins. In addition to cyclin A and CDK2, other cell cycle controllers such as pRb, pRb2/p130, p107, and p300 are also cellular targets for E1A proteins (Moran 1993). Further evidence linking cell cycle control to malignant transformation is provided by other DNA tumor viral oncoproteins such as SV40 T antigen and human papilloma virus (HPV) E7, which transform cells via interaction with a similar set of cellular proteins. pRb, the product of the retinoblastoma tumor suppressor gene, is the most important protein regulating the passage through R and hence the progression into the S phase (Weinberg, 1995). pRb connects many extracellular stimuli to the transcriptional mechanism. In fact, it controls the expression of many genes necessary for G_1/S phase progression, and its function is regulated by different components of the cell cycle clock. pRb undergoes a number of post-translational modifications (namely, phosphorylative events) in concurrence with R transition. Before G_1 phase progression is initiated by mitogen stimulation, pRb is underphosphorylated, a form known to be active in repressing cell cycle progression. As the cells progress into G_1/S phase, pRb becomes phosphorylated, and this phosphorylation causes the inactivation of its growth inhibitory function. pRb then maintains this hyperphosphorylated configuration throughout the remainder of the cell cycle, becoming underphosphorylated once again upon emergence from M (Weinberg, 1995).

According to the pattern of pRb phosphorylation and its phosphorylation-dependent growth suppressive properties, pRb might function as the guardian of the G_1/S phase transition at the R-point gate. In response to signals, such as mitogens, that induce its phosphorylation, pRb permits the cell to proceed into late G_1.

Which are the mediators of signals influencing pRb phosphorylative status? Phosphopeptide analysis of pRb indicates the existence of about dozen distinct sites of phosphorylation on serine and threonine residues (Knudsen and Wang, 1996). These sites are targeted by cyclin/CDKs specifically induced during different stages of G_1/S phase progression. Thus, cyclin/CDK-mediated phosphorylation of pRb is likely to be the mechanism by which the growth suppressive function of pRb is turned off during the G_1/S phase transition and in the following phases of the cell cycle. Cyclins of the D class (D1, D2, and D3) are the first cyclins to be expressed upon mitogenic stimulation of quiescent cells. Synthesis of these cyclins in many cell types is growth factor dependent and occurs in early G_1 (Weinberg, 1995). Once these cyclins have been accumulated, they form complexes with CDK4 and CDK6 to regulate their function. Accumulating evidence indicates that cyclin D/CDK4/6 are most prominently implicated in the phosphorylation of pRb (Dowdy et al., 1993). D-type cyclins are unique in their ability to physically interact with pRb, recruiting CDKS to their substrate (Matsushime et al., 1994; Strauss et al., 1995). Therefore, D cyclins are perceived as a link between the cell cycle machinery and growth factor–induced signaling (Lukas et al., 1996). However, the effective difference is still unclear between the three known D cyclins (D1, D2, and D3) in binding to and determining both phosphorylative status and function of pRb. As the cell proceeds though the R point into the mid-to-late G_1, other cyclins, such as cyclins E and A, form complexes with CDK2 that which specifically target pRb phosphorylation sites to maintain pRb phosphorylation (Bartek et al., 1996). Sequential activation of cyclin E/CDK2 and cyclin A/CDK2 at late G_1 and S phase is also a determinant for initiation of DNA replication (Krude et al., 1997). This model of G_1/S phase regulation (Fig. 2.6) defines cyclins and CDKS as proteins possibly implicated in the genesis of tumors, or *tumor promoters.* In fact, deregulated expression and/or activity of these proteins would impair the growth-inhibitory activity of pRb and lead to a state of uncontrolled cell proliferation. On the other hand, loss of function of the physiological inhibitory constraint of cyclins/CDKS, i.e., CDI, or of their target, pRb, would produce the same effect. Therefore, CDI and pRb are considered *tumor suppressors,* that is, the natural counterpart of tumor promoters. Further linking pRb to the pathway to cellular transformation is the recent finding of its involvement in the ras signaling pathway (Peeper et al., 1997). Inactivation of ras results in a lack of cyclin D protein accumulation and an increase in hypophosphorylated pRb. Given ras's well-known connection to cancer, these findings boost the importance of these cell cycle proteins in keeping a cell from becoming cancerous.

Studies from mice deficient in CDI demonstrate that loss of function of p16, the specific cyclin D activity inhibitor, associates with high rates of tumorigenesis (Serrano et al., 1996), whereas p21 and p27 null mice

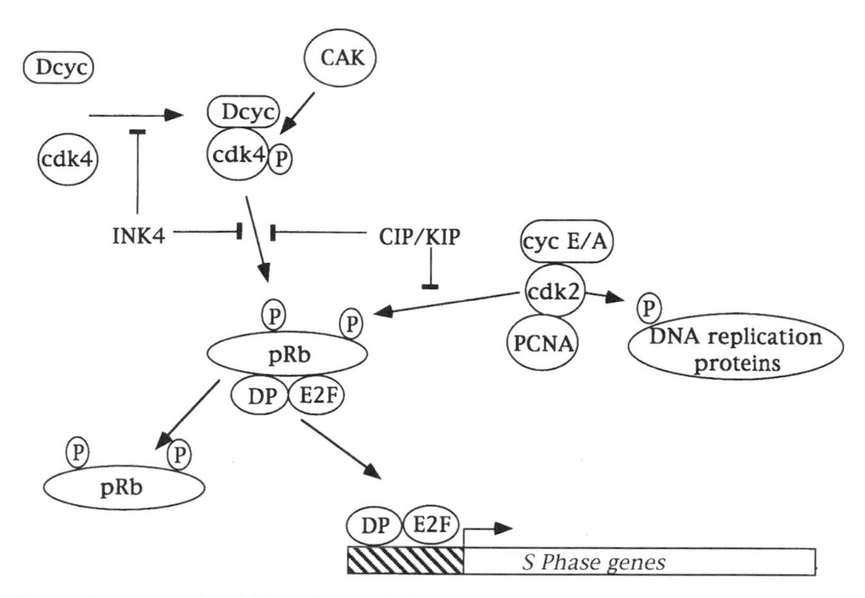

Figure 2.6. How the INK and Cip/Kip inhibitors act to halt cell cycle progression.

display, at most, a defective response following DNA-damaging agents and hyperplasia, respectively (Deng et al., 1995; Fero et al., 1996; Kiyokava et al., 1996; Nakayama et al., 1996). In addition, in humans, some tumor types, such as sarcomas and lymphomas, have frequent p16 mutations, especially the skin-specific cancer melanoma. These observations led to the hypothesis that a specific pathway, involving cyclin D, CDK4/6, p16, and pRb, plays a crucial role during the R point in regulating the entry of the cell in a mitogen-independent cell cycle progression and DNA synthesis. Disruption of such a pathway or altered function of their components results not only in reduced growth control but also in an acceleration in the rate of accumulation of genetic alterations and hence may be required for the genesis of human cancers. Nevertheless, it has been recently shown that targeted disruption of mouse p57 gene and generation of mice devoid of p57 expression reveal alterations in cell proliferation and differentiation, with several abnormalities, including abdominal muscle defects, cleft palate, endochondral bone ossification defects, renal medullary dysplasia, adrenal cortical hyperplasia and cytomegaly, lens cell hyperproliferation, and apoptosis (Zhang et al., 1997; Yan et al., 1997). Interestingly, many of these phenotypes are found in the Beckwith-Wiedeman syndrome, an hereditary disorder in which a rearrangement at the 11p15 site (where p57 resides) has been observed. Although this syndrome predisposes to the development of several childhood tumors, a direct link between neoplastic development and p57 absence has not been provided by p57 null mice (Yan et al., 1997).

pRb exerts its antiproliferative function by regulating a number of downstream effectors (Kouzarides, 1995). The best-characterized pRb targets are members of the E2F/DP family of transcription factors, generically referred as E2F. Five E2F species (E2F1–5) and three DP family

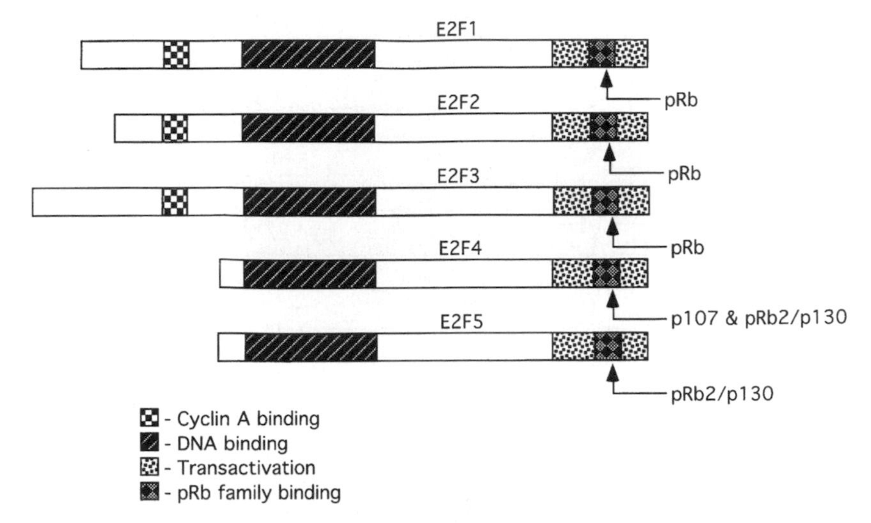

Figure 2.7. Schematic of E2F family members.

members (DP1–3) have been identified and characterized (La Thangue, 1996) (Fig. 2.7). E2F and DP members heterodimerize in various combinations to bind to a consensus sequence (E2F site) present in the promoter of genes required for progression from G_1 to S, including c-myc, N-myc, B-myb, cdc2, DHFR, thymidine synthetase, thymidine kinase, and DNA polymerase α (Adams and Kaelin, 1995). Extensive studies have shown that E2F/DP free heterodimers are active in promoting the transcription of downstream genes and that the interaction with unphosphorylated pRb represses activation of E2F-responsive genes (Chellappan et al., 1991; Weintraub et al., 1992; Nevins, 1992; Helin et al., 1993). pRb preferentially binds a subset of E2F members, including E2F1–3, via a domain (termed the Rb "pocket") also required for binding to the viral oncoproteins adenovirus E1A, SV40 T antigen and hpV E7 (Nevins, 1992; Weinberg, 1995). Both cellular proteins (E2Fs) and oncoviral products bind preferentially the un(der)phosphorylated form of pRb, and overexpression of either E2F1 or viral oncoproteins in quiescent cells is able to activate E2F-dependent transcription, leading to inappropriate S phase entry that is eventually followed by apoptotic cell death (Johnson et al., 1993; Qin et al., 1994; Shan and Lee, 1994). Therefore, overexpression of E2F or disruption of pRb/E2F complexes via sequestration of pRb by oncoviral proteins is sufficient to activate E2F-dependent transcription. Nevertheless, several observations have shown that Rb binding to E2F is required for pRb-mediated repression of transcription rather than merely representing inactivation or sequestration of E2F (Weintraub et al., 1995). Thus, at least three functional states of E2F-responsive genes can be envisioned: one repressed by the pRb/E2F/DP complex, another activated by free E2F/DP heterodimers, and a basal state where the transcription is E2F independent (Weintraub et al., 1995; Tommasi and Pfeiffer, 1995; Zwicker et al., 1996). According to this model, E2F1 activity is critical to drive the cell into S phase, but it is also important for the cell growth suppressive activity of pRb. Inactivation of the E2F1 gene in the mouse germline results in hyperplasia and even neoplasia instead of the expected proliferative failure (Field et al., 1996; Yamasaki et al., 1996). These results lead to the speculation that E2F1 can exert an important role in repressing the transcription of growth promoting genes by recruiting pRb. An alternative explanation of the E2F1 null mice phenotype is that E2F1 can in vivo modulate the balance between cell survival, death, and postmitotic differentiation in certain tissues. In cells induced to progress into G_1/S phase by mitogens, E2F1 participation in promoting S phase entry is restricted to a window late in G_1, when E2F1 stimulates the expression of genes required for DNA synthesis. Moreover, active E2F1, released from pRb, fails to induce S phase in the presence of dominant-negative mutants of CDK2 and CDK3, suggesting that CDK3 also contributes to activation of E2F (Hofmann et al., 1996). As cells have had access to S phase, the E2F1 function is no longer required, and its DNA-binding capacity is regulated by the S phase–specific cyclin A. Indeed, cyclin A/CDK2 complexes target E2F1/DP1 heterodimers and inhibit their binding to DNA by direct phosphorylation of DP1 (Krek et al., 1994). Thus, the function of E2F1 is downregulated by factors (cyclin A/CDK2) that have been induced by E2F1 itself, and this autoregulatory loop may serve to prevent deregulated E2F1 function. Artificial elimination of this cyclin A function (or by extension excessive accumulation of E2F1) results in S phase arrest followed by apoptosis (Krek et al., 1994). E2F1 function also depends on its expression levels. Higher expression of E2F1 correlates with greater potential for both cell killing and transformation. E2F1 synthesis fluctuates during the cell cycle, with almost undetectable levels in cells emerging from G_0 and relative accumulation in mid-to-late G_1 (Moberg et al., 1996). In this regard it is noteworthy that E2F1 abundance results both from its transcription and from its degradation. It has been recently demonstrated that a function of pRb is in protecting E2F1 from degradation by the ubiquitin–proteosome pathway (Hofmann et al., 1996; Hateboer et al., 1996). In this context, pRb binding to E2F1 in G_1 would stabilize a pRb/E2F1/DP1 complex that allows pRb itself to express its control on the G_1/S phase progression. This complex picture of effects has led to the uncertainty of whether E2F1 is defined as a tumor promoter or a tumor suppressor.

The pRb-bound E2F species does not account for the entire E2F activity during the cell cycle. In fact, the most predominant E2F species in quiescent cells are E2F4 and E2F5 (Moberg et al., 1996). At this stage of the cell cycle, E2F4 is found in a complex with a pRb-related protein, pRb2/p130. pRb-related proteins are the products of the p107 and pRb2/p130 genes (Whyte, 1995; Paggi et al., 1996). These proteins show related structures and similar biochemical properties to pRb (Fig. 2.8). Much of the sequence similarity among pRb, p107, and pRb2/p130 resides in the domain that mediates the interaction with E2F/DP members and viral oncoproteins (Mayol et al., 1993; Paggi et al., 1996). Like pRb, p107 and pRb2/p130 interact with and inhibit the transcriptional activity of E2F/DP heterodimers, and oncoviral disruption of these complexes

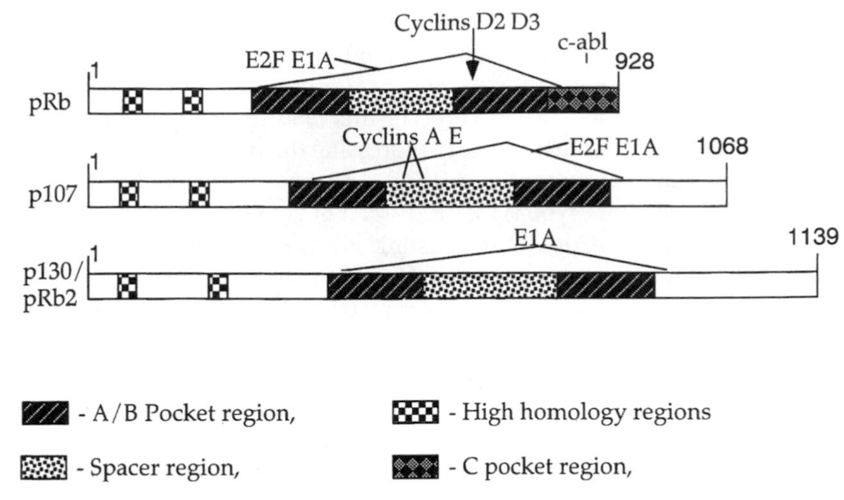

Figure 2.8. Schematic of pRb family members.

reinduce E2F site–dependent transcription and at some extent cell cycle progression (Whyte, 1995). Both p107 and pRb2/p130 are also putative substrates of G_1 cyclin/CDK substrates (Baldi et al., 1995; Claudio et al., 1996). Although both p107 and pRb2/p130 have growth suppressive properties, these proteins appear to function in growth-signaling pathways distinct from those involving pRb (Vairo et al., 1995). As an example, ectopic expression of either p107 or pRb2/p130 suppresses proliferation in cell lines where pRb overexpression is uneffective in blocking cell cycle progression (Zhu et al., 1993; Claudio et al., 1994). This functional difference is further supported by distinct properties shown by pRb, p107, and pRb2/p130. First, p107 and pRb/p130 interact with a subset of E2F proteins different from those of pRb. In fact, p107 and pRb2/p130 exclusively bind E2F4 and E2F5, whereas pRb specifically binds E2F1–3, although an E2F4/Rb complex is detectable during G_1/S phase transition, suggesting that pRb might play the role of universal regulator of E2F/DP member function (Moberg et al., 1996). Second, unlike pRb, p107 and pRb2/p130 associate stably with the G_1-specific cyclin E/CDK2 and cyclin A/CDK2 kinases to form E2F-containing complexes still able to bind E2F sites (Whyte, 1995). These complex formations coincide with the timing of appearance of cyclin E and cyclin A during G_1/S phase progression and is inhibited by increased expression of the CDI p21 and p27 (Shiyanov et al., 1996; Zerfass-Thome et al., 1997). The structural basis for CDI disruption of E2F complexes is provided by the high homology in the region of p107 and p21 required for cyclin/CDK2 binding (Zhu et al., 1995). Although the specific function of these cyclin/CDK-containing E2F complexes remains undefined, it is possible that cyclins E and A complexed to CDK2 are recruited by E2F complexes at the DNA level in late G_1 and S phase. At that time, cyclin E/CDK2 and cyclin A/CDK2 synergistically trigger initiation of DNA replication (Krude et al., 1997). Finally, analysis of pocket protein knockout mice reveals that, unlike pRb knockout mouse embryos, which die in utero, homozygous knockouts of p107 and pRb2/p130 had little discernible effects on development or on the growth of mouse embryo fibroblasts in culture, suggesting that a functional redundance may exist among these proteins. Furthermore, mice concomitantly nullizygous for both p107 and pRb2/p130 exhibited deregulated chondrocyte growth and neonatal lethality, but not increased tumor incidence, as opposed to heterozygous pRb +/− mice, which develop several cancers (Cobrinik et al., 1996). These results suggest that, although in certain specific settings p107 and pRb2/p130 perform growth regulatory functions that are not fulfilled by pRb, pRb is unique in the pocket protein family as a tumor suppressor protein. In accordance with these findings, pRb is frequently lost or mutated in many human cancers, whereas p107 does not appear to be targeted by tumor promoting mutations (Weinberg, 1995). Preliminary data, however, suggest that pRb2/p130 may be lost in a number of transformed cell lines, such as the HONE-1 nasopharyngeal carcinoma line (Claudio et al., 1994; Cinti and Giordano, unpublished data). Interestingly, an analysis of the levels of expression of E2F-responsive genes in cells lacking pRb, p107, or pRb2/p130 has been recently carried out, showing that these pocket proteins specifically regulate distinct sets of genes involved in the regulation of the cell cycle, with p107 and pRb2/p130 exerting overlapping functions (Hurford et al., 1997). All these data suggest that pRb, p107, and pRb2/p130 cooperate to confine the activation of E2F, and therefore the expression of E2F-responsive genes, to precise stages of the cell cycle. A complex picture of E2F-dependent transcriptional regulation therefore occurs along the cell cycle (Fig. 2.9) (Paggi et al., 1996), involving all of the pocket proteins: In G_0, E2F4 and E2F5 are associated with pRb2/p130 in a transcriptionally inactive complex. Upon mitogenic stimulation, pRb2/p130-released and newly synthesized E2F4 accumulates in free form and initiates a cascade of events leading to the expression of other E2F/DP members (Paggi et al., 1996). E2F/DP heterodimer activity is then controlled at the G_1/S phase transition by the interplay between p107 and pRb, which are sequentially activated and phosphorylate these pocket proteins to hamper the formation of E2F-containing complexes. It has been recently shown that a further level of E2F/DP regulation might occur at the cytoplasmic level. The observation that E2F4 and E2F5, as well as DP1, are devoid of a nuclear localization signal (nls) (Magae et al., 1997; de la Luna et al., 1997) prompted many investigators to study the subcellular localization of these proteins and the mechanism(s) responsible for their nuclear import. Different groups (Magae et al., 1997; de la Luna et al., 1997; Lindeman et al., 1997; Muller et al., 1997; Verona et al., 1997; Allen et al., 1997; Puri et al., 1998) have shown that DP1, E2F4, and E2F5 subcellular localization is cell cycle regulated, with a nuclear or cytoplasmic distribution that changes depending on the specific cell cycle phase and the cell type studied. As for E2F4, it is mostly nuclear in G_0/G_1 and accumulates in the cytoplasm upon mitogenic stimulation of G_1/S phase progression. Accordingly, when these proteins lacking the nls are ectopically expressed, they accumulate in the cyto-

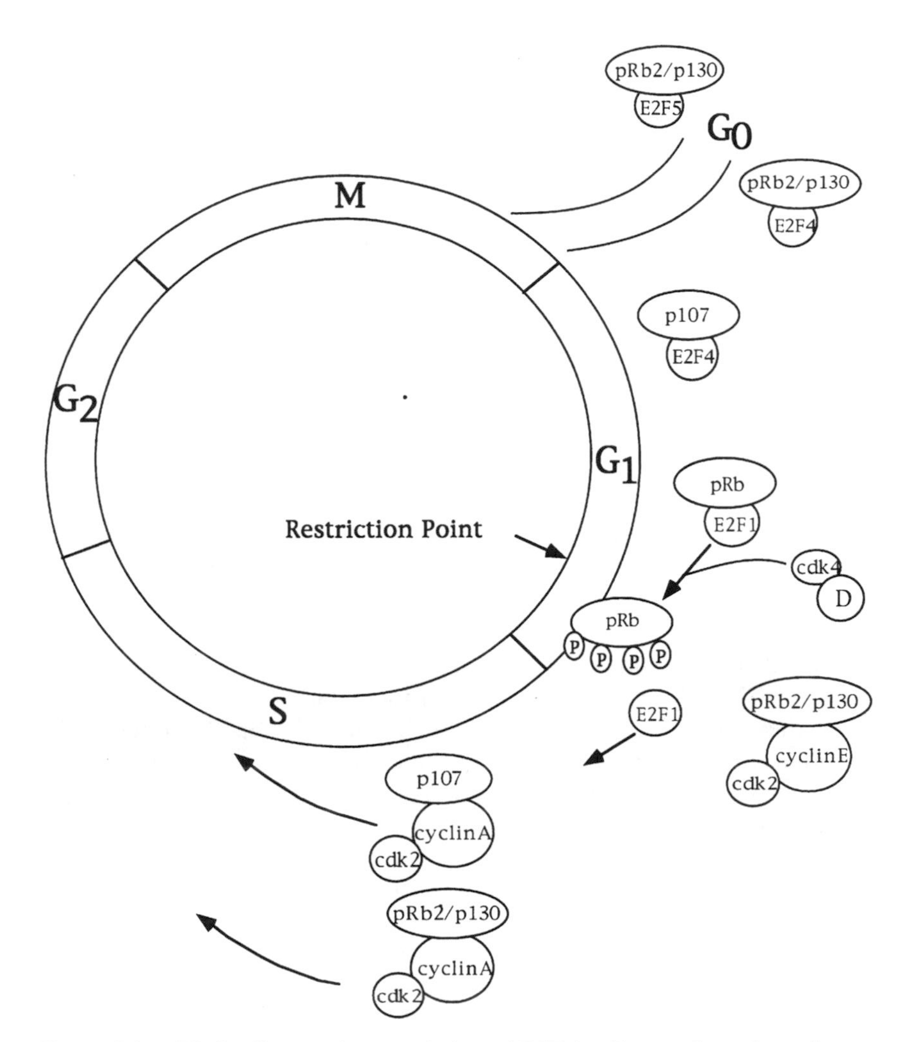

Figure 2.9. pRb family member regulation of E2F family members throughout the cell cycle.

plasm and do not affect the cell cycle progression. While E2F1 co-expression is sufficient to induce DP1 nuclear translocation, E2F4 and E2F5 nuclear accumulation can be efficiently induced by co-expression of "pocket" proteins or by some spliced variants of DP3 that have the nls. In keeping with this ideas the effect on the cell cycle of E2F4/5 nuclear accumulation depends on the specific partner that provides them the nls, with "pocket" protein–induced nuclear translocation that correlates with cell growth suppression, while DP3-mediated E2F4/5 nuclear entry produces a mitogenic effect (Allen et al., 1997; Puri et al., 1998). However, because most of these studies are based on protein overexpression, the physiological regulation and the relevance of this cytoplasmic pathway remain to be defined.

The functional difference between pRb and the other pocket proteins can also reside in the peculiar property of pRb to regulate gene transcrip-

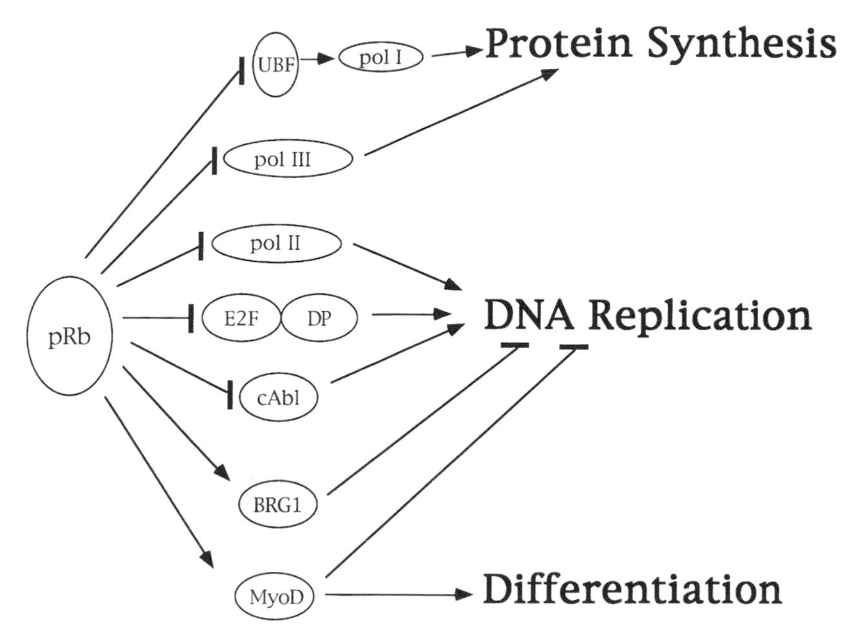

Figure 2.10. The many functions of the Rb protein.

tion via interaction with a number of other proteins (Fig. 2.10). Briefly, pRb can repress the activity of all three RNA polymerases (pol I, pol II, and pol III), thereby globally influencing cellular biosynthesis (Kouzarides, 1995). In addition, pRb can interact, through a domain in the C terminus outside of the A/B pocket, with cAbl, a tyrosine kinase that can enhance transcription during S phase, and phosphorylation of pRb correlates with both release of cAbl and its activation (Welch and Wang, 1993). Finally, it has been shown that pRb can interact with the differentiation-specific proteins and tumor suppressors MyoD (Gu et al., 1993) (see below) and BRG1 (Dunaief et al., 1994) and that these interactions may be relevant for cell cycle arrest and terminal differentiation.

Another tumor promoter is the protooncogene c-myc. Protooncogenes are proteins that play a critical role in cell growth control and that are abnormally activated in certain tumors by structural and functional alterations. On the basis of their cellular location and functional properties they are classified as growth factors, growth factor receptors, signal transducers, protein kinases, or transcription factors. Among the transcription factors, c-myc is one of the most widely studied protooncogenes (Henriksson and Luscher, 1996). A number of human tumors display genetic alterations at the c-myc locus. c-myc protein (cMyc), in conjunction with its dimerization partner Max, is capable of inducing activation of G_1 cyclin/CDK complexes and S phase entry in the absence of mitogen stimulation. In addition, cMyc has been implicated in the induction of apoptosis by growth factor deprivation. However, few transcriptional targets of cMyc have been identified. More recently, it has been shown that cdc25 phosphatases are a target of cMyc and can represent the link between cMyc function and the cell cycle machinery (Galaktionov et al., 1996). In fact, the Myc/Max heterodimer binds to Myc-binding sites in the cdc25A gene to activate cdc25A transcription. cdc25A induction plays an essential function in G_1/S phase transcription. The expression of this phosphatase peaks as early as 3 hours after mitogen stimulation in quiescent fibroblasts and activates CDKs by dephosphorylation on specific tyrosine residues. This coincides with c-myc induction and cMyc-dependent stimulation of cyclin D/CDK4 and cyclin E/CDK2 in the same cells, which also involves ras cooperation (Leone et al., 1997). Therefore, cMyc-dependent cdc25A expression following growth factor stimulation might induce CDK-dependent pRb phosphorylation and mediate the oncogenic activity of cMyc.

CELL CYCLE CHECKPOINTS FROM YEASTS TO MAMMALS

As mentioned, cell cycle checkpoints are regulatory pathways that control the orderly and timely succession of cell cycle transitions and ensure that DNA replication and chromosome segregation are completed with high fidelity. In addition, checkpoints respond to DNA damage by arresting the cell cycle, thus providing time for repair, and by inducing transcription of genes that facilitate repair. Loss of checkpoint control is thought to result in genomic instability and may be implicated in cell transformation (Hartwell, 1992). Many components of these checkpoint pathways are shared among all eukaryotes (Table 2.3).

S Phase and DNA Damage Checkpoints in Yeasts

Many genes that are required for the S phase checkpoint (dependence of mitosis on completion of DNA replication) and the DNA damage checkpoint (dependence of mitosis on undamaged DNA) have been identified in the yeasts *Sac. cerevisiae* and *Sch. pombe* (Table 2.3). Many of these checkpoint genes have recently been sequenced, providing new information on how checkpoints work and reveal similarities and differences between budding yeast and fission yeast, presumably due to their different modes of cell division (Elledge, 1996; Stewart and Enoch, 1996) (Fig. 2.11A,B).

At the center of checkpoint control are evolutionarily conserved checkpoint kinases belonging to the phosphatidylinositol-3 (PI_3) family: Mecl(Sc) in *Sac. cerevisiae,* Rad3+(Sp) in *Sch. pombe,* and the ATM (ataxia telangiectasia mutated) gene product in humans (Enoch and Norbury, 1995; Stewart and Enoch, 1996). A number of related genes play a role in DNA damage checkpoints in flies and mice (Zakian, 1995). Mecl(Sc) and Rad3(Sp) are required for both the S phase and the DNA damage checkpoints. In *Sac. cerevisiae,* Mecl(Sc) is also required, together with the Rad53(Sc) gene, for a newly described checkpoint that slows the rate of DNA replication in the presence of alkylating agent–induced DNA damage (Paulovitch and Hartwell, 1995). The recent isolation and characterization of the gene encoding the catalytic

TABLE 2.3. DNA Replication and Damage Checkpoint Genes in Yeasts and Mammalians

Organism			
Sac. cerevisiae	*Sch. pombe*	Mammals	Activity
Rad9			
Rad24	rad17		RF-C related
Rad17	rad1		Nuclease
Mec3			
Mec1	rad3	ATRPI-3	Kinase
TEL1		ATMPI-3	Kinase
Rad53	cds1		Kinase
Pol II	cdc30	pol ε	polymerase
DPB11	CUT5		
Chk1[1]	chk1		Kinase
PDS1			
DUN1			Kinase
BMH1,2[1]	Rad24, 25	14-3-3	
Rad26			
Rad9		HRAD9	
		p53	Transcription
		p21	CDK inhibitor

[1]Genes in *Sac. cerevisiae* that have not yet been proven to have checkpoint functions.

subunit of the DNA-dependent protein kinase (DNA-PK) showed that it encodes a protein with similarity to a subgroup of PI_3 kinases (Hartley et al., 1995). DNA-PK is required for immunoglobulin gene rearrangement, recombination, and repair of radiation-induced double-stranded breaks (DSBs) (Jeggo et al., 1995; Weaver, 1995). Neither the DNA-PK holoenzyme nor the catalytic subunit on its own possesses lipid kinase activity. Two regulatory subunits, Ku70 and Ku80, which bind to DSBs, are required for DNA-PK to function as a serine/threonine protein kinase. The mode of activation of DNA-PK by DSBs may serve as a model for the activation of checkpoint kinases. DNA damage or S phase arrest may be accompanied by the formation of DNA structures that are able to recruit the checkpoint kinases into catalytically active complexes (Stewart and Enoch, 1996).

Other pairs of checkpoint genes that are conserved between *Sac. cerevisiae* and *Sch. pombe* are Rad17(Sc)/Rad1+(Sp) (encoding putative 3′-5′ exonucleases) (Lydall and Weinert, 1995) and Rad24(Sc)/Rad17+(Sp) (displaying a limited homology to the human replication factor C (RF-C), which is required for DNA polymerase δ and ε binding to primed DNA) (Griffiths et al., 1995). Interestingly, the mammalian tumor suppressor protein p53, which is implicated in many cell cycle checkpoints, has been recently demonstrated to possess 3′–5′ exo-

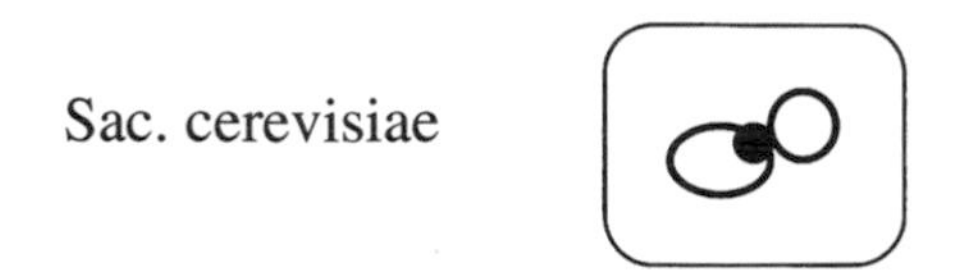

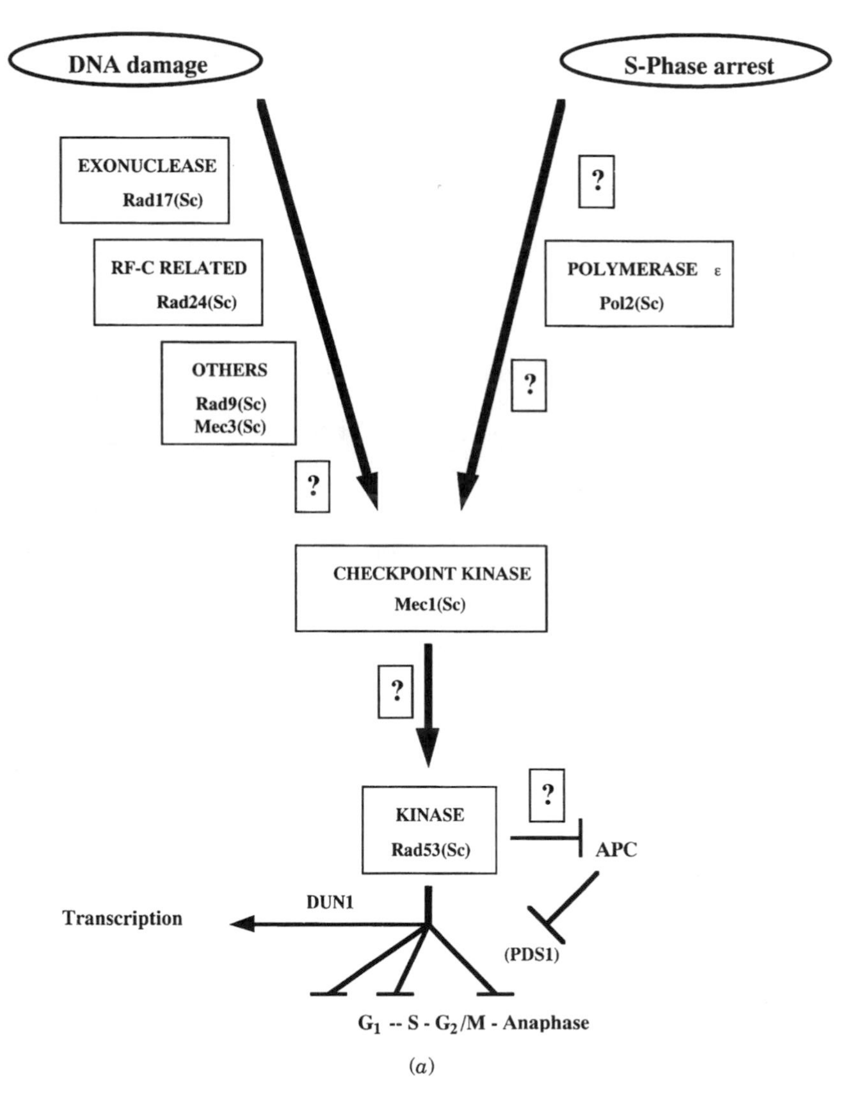

(*a*)

Figure 2.11. DNA damage checkpoint pathways in **(A)** *Sac. cerevisiae,* **(B)** *Sch. pombe,* and **(C)** mammalian systems.

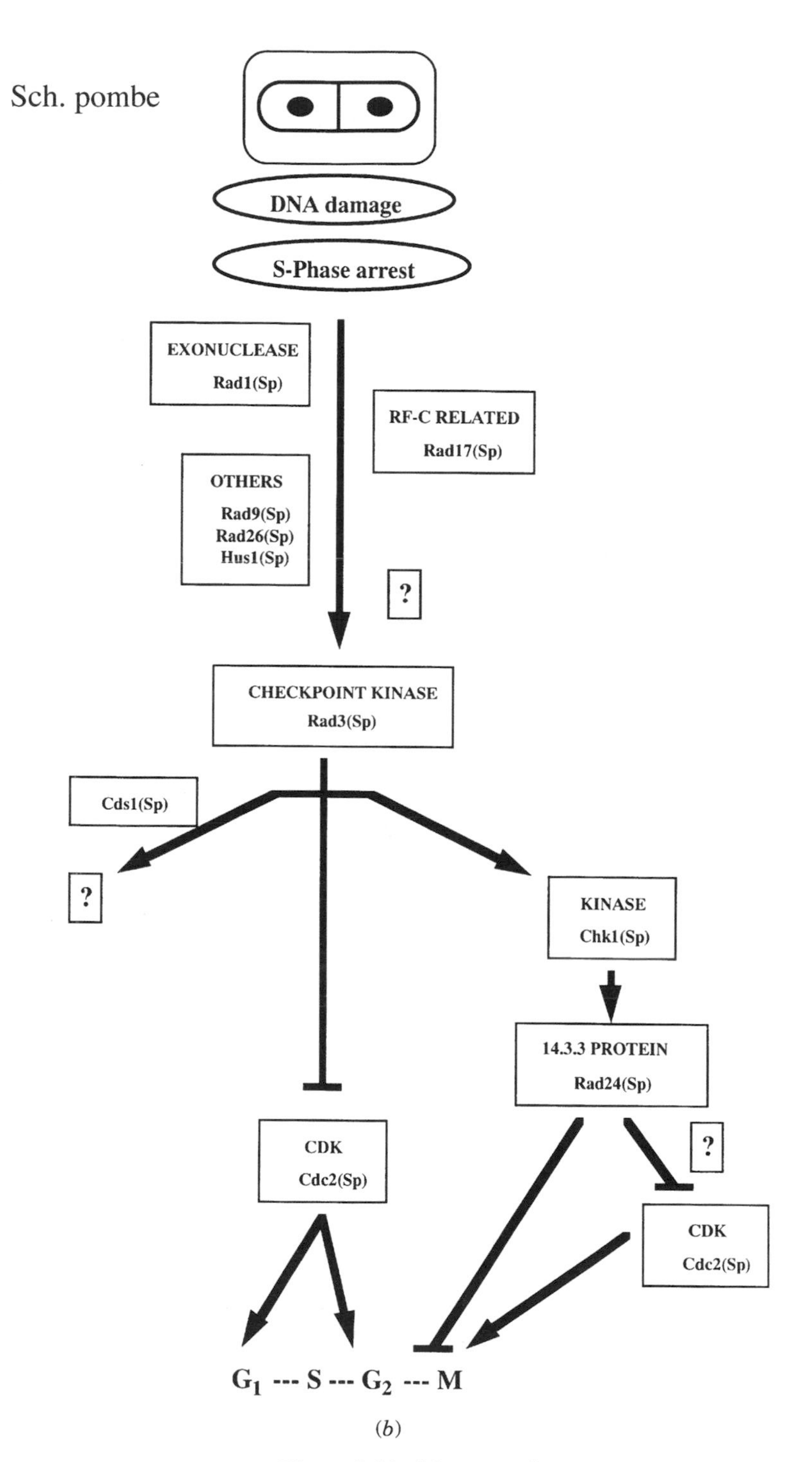

(*b*)

Figure 2.11. (*Continued*)

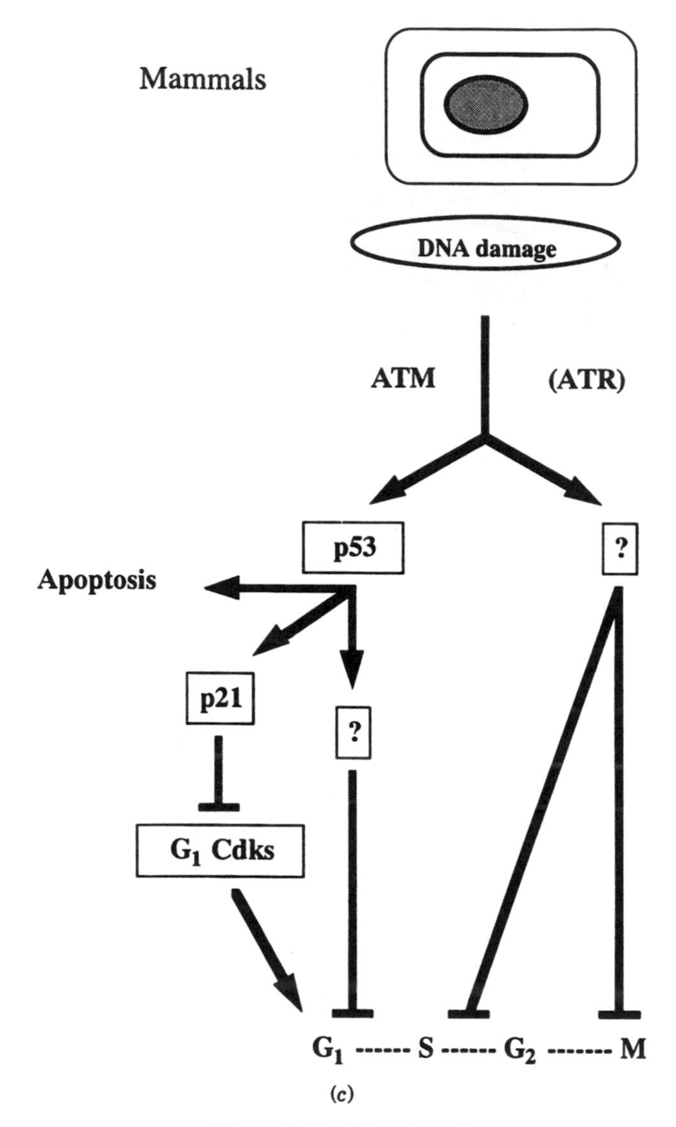

(*c*)

Figure 2.11. (*Continued*)

nuclease activity (Mummenbrauer et al., 1996). Whether these proteins are intrinsic components of the checkpoint kinase (functioning like Ku70 and Ku80) or are involved in sensing primary DNA lesions and converting them into DNA structures that activate the checkpoint kinases is not clear at this time. If DNA structures are involved, single-stranded DNA is the best candidate (Lydall and Weinert, 1995). DNA damage–induced DSBs are rapidly converted into molecules with long 3′ tails. Rad24(Sc) is involved in the formation of single-stranded DNA, and, while it is likely that Rad17(Sc) functions similarly, nothing is known about the equivalent *Sch. pombe* proteins Rad1(Sp) and Rad17(Sp). Single-stranded DNA may also activate the S phase checkpoint as well, as large regions of ssDNA have been detected in mammalian cells replicating their DNA. Data from both yeasts suggest that a functional replication complex is required (Stewart and Enoch, 1996). However, the S phase checkpoint in *Sch. pombe* requires both Rad1(Sp) and Rad17(Sp) (Enoch et al., 1992), while Rad17(Sc) and Rad24(Sc) are not required in *Sac. cerevisiae* (Weinert et al., 1994). This implies either that in *Sac. cerevisiae* different proteins generate single-stranded DNA in S phase arrest and DNA damage or that in *Sac. cerevisiae* the S phase checkpoint is activated by mechanisms other than single-stranded DNA. The link between the DNA replication machinery and the S phase checkpoint is supported by the observation that in *Sac. cerevisiae* cells with C-terminal truncations of the catalytic subunit of DNA polymerase α or pol II, are defective for the S phase checkpoint but not for the DNA damage checkpoint and can replicate their DNA correctly (Navas et al., 1995).

Targets of the checkpoint kinases Mec1(Sc) and Rad3(Sp) have been recently identified. In *Sac. cerevisiae* the Rad53(Sc) protein kinase is activated by trans-phosphorylation in a Mec1-dependent manner in response to both DNA damage and S phase arrest (Sanchez et al., 1996; Sun et al., 1996). However, phosphorylation of Rad53(Sc) is observed upon DNA damage in Mec1(Sc) mutants by overexpressing Tel1, suggesting that Tel1 and Mec1(Sc) have redundant functions in the activation of Rad53(Sc) (Sanchez et al., 1996; Sun et al., 1996). The in vivo targets of activated Rad53(Sc), however, have not been identified. A Rad53(Sc) homologue in *Sch. pombe,* Cds1(Sp), has been recently discovered, but its function may be different because Cds1(Sp) mutants have a phenotype that differs from those of other *Sch. pombe* checkpoint mutants (Murakami and Ikayama, 1995). In *Sch. pombe,* Rad3(Sp) phosphorylates different targets in response to DNA damage and S phase arrest. The Chk1(Sp) putative protein kinase is activated, by autophosphorylation, after DNA damage in a Rad1+(Sp), Rad3+(Sp), Rad9+(Sp), Rad17+(Sp), and Rad26+(Sp) dependent manner (Walworth and Bernards, 1996). The fission yeast Chk1 gene is not required for normal growth, but plays an essential role for yeasts to survive radiation exposure (Walworth and Bernards, 1996). It encodes a Ser/Thr protein kinase that becomes phosphorylated after exposure to ultraviolet light, radiation and alkylating agents. After DNA damage, Chk1 is essential for G_2 arrest induction. A proposed mechanism for Chk1 G_2 arrest in response to ultraviolet light involves Chk1-mediated phosphorylation of wee1, which in turn inhibits cdc2 activity by phosphorylation at Tyr-15 (O'Connel et al., 1997). The G_2/M checkpoint induced by radiation, however, can occur even in the absence of wee1 or mik1 (the two cdc2 kinases) (Furnari et al., 1997). In this case, cdc25, the phosphatase for cdc2, has been implicated as a downstream target for Chk1 activity. Indeed, Chk1 forms a complex with and phosphorylates cdc25 at Ser-216, an event that occurs during S and G_2 phases but not in metaphase (Furnari et al., 1997; Peng et al., 1997). This Chk1/cdc25 pathway is conserved through evolution, because human Chk1 has been cloned and resides in the nucleus, having the ability to phosphorylate three human homologues of cdc25 (cdc25A, -B, -C) (Sanchez et al., 1997). Interestingly, overproduction of a Ser-216–mutated cdc25 in a human cell line (HeLa cells) abrogates radiation-induced G_2 arrest (Peng et al., 1997). Genetic studies indicate that Chk1(Sp) interacts with the 14-3-3 adaptor proteins encoded by Rad24+(Sp), which is specific to the DNA damage checkpoint as well, playing an essential role in the G_2/M checkpoint response to the DNA damage (Ford et al., 1994). In fact, phosphorylation of cdc25 at Ser-216 does not alter its phosphatase activity, but instead generates a phosphoepitope motif that mediates the direct binding to 14-3-3 (Sanchez et al., 1997; Peng et al, 1997). Thus, genetic studies in *Sch. pombe* have defined a G_2/M checkpoint pathway activated by DNA damage and involving a Rad3-dependent activation of Chk1. This latter phosphorylates cdc25 to induce cdc25/14-3-3 association, which in turn keeps cdc25 phosphatase sequestered in an inactive complex, thereby preventing it from dephosphorylating and activating cdc2. Inactivation of cdc2 will finally prevent G_2/M phase transition. This part of the checkpoint pathway is yeast-species specific: *Sac. cerevisiae* does not have chk1+(Sp)–related genes, and the 14-3-3 protein encoded by the *Sac. cerevisiae* BMH1 gene does not work in checkpoint controls. The ultimate target of the S phase checkpoint in *Sch. pombe* is the inhibitory phosphorylation of Tyr-15 on the cdc2(Sp) kinase (Enoch and Nurse, 1990). Again, this is yeast specific, and in *Sac. cerevisiae* tyrosine phosphorylation of cyclin-dependent kinases is not involved in either DNA damage or S phase checkpoints (Amon et al., 1992). In higher eukaryotes, CDK tyrosine phosphorylation is required for checkpoint control as in *Sch. pombe* (Terada et al., 1995). In conclusion, similarities and differences have been unraveled in the molecular organization of the DNA damage and the S phase checkpoints in the budding and fission yeasts.

Mammalian DNA Damage Checkpoints

Mammalian cells display a cell cycle response to DNA damage similar to that of yeasts, but, in addition, they can also activate the cell death pathway (Fig. 2.11C). Indeed, for a multicellular organism, the elimination of damaged cells is an acceptable strategy in view of the final goal of the survival of the organism as a whole. Apoptotic responses are discussed elsewhere in more detail. The G_1 DNA damage checkpoint is controlled by three genes: mutated in ataxia telangiectasia (ATM) (Savitsky et al., 1995), p53 (Ko and Prives, 1996), and p21 (Brugarolas et al., 1995; Deng et al., 1995). The tumor suppressor gene p53 is the most widely studied. It encodes a transcription factor that is activated in response to DNA damage and perturbation of the nucleotide pools. Cells deficient in p53 are unable to arrest in G_1 in response to γ-irradiation and show reduced apoptosis. At least part of p53's ability to induce G_1 arrest is linked to the transcriptional activation of the gene encoding the S phase CDK inhibitor p21 (Ko and Prives, 1996; Sherr, 1996). However, mice lacking p21 have a less severe defect in G_1 arrest than p53 $-/-$ animals, and the existence of a second p53-dependent G_1 arrest pathway is likely. Tyrosine phosphorylation of CDK4 is required for G_1

arrest in response to ultraviolet irradiation, which is a good candidate as a final target for this putative p21 independent pathway (Terada et al., 1995). p21 controls checkpoint function in human cells as well (Waldman et al., 1995). Failure to destroy the Sicl inhibitor in yeast cells may represent the mammalian functional counterpart of p21 inhibition. p53 increases both its stability and its activity as a transcription factor increase in response to DNA damage.

Despite intensive study, the mechanism by which p53 senses DNA damage remains elusive. Because cells lacking ATM have a reduced and delayed p53 response to DNA damage (Kastan et al., 1992), and, in view of its relationship with the yeast proteins Mecl and Rad3, ATM is thought to act upstream p53 and to contribute to its activation. A recent report has suggested that radiation-induced and ATM-dependent activation of p53 is mediated by the phosphorylation of p53 at Ser-15, possibly by DNA-PK, and that this phosphorylation can inhibit the binding with the p53 repressor MDM2 (Shieh et al., 1997). Nevertheless, ATM mutants still apoptose in a p53-dependent manner in response to DNA damage, and ATM-independent mechanisms for p53 activation await to be defined. In fact, Ser-15 phosphorylation (Shieh et al., 1997) and activation (Xu and Baltimore, 1996) of p53 occur in ATM-deficient cells as well. On the other hand, functional ATM is required for activation of cAbl tyrosine kinase in response to radiation. Upon irradiation, cAbl phosphorylation at Ser-465 by the PI_3 kinase domain of ATM activates cAbl tyrosine kinase (Baskaran et al., 1997; Shafman et al., 1997). DNA-PK has also been reported to function as an activator of cAbl (Kharabanda et al., 1997). Although cAbl can phosphorylate the C-terminal repeat domain of RNA pol II (Baskaran et al., 1996), thereby influencing the transcription of some genes in response to DNA damage, other physiological downstream targets for cAbl tyrosine kinase are still unknown, and the matter of whether cAbl may cooperate with p53 in the induction of cell cycle arrest (Wen et al., 1996) remains controversial. The observation that both ATM and p53 are frequently mutated in human cancers strongly implicates the DNA checkpoint in the prevention of cancer (Elledge, 1996; Sherr, 1996). However, although it is believed that the loss of cell cycle arrest in response to various situations leads to genomic instability and cancer, this is not formally proven. Indeed, when the ability to arrest the cell cycle is specifically affected without a broader interference with the entire signaling pathway, i.e., loss of p21 in p21 knockout mice, no increase in cancer development has been observed, and p21 mutations are very rare in human cancers (Brugarolas et al., 1995; Deng et al., 1995). A number of general questions remain. The first is how checkpoint pathways integrate with the normal control of cell proliferation and development. Inappropriate expression of certain proteins affecting cell proliferation (the protooncogene c-myc, the cell cycle regulator E2F1, the viral proteins E1A from adenovirus, (HPV E7, and pX of the human hepatitis B virus) can activate the p53-dependent checkpoint pathway (Evan, 1995; Chirillo et al., 1997). In addition, if checkpoints are involved in cancer prevention in vivo, which are the contributions of DNA repair, cell cycle arrest, and apoptosis? Indeed, the exploitation of checkpoint-mediated apoptosis pathways and sensitization to anticancer drugs, due to loss of checkpoint function, are currently main targets of many chemotherapeutic strategies.

The Spindle Assembly Checkpoint

In most cells the presence of unattached chromosomes or defects in spindle assembly activates an intracellular signaling pathway known as the *spindle assembly checkpoint* that blocks anaphase onset and stabilizes the APC substrates (Elledge, 1996). The process monitored by this checkpoint is complex: Spindle assembles; chromosomes attach to the spindle via a specialized protein structure that forms on the centromeres—the kinetochore; kinetochores of sister chromatids bind to spindle fiber on opposite spindle poles; chromosomes reach the metaphase plate. The integrity of this system is required for keeping chromosomal segregation fidelity, and its loss may contribute to aneuploidy in cancerous cells. Both the molecular identity of sensors and the nature of the signals that are sensed by the checkpoint are still under investigation. The best candidates for signals are lack of chromosome attachment to the spindle and the absence of the tension generated on the attached chromosomes (Elledge, 1996; Wells, 1996). The known transducers in the spindle assembly checkpoint—MAD1, MAD2, and MAD3 (mitotic arrest defective) (Li and Murray, 1991) and BUB1, BUB2, and BUB3 (budding uninhibited by benimidazole) (Hoyt et al., 1991)—have been isolated in screens for the failure to arrest in the presence of microtubule depolymerizing drugs. MAD and BUB mutants display an increase in spontaneous chromosome loss. The spindle assembly checkpoint signals through two kinases, BUB1 and MPS1, the product of the MPS1 gene first identified as a gene required for spindle pole assembly. MPS1 overexpression arrests the cell cycle (Hardwick and Murray, 1995), and this is dependent on the MAD/BUB genes. MPS1 represents the initial signaling event and phosphorylates MAD1 (Hardwick and Murray, 1995). MAD2 cooperates with MPS1, and MAD3 and BUB1 are placed downstream. Recent results indicate that the spindle assembly checkpoint signaling pathway targets the APC (Elledge, 1996). The checkpoint may either inhibit the activation of the APC, thus stabilizing its substrates, or modify and protect factors like CUT2 and PDS1 from the APC until chromosomes are properly attached to the mitotic spindle. In normal mitosis the intrinsic lag period before APC activation is sufficient, and the checkpoint may intervene only when the fidelity of the segregation system needs to be enhanced.

CELL CYCLE AND DIFFERENTIATION

Based on the potential for growth and differentiation, animal tissues can be classified into three groups: renewing, stable, and static. Differences among them reside in their different cell cycle responses to mitogenic stimuli. Cells from renewing tissues are periodically replaced by precursors, which are produced by continual proliferation of their progenitor cells, as observed in epidermidis and mucosal epithelia in the digestive tract. Stable tissues, represented by glandular and connective cell types, maintain certain functions but retain proliferative potential as well. Hepatocytes are an example of functional parenchymal cells that exhibit dormant growth potential. Finally, terminally differentiated cells from static tissues, such as neurons, adipocytes, and skeletal and cardiac muscle cells, lose their proliferative potential and maintain their differentiated functions throughout life. These cells irreversibly withdraw from the cell cycle and are unable to progress into the G_1/S phase and undergo DNA synthesis in response to mitogens. Thus, in static tissue, differentiation and cell multiplication are mutually exclusive processes. The biological significance of such growth regulation is given by the possibility that, once the differentiation program is completed, even one cell dividing when it should not can lead to cancer. If the problem were widespread, correct formation of tissues would be disrupted. This peculiar property of terminally differentiated cells also renders static tissues particularly susceptible to irreversible damage due to cell loss (i.e., necrosis and apoptotic cell death resulting from ischemia or other damaging stimuli). Indeed, as well exemplified in cardiac or brain infarction, damaged cardiomyocytes or neurons cannot be repaired or replaced by new progenitor cells. Conversely, the irreversible cell cycle exit of terminally differentiated cells makes them unable to transform into neoplastic cells. In fact, no tumors can arise from muscle fibers, cardiomyocytes, neurons, or adipocytes.

An amenable paradigm for terminal differentiation is offered by in vitro myogenesis (Yaffe and Saxel, 1977). The induction of mononucleated myoblast fusion into fully differentiated multinucleated myotubes recapitulates an highly ordered process of temporally separable events: upregulation of muscle-specific regulatory transcription factors (commitment), induction of proteins that irreversibly inhibit cell cycle progression (permanent cell cycle withdrawal), fusion of mononucleated cells into multinucleated myotubes (phenotypic differentiation), and expression of muscle-specific structural and contractile proteins (biochemical differentiation) (Andrés and Walsh, 1996). The final result is the formation of multinucleated myotubes with contractile properties that are unable to divide in response to growth factors. Interestingly, muscle-specific regulatory factors, such as the basic helix-loop-helix (bHLH) protein MyoD, can play a critical role in inhibiting cell proliferation, in addition to inducing biochemical differentiation (Crescenzi et al., 1990; Sorrentino et al., 1990). How is MyoD or other myogenic proteins able to trigger a program leading to an irreversible cell cycle withdrawal? Recent studies have focused on the control of cyclin/CDK activity and Rb family member functions during myogenic differentiation. All of the progress in this field stems from the first observation that the pRb functional integrity is required for the formation of multinucleated myotubes refractory to mitogenic signals (Gu et al., 1993). These studies demonstrated that pRb cooperates with MyoD and other bHLH myogenic factors, possibly by a physical interaction, and that either the functional inactivation of pRb through viral oncoproteins or the absence of pRb results in the impairment of both myogenic and cell cycle suppressive properties of bHLH regulatory proteins. Furthermore, experiments using muscle cells derived from pRb knockout mice (Schneider et al., 1994) suggest that a functional redundance between pRb and its relatives p107 and pRb2/p130 may exist to some extent because multinucleated myotubes also form in the absence of pRb. However, these pRb −/− multinucleated cells, which display p107 levels higher than normal myotubes, respond to mitogens with cell cycle re-entry and DNA synthesis. Knowing the nature of pRb, whose function is regulated by post-transcriptional modification (i.e., phosphorylation by CDK inhibits its growth suppressive properties) (Weinberg, 1995), it has been a natural consequence to search for mechanisms responsible for keeping pRb, and eventually p107 and pRb/p130, in their underphosphorylated form. Results obtained in the last 2 years have defined an interplay between myogenic factors and proteins important to cell cycle control that contribute to regulate cell cycle arrest in differentiating myocytes. It has been proposed by different groups that the antiproliferative property of MyoD might reside in its ability to induce the expression of the CDI p21 in a p53-independent manner, thereby inhibiting CDK activity during muscle differentiation (Halevy et al., 1995; Parker et al., 1995; Guo et al., 1995). The final outcome of this pathway is the maintenance of pRb (and the other pocket proteins) in an underphosphorylated form. The induction of p21 by MyoD also prevents the formation of E2F complexes containing the kinases cyclin E/CDK2 and A/CDK2 (Puri et al., 1997b), those that are possibly involved in the initiation of DNA synthesis (Krude et al., 1997). A further potentiation of this functional loop is achieved by MyoD-dependent induction of pRb gene expression in differentiating myocytes (Martelli et al., 1994). However, it is noteworthy that cells from embryos in which the MyoD gene had been inactivated undergo terminal differentiation without any abnormalities in the program of permanent cell cycle arrest (Parker et al., 1995). The normal induction of p21 levels and the conserved growth suppressive/myogenic function of pRb in MyoD −/− muscle progenitors, indicates that functional redundancy must exist among bHLH proteins. Because Myf5 has the same range of activities as MyoD in both embryos and cultured cell lines (Weintraub, 1993), it is likely that Myf5 can fulfill the antiproliferative activities of MyoD in MyoD knockout mice. This evidence demonstrates that the interplay between myogenic bHLH transcription factors CDI and pRb contributes to induce irreversible cell cycle arrest in differentiating myocytes by counteracting the proliferative potential of cyclin/CDKs and the pRb targets, E2F/DP family members. According to this model, in myocytes undergoing terminal differentiation, irreversible cell cycle exit also coincides with the downregulation of all the cyclins, with the exception of cyclin D3, and the inactivation of CDK function (Cardoso et al., 1993; Rao et al., 1994, 1995; Guo et al., 1995), which leads to the expression of unphosphorylated pRb and pRb2/p130 at high levels (Kiss et al., 1995) and the suppression of E2F/DP activity (Shin et al., 1995; Corbeil et al., 1995). Recently, another potential target for Rb and pRb2/p130 to initiate and establish cell cycle arrest during terminal differentiation has been identified in the transcriptional repressor HBP1. It is a protein belonging to the sequence-specific HMG transcription factor family, which interacts selectively with pRb and pRb2/p130. HBP1 functions as a transcriptional repressor of the cMyc promoter through E2F site and, when overexpressed, leads to efficient cell cycle arrest (Tevosian et al., 1997). The overexpression of either cyclin D1 (Rao et al., 1994; Skapek et al., 1995) or cyclin E2F1 (Wang et al., 1995) results in the inhibition of muscle differentiation. Likewise, the persistent activation of Id, a bHLH dominant-negative protein that impairs the DNA-binding activity of these myogenic factors, may inhibit irreversible cell cycle exit in myocytes even in the presence of high p21 levels (Benezra et al., 1990; Kurabayashi et al., 1994; Puri et al., 1997c). Taken together, these observations are consistent with a regulation of the cell cycle in differentiating myocytes exerted by environmental mitogenic and antimitogenic stimuli (Fig. 2.12A–F).

In the presence of a high concentration of certain growth factors, the proliferative potential of muscle progenitors is supported and terminal differentiation is prevented by a sustained expression and activity of cyclins/CDKs and Id. A decrease in mitogenic growth factors and/or the presence of positive regulators of myogenic differentiation would determine the induction of conditions permissive for either myogenic or antiproliferative activities of bHLH factors, thereby triggering the activation of the CDI/pRb pathway. Thus the balance of mitogenic and antimitogenic signals ultimately dictates the conditions for establishment of an irreversible cell cycle arrest and terminal differentiation in many tissues. Although this model would emphasize the role of cyclin D1, whose expression is typically enhanced by extracellular mitogens, and its associated kinases in the phosphorylation of specific substrates, such as pRb and bHLH myogenic regulators, the effective importance of this loop in the suppression of terminal differentiation remains undefined.

Interestingly, during neuronal differentiation, the activity of CDK5 and its regulatory counterpart p35 increases in postmitotic neurons of the central nervous system (Nikolic et al., 1996). Despite its regulatory function for CDK5, p35 does not exhibit homology with any known cyclin. This protein is expressed exclusively in differentiated neurons, and CDK5/p35 kinase activity closely parallels the extent of corticogenesis when all the other cyclin/CDK complexes are inactive. According to this pattern of expression and activity, it has been demonstrated that CDK5/p35 kinase activity is essential for neurite outgrowth during neuronal differentiation. Furthermore, CDK5 activation by p35 is not inhibited by the CDI p27, which accumulates at high levels in cortical postmitotic neurons and efficiently recognizes and inactivates other cyclin/CDK complexes, including cyclinD/CDK5. This demonstrates

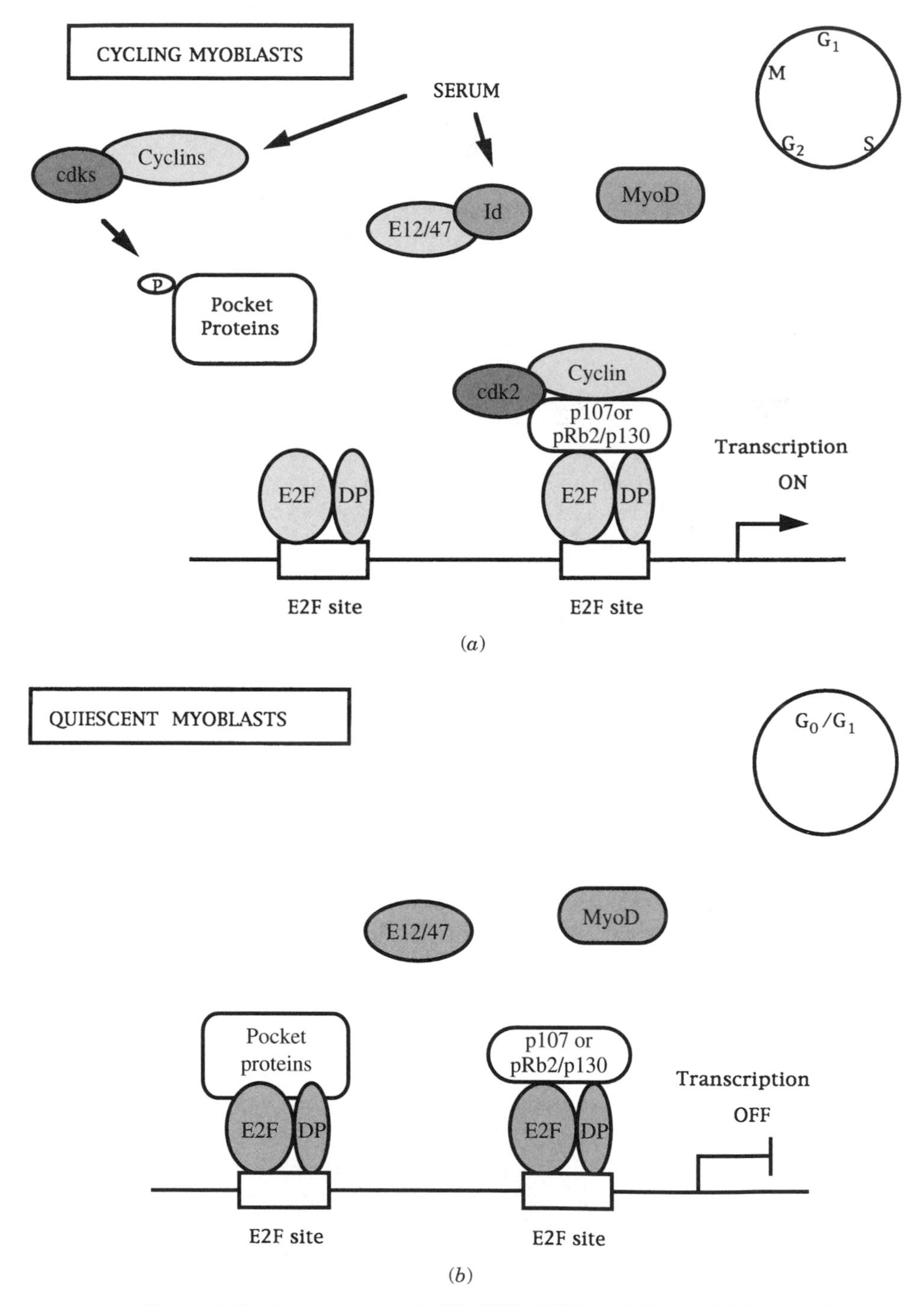

Figure 2.12. Arrangement of pRb, E2F, CDK, and CDK inhibitor family members during myogenesis at the points of **(A)** cycling, **(B)** quiescence, **(C)** quiescence with serum added, **(D)** differentiation, **(E)** differentiation with serum added, and **(F)** terminal differentiation.

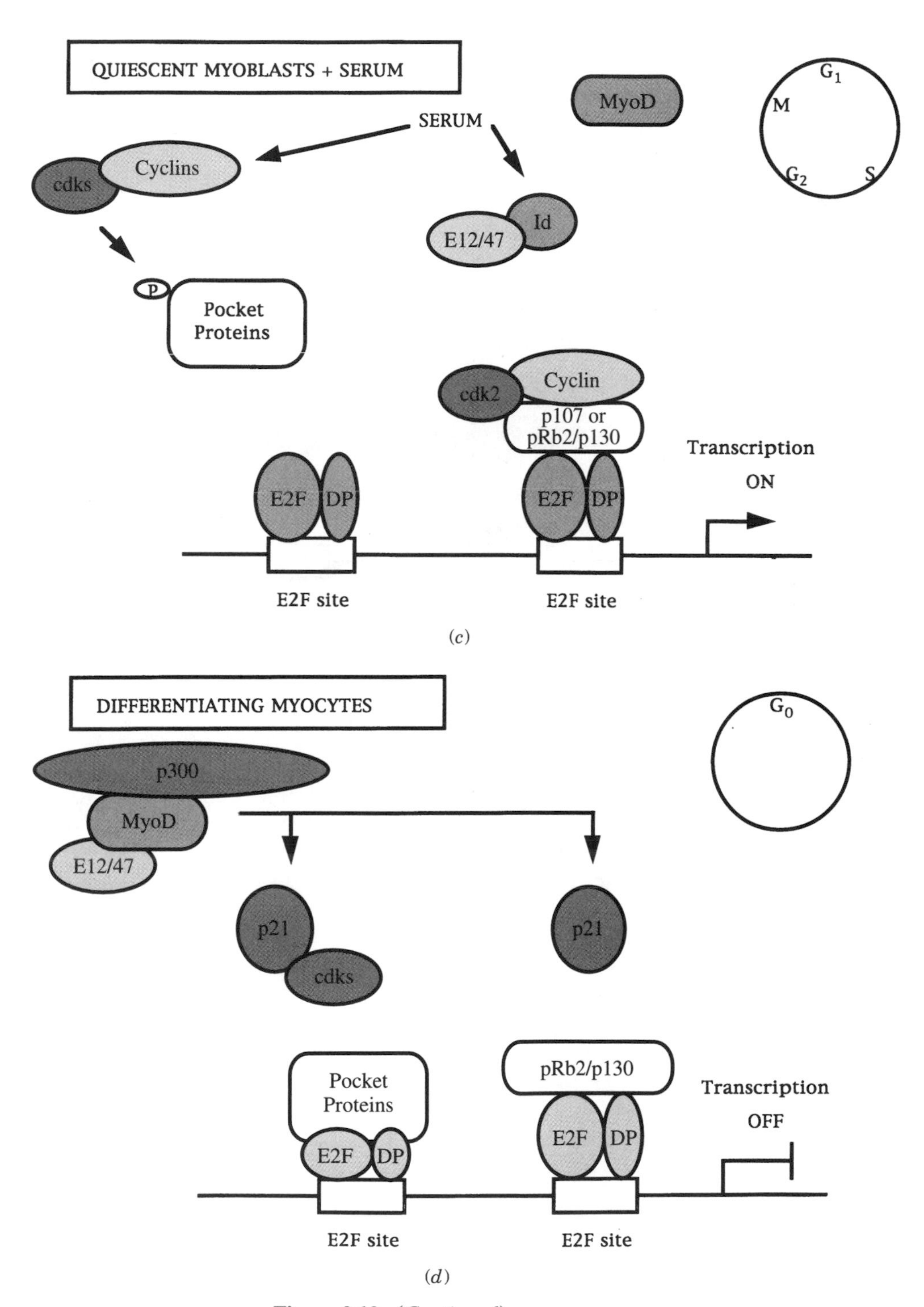

Figure 2.12. (*Continued*)

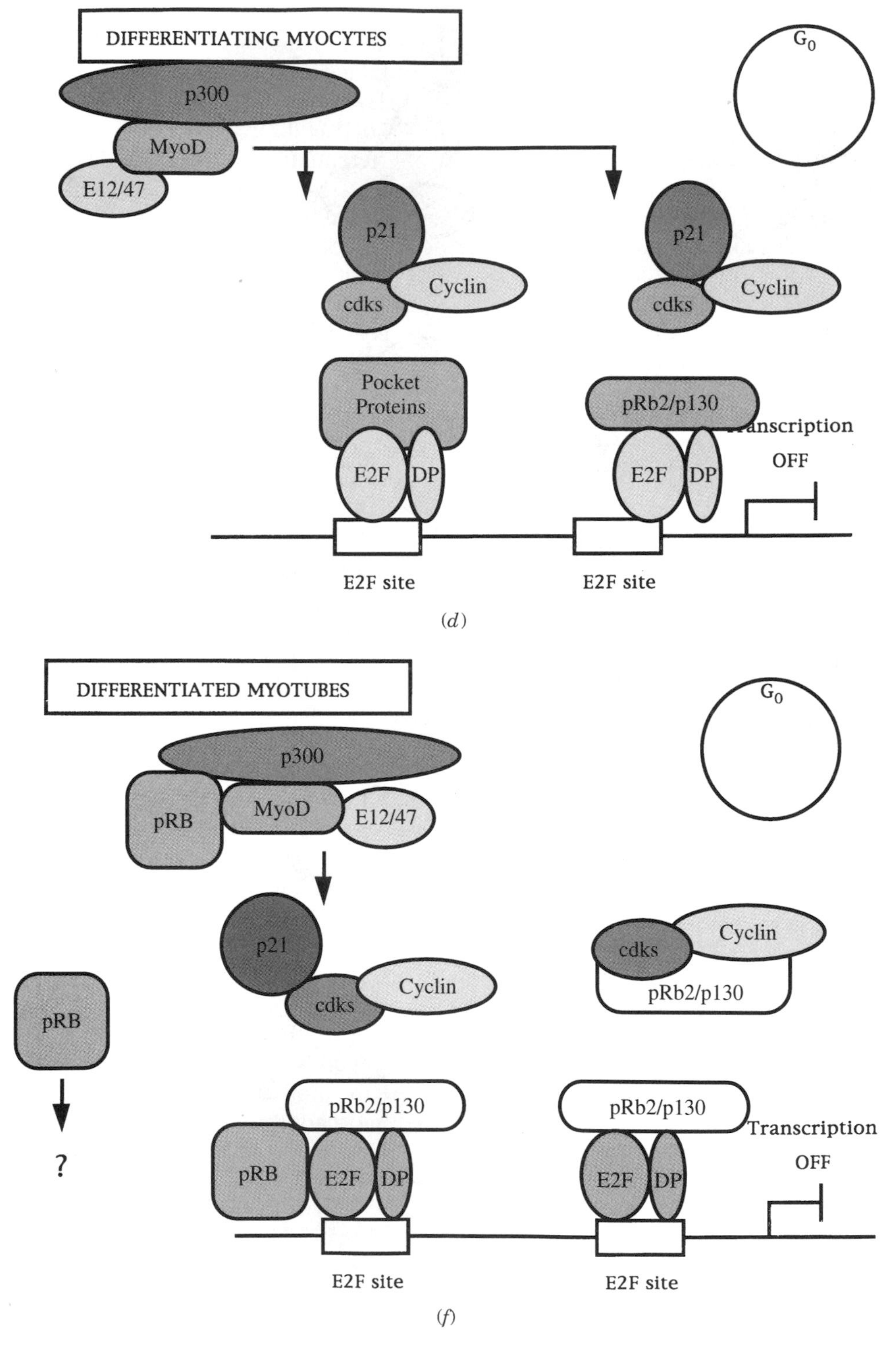

(*d*)

(*f*)

Figure 2.12. (*Continued*)

that, in some cell types, specific CDK activity can be required for induction of cell cycle arrest. The functional significance of this activity and the specific targets remain to be uncovered. Likewise, CDK5 kinase activity is required for normal somitic muscle development and CDK5 has an essential role in promoting expression of bHLH during myogenesis in vivo, although in the case of muscle differentiation CDK5 has nuclear localization (Philpott et al., 1997).

As for adipogenesis, several transcription factors collaborate to stimulate fat cell differentiation. Among them, the peroxisome proliferator-activated receptor (PPARg) seems to play in adipocytes the same cell suppressive function as MyoD in muscle cells. A possible mechanism for PPARg-dependent cell cycle arrest resides in its ability to inhibit E2F/DP DNA binding, possibly via activation of the PP2A phosphatase and consequent dephosphorylation of E2F/DP members (Altiok et al., 1997).

A p53-independent p21 (or eventually other CDI) upregulation and the maintenance of pRb (and the other pocket proteins) in its underphosphorylated form has also been demonstrated to occur as an essential regulatory mechanism during terminal differentiation in other cell types such as keratinocytes, melanocytes, hemopoietic cell lines, hepatocytes, and intestinal epithelial cells (Jiang et al., 1994, 1995; Steinman et al., 1994; Missero et al., 1995; Missero and Dotto, 1996; MacLeod et al., 1995). However, mice deficient in p21 function do not exhibit significant abnormalities in the development of most tissues (Deng et al., 1996), suggesting that different universal CDIs (i.e., p27 and/or p57) can have overlapping functions in driving cell cycle arrest in differentiating cells of many organs. Therefore, it appears that CDI and pRb participate as a part of an essential regulatory mechanism controlling the cell cycle in cells committed to differentiate inside specialized tissues.

Regarding the specific role of pRb family members p107 and pRb2/p130, the generation of mice homozygous for null alleles of both p107 and pRb2/p130 revealed that pRb integrity is sufficient for differentiation of many tissues, whereas p107 and pRb2/p130 may have a more prominent albeit redundant function specifically in chondrocyte development (Lee et al., 1996; Cobrinik et al., 1996).

The pivotal role of pRb and the pocket proteins in the control of the cell cycle during differentiation is also confirmed by the finding that oncoviral proteins known to target the pocket proteins, such as adenovirus E1A and SV40 large T antigen, interfere with the program of permanent cell cycle arrest induction and biochemical/morphological differentiation of skeletal muscle cell, cardiomyocytes, keratinocytes, adipocytes, and melanocytes (Webster et al., 1988; Gu et al., 1993; Crescenzi, et al., 1995; Kirsherbaum et al., 1995; Missero et al., 1995; Yavuzer et al., 1995). Interestingly, it has been recently demonstrated that E1A's ability to suppress terminal differentiation in almost all these cell types depends on the integrity of a region that interacts with p300, a large protein belonging to a family of transcriptional coactivators including the CREB binding protein CBP (Mymryk et al., 1992; Caruso et al., 1993; Kirsherbaum et al., 1995; Missero et al., 1995; Yavuzer et al., 1995). More extensive studies have shown that p300 is a specific coactivator for MyoD (Yuan et al., 1996) and other myogenic proteins (Sartorelli et al., 1997), and its function is required for MyoD-dependent p21 transcription and efficient cell cycle arrest in myocytes induced to differentiate (Eckner et al., 1996; Puri et al., 1997a). The ability of p300 to coactivate MyoD-dependent cell cycle arrest in differentiating myocytes is tightly dependent on the recruitment by p300 of an acetyltransferase, PCAF, which contributes to efficiently stimulate p21 transcription. Complex formation between the sequence-specific transcription factor MyoD and the coactivators p300 and PCAF is critical for induction of myogenic terminal differentiation, because disruption of such a complex by viral proteins (E1A and T antigen) is associated with their antidifferentiative potential (Puri et al., 1997d).

Because E1A's potential to reactivate cell cycle progression and DNA synthesis in terminally differentiated multinucleated myotubes is not impaired by mutations affecting its p300-binding region, but is largely dependent on its ability to interfere with pRb function (Puri et al., 1997a), it is likely that p300's role is restricted to the induction of conditions required for cell cycle arrest (i.e., CDI upregulation) in differentiating cells, and pRb plays an essential role in preventing cell cycle reactivation in terminally differentiated myotubes.

CONCLUSION

We have reviewed the current knowledge of cell cycle protein involvement in a number of different cellular processes. As opposed to previous roles assigned to these proteins, which were solely to drive the cell into DNA synthesis or mitosis or to inhibit either, the cell cycle machinery has been found also to be absolutely essential for embryogenesis, cellular differentiation, sensing of DNA damage, monitoring of the assembly of chromosomes during mitosis, and a number of other phenomena. Because the proteins involved in control of cellular division are the closest and final decision-makers in the process of cell replication, it is important to know as much about them as possible, as many diseases that strike mammals revolve around the deregulation of these processes. Undoubtedly, further understanding of the functions of these proteins in the future will allow investigators to devise better assessments and treatments of altered cell cycle sicknesses such as cancer.

REFERENCES

Adams PD, Kaelin WG Jr (1995): Transcriptional control by E2F. Semin Can Biol 6:99–108.

Allen KE, de la Luna S, Kerkhoven RM, et al. (1997): Distinct mechanism of nuclear accumulation regulates the functional consequence of E2F transcription factors. J Cell Sci 110:2819–2831.

Altiok S, Xu M, Spiegelman BM (1997): PPARg induces cell cycle withdrawal: Inhibition of E2F/DP DNA binding activity via down-regulation of PP2A. Genes Dev 11:1987–1998.

Amon A, Irniger S, Nasmyth K (1994): Closing the cell cycle in yeast. G_2 cyclin proteolysis initiated at mitosis persists until the activation of the G_1 cyclins at the next cycle. Cell 77:1037–1050.

Amon A, Surana U, Muroff I, Nasmyth K (1992): Regulation of $p34^{cdc28}$ tyrosine phosphorylation is not required for entry into mitosis in *S. cerevisiae.* Nature 355:368–371.

Amon A, Tyers M, Futcher B, Nasmyth K (1993): Mechanisms that help the yeast cell cycle clock tick: G_2 cyclins transcriptionally activate G_2 cyclins and repress G_1 cyclins. Cell 74:993–1007.

Andrés V, Walsh K (1996): Myogenin expresion, cell cycle withdrawal, and phenotypic differentiation are temporally separable events that precede cell fusion upon myogenesis. J Cell Biol 132:657–666.

Atherton-Fessler S, Parker LL, Geahlen RL, Piwnica-Worms H (1993): Mechanisms of p34cdc2 regulation. Mol Cell Biol 13:1675–1685.

Bai C, Sen P, Hofmann K, et al. (1996): SKP1 connects cell cycle regulators to the ubiquitin proteolysis machinery through a novel motif, the F-box. Cell 86:263–274.

Baldi A, De Luca A, Claudio PP, Baldi F, Giordano GG, Tommasino M, Paggi MG, Giordano A (1995): The RB2/p130 gene product is a nuclear protein whose phosphorylation is cell cycle regulated. J Cell Biochem 59:402–408.

Bartek J, Bartkova J, Lukas J (1996): The retinoblastoma protein pathway and the restriction point. Curr Opin Cell Biol 8:805–814.

Bartel B, Wunning I, Varshavsky A (1990): The recognition component of the N-end rule pathway. EMBO J 9:3179–3189.

Baskaran R, Chiang GG, Wang JYJ (1996): Identification of a binding site in cAbl tyrosine kinase for the C-terminal repeated domain of RNA polymerase II. Mol Cell Biol 16:3361–3369.

Baskaran R, Wood L, Whitaker L, et al. (1997): Ataxia teleangectasia mutant protein activates cAbl tyrosine kinase in response to ionizing radiation. Nature 387:516–519.

Beach D, Durkacz B, Nurse P (1982): Functionally homologous cell cycle control genes in budding and fission yeast. Nature 300:706–709.

Benezra R, Davis RL, Lockshon D, Turner DL, Weintraub H (1990): The protein Id: A negative regulator of helix-loop-helix DNA binding proteins. Cell 61:49–59.

Blondel M, Mann C (1996): G_2 cyclins are required for the degradation of G_1 cyclins in yeast. Nature 384:279–282.

Booher RN, Alfa CE, Hyams JS, Beach DH (1989): The fission yeast cdc2/cdc13/suc1 protein kinase: Regulation of catalytic activity and nuclear localization. Cell 58:485–497.

Booher RN, Beach DH (1988): Involvement of $cdc13^+$ in mitotic control in *Schizosaccaromyces pombe:* Possible interaction of the gene product with microtubules. EMBO J 7:2321–2327.

Brandeis M, Hunter T (1996): The proteolysis of mitotic cyclins in mammalian cells persists from the end of mitosis until the onset of S-phase. EMBO J 15:5280–5289.

Brugarolas J, Chandrasekaran C, Gordon J, et al. (1995): Radiation induced cell cycle arrest compromised by p21 deficiency. Nature 377:552–557.

Cardoso MC, Leonhardt H, Nadal-Ginard B (1993): Reversal of terminal differentiation and control of DNA replication: Cyclin A and cdk2 specifically localize at subnuclear sites of DNA replication. Cell 74:976–992.

Caruso M, Martelli F, Giordano A, Felsani A (1993): Regulation of MyoD gene transcription and protein function by the transforming domains of the adenovirus E1A oncoprotein. Oncogene 8:267–278.

Chellappan SP, Hiebert S, Mudri M, et al. (1991): The E2F transcription factor is a cellular target for the Rb protein. Cell 65:1053–1061.

Chen J, Jackson PK, Kirschner MW, Dutta A (1995): Separate domains of p21 involved in the inhibition of cdk kinase and PCNA. Nature 374:386–388.

Chirillo P, Pagano S, Natoli G, et al. (1997): The hepatitis B virus X gene induces p53 mediated programmed cell death. Proc Natl Acad Sci USA 94:8162–7.

Ciechanover A (1994): The ubiquitin–proteasome proteolytic pathway. Cell 79:13–22.

Claudio PP, De Luca A, Howard CM, et al. (1996): Functional analysis of pRb2/p130 interaction with cyclins. Cancer Res 56:2003–2008.

Claudio PP, Howard C, Baldi A, et al. (1994): p130/pRb has growth suppressive properties similar to yet distinctive to those of retinoblastome family members pRb and p107. Cancer Res 54:5556–5560.

Cobrinik D, Lee M-H, Hannon G, et al. (1996): Shared role of the pRb-related p130 and p107 proteins in limb development. Genes Dev 10:1633–1644.

Coleman TR, Tang Z, Dunphy WJ (1993): Negative regulation of the wee1 protein kinase by direct action of the nim1/cdr1 mitotic inducer. Cell 72:919–929.

Corbeil HB, Whyte P, Branton PE (1995): Characterization of transcription factor E2F complexes during muscle and neuronal differentiation. Oncogene 11:909–920.

Correa-Bordes S, Nurse P (1995): $p25^{rum1}$ orders S phase and mitosis by acting as an inhibitor of the $p34^{cdc2}$ mitotic kinase. Cell 83:1001–1009.

Crescenzi M, Fleming TP, Lassar AB, et al. (1990): MyoD induces growth arrest independent of differentiation in normal and transformed cells. Proc Natl Acad Sci USA 87:8442–8446.

Crescenzi M, Soddu S, Tato F (1995): Mitotic cycle reactivation in terminally differentiated cells by adenovirus infection. J Cell Physiol 162:26–35.

de la Luna S, Burden MJ, Lee C-W, et al. (1997) Nuclear accumulation of the E2F heterodimer regulated by subunit composition and alternative splicing of a nuclear localization signal. J Cell Sci 109:2443–2452.

Deng C, Zhang P, Harper P, et al. (1995): Mice lacking p21 CIP1/WAF1 undergo normal development, but are defective in the G_1 checkpoint control. Cell 82:675–684.

de Nooji JC, Letendre MA, Hariharan IK (1996): A cyclin-dependent kinase inhibitor, Dacapo, is necessary for timely exit from the cell cycle during *Drosophila* embryogenesis. Cell 87:1237–1248.

Deng C, Zhang P, Harper JW, Elledge S, Leder P (1995): Mice lacking p21/CIPL/WAF1 undergo normal development, but are defective MGI check point control. Cell 82:675–84.

Desai D, Gu Y, Morgan DO (1992). Activation of human cyclin-dependent kinases in vitro. Mol Cell Biol 3:571–582.

Desai D, Wessling HC, Fisher RP, Morgan DO (1995): Effects of phosphorylation by CAK on cyclin binding CDC2 and CDCK2. Mol Cell Biol 15:345–350.

Deshaies RJ, Chau V, Kirschner M (1995): Ubiquitination of the G_1 cyclin Cln2p by a Cdc34-dependent pathway. EMBO J 14:303–312.

Devault AA, Martinez M, Fesquet D, et al. (1995): MAT 1 (menage-à-trois) a new RING finger protein subunit stabilizing cyclin H-cdk7 complexes in starfish and *Xenopus* CAK. EMBO J 14:5627–5636.

Dowdy SF, Hinds PW, Louis K, et al. (1993): Physical interaction of the retinoblastoma protein with human D cyclin. Cell 73:499–511.

Draetta G, Beach DH (1988): Activation of cdc2 protein kinase during mitosis in human cells: Cell cycle dependent phosphorylation and subunit rearrangement. Cell 54:17–26.

Draetta G, Brizuela L, Patashkin J, Beach D (1987): Identification of p34 and p13, human homologs of cell cycle regulators of fission yeast encoded by $cdc2^+$ and $suc1^+$. Cell 50:319–325.

Draetta G, Luca F, Westendorf J, Brizuela L, Ruderman J, Beach D (1989): Cdc2 protein kinase is complexed with both cyclin A and b: Evidence for proteolytic inactivation of MPF. Cell 56:829–838.

Dunaief JL, Strober BE, Guha S, et al. (1994): The retinoblastoma protein and BRG1 form a complex and cooperate to induce cell cycle arrest. Cell 79:119–130.

Eckner R, Yao TP, Oldread E, Livingston DM (1996): Interaction and functional collaboration of p300/CBP and bHLH proteins in muscle and B cell differentiation. Genes Dev 10:2478–2490.

El Deiry, WS, Tokino T, Velculescu VE, et al. (1993): WAF1 a potential mediator of p53 tumor suppression. Cell 75:817–825.

Elledge S (1996): Cell cycle checkpoints: Preventing an identity crisis. Science 274:1664–1772.

Elledge SJ, Winston J, Harper JW (1996): A question of balance: The role of cyclin–kinase inhibitors in development and tumorigenesis. Trends Cell Biol 6:388–392.

Enoch TA, Carr M, Nurse P (1992): Fission yeast genes involved in coupling mitosis to completion of DNA replication. Genes Dev 6:2035–2046.

Enoch T, Norbury C (1995): Cellular responses to DNA damage: Cell cycle checkpoints, apoptosis and the roles of p53 and ATM. Trends Biochem Sci 20:426–430.

Enoch T, Nurse P (1990): Mutation of fission yeast cell cycle control genes abolishes dependence of mitosis on DNA replication. Cell 60:665–673.

Epstein CB, Cross FR (1992): CLB5: A novel B cyclin from budding yeast with a role in S-phase. Genes Dev 6:1695–1706.

Evan GI (1995): p53-dependent checkpoint pathways. Curr Opin Cell Biol 7:825–831.

Evans T, Rosenthal ET, Youngblom J, Diste, D, Hunt T (1983): Cyclin: A protein specified by maternal mRNA in sea urchin eggs that is destroyed at each cleavage division. Cell 33:389–396.

Featherstone C, Russell P (1991): Fission yeast $p107^{wee1}$ mitotic inhibitor is a tyrosine/serine kinase. Nature 349:808–811.

Feaver WJ, Gileadi O, Li Y, Kornberg RD (1994): Relationship of CDK-activating kinases and RNA polymerase II CTD kinase TFIIH/TFIIK. Cell 79:1103–1109.

Fero ML, Rivkin M, Tasch M, et al. (1996): A syndrome of multiorgan hyperplasia with feature of gigantism, tumorigenesis, and female sterility in p27Kip1-deficient mice. Cell 85:733–744.

Field SJ, Tsai FY, Kuo F, et al. (1996): E2F-1 functions in mice to promote apoptosis and suppress proliferation. Cell 85:549–561.

Firpo EJ, Koff A, Solomon MJ, Roberts JM (1994): Inactivation of a cdk2 inhibitor during interleukin2-induced ptoliferation of human T lymphocytes. Mol Cell Biol 14:4889–4901.

Fisher DL, Nurse P (1995): Cyclins of the fission yeast *Schizosaccharomyces pombe.* Semin Cell Biol 6:73–78.

Fisher DL, Nurse P (1996): A single fission yeast mitotic cyclin B–p34cdc2 kinase promotes both S phase and mitosis in the absence of G_1 cyclins. EMBO J 15:850–860.

Fisher RP, Jin P, Chamberlin HM, Morgan DO (1995): Alternative mechanism of CAK assembly require an assembly factor or an activated kinase. Cell 83:47–57.

Fisher RP, Morgan DO (1994): A novel cyclin associates with MO15–CDK7 to form the CDK activating kinase. Cell 78:713–724.

Ford JC, Al-Khodairy F, Fotou E, Sheldrick KS, Griffiths DJF, Carr AM (1994): 14-3-3 Protein homologs required for the DNA damage checkpoint in fission yeast. Science 265:533–535.

Funabiki H, Yamano H, Kumada K, Nagao K, Hunt T, Yanagida M (1996): Cut2 proteolysis required for sister-chromatid separation in fission yeast. Nature 381:438–441.

Furnari B, Thind N, Russel P (1997): Cdc25 mitotic inducer targeted by Chk1 DNA damage checkpoint kinase. Science 277:1495–1497.

Galaktionov K, Beach D (1991): Specific activation of cdc25 tyrosine phosphatases by B-type cyclins: Evidence for multiple roles of mitotic cyclins. Cell 67:1181–1194.

Galaktionov K, Chen X, Beach D (1996): Cdc25 cell-cycle phosphatase as a target of c-myc. Nature 382:511–517.

Gautier J, Solomon MJ, Booher RN, Bazan JF, Kirschner MW (1991): Cdc25 is a specific tyrosine phosphatase that directly activates p34cdc2. Cell 67:197–211.

Ghiara JB, Richardson HE, Sugimoto K, et al. (1991): A cyclin B homolog in *S. cerevisiae:* Chronic activation of the cdc28 protein kinase by cyclin prevents exit from mitosis. Cell 65:163–174.

Giordano A, Lee JH, Scheppler JA, Herrmann C, Harlow E, Deuschle U, Beach D, Franza BR Jr (1991a): Cell cycle regulation of histone H1 kinase activity associated with the adenoviral protein E1A. Science 253:1271–1275.

Giordano A, McCall C, Whyte P, Franza BR Jr (1991b): Human cyclin A and the retinoblastoma protein interact with similar but distinguishable sequences in the adenovirus E1A gene product. Oncogene 6:481–485.

Giordano A, Whyte P, Harlow E, Franza BR Jr, Beach D, Draetta G (1989): A 60 kDa cdc2-associated polypeptide complexes with the E1A proteins in adenovirus-infected cells. Cell 58:981–990.

Girard F, Strausfeld U, Fernandez A, Lamb NJC (1991): Cyclin A is required for the onset of DNA replication in mammalian fibroblasts. Cell 67:1169–1179.

Glotzer M, Murray AW, Kirscner MW (1991): Cyclin is degraded by the ubiquitin pathway. Nature: 349:132–138.

Goebl MG, Yochem J, Jentsch S, McGrath JP, Varshavsky A, Byers B (1988): The yeast cell cycle gene cdc34 encodes a ubiquitin-conjugating enzyme. Science 241:1331–1334.

Gould KL, Nurse P (1989): Tyrosine phosphorylation of the fission yeast $cdc2^+$ protein kinase regulates entry into mitosis. Nature 342:39–45.

Grana X, Claudio PP, DeLuca A, Sang N, Giordano A (1994): PISSLRE, a human novel CDC2-related protein kinase. Oncogene 9:2097–2103.

Griffiths DJF, Garbet NG, McCready S, Lehmann AR, Carr AM (1995): Fission yeast Rad17: A homologue of budding yeast RAD24 that shares regions of sequence similarity with DNA polymerase accessory proteins. EMBO J 14:5812–5823.

Gu W, Schneider JW, Condorelli G, et al. (1993): Interaction of myogenic transcription factors and the retinoblastoma protein mediates muscle cell commitment and differentiation. Cell 72:309–324.

Guan K-L, Jenkins CW, Li Y, et al. (1994): Growth suppression by p18, a $p16^{INK4b\text{-}MTS1}$ and $p14^{INK4b\text{-}MTS2}$-related CDK6 inhibitor, correlates with wild-type pRb function. Genes Dev 8:2939–2952.

Guo K, Wang G, Adres S, et al. (1995): MyoD induced expression of p21 inhibits cyclin-dependent kinase activity upon myocyte terminal differentiation. Mol Cell Biol 15:3823–3829.

Haese GJD, Walworth N, Carr AM, Gould KL (1996): The wee1 protein kinase regulates T14 phosphorylation of fission yeast Cdc2. Mol Biol Cell 6:371–385.

Hagan I, Hayles J, Nurse P (1988): Cloning and sequencing of the cyclin-related $cdc13^+$ gene and a cytological study of its role in fission yeast mitosis. J Cell Sci 91:587–595.

Halevy OB, Novitch G, Spicer DB, et al. (1995): Correlation of terminal cell cycle arrest of skeletal muscle with induction of p21 by MyoD. Science 267:1018–1021.

Hannon GJ, Beach D (1994): $p15^{INK4b}$ is a potential effector of TGFβ-induced cell cycle arrest. Nature 371:257–261.

Hardwick KG, Murray AW (1995): Mad1p, a phosphoprotein component of the spindle assembly checkpoint in budding yeast. J Cell Biol 131:709–720.

Harper JW, Adami GR, Wei K, et al. (1993): The p21 cdk-interacting protein Cip1 is a potent inhibitor of G_1 cyclin-dependent kinases. Cell 75:805–816.

Hartley KO, Gell D, Smith CGM, et al. (1995): DNA-dependent protein kinase catalytic subunit: A relative of phosphatidylinositol 3-kinase and the ataxia telangiectasia gene product. Cell 82:849–856.

Hartwell LH (1992): Defects in a cell cycle checkpoint may be responsible for the genomic instability of cancer cells. Cell 71:543–546.

Hartwell LH, Culotti J, Pringle JR, Reid BJ (1974): Genetic control of the cell division cycle in yeast. Science 183:46–51.

Hartwell LH, Weinert TA (1989): Checkpoints: Controls that ensure the order of cell cycle events. Science 246:629–634.

Hateboer G, Kerkhoven RM, Shvarts A, Bernards R, Beijersbergen R (1996): Degradation of E2F by the ubiquitin–proteasome pathway: Regulation by retinoblastoma family proteins and adenovirus transforming proteins. Genes Dev 10:2960–2970.

Hayles J, Fisher D, Woolard A, Nurse P (1994): Temporal order of S phase and mitosis in fission yeast is determined by the state of the $p34^{cdc2}$–mitotic B cyclin complex. Cell 78:813–822.

Helin K, Harlow E, Fattaey A (1993): Inhibition of E2F1 transactivation by direct binding of the retinoblastoma protein. Mol Cell Biol 13:6501–6508.

Hengst L, Reed SI (1996): Translational control of p27Kip1 accumulation during the cell cycle. Science 271:1861–1864.

Henriksson M, Luscher B (1996): Proteins of the Myc network: Essential regulators of cell growth and differentiation. Adv Cancer Res 68:109–180.

Hershko A, Ganoth D, Sudakin V, et al. (1994): Components of a system that ligates cyclin to ubiquitin and their regulation by the protein kinase cdc2. J Biol Chem 269:4940–4946.

Hofmann F, Martelli F, Livingston DM, Wang Z (1996): The retinoblastoma gene product protects E2F-1 from degradation by the ubiquitin–proteasome pathway. Genes Dev 10:2949–2959.

Holloway SL, Glotzer M, King RW, Murray AW (1993): Anaphase is initiated by proteolysis rather than by the inactivation of maturation-promoting factor. Cell 73:1393–1402.

Hoyt MA, Lotis L, Roberts BT (1991): *S. cerevisiae* genes required for cell cycle arrest in response to loss of microtubule function. Cell 66:507–517.

Hoyt MA (1997): Eliminating all obstacles: Regulated proteolysis in the eukariotic cell cycle. Cell 91:149–151.

Huibregtse JM, Scheffner M, Beauderon S, Howley PM (1995): A family of proteins structurally and functionally related to E6-AP ubiquitin protein ligase. Proc Natl Acad Sci USA 92:2563–2567.

Hurford RK Jr, Cobrinic D, Lee MH, Dyson N (1997): pRd and p107/p130 are required for the regulated expression of different sets of E2f responsive genes. Genes Dev 11:1447–63.

Iavarone A, Massague J (1997): Repression of the Cdk activator Cdc25A and cell cycle arrest by cytokine TGF-b in cells lacking the CDK inhibitor p15. Nature 387:417–421.

Irninger S, Piatti S, Michaelis C, Nasmyth K (1995): Genes involved in sister chromatid separation are needed for B-type cyclin proteolysis in budding yeast. Cell 81:269–277.

Izumi T, Walker DH, Maller JL (1992): Periodic changes in phosphorylation of the *Xenopus* cdc25 phosphatase regulate its activity. Mol Cell Biol 3:927–939.

Jaffrey PD, Russo AA, Polyak K, et al. (1995): Mechanism of cdk activation revealed by the structure of a cyclin A–cdk2 complex. Nature 376:313–320.

Jeggo PA, Taccioli GE, Jackson SP (1995): Menages a trois: Double strand break repair, V(D)J recombination and DNA-PK. Bioessays 17:949–957.

Jiang H, Fisher PB (1993): Use of a sensitive and efficient subtraction hybridization protocol for the identification of genes differently regulated during the induction of differentiation in human melanoma cells. Mol Cell Differ 1:285–299.

Jiang H, Lin J, Su Z-Z, et al. (1994): Induction of differentiation in human promyelocytic HL-60 leukemia cells activates p21, WAF1/CIP1, expression in the absence of p53. Oncogene 9:3397–3406.

Jiang H, Lin J, Su Z-Z, et al. (1995): The melanoma differentiation–associated gene mda-6, which encodes the cyclin-dependent kinase inhibitor p21, is differently expressed during growth, differentiation and progression in human melanoma cells. Oncogene 11:1855–1864.

Johnson DG, Schwarz JK, Cress WD, Nevins JR (1993): Expression of E2F1 induces quiescent cells to enter S phase. Nature 365:349–352.

Kastan MB, Zhan Q, El-Deiry WS, et al. (1992): A mammalian cell cycle checkpoint pathway utilizing p53 and GADD45 is defective in ataxia-telangiectasia. Cell 71:587–597.

Kharabanda S, Pandey P, Jin S, et al. (1997): Functional interaction between DNA–PK and cAbl in response to DNA damage. Nature 386:732–735.

King RW, Deshaies RJ, Peters J-M, Kirschner MW (1996a): How proteolysis drives the cell cycle. Science 274:1652–1659.

King RW, Glotzer M, Kirschner MW (1996b): Mutagenic analysis of the destruction signal of mitotic cyclins and structural characterization of ubiquitination intermediates. Mol Biol Cell 7:1343–1357.

King RW, Peters J-M, Tugendreich S, Rolfe M, Hieter P, Kirscner MW (1995): A 20S complex containing CDC27 and CDC16 catalyzes the mitosis specific conjugation of ubiquitin to cyclin B. Cell 81:279–288.

Kipreos ET, Lander LE, Wing JP, Hew W, Hedgecock EM (1996). Cul-1 is required to cell cycle exit in C. Elegans and identifies a novel gene family. Cell 85:829–39.

Kirsherbaum LA, Schneider MD (1996): Adenovirus E1A represses cardiac gene transcription and reactivates DNA synthesis in ventricular myocytes, via alternative pocket proteins and p300 binding domains. J Biol Chem 270:7791–7794.

Kirshenbaum LA, Schneider MD (1995): Adenovirus E1A represses cardiac gene transcription and reactivates DNA synthesis in ventricular myocytes, via alternative pocket protein- and p300-binding domains. J Biol Chem 270:7791–4.

Kiss M, Gill RM, Hamel P (1995): Expression and activity of the Retinoblastoma protein (pRb)–family proteins, p107 and p130, during L6 myoblasts differentiation. Cell Growth Differ 6:1287–1298.

Kiyokava H, Kineman RD, Manova-Todorova KO, et al. (1996): Enhanced growth of mice lacking the cyclin-dependent kinase inhibitor function of p27 Kipl. Cell 85:721–732.

Klotzbucher A, Stewart E, Harrison D, Hunt T (1996): The destruction box of cyclin A allows B-type cyclins to be ubiquitinated, but not efficiently detroyed. EMBO J 15:3053–3064.

Knude T, Jackman M, Pines J, Laskey RA (1997): Cyclin/cdk-dependent initiation of DNA replication in a human cell-free system. Cell 88:109–119.

Knudsen ES, Wang JYJ (1996): Differential regulation of retinoblastoma protein function by specific Cdk phosphorylation sites. J Biol Chem 271:8313–8320.

Ko LJ, Prives C (1996): p53: Puzzle and paradigm. Genes Dev 10:1054–1072.

Koch C, Nasmyth K (1994): Cell cycle transcription in yeast. Curr Opin Cell Biol 6:451–459.

Koch C, Schleiffer A, Ammeter G, Nasmyth K (1996): Switching transcription on and off and during the yeast cell cycle: Cln/Cdc28 kinases activate bound transcription factor SBF (Swi4/Swi6) at Start, whereas Clb/Cdc28 kinases displace it from the promoter in G2. Genes Dev 10:129–141.

Koff A, Giordano A, Desai D, Yamashita K, Harper JW, Elledge S, Nisitimoto T, Morgan DO, Franza BR, Roberts JM (1993): Formation and activation of a cyclin E-cdk2 complex during the G1 phase of the human cell cycle. Science 257:1609–94.

Kornitzer D, Raboy B, Kulka RG, Fink GR (1994): Regulated degradation of transcription factor Gcn4. EMBO J 13:6021–6030.

Kouzarides T (1995): Transcriptional control by the retinoblastoma protein. Semin Cancer Biol 6:91–98.

Krek W, Ewen ME, Shirodkar S, et al. (1994): Negative regulation of the growth-promoting transcription factor E2F-1 by a stably bound cyclin A–dependent protein kinase. Cell 78:161–172.

Krek W, Nigg EA (1991a): Differential phosphorylation of vertebrate p34cdc2 kinase at the G_1/S and G_2/M transition of the cell cycle; identification of major phosphorylation sites. EMBO J 10:305–316.

Krek W, Nigg EA (1991b): Mutations of p34cdc2 phosphorylation sites induce premature mitotic events in HeLa cells: Evidence for a double block to p34cdc2 kinase activation in vertebrates. EMBO J 10:3331–3334.

Krude T, Jackman M, Pines J, Laskey RA (1997): Cyclin/Cdk-dependent initiation of DNA replication in a human cell-free system. Cell 88:109–119.

Kumagai A, Dumphy WG (1991): The cdc25 protein controls tyrosine dephosphorylation of the cdc2 protein in a cell free system. Cell 64:903–914.

Kurabayashi M, Jeyaseelan R, Kedes L (1994): Doxorubicin represses the function of the myogenic helix-loop-helix transcription factor MyoD. J Biol Chem 269:6031–6039.

Lane ME, Sauer K, Wallace K, et al. (1996): Dacapo, a cyclin-dependent kinase inhibitor, stops cell proliferation during *Drosophila* development. Cell 87:1225–1236.

Lanker S, Valdovieso MH, Wittemberg C (1996): Rapid degradation of the G_1 cyclin Cln2 induced by CDK-dependent phosphorylation. Science 271:1597–1600.

LaBaer J, Garrett MD, Stevenson LF, Slingerland JM, Sandhu C, Chou HS, Fattaey A, Harlow E (1997): New functional activities for the p21 family of CDK inhibitors. Genes Dev 11:847–62.

La Thangue NB (1996): E2F and the molecular mechanism of early cell-cycle control. Biochem Soc Trans 24:55–59.

Lee MG, Nurse P (1987): Complementation used to clone a human homolog of the fission yeast cell cycle control gene $cdc2^+$. Nature 327:31–35.

Lee M-H, Reynisdottìr I, Massague J (1995): Cloning of $p57^{KIP2}$, a cyclin-dependent kinase inhibitor with unique domain structure and tissue distribution. Gene Dev 9:639–649.

Lee M-H, Williams BO, Mulligan G, et al. (1996): Targeted disruption of p107: Functional overlap between p107 and pRb. Genes Dev 10:1621–1632.

Lehner CF, O'Farrell PH (1989): Expression and function of *Drosophnila* cyclin A during embryonic cell cycle progression. Cell 56:957–968.

Leone G, De Gregori J, Sears R, et al. (1997): Myc and Ras collaborate in inducing accumulation of active cyclinE/cdk2 and E2F. Nature 387:422–426.

Lew DJ, Kornbluth S (1996): Regulatory roles of cyclin dependent kinase phosphorylation in cell cycle control. Curr Opin Cell Biol 8:795–804.

Li R, Murray AM (1991): Feedback control of mitosis in budding yeast. Cell 66:519–531.

Li S, MacLachlan TK, De Luca A, Giordano A (1995): The cdc2-related kinase PISSLRE is essential for cell growth and is involved in G_2 phase of the cell cycle. Cancer Res 55:3395–3399.

Li Y, Jenkins CW, Nichols MA, Xiong Y (1994a): Cell cycle expression and p53 regulation of the cyclin-dependent kinase inhibitor p21. Oncogene 9:2261–2268.

Li Y, Nichols MA, Shay JW, Xiong Y (1994b): Transcriptional repression of the D-type cyclin-dependent kinase inhibitor p16 by the retinoblastoma susceptibility gene product. Cancer Res 54:6078–6082.

Lindeman G, Gaubatz S, Livingston D, et al. (1997): The subcellular localization of E2F4 is cell cycle dependent. Proc Natl Acad Sci USA 94:5095–5100.

Luca FC, Shibuya EK, Dohrmann CE, Rudermann JV (1991): Both cyclin A Δ60 and B Δ97 are stable and arrest cells in M phase but only cyclin B Δ97 turns on cyclin destruction. EMBO J 10:4311–4320.

Lukas J, Bartkova J, Bartek J (1996): Convergence of mitogenic signalling cascades from diverse classes of receptor at the cyclin D–cyclin-dependent kinase-pRb-controlled G1 checkpoint. Mol Cell Biol 16:6917–6925.

Lundgren K, Walworth N, Booher R, et al. (1991): mik1 and wee1 cooperate in the inhibitory tyrosine phosphorylation of cdc2. Cell 64:1111–1122.

Luo Y, Hurwitz J, Massague J (1995): Cell cycle inhibition mediated by functionally independent CDK and PCNA inhibitory domains of p21. Nature 375:159–161.

Lydall D, Weinert T (1995): Yeast checkpoint genes in DNA damage processing: Implications for repair and arrest. Science 270:1488–1491.

MacLachlan TK, Sang N, Giordano A (1995): Cyclins, cdks and cdk inhibitors: Implications in cell cycle control and cancer. Crit Rev Eukaryot Gene Expression 4:32–62.

MacLeod KF, Sherry N, Hannon G, et al. (1995): p53-dependent and independent expression of p21 during cell growth, differentiation and DNA damage. Genes Dev 9:935–944.

Magae J, Wu C-H, Illenye S, et al. (1997): Nuclear localization of DP and E2F transcription factors by heterodimeric partners and retinoblastoma protein family members. J Cell Sci 109:1717–1726.

Makela TP, Tassan JP, Nigg EA, Frutiger S, Hughes GJ, Weinberg RA (1994): A cyclin associated with the CDK-activating kinase MO15. Nature 371:254–257.

Martelli F, Cenciarelli C, Santarelli G, et al. (1994): MyoD induces retinoblastoma gene expression during myogenic differentiation. Oncogene 9:3579–3590.

Martin-Castellano C, Labib K, Moreno S (1996): B-type cyclins regulate G_1 progression in fission yeast in opposition to the $p25^{rum1}$ cdk inhibitor. EMBO J 15:839–849.

Mathias N, Johnson SL, Winey M, et al. (1996): Cdc53 acts in concert with Cdc4p and Cdc34p to control the G1-to-S-phase transition and identifies a conserved family of proteins. Mol Cell Biol 16:6634–6643.

Matsuoka M, Edwards M, Bai C, et al. (1995): $p57^{KIP2}$, a structurally distinct member of the p21^{CIP1}cdk inhibitor family, is a candidate tumor suppressor gene. Genes Dev 9:650–662.

Matsushine H, Quelle DE, Shurleff SA, et al. (1994) D-type cyclin dependent kinase activity in mammalian cells. Mol Cell Biol 14:2066–2076.

Mayol X, Graña X, Baldi A, Sang N, Hu Q, Giordano A (1993): Cloning of a new member of the retinoblastoma gene family (pRb2) which binds to the E1A transforming domain. Oncogene 8:2561–2566.

Mendenhall MD (1993): An inhibitor of p34CDC28 protein-kinase activity in *Saccharomyces cerevisiae.* Science 259:216–219.

Minshull J, Blow JJ, Hunt T (1989): Translation of cyclin mRNA is necessary for extracts of activated *Xenopus* eggs to enter mitosis. Cell 56:947–956.

Missero C, Colautti E, Eckner R, et al. (1995): Involvement of the cell cycle inhibitor Cip1/WAF1 and the E1A associated p300 protein in terminal differentiation. Proc Natl Acad Sci USA 92:5451–5455.

Missero C, Dotto A (1996): p21 WAF1/Cip1 and terminal differentiation control of normal epithelia. Mol Cell Differ 4:1–16.

Moberg K, Starz MA, Lees J (1996): E2F-4 switches, from p130 to p107 and pRb in response to cell cycle re-entry. Mol Cell Biol 16:1436–1449.

Mondesert O, McGowan C, Russell P (1996): Cig2, a B-type cyclin, promotes the onset of S in *Schizosaccharomyces pombe.* Mol Cell Biol 16:1527–1533.

Moran E (1993): DNA tumor virus transforming proteins and the cell cycle. Curr Opin Genet Dev 3:63–70.

Moreno S, Nurse P (1994): Regulation of progression through the G_1 phase of the cell cycle by the rum1$^+$ gene. Nature 367:236–242.

Morgan DO (1995): Principles of CDK regulation. Nature 374:131–134.

Muller H, Moroni MC, Vigo E, et al. (1997): Induction of S-phase entry by E2F transcription factors depends on their nuclear localization. Mol Cell Biol 17:5508–5520.

Mummenbrauer T, Janus F, Muller B, Wiessmuller L, Deppert W, Grosse F (1996): p53 protein exhibits 3′- to 5′-exonuclease activity. Cell 85:1089–1099.

Murakami H, Okayama H (1995): A kinase from fission yeast responsible for blocking mitosis in S phase. Nature 374:817–819.

Murray AW (1992): Creative blocks: Cell cycle checkpoints and feed-back controls. Nature 359:599–604.

Murray AW, Kirschner MW (1989): Cyclin synthesis drives the early embryonic cell cycle. Nature 339:275–280.

Murray AW, Solomon MJ, Kirschner MW (1989): The role of cyclin synthesis and degradation in the control of the maturation promoting factor activity. Nature 339:280–286.

Mymryk JS, Lee RWH, Bayley ST (1992): Ability of the adenovirus 5 E1A protein to suppress differentiation of BC3H1 myoblasts correlates with their binding to a 300 kDa cellular protein. Mol Biol Cell 3:1107–1115.

Nakanishi M, Robetorge RS, Adami GR, et al. (1995): Identification of the active region of the DNA synthesis inhibitor gene p21. EMBO J 14:555–563.

Nakayama K, Ishida N, Shirane M, et al. (1996): Mice lacking p27Kip1 display increased body size, multiple organ hyperplasia, retinal dysplasia and pituitary tumors. Cell 85:707–720.

Nasmyth K (1993): Control of the yeast cell cycle by the cdc28 protein kinase. Curr Opin Cell Biol 5:166–179.

Nasmyth K (1996): Viewpoint: Putting the cell cycle in order. Science 274:1643–1645.

Navas TA, Zhou Z, Elledge SJ (1995): DNA polymerase ε links the DNA replication machinery to the S phase checkpoint. Cell 80:29–39.

Nevins JR (1992): E2F: A link between the Rb tumor suppressor protein and viral oncoproteins. Science 258:424–429.

Nikolic M, Dudek H, Kwon YT, et al. (1996): The cdk5/p35 kinase is essential for neurite outgrowth during neuronal differentiation. Genes Dev 10:816–825.

Noda A, Ning Y, Venable SF, et al. (1994): Cloning of senescent cell-derived inhibitors of DNA synthesis using an expression screen. Exp Cell Res 211:90–98.

Norbury C, Blow J, Nurse P (1991): Regulatory phosphorylation of the $p34^{cdc2}$ protein kinase invertebrates. EMBO J 10:3321–3329.

Nourse J, Firpo E, Flanagan WM, et al. (1994): Interleukin 2–mediated elimination of p27 cyclin dependent inhibitor prevented by rapamycin. Nature 372:570–573.

Nugroho TT, Mendenhall MD (1994): An inhibitor of yeast cyclin-dependent kinase plays an important role in ensuring the genomic integrity of daughter cells. Mol Cell Biol 14:3320–3328.

Nurse P (1994): Ordering S phase and M phase in the cell cycle. Cell 79:547–550.

Nurse P, Thuriaux P (1980): Regulatory genes controlling mitosis in the fission yeast *Schizosaccaromyces pombe.* Genetics 96:627–637.

Nurse P, Thuriaux P, Nasmyth K (1976): Genetic control of the cell division cycle in the fission yeast *Schizosaccaromyces pombe.* Mol Gen Genet 146:167–178.

O'Connel M, Raleigh J, Verkade H, et al. (1997): Chkl is a weel kinase in the G_2 DNA damage checkpoint inhibiting cdc2 by Y15 phosphorylation. EMBO J 16:545–554.

Ohsumi K, Katagiri C, Kishimoto T (1993): Chromosome condensation in *Xenopus* mitotic extracts without histone H1. Science 262:2033–2035.

Ohta K, Shijna N, Okamura E, et al. (1993): Microtubule enucleating activity of centrosomes in cell free extracts from *Xenopus* eggs: Involvement of phosphorylation and accumulation of pericentriolar material. J Cell Sci 104:125–137.

Okazaki K, Nishizawa M, Furuno N, Yasuda H, Sagata N (1992): Differential occurrence of CSF-like activity and transforming activity of Mos during the cell cycle in fibroblasts. EMBO J 11:2447–2456.

Pagano M, Pepperkok R, Verde F, Ansorge W, Draetta G (1992): Cyclin A is required at two points in the human cell cycle. EMBO J 11:962–972.

Pagano M, Tam SW, Theodoras AM, et al. (1995): Role of the ubiquitin–proteasome pathway in regulating abundance of the cyclin-dependent kinase inhibitor p27. Science 269:682–685.

Paggi MG, Baldi A, Bonetto F, Giordano A (1996): Retinoblastoma protein family in cell cycle and cancer—A review. J Cell Biochem 62:418–430.

Pardee AB (1989): G_1 events and regulation of cell proliferation. Science 246:603–608.

Parker LL, Walter SA, Young PG, Piwnica-Worms H (1993): Phosphorylation and inactivation of the mitotic inhibitor Weel by the niml/cdrl kinase. Nature 363:736–738.

Parker S, Eichele BG, Zhang P, et al. (1995): p53-Independent expression of p21 in muscle and other terminally differentiated cells. Science 267:1024–1027.

Parry D, Bates S, Mann DJ, Peters G (1995): Lack of cyclinD–cdk complexes in pRb-negative cells correlates with high levels of $p16^{INK4-MTS1}$ tumor suppressor gene product. EMBO J 14:503–511.

Patra D, Dunphy WG (1996): Xe-p9, a *Xenopus* SUC1/CSK homolog, has multiple essential roles in cell cycle control. Genes Dev 10:1503–1515.

Paulovich AG, Hartwell LH (1995): A checkpoint regulates the rate of progression through S phase in *S. cerevisiae* in response to DNA damage. Cell 82:841–847.

Peeper DS, Upton TM, Ladha MH, Neuman E, Zalvide J, Bernards R, DeCaprio JA, Ewen MA (1997): Ras signalling linked to cell-cycle machinery by the retinoblastoma protein. Nature 386:177–181.

Peng C, Graves P, Thoma R, et al. (1997): Mitotic and G_2 checkpoint control: Regulation of 14-3-3 protein binding by phosphorylation of Cdc25 on Serine 216. Science 277:1501–1505.

Philpott A, Porro E, Kirschner MW, et al. (1997): The role of cyclin-dependent kinase 5 and a novel regulatory subunit in regulating muscle differentiation and patterning. Genes Dev 11:1409–1421.

Pines J (1996): Cyclin from sea urchins to HeLas: Making the human cell cycle. Biochem Soc Trans 24:15–33.

Pines J, Hunter T (1990): Human cyclin A is adenovirus E1A–associated protein p60 and behaves differently from cyclin B. Nature 346:760–763.

Plon SE, Leppig KA, Do HN, Groudine M (1993): Cloning of the human homolog of the CDC34 cell cycle gene by complementation in yeast. Proc Natl Acad Sci USA 90:10484–8.

Polyack K, Kato JY, Solomon MJ, et al. (1994a): $p27^{Kip1}$, a cyclin–cdk inhibitor, links transforming growth factor-β and contact inhibition to cell cycle arrest. Genes Dev 8:9–22.

Polyack K, Lee M-H, Erdjument-Bromage H, et al. (1994b): Cloning of $p27^{Kip1}$ a cyclin-kinase inhibitor and a potential mediator of extracellular antimitogenic signals. Cell 78:59–66.

Puri PL, Avantaggiati ML, Balsano C, et al. (1997a): p300 is required for MyoD-dependent cell cycle arrest and muscle specific gene transcription. EMBO J 16:369–383.

Puri PL, Balsano C, Burgio VL, et al. (1997b): MyoD prevents cyclinA/cdk2 containing E2F complexes formation in terminally differentiated myocytes. Oncogene 14:1171–1184.

Puri PL, Cimino L, Fulco M, et al. (1998): Regulation of E2F4 mitogenic activity during terminal differentiation by its heterodimerization partners for nuclear localization. Cancer Res 58:1325–31.

Puri PL, Medaglia S, Cimino L, et al. (1997c): Uncoupling of p21 induction and MyoD activation results in the failure of irreversible cell cycle arrest in doxorubicin-treated myocytes. J Cell Biochem 66:27–36.

Puri PL, Sartorelli V, Yang X-J, et al. (1997d): Differential roles of p300 and PCAF acetyltransferases in muscle differentiation. Mol Cell 1:35–45.

Quelle DE, Zindy F, Ashmun RA, Sherr CJ (1995): Alternative reading frame of the INK4a tumor suppressor gene encode two unrelated proteins capable of inducing cell cycle arrest. Cell 83:993–1000.

Qin XQ, Livingston DM, Kaelin WG, Adams P (1994): Deregulated E2F1 expression leads to S-phase entry and p53-mediated apoptosis. Proc Natl Acad Sci USA 91:10918–10922.

Rao SS, Chu C, Kohtz DS (1994): Ectopic expression of cyclin D1 prevents activation of gene transcription by myogenic basic helix-loop-helix regulators. Mol Cell Biol 14:5259–5267.

Rao SS, Chu C, Kohtz DS (1995): Positive and negative regulation of D-type cyclin expression in skeletal myoblasts by basic fibroblasts factor b. A role for cyclin D1 in control of myoblasts differentiation. J Biol Chem 270:4093–4100.

Reinisdottir I, Massague J (1997): The subcellular location of p15 and p27 coordinate their inhibitory interaction with cdk4 and cdk2. Genes Dev 11:492–503.

Richardson HE, Wittemberg C, Cross F, Reed SI (1989): An essential G_1 function for cyclin-like proteins in yeast. Cell 58:1127–1133.

Russell P, Nurse P (1986): Cdc25 functional as an inducer in the mitosis control of fission yeast. Cell 45:559–567.

Russell P, Nurse P (1987): Negative regulation of mitosis by wee1$^+$, a gene encoding a protein kinase homolog. Cell 45:145–153.

Russo AA, Jeffrey PD, Patten AK, et al. (1996): Crystal structure of the p27^{Kip1} cyclin-dependent-kinase inhibitor bound to the cyclin A-Cdk2 complex. Nature 382:325–331.

Sanchez Y, Desany BA, Jones WJ, Wang B, Elledge SJ (1996): Regulation of RAD53 by the ATM-like kinase MEC1 and TEL1 in yeast cell cycle checkpoint pathways. Science 271:357–360.

Sanchez Y, Wong C, Toma R, et al. (1997): Conservation of the Chk1 checkpoint pathway in mammals: Linkage of DNA damage to Cdk regulation though Cdc25. Science 277:1497–1501.

Sartorelli V, Huang J, Hamamori Y, Kedes L (1997): Molecular mechanisms of myogenic coactivation by p300: Direct interaction with the activation domain of MyoD and with the MADS of MEF2C. Mol Cell Biol 17:1010–1026.

Sauer K, Knoblich JA, Richardson H, Lehner CF (1995): Distinct modes of cyclin E/cdc2c kinase regulation and S phase control in mitotic and endoreduplication cycles of *Drosophila* embryogenesis. Genes Dev 9:1327–1339.

Savitsky K, Bar-Shira A, Gilad S, et al. (1995): A single ataxia-teleangiectasia gene with a product similar to PI-3 kinase. Science 268:1749–1753.

Schneider BL, Yang QH, Futcher B (1996): Linkage of replication to Start by the Cdk inhibitor Sic1. Science 272:560.

Schneider JW, Gu W, Zhu L, et al. (1994): Reversal of terminal differentiation mediated by p107 in Rb−/− muscle cells. Science 264:1467–1470.

Schwob E, Bohm T, Mendenhall MD, Nasmyth K (1994): The B-type cyclin kinase inhibitor p40^{SIC1} controls the G_1 to S transition in *S. cerevisiae.* Cell 79:233–244.

Schwob E, Nasmyth K (1993): CLB5 and CLB6, a new pair of B cyclins involved in DNA replication in *Saccharomyces cerevisiae.* Genes Dev 7:1160–1175.

Serizawa H, Makela TP, Conaway JW, et al. (1995): Association of cdk-activating kinase subunit with transcription factor TFIIH. Nature 374:280–282.

Serrano M, Gomaz-Lahoz E, DePinho E, et al. (1995): Inhibition of Ras-induced proliferation and cellular transformation of p16^{INK4}. Science 267:249–252.

Serrano M, Lee HW, Chin L, et al. (1996): Role of the INK4a locus in tumor suppression and cell mortality. Cell 85:27–37.

Seufert W, Futcher B, Jentsch S (1995): Role of a ubiquitin-conjugating enzyme in degradation of S-phase and M-phase cyclins. Nature 373:78–81.

Shafman T, Khanna K, Kedar P, et al. (1997): Interaction between ATM protein and cAbl in response to DNA damage. Nature 387:520–523.

Shan B, Lee W-H (1994): Deregulated expression of E2F1 induces S-phase entry and leads to apoptosis. Mol Cell Biol 14:8166–8173.

Sheikh MD, Li X, Chen J, et al. (1994): Mechanism of regulation of WAF1/Cip1 gene expression in human breast carcinoma: Role of p53-dependent and independent signal transduction pathways. Oncogene 9:3407–3415.

Sherr CJ (1996): Cancer cell cycles. Science 274:1672–1677.

Sherr CJ, Roberts JM (1995): Inhibitors of mammalian G1 cyclin-dependent kinases. Genes Dev 9:1149–1163.

Shieh S-Y, Ikeda M, Taya Y, et al. (1997): DNA damage–induced phosphorylation of p53 alleviates inhibition by MDM2. Cell 91:325–334.

Shin EK, Shin A, Paulding C, et al. (1995): Multiple changes in E2F function and regulation occur upon muscle differentiation. Mol Cell Biol 15:2252–2262.

Shiyanov P, Bagchi S, Adami G, et al. (1996): p21 disrupts the interaction between cdk2 and the E2F–p130 complex. Mol Cell Biol 16:737–744.

Skapek SX, Rhee J, Spicer DB, Lassar A (1995): Inhibition of myogenic differentiation in proliferating myoblasts by cyclin D1–dependent kinase. Science 267:1022–1024.

Solomon MJ, Glotzer M, Lee TH, Philippe M, Kirschner MW (1990): Cyclin activation of p34cdc2. Cell 63:1013–1024.

Sorrentino V, Pepperkok R, Davis RL, et al. (1990): Cell proliferation inhibited by MyoD independent of differentiation. Nature 345:813–815.

Steinman RA, Hoffman B, Iro A, et al. (1994): Induction of p21 (WAF-!/CIP1) during differentiation. Cloning of a new member of the retinoblastoma gene family (pRb2) which bind to the E1A transforming domain. Oncogene 8:2561–2566.

Stern B, Nurse P (1996): A quantitative model for the cdc2 control of S phase and mitosis in fission yeast. Trends Genet 12:345–350.

Strauss M, Lukas J, Bartek J (1995): Unrestricted cell cycling and cancer. Nat Med 1:1245–1246.

Stewart E, Enoch T (1996): S-phase and DNA-damage checkpoints: A tale of two yeasts. Curr Opin Cell Biol 8:781–786.

Stewart E, Kobayashi H, Harrisoy D, Hunt T (1994): Destruction of Xenopus cyclins A and B2, but not B1, requires binding to p34 cdc2. EMBO J 13:584–94.

Sudakin V, Ganoth D, Dahan A, et al. (1995): The cyclosome, a large complex containing cyclin-selective ubiquitin ligase activity, targets cyclins for destruction at the end of mitosis. Mol Biol Cell 6:185–197.

Sun Z, Fay DS, Marini F, et al. (1996): Spk1/Rad53 is regulated by Mec1-dependent protein phosphorylation in DNA replication and DNA damage checkpoint pathways. Genes Dev 10:395–406.

Surana U, Amon A, Dowzer C, McGrew J, Byers B, Nasmyth K (1993): Destruction of the CDC28/CLB mitotic kinase is not required for the metaphase to anaphase transition in budding yeast. EMBO J 12:1969–1978.

Tam SW, Shay JW, Pagano M (1994): Differential expression and cell cycle regulation of the cyclin-dependent kinase 4 inhibitor p16. Cancer Res 54:5816–5820.

Tassan JP, Jaquenoud M, Fry AM, et al. (1995): In vitro assembly of a functional human cdk7–cyclin H complex requires MAT1, a novel 36 kDa RING finger protein. EMBO J 14:5608–5617.

Terada Y, Tatsuka M, Jinno S, Okayama H (1995): Requirement for tyrosine phosphorylation of Cdk4 in G1 arrest induced by ultraviolet irradiation. Nature 376:358–362.

Tevosian SG, Shih HH, Mendelson KJ, et al. (1997): HBP1: A HMG box transcriptional repressor that is targeted by the retinoblastoma family. Genes Dev 11:383–396.

Tommasi S, Pfeiffer GP (1995): In vivo structure of the human cdc2 promoter: Release of a p130–E2F-4 complex from sequences immediately upstream of the transcription initiation site coincides with the induction of csc2 expression. Mol Cell Biol 15:6901–6913.

Toyoshima H, Hunter T (1994): p27, a novel inhibitor of G_1 cyclin/cdk protein kinase activity is related to p21. Cell 78:67–74.

Tsai LH, Harlow E, Meyerson M (1991): Isolation of the human cdk2 gene that encodes the cyclin A– and adenovirus E1A–associated p33 kinase. Nature 353:174–177.

Tyers M, Tokiwa G, Futcher B (1993): Comparison of the *Saccharomyces cerevisiae* G_1 cyclins: Cln3 may be an upstream activator of Cln1, Cln2 and other cyclins. EMBO J 12:1955–1868.

Vaga SG, Hannon GJ, Beach D, Stillman B (1994): The p21 inhibitor of cyclin-dependent kinases controls DNA replication by interaction with PCNA. Nature 369:574–578.

Vairo G, Livingston DM, Ginsberg D (1995): Functional interaction between E2F4 and p130: Evidence for distinct mechanism underlying growth suppression by different retinoblastoma protein family members. Genes Dev 9:869–881.

Vander Velden HM, Lohka MJ (1993): Mitotic Arrest caused by the amino terminus of Xenopus cycle B2. Mol. Cell. Biol. 13:1480–8.

Verona R, Moberg K, Estes S, et al (1997): E2F activity is regulated by cell cycle–dependent changes in subcellular localization. Mol Cell Biol 17:7268–7282.

Waldman T, Kinzler KW, Vogelstein B (1995): p21 is necessary for the p53-mediated G_1 arrest in human cancer cells. Cancer Res 55:5187–5190.

Waldman T, Lengauer C, Kinzler KW, Vogelstein B (1996): Uncoupling of S phase and mitosis induced by anticancer agents in cells lacking p21. Nature 381:713–716.

Walker DH, DePaoli-Roach AA, Maller JL (1992): Multiple roles for protein phosphatase 1 in regulating the *Xenopus* early embryonic cell cycle. Mol Cell Biol 3:687–698.

Walworth NC, Bernards R (1996): Rad-dependent response of the chk1-encoded protein kinase at the DNA damage checkpoint. Science 271:353–356.

Wang J, Helin K, Jin P, Nadal-Ginard B (1995): Inhibition of in vitro myogenic differentiation by cellular transcription factor E2F1. Cell Growth Differ 6:1299–1306.

Weaver DT (1995): What to do at an end: DNA double-strand-break repair. Trends Genet 11:388–392.

Webster KA, Muscat G, Kedes L (1988): Adenovirus E1A products suppress myogenic differentiation and inhibit transcription from muscle specific promoters. Nature 332:553–557.

Weinberg RA (1995): The retinoblastoma protein and cell cycle control. Cell 81:323–330.

Weinert TA, Kiser GL, Hartwell LH (1994): Mitotic checkpoint genes in budding yeast and dependence of mitosis on DNA replication and repair. Genes Dev 8:652–665.

Weintraub H (1993) The MyoD family and myogenesis: Redundancy, network and thresholds. Cell 75:1241–1244.

Weintraub SJ, Chow KNB, Luo RX, et al. (1995): Mechanism of active transcriptional repression by the retinoblastoma protein. Nature 375:812–815.

Weintraub SJ, Prater CA, Dean DC (1992): Retinoblastoma protein switches the E2F site from positive to negative elements. Nature 358:259–261.

Welch PJ, Wang JY (1993): A C-terminal protein-binding domain in the retinoblastoma protein regulates nuclear cAbl tyrosine kinase activity in the cell cycle. Cell 5:779–790.

Wells W (1996): The spindle assembly checkpoint: Aiming for a perfect mitosis, every time. Trends Cell Biol 6:228–234.

Wen S-T, Jackson PK, Van Etten RA (1996): The cytostatic function of cAbl is controlled by multiple nuclear localization signal and requires the p53 and Rb tumor suppressor gene products. EMBO J 15:1583–1595.

Whyte P (1995): The retinoblastoma protein and its relatives. Semin Cancer Biol 6:83–90.

Willems AR, Lanker S, Patton EE, et al. (1996): Cdc53 targets phosphorylated G_1 cyclins for degradation by the ubiquitin proteolytic pathway. Cell 86:453–463.

Wu L, Nurse P (1993): Nim1 kinase promotes mitosis by inactivating Wee1 tyrosine kinase. Nature 363:738–741.

Xiong Y, Zhang H, Beach D (1992): D-type cyclins associate with multiple protein kinases and the DNA replication and repair factor PCNA. Cell 71:505–514.

Xiong Y, Hannon GJ, Zhang H, et al. (1993a): p21 is a universal inhibitor of cyclin kinases. Nature 366:701–704.

Xiong Y, Zhang H, Beach D (1993b): Subunit rearrangement of the cyclin-dependent kinases is associated with cellular transformation. Genes Dev 7:1572–1583.

Xu Y, Baltimore D (1996): Dual role of ATM in cellular response to radiation in cell growth control. Genes Dev 10:2401–2410.

Yaffe D, Saxel O (1977): Serial passaging and differentiation of myogenic cells isolated from dystrophic mouse muscle. Nature 270:725–727.

Yaglom J, Linskens H, Sadis S, Rubin DM, Futcher B, Finley D (1995): $p34^{CDC28}$ mediated control of Cln3 cyclin degradation. Mol Cell Biol 15:731–741.

Yamamoto A, Guacci V, Koshland D (1996): J Cell Biol 133:85.

Yamasaki L, Jacks T, Bronson R, et al. (1996): Tumor induction and tissue atrophy in mice lacking E2F-1. Cell 85:537–548.

Yan Y, Frisen J, Lee M-H, et al. (1997) Ablation of the CDK inhibitor p57Kip2 results in increased apoptosis and delayed differentiation during mouse development. Genes Dev 11:973–983.

Yavuzer U, Keenan E, Lowings E, et al. (1995): The microphthalmia gene product interacts with the retinoblastoma protein in vitro and is a target for deregulation of melanocyte-specific transcription. Oncogene 10:123–134.

Yu H, King RW, Peters J-M, Kirshner MW (1996): Curr Biol 6:455.

Yuan W, Condorelli G, Caruso M, et al. (1996): Human p300 protein is a coactivator for the transcription factor MyoD. J Biol Chem 271:9009–9013.

Zachariae W, Nasmyth K (1996): TPR proteins required for anaphase progression mediate ubiquitination of mitotic B type cyclins in yeasts. Mol Biol Cell 7:791–801.

Zakian VA (1995): ATM-related genes: What do they tell us about functions of the human gene? Cell 82:685–687.

Zerfass-Thome K, Schulze A, Zwersche W, et al. (1997): p27 blocks cyclin E–dependent transactivation of cyclin A gene expression. Mol Cell Biol 17:407–415.

Zhang H, Hannon GJ, Beach D (1994): p21-containing cyclin kinases exist in both active and inactive states. Genes Dev 8:1750–1758.

Zhang H, Xiong Y, Beach D (1993): Proliferating cell nuclear antigen and p21 are components of multiple cell cycle kinase complexes. Mol Cell Biol 4:897–906.

Zhang P, Liegeois NJ, Wong C, et al. (1997): Altered cell differentiation and proliferation in mice lacking p57Kip2 indicates a role in Beckwith-Wiedemann syndrome. Nature 387,151–158.

Zhu L, Harlow E, Dynlacht BD (1995): p107 uses a p21 CIP1–related domain to bind cyclin/cdk2 and regulate interactions with E2F. Genes Dev 9:1740–1752.

Zhu L, van den Heuvel S, Helin K, et al. (1993): Inhibition of cell proliferation by p107, a relative of the retinoblastoma protein. Genes Dev 7:1111–1125.

Zwicker J, Liu N, Engeland K, et al. (1996): Cell cycle regulation of E2F site occupation in vivo. Science 271:1595–1596.

CHAPTER 3

REGULATION OF DNA REPLICATION AND S PHASE

G. PREM VEER REDDY
Department of Urology, Henry Ford Health System, Detroit, MI 48202

INTRODUCTION

The onset of DNA replication during the cell cycle is considered to be the hallmark of cell entry into S phase. Accurate duplication of the DNA during each cycle of cell division is fundamental to the propagation of life itself. This most basic biosynthetic process of DNA replication has to ensure that the DNA in the cells is duplicated fully during S phase of the cell cycle so that the genetic information is transmitted in its entirety from one cell generation to the next. In addition to ensuring complete and timely duplication of DNA, the replicative process must also restrict itself from re-replicating any portion of nuclear DNA within a single cell cycle. Any failure to duplicate nuclear DNA accurately, completely, timely, or only once within a cell cycle is debilitating to cells, often leading to their death or promiscuous proliferation as in cancers.

Early studies with prokaryotes have been instrumental in identifying and characterizing several of the enzymes and proteins fundamental to DNA replication (Kornberg and Baker, 1992). However, an understanding of the regulation of DNA replication in eukaryotes is complicated by the structural organization of DNA into nucleosomes and chromosome scaffolds. A complex architecture resulting from the close association of DNA with tightly bound proteins (histones) in eukaryotes could alter the ability of specific DNA sequences in a chromosome to serve as the origins for DNA replication. This is in contrast to the potential of specific sequences on bare DNA of eukaryotes to initiate DNA synthesis both in vivo and in vitro. A limited duration of S phase in which a large amount of DNA has to duplicate in eukaryotes also necessitates simultaneous replication of their DNA at multiple sites on each chromosome. For instance, instead of the usual 8–10 hours of S phase, a single replication

The Molecular Basis of Cell Cycle and Growth Control, Edited by G.S. Stein, R. Baserga, A. Giordano, and D.T. Denhardt
ISBN 0-471-15706-6, pages 80–154.

fork extending at an observed rate of 5 kb per minute on each chromosome would require over 15 days to fully duplicate $\simeq 3 \times 10^6$ kb DNA in 23 chromosomes during a single S phase in a human cell. Replication at multiple sites on a chromosome must be coordinated with transcriptional activity to facilitate simultaneous expression of genes necessary for subsequent steps in cell division and for other functions vital to the cells. Furthermore, the replication must be accompanied by the duplication of nuclear proteins necessary for restoring the chromosome architecture that regulates specific patterns of gene expression inherent to the cell types.

While the process of DNA replication itself has received much deserved attention in recent years, the role of deoxynucleotides and the enzymes required for their synthesis in regulation of DNA replication in eukaryotes have not been fully appreciated. To sustain a maximal rate of DNA replication with a high fidelity characteristic of multicellular organisms, there must be a continuous and balanced supply of all four deoxynucleoside triphosphates (dNTPs) to the sites of replication. A low steady-state level of dNTP pools observed in proliferating cells also necessitates their compartmentalization in the microvicinity of DNA replication. Furthermore, while most of the enzymes of dNTP synthesis are present in the cells continuously throughout the cell cycle, their activity inside the cells remains undetectable until the onset of DNA replication in S phase (Reddy, 1982). These observations raise several issues fundamental to the regulation of DNA replication and the entry of cells into S phase. For example, how can dNTPs be compartmentalized at the sites of DNA replication when no physical barriers are present inside the cells to limit their free diffusion throughout the nucleus and the cytoplasm? What limits the enzymes of dNTP synthesis from being catalytically active inside the cells during other phases of the cell cycle when DNA is not replicating? Studies to address these and other issues relating to deoxynucleotide metabolism during the transition of cells from G_1 into S phase have been reviewed in detail elsewhere (Reddy and Fager, 1993).

Much of our recent understanding of the regulation of DNA replication and S phase in eukaryotes has emerged from the convergence of observations from elegant studies in yeast (by geneticists and molecular biologists) and in frog eggs (by biochemists and physiologists). Although the foundation for these studies was laid some 30 years ago, most knowledge in this field has accumulated only within the last 10 years. During this period, we have witnessed the identification and characterization of cyclins and cyclin-dependent kinases (CDKs) associated with the entry into and progression through S phase, origins of DNA replication, proteins involved in the initiation of DNA replication, proteins that prevent DNA from replicating more than once per cell cycle, and the enzyme telomerase, which acts to protect telomeric DNA against a progressive loss during each round of DNA replication. Although these discoveries have come essentially from studies with unicellular organisms and eggs, important details of their involvement in regulation of DNA replication and S phase seems to be universal to all proliferating cells from higher organisms. How each of these discoveries contributed to an understanding of the range of complex mechanisms devised by the cells to ensure precise duplication of total nuclear DNA within the confines of S phase are presented in this chapter.

CELL CYCLE REGULATORY PROTEINS THAT CONTROL PROGRESSION OF CELLS FROM G_1 INTO S PHASE

Cell Fusion Studies Revealing the Presence of Regulatory Factors Controling the Entry of Cells Into S Phase

Mammalian cell fusion experiments by Rao and Johnson in the early 1970s led to many elegant studies of the molecular events contributing to control of DNA replication during the cell cycle in eukaryotes. They, for the first time, examined the ability of intracellular components of cells in one phase of the cell cycle to affect the ability of the cells in other phases of the cell cycle to enter S phase and mitosis (Rao and Johnson, 1970). Following the fusion of synchronized HeLa cells in S phase with those in G_1 or G_2 phase, they were able to demonstrate that, while the cells in G_1 phase can be made to enter abruptly into S phase by a rapid induction of DNA synthesis, the cells in G_2 phase cannot be made to re-replicate DNA before they undergo mitosis. G_1 cells fused to S phase cells not only entered into S phase abruptly but also transited G_2 and M phases much earlier than those that had not been fused or had been fused to G_1 or G_2 phase cells. On the other hand, neither G_1 nor G_2 phase cells could prevent S phase cells from completing DNA replication and their subsequent passage through G_2 and M phases.

Taken together, these observations unveiled several insights into the control of DNA replication during cell cycle: (1) The onset of DNA replication in G_1 cells requires specific factors whose expression and/or activation is restricted to the cells that have committed to S phase; otherwise, the DNA itself is competent to replicate at any point in G_1 (2) G_1 or G_2 cells do not contain any inhibitors that are capable of preventing S phase cells from completing DNA replication and passing through G_2 and M phases; thus, the commitment of cells to enter into S phase is the rate-limiting step in their ability to transit through a full cycle of cell division. (3) DNA replicated once during S phase becomes inaccessible to the factors in S phase cells for its re-replication at any time prior to nuclear division, suggesting that the nuclear factors necessary for the initiation of DNA replication are being turned over with one round of replication during S phase, and the restoration of these factors again in the nucleus would require its breakdown during mitosis. Two of the most fundamental questions in the control of the cell cycle then became (1) What are the factors in S phase cells that are capable of initiating DNA replication in G_1 phase cells? (2) What are the nuclear factors that render DNA incapable of re-replicating following one round of its replication during S phase?

Maturation Promoting Factor

Maturation-promoting factor (MPF) was initially identified in mature frog oocytes as a substance capable of completing meiosis and inducing repeated mitotic cycles in immature oocytes that are arrested in metaphase of meiosis II (Masui and Markert, 1971; Ecker and Smith, 1971). A similar factor capable of inducing mitosis was also indicated from fusion studies with *Physarum polycephalum* (Rusch et al., 1966) and HeLa cells (Rao and Johnson, 1970). Cell-free extracts prepared from *Xenopus* eggs were shown to support periodic DNA synthesis in sperm pronuclei (Hutchison et al., 1987). This DNA synthesis in *Xenopus* egg extracts was found to be sensitive to the inhibitors of protein synthesis, indicating periodic changes in the level of key cellular components required for the induction of DNA synthesis. Based on its ability to induce chromosome condensation and nuclear membrane breakdown, MPF capable of inducing DNA replication was purified from *Xenopus* extracts (Lohka et al., 1988). Purified MPF consisted of two subunits, one of them being 34 kD with serine/threonine protein kinase and the other a 45 kD substrate. The 34 kD subunit with kinase activity is identified as a *Xenopus* homologue of cdc2 protein kinase and the 45 kD subunit as a *Xenopus* B-type cyclin, which were by then implicated to be involved in the control of mitosis in yeast (see below). Although an MPF-like substance capable of promoting S phase, S phase–promoting factor (SPF), has not been purified for the lack of a suitable assay, yeast genetics and in vitro DNA replication systems employing egg extracts have allowed the identification of cellular components representative of SPF. These factors have come to be known as *S phase cyclins* and cyclin-dependent kinases (CDKs).

Identification of Cell Cycle Regulatory Proteins

Genetic studies of Hartwell and coworkers with brewer's or budding yeast, *Saccharomyces cerevisiae,* in early 1970s led to the identification of genes whose products are critical for a sequential progression of cells from one stage to the next in the cell division cycle (Hartwell et al., 1970; Hartwell, 1971; Hereford and Hartwell, 1974). These essential genes are now collectively known as *cell-division cycle* (cdc) genes. These studies also demonstrated that the initiation of certain steps in the cell-division cycle is dependent on the completion of one or more preceeding steps. For instance, onset of mitosis and nuclear division is dependent on the completion of DNA replication, and the completion of mitosis is dependent on the assembly of the mitotic spindle.

Nurse and colleagues extended these studies to identify genes controlling the cell-division cycle of the fission yeast *Schizosaccharomyces pombe,* whose cell cycle resembles more closely that of mammalian somatic cells. *Sch. pombe* attains a critical mass during interphase in order for its entry into S phase and divides into two daughter cells of equal size following mitosis (Nurse, 1975; Nurse and Thuriaux, 1977;

Nasmyth et al., 1979). Studies with fission yeast lead to the identification of a number of genes involved in DNA synthesis, mitosis, nuclear division, and cell plate formation (Nurse et al., 1976; Nasmyth and Nurse, 1981). One of these genes, cdc2, turned out to be of particular importance as we came to know of its central role in the entry of cells into both S phase and mitosis (Nurse and Thuriaux, 1980; Simanis et al., 1987). While certain mutations in this gene prevented cells from entering into mitosis, other mutations, in contrast, caused the cells to undergo mitosis earlier than usual. Complimentation essays revealed that the mutations in the cdc2 gene can be rescued by $cdc28^{+}$, which has been identified previously as a member of the cdc family in budding yeast, thus making cdc2 in *Sch. pombe* a functional homologue of cdc28 in *Sac. cerevisiae* (Beach et al., 1982). Further extension of these studies to rescue cdc2 mutations in fission yeast by expressing a human cDNA library lead to the identification of a human analogue of cdc2. The predicted protein sequence of a human homologue and that of the yeast cdc2 gene were found to be very similar (Lee and Nurse, 1987). Because of these functional, structural, and apparent molecular weight similarities, these proteins controlling mitosis in different species are now commonly referred to as $p34^{cdc2}$ or cdc2p.

Cyclins and CDKs in Yeast

In yeast, the cdc2p at START, a point in G_1 when commitment to the initiation of DNA replication (S phase) is made, and at the beginning of mitosis is controlled by its interaction with a number of other proteins and its phosphorylation state. In *Sch. pombe,* three B-type cyclins encoded by cdc13, cig1, and cig2 interact individually with cdc2p to activate its protein kinase at specific points in cell cycle. While cdc13 activation of cdc2p is essential for the entry of cells into mitosis (Moreno et al., 1989; Booher et al., 1989), cig2 plays a major role in activating cdc2p in G_1 phase during the transition of cells from G_1 into S phase (Martin-Castellanos et al., 1996; Mondesert et al., 1996). However, cdc13 is also capable of substituting for cig2 to allow the progression of cig2-deficient mutants from G_1 into S phase, albeit with some delay in G_1 (Fisher and Nurse, 1996; Martin-Castellanos et al., 1996). Furthermore, the onset of S phase in *Sch. pombe* could not be blocked completely without deleting all three cyclins (cdc13, cig1, and cig2), suggesting that cig1 may also play a minor role in the G_1 to S transition (Fisher and Nurse, 1996).

An intriguing observation of genetic studies in yeast is that the inactivation of the cdc13-dependent cdc2p kinase, as observed in cdc13 and cdc2 temperature-sensitive ($cdc13^{ts}$ and $cdc2^{ts}$) mutants, a cdc13 deletion ($cdc13\Delta$) mutant, or a rum 1 (a specific inhibitor for cdc13/cdc2p)–overexpressing mutant, while causing a block in mitosis as expected, actually allows the cells to undergo repeated re-replication of DNA (Hayles et al., 1994; Moreno and Nurse, 1994; Correa-Bordes and Nurse, 1996; Jallepalli and Kelly, 1996). These studies point to an additional role of mitotic cyclin/cdc2p kinase in preventing re-replication of DNA and ensuring that there is only one S phase per cell cycle. Although three different cyclins (cdc13, cig1, and cig2) are known to specify the function of a single protein kinase, cdc2p, to allow the progression of *Sch. pombe* from one phase to the next in cell cycle, it seems that a single cyclin, cdc13, interacting with cdc2p is capable of determining whether cells can enter into S phase or mitosis and also prevents the cells from re-replicating their DNA before undergoing mitosis. Regulator(s) and/or the mechanism(s) that confer these three fundamentally different functions to the cdc13/cdc2p complex during cell cycle in *Sch. pombe* has been a subject of intense investigation in recent years (Stern and Nurse, 1996; Tyson et al., 1996).

In the budding yeast *Sac. cerevisiae,* cdc28p performs all basic functions that cdc2p is known to carry out in fission yeast. The B-type cyclins that regulate cdc28p kinase activity during the cell cycle in budding yeast are encoded by CLB genes 1 to 6 (clb1 to clb6). The cyclins encoded by CLB5 and CLB6 (clb5 and clb6) are involved in triggering DNA replication at the onset of S phase (Schwob and Nasmyth, 1993). Those encoded by CLB3 and CLB4 (clb3 and clb4) induce the formation of mitotic spindle (Fitch et al., 1992; Richardson et al., 1992), and CLB1 and CLB2 (clb1 and clb2) trigger nuclear division (Surana et al., 1991). CLB5 and CLB6 mutations lead to a delay, but not to a total block, in the entry of cells into S phase (Schwob and Nasmyth, 1993; Kuhne and Linder, 1993), suggesting that other cyclins (clb1 through clb4) that accumulate later in the cell cycle may also be involved in the induction of S phase. In budding yeast, there are at least three other cyclins that are distinct from B-type cyclins and are known to play a crucial role in progression of cells through G_1. These cyclins, designated as CLNs, in association with cdc28p, activate clb cyclins required for the entry of cells into S phase (Dirick et al., 1995; Schwob et al., 1994). Cln cyclins seem to be dispensable in this role as G_1 arrest in mutant cells lacking all three cln cyclins can be rescued by inducing the expression of clb5 or even clb2 from the GAL1–10 promoter (Schwob and Nasmyth, 1993; Amon et al., 1994). It is, however, intriguing to note that the function of both the cln cyclins and clb cyclins is being mediated through their interaction with a single protein kinase, cdc28p.

Cyclins and CDKs in Mammalian Cells

In mammalian cells, as in *Ascomycetes,* a range of cyclin subunits activate cyclin-dependent kinases (CDKs) to trigger their transition from one phase to the next during cell cycle. However, the specific cyclins and CDKs involved in the transition of mammalian cells through G_1 and into S phase are different from those required for similar processes in yeast. For instance, cdc2p/cdc28p kinase, which is common for induction of both the S phase and mitosis in yeast, is required only for mitosis in mammalian cells and is designated as CDK1 (Riabowol et al., 1989; Hamaguchi et al., 1992). There are at least three other CDKs known to be involved in the transition of mammalian cells through G_1 and into S

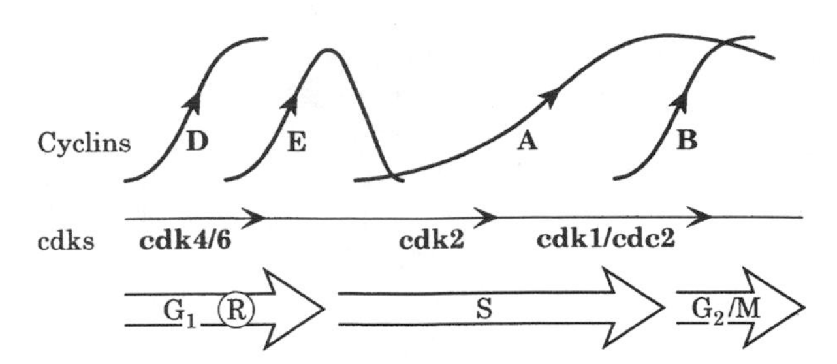

Figure 3.1. Periodic changes in some of the cyclins as mammalian cells progress from G_1 into S phase. The protein level of cyclin-dependent kinases (CDKs) is suggested to remain relatively stable throughout the cell cycle.

phase: CDK4 and CDK6 for progression through early to late G_1 (Sherr, 1994) and CDK2 for entry into and progression through S phase (Fang and Newport, 1991; Paris et al., 1991).

The D-type cyclins in association with CDK4 or CDK6 are suggested to be involved in progression of mammalian cells not only through G_1 but also in triggering their entry into S phase (Baldin et al., 1993; Quelle et al., 1993; Sherr, 1993). Overexpression of cyclin D1, in either cycling or quiescent cells stimulated to enter into a proliferative cycle, resulted in a significant reduction in the G_1 phase of their cell cycle (Resnitzky et al., 1994; Quelle et al., 1993). Although the cells overexpressing cyclin D1 in these studies exhibited a reduced G_1 phase, their overall rate of proliferation remained essentially the same as that of normal cells. Similarly, cyclin D1 in transgenic mice did not alter the proliferative rate of the cells, but, in combination with other oncogenes, was able to induce tumorigenesis (Bodrug et al., 1994; Lovec, 1994). In comparison to the observations made in a variety of cells overexpressing cyclin D1, cells lacking a cyclin D1 gene are found to undergo cell division and complete a normal prenatal development (Sicinsk et al., 1995). These observations in embryonic cells raised the possibility that either cyclin D1 is not directly responsible for the transition of mammalian cells into S phase or, as in yeast, other cyclins expressed later in the cell cycle are able to fulfill the function of cyclin D1 to trigger cell entry into S phase.

Cyclins E and A, in association with CDK2, are known to be more directly involved in the entry and progression of cells through S phase. Both of these proteins increase to a maximum level at the time when cells pass through S phase (Fig. 3.1); the level of cyclin E increasing at the beginning of S phase (Koff et al., 1992; Dulic et al., 1992) and that of cyclin A increasing during S and G_2 phases (Pines and Hunter, 1990; Tsai et al., 1991). Experiments to examine the role of these two cyclins in DNA replication have largely been carried out in *Drosophila* embryos and *Xenopus* egg extracts. *Drosophila* embryos defective in cyclin E arrest in G_1 and cease to replicate DNA following the 16th cycle of mitotic division when maternal cyclin E transcripts suddenly disappear. An ectopic expression of cyclin E immediately following their exit from mitosis in the 16th cycle allowed these cells to overcome G_1 arrest and enter into S phase (Knoblich et al., 1994). A similar induction of cyclin

E in quiescent mammalian cells allowed their progression into S phase with a modest reduction in the time required for their transit through G_1 (Ohtsubo and Roberts, 1993; Resnitzky et al., 1994). This requirement for cyclin E in induction of mammalian cells to enter into S phase could not be alleviated by the presence of other cyclins, indicating an absolute requirement of cyclin E for cellular commitment to S phase.

Among all the known cyclins in mammalian cells only cyclin A, with some structural and functional resemblance to B-type cyclins, is found to regulate more than one step in the mammalian cell cycle. It has been implicated in the control of S phase (Girard et al., 1991; Pagano et al., 1992; Zindy et al., 1992; Strausfeld et al., 1996), as well as mitosis (Minshull et al., 1989; Lehner and O'Farrell, 1989; Strausfeld et al., 1996), and also in preventing nuclear DNA from replicating more than once per cell cycle (Sauer et al., 1995). Periodic variations in the level of cyclin A/cdk2 activity during the cell cycle seem to determine its functional specificity to induce either S phase or mitosis; its low level promotes the passage through S phase, and its high level causes the induction of mitosis (Strausfeld et al., 1996). In contrast to these observations in studies with mammalian cultured cells, cyclin A does not seem to be required for DNA replication in *Xenopus* egg extracts and during certain divisions of *Drosophila* embryogenesis (Knoblich and Lehner, 1993; Lehner and O'Farrell, 1990). This makes cyclin A relatively less important than cyclin E in the induction of S phase at least during embryonal development in higher eukaryotes. However, the labile nature of both of these cyclins and the increase in the level of their activity at the beginning of S phase (Dou et al., 1993) is consistent with their role in allowing cells to pass the restriction (R) point, the equivalent of the START site in the yeast cycle, when the cells become committed to S phase (Pardee, 1974, 1989).

Cyclin/CDK Inhibitors

Periodic changes in cyclins alone do not fully account for the stringent temporal order of cell cycle progression in yeast and mammalian cells. There is a second tier of regulation being exerted by a family of low molecular weight proteins composed of ankyrin repeats that inhibit the activity of CDKs. In recent years, these cdk inhibitors (CDIs) have emerged as major players in regulating cell cycle progression from G_1 into S phase and also in preventing DNA from re-replicating prior to the nuclear division. In *Sac. cerevisiae,* although the B-type cyclin (clb5/clb6)–cdc28p complexes are present in G_1 phase, they are incapable of inducing the entry of cells into S phase because of their inhibition by the Sic1 protein (Schwob et al., 1994; Nugroho and Mendenhall, 1994). The Sic1p level in the cells increases transiently following mitosis and reaches its highest level in early G_1, blocking (clb5/clb6)–cdc28p activity. Shortly before S phase it undergoes a rapid ubiquitin-dependent proteolysis permitting (clb5/clb6)–cdc28p kinase to be active (Donovan et al., 1994; Schwob et al., 1994). At least three proteins, cdc34p (Goebl et al., 1988), cdc4p (Mathias et al., 1996), and SKP1p (Bai et al., 1996), are known to be involved in ubiquitin-dependent proteolysis (see below) of Sic1p at the START. Temperature-sensitive mutations in any of these genes leads to the accumulation of Sic1p at nonpermissive temperatures, resulting in the arrest of cells in late G_1 with 1N DNA content. Deletion of Sic1 gene in these mutants allows them to initiate DNA replication at nonpermissive temperature (Schwob et al., 1994). Sic1 deletion also uncouples initiation of DNA replication from other cell cycle events, such as budding, and allows an immediate initiation of DNA replication following mitosis (Schneider et al., 1996). Furthermore, the only nonredundant essential function of G_1 cln cyclins in allowing the transition of cells from G_1 into S phase is to target Sic1p for ubiquitin-dependent proteolysis. This is indicated by the fact that the lethality of a cyclin triple (CLN1, CLN2, and CLN3) mutation can be suppressed by Sic1 deletion (Schneider et al, 1996). Thus the degradation of Sic1p represents a crucial step in overall commitment of *Sac. cerevisiae* at the START to initiate DNA replication and enter into S phase.

A similar mechanism involving CDIs in the control of onset of DNA replication and S phase in *Sch. pombe* is indicated from the identification of the $rum1^+$ gene whose overexpression resulted in several rounds of DNA replication in the absence of mitosis (Moreno and Nurse, 1994). Overexpression of $rum1^+$ resulted in a significant drop in cdc13/cdc2p kinase activity. Deletion of $rum1^+$ eliminated the G_1 interval required for attaining critical cell mass before entering into S phase. The $rum1^+$ deletion also prevented the G_1 arrest that normally occurs following nitrogen starvation in wild-type cells, indicating a $rum1^+$ requirement for G_1 arrest at START (Moreno and Nurse 1994). While rum^+ protein is found to inhibit cdc13/cdc2p kinase (Correa-Bordes and Nurse, 1996), its regulation during the cell cycle and its involvement in the regulation of cdc13/cdc2p kinase activity in G_1 remains to be identified.

An initial understanding of CDI involvement in regulation of S phase onset in mammalian cells has come from the work of Koff et al. (1993). They found that the extracts of the cells arrested in G_1 by transforming growth factor-β (TGF-β) contained normal amounts of cyclin E and CDK2 but failed to exhibit cyclin E/CDK2 activity. These studies revealed the presence of an inhibiting factor in extracts of TGF-β–treated cell that is capable of blocking cyclin E/CDK2 activity in the extracts of untreated cells. This inhibitory factor is identified to be a heat-stable protein with an apparent molecular weight of 27 kD that exists in an inactive form in untreated cells and is referred to as $p27^{kip1}$ (Polyak et al., 1994). $p27^{kip1}$ levels are found to decline as quiescent macrophage cells (Kato et al., 1994) and T cells (Firpo et al., 1994) are induced to enter into S phase following growth factor/cytokine stimulation.

The advent of two-hybrid screening technology lead to the identification of at least three other CDIs, designated as INK4, Pic1, and cip2/CDI1 in mammalian cells (Fig. 3.2). Although each of these CDIs was isolated by using specific CDKs as "baits," in this technique they have the potential to bind and inactivate other CDKs interacting with a variety of cyclins. For instance, INK4 isolated by using CDK4 as a bait (Serrano

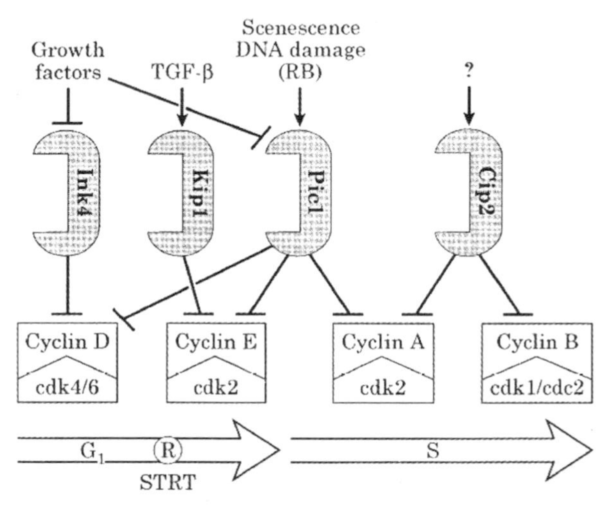

Figure 3.2. Regulation and cyclin/CDK targets of some of the cyclin-dependent kinase inhibitors (CDIs) that are known to play a role in progression of mammalian cells through G_1 and S phases.

et al., 1993) inactivates both CDK4 and CDK6 associated with cyclin D (Xiong et al., 1993; Hannon and Beach, 1994; Guan et al., 1994) and arrests the cells in G_1. Cyclin D/CDK4- or CDK6-dependent phosphorylation of retinoblastoma protein (Rb) is essential for the progression of cells through G_1 (Quelle et al., 1993; Resnitzky et al., 1994). INK4 fails to induce G_1 arrest in the cells that lack functional Rb (Lukas et al., 1995; Koh et al., 1995), suggesting INK4 involvement in the control of cyclin D–CDK4/CDK6–dependent phosphorylation of Rb required for the progression of cells through G_1 and into S phase. INK4 regulation of these processes seems to play an important role in the proliferation of normal cells as a number of primary tumors and tumor cell lines are found to contain mutations in its gene (Hunter and Pines, 1994).

Similarly, cip1 isolated by its binding to CDK2 in a two-hybrid screening system (Harper et al., 1993), also known as WAF1 (El-Derry et al., 1993), Sdi1 (Noda et al., 1994), Cap20 (Gu et al., 1993), and Pic1 (Hunter, 1993), inhibits the kinase activity of cyclin A/CDK2, cyclin E/CDK2, and cyclin D1/CDK4 strongly and that of cyclin B/cdc2 much more weakly. It is found associated with cyclins A, E, and D1 under in vitro conditions (Xiong et al., 1993). cip1/Pic1 induction is linked more directly to the block in the entry of x-ray–treated cells into S phase. DNA damage caused by irradiation is shown to activate wild-type (but not mutant) p53, which in turn binds to cip1/Pic1 promoter, leading to its transcription. An increase in abundance of cip1/Pic1 following p53 activation blocks the entry of irradiated cells into S phase by inactivating cyclin E/CDK2 and/or cyclin D/CDK4 (Deng et al., 1995). However, p53 activation is not the only mechanism by which Pic1 expression is induced. Actually, in senescent cells an increase in cip1/Pic1 levels is observed in direct contrast to the decrease in p53 levels and in muscle, and other terminally differentiated cells growth factors seem to be involved in regulation of its expression (Parker et al., 1995).

One other CDI, referred to as cip2 or CDI1, has been isolated by virtue of its binding to CDK2 or cdc2 in two-hybrid screening systems (Gyuris et al., 1993; Harper et al., 1993). This is a 24 kD protein with structural and functional similarities to cdc25p phosphatase. cip2 binds tightly to CDK2 and cdc2 but not to CDK4. Overexpression of wild-type, but not mutant, cip2/CDI1 in HeLa or yeast cells leads to their arrest in G_1, and in normally cycling cells its mRNA and protein reach a maximum level in late G_1 (Gyuris et al., 1993), suggesting a negative regulatory role for this protein also in the control of cell entry into S phase.

Thus, most of the CDIs identified to date exhibit a broad specificity in their binding to CDKs. This has made it difficult to assign the negative regulatory function exclusively to any one particular CDI in the control of S phase onset in mammalian cells. Furthermore, even when a potential role in blocking S phase is assigned to a particular CDI, it is hard to establish whether its regulation during cell cycle contributes to the G_1/S checkpoint control or to the growth factor–induced signal transduction pathways governing G_1/S transition at the restriction (R) point in cell cycle. However, the observations that INK4 and cip1/Pic1 in transformed cells with normal checkpoint controls do not respond to growth inhibitory signals, such as TGF-β and cell–cell contact, and that they fail to inhibit G_1 cyclin/CDK activity indicate possible involvement of these two CDIs in signal transduction pathways controlling the transition of cells from G_1 into S phase (Nasmyth and Hunt, 1993).

Proteolysis in Progression of Cells from G_1 Into S Phase

Degradation of these CDIs at the end of G_1 by a ubiquitin-dependent mechanism is an essential step in the onset of DNA replication (Fig. 3.3). In mammalian cells, as in yeast, the CDIs are marked for degradation by a ubiquitin-conjugating enzyme system consisting of E1 (ubiquitin-activating enzyme), E2 (ubiquitin-conjugating enzyme), and E3 (ubiquitin-ligating enzyme) (Ciechanover, 1994). While E1 enzymes initiate the first step in the reaction, a variety of E2 enzymes in conjunction with E3 enzymes seems to determine the specificity for the proteins targeted for ubiquitination. Once the target proteins are ubiquitinated, they are readily degraded by 26S proteosomes in an ATP-dependent reaction (Hilt and Wolf, 1996). This ubiquitin-dependent proteolysis is also responsible for the destruction of cyclins contributing to periodic changes in their levels during cell cycle (Glotzer et al., 1991; Luca et al., 1991; Deshaies et al., 1995; Yaglom et al., 1995; Seufert et al., 1995). The presence of cis-acting signals, such as the "destruction box" consisting of short stretches of highly conserved amino acids or "PEST" sequences consisting of regions rich in proline, aspartic acid, glutamic acid, serine, and threonine, on cyclins targets them for ubiquitin-dependent proteolysis. It is not known whether any such cis-acting signals are present on

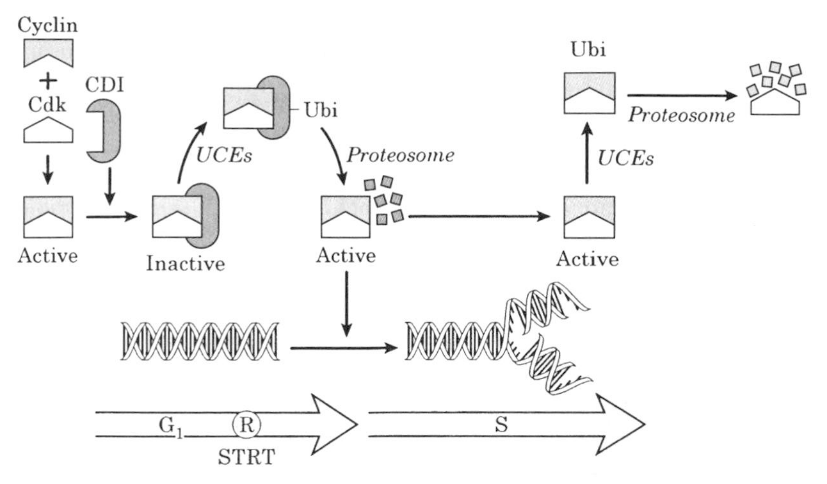

Figure 3.3. Proposed role of proteolysis in regulation of DNA replication and the entry of cells into S phase. For initiation of replication, cyclin-dependent kinase inhibitors (CDIs) are marked for destruction by ubiquitin-conjugating enzymes (UCEs). Ubiquitinated CDI is then degraded by a 26S proteosome complex, allowing cyclin/CDKs to activate initiation of DNA replication and entry of cells into S phase. Once DNA replication is initiated, cyclins are similarly degraded by UCEs and proteosome to prevent reinitiation of replication within the same S phase and to allow cells to progress through G_2 and M phases.

CDIs to make them susceptible to ubiquitination. Specific phosphorylated states of these proteins also seem to determine their susceptibility to ubiquitination. CLN2 and CLN3 are found phosphorylated before ubiquitination, and such a phosphorylated state seems to be necessary for their proteolysis because nonphosphorylated mutants of CLN2 and CLN3 were stable in vitro (Deshaies et al., 1995; Yaglom et al., 1995). Similarly, phosphorylation of CDIs by G_1 cyclin CDK complexes during late G_1 may trigger their ubiquitin-dependent degradation, thereby allowing S phase cyclin CDKs to activate the progression of cells into S phase. Consistent with this possibility is the observation that the S phase block in G_1 cyclin mutants is circumvented by Sic1 deletion in yeast (Schneider et al., 1996; Tyers, 1996), indicating that the sole function of G_1 cyclins in regulating the entry of cells into S phase could be to induce Sic1p phosphorylation, thereby making it a target for ubiquitin-dependent proteolysis.

ROLE OF CYCLIN/CDK IN REGULATION OF DNA REPLICATION AND THE ONSET OF S PHASE

Cyclin/CDK-Mediated Expression of the Enzymes Associated With DNA Synthesis

While the specific substrates of cyclin CDKs that are directly involved in DNA replication remain to be identified, active cyclin CDK complexes are known to stimulate the expression of a number of enzymes and proteins associated with DNA replication. This is accomplished by changing the phosphorylated state of Rb and Rb-like proteins (p130, p107), which control the activity of a family of heterodimeric transcriptional regulators called E2Fs (La Thange, 1994). E2Fs induce expression of a number of cell cycle and DNA synthesis regulatory genes by binding to their promoter sequences. These genes include those encoding thymidine kinase, dihydrofolate reductase, thymidylate synthase, DNA polymerase α, cdc2, cyclin E, cyclin A, and c-myc (Means et al., 1992; Ogris et al., 1993; Pearson et al., 1991; Dalton, 1992; Lam and Watson, 1993; Mudryj et al., 1990; Reed et al., 1992; Geng et al., 1996; Karlseder et al., 1996; Sherr, 1996; Dou et al., 1992). In a hypophosphorylated state Rb represses the expression of these genes by binding and inactivating E2F/DP-1 heterodimeric transcription factors. Phosphorylation of Rb causes Rb to be released from E2F/DP-1 complexes, allowing E2F/DP-1 complexes to be transcriptionally active. Cyclin D/CDK4 (Lukas et al., 1994; Resnitzky et al., 1993; Quelle et al., 1993; Baldin et al., 1993) in G_1 and cyclin E/CDK2 (Hinds et al., 1992; Beijersbergen et al., 1995) during the transition of cells from G_1 into S are known to induce phosphorylation of Rb, leading to E2F/DP-1–dependent induction of the enzymes and proteins regulating the onset of DNA replication. Because the expression of both cyclin E and E2F itself is under the control of E2F trans-activation, once their expression is initiated by cyclin D1/CDK4/6–dependent phosphorylation of Rb in late G_1, they orchestrate the induction of their own expression by positive feedback regulation of E2F during G_1/S transition (Weinberg, 1995; Johnson et al., 1994). This would allow the expression of enzymes and proteins associated with DNA synthesis in cells to switch from being induced by mitogen-dependent cyclin D/CDK4/6 activation in G_1 to mitogen-independent cyclin E/CDK2 activation during G_1/S transition. Such a switch to growth factor–independent induction of the enzymes and proteins associated with DNA synthesis would ensure an irreversible commitment of cells to progress through S phase and, normally, through the remainder of cell cycle.

In addition to cyclin D– and E–dependent kinases, cyclin A–dependent kinase is also capable of phosphorylating Rb to allow E2F trans-activation of genes. This is indicated from the observation that the ectopic expression of cyclin A induces Rb phosphorylation and premature entry of cells into S phase (Hinds et al., 1992; Resnitzky et al., 1995). Both cyclin E and cyclin A are found associated with E2F complexes bound to DNA during late G_1 and S phases, respectively (Mudryj et al., 1991; Bandara et al., 1991; Lees et al., 1992). Cyclin A associated with E2F/DNA complexes is able to recruit catalytic kinase subunit CDK2 to phosphorylate Rb or p107 bound to E2F (Bandara et al., 1992; Cao et al., 1992; Devoto et al., 1992). Such cyclin A/CDK2 activity also seems to be critical for the entry of cells into S phase because inhibition of cyclin A function prevents the cells from entering into S phase (Girard et al., 1991; Pagano et al., 1992; Zindy et al., 1992). In addition to being able to promote the expression of genes necessary for the entry of cells into S phase by phosphorylating Rb, cyclin A/CDK2 is also implicated in suppressing gene expression at the end of S phase. Cyclin A/CDK2 is shown to bind stably to E2F/DP-1 DNA complexes and phosphorylate DP-1. Such phosphorylation causes E2F/DP-1 to be released from DNA, thereby inhibiting the expression of E2F-target genes (Krek et al., 1994). These opposing dual roles of cyclin A/CDK2 in gene expression seem to be facilitated by the periodic changes in cyclin A levels: at low levels, in the beginning of S phase, promoting gene expression by phosphorylating Rb, and at high levels, toward the end of S phase, suppressing E2F-target gene expression by phosphorylating its DNA-binding subunit DP-1. Thus the gene expression facilitated by periodic changes in the levels of cyclins E and A during late G_1 and early S phase represents an important step in the ability of cells to enter into S phase and initiate DNA replication.

Cyclin/CDK Regulation of Proteins Required for the Initiation of DNA Replication

An induction of enzymes and proteins associated with DNA synthesis itself may not be the final step in triggering the initiation of DNA replication at the onset of S phase, because *Xenopus* egg extracts containing all the enzymes and proteins necessary for DNA synthesis can be blocked from initiating DNA replication by immunodepletion or CDI affinity depletion of cyclins (Jackson et al., 1995) or their associated CDKs (Blow and Nurse, 1990; Fang and Newport, 1991; Strausfeld et al., 1994; Jackson et al., 1995). These observations point to a more direct role of cyclin/CDKs in activating key component(s) of DNA replication machinery that are involved in triggering the initiation of DNA replication. Although no such S phase cyclin/CDK substrate that is specific in driving the cells from G_1 into S phase has yet been identified, mitotic cyclin/CDK in yeast is shown to be directly involved in suppressing the level and function of an essential protein required for the initiation of DNA replication (Jellepalli and Kelly, 1996). cdc6 protein (cdc6p) in *Sac. cerevisiae* and its related cdc18 protein (cdc18p) in *Sch. pombe* play a central role not only in initiation of DNA replication but also in ensuring that the DNA replication occurs in each cycle before a cell enters mitosis. In the absence of these proteins in late G_1, cells enter mitosis prematurely without first replicating their DNA, and consequently they die because of the reduction in genetic makeup of the daughter cells (Kelly et al., 1993; Piatti et al., 1995). Both of these proteins are extremely labile with half-lives of less than 5 minutes and are synthesized de novo in late G_1 (Nishitani and Nurse, 1995; Muzi-Falconi et al., 1996; Piatti et al., 1995) and also in late G_2 (Piatti et al., 1995). Binding of these proteins to the origin recognition complex (ORC) at the DNA replication initiation sites (origins of replication) on the chromosome seems to be the rate-limiting step for the initiation of DNA replication (Liang et al., 1995; Leatherwood et al., 1996; Coleman et al., 1996) (see below). cdc18p/cdc6p bound to ORC at the origins is inactivated immediately following the initiation of DNA replication so that it is prevented from reinitiating DNA replication before mitosis occurs. Mitotic cyclin/CDKs are found to be responsible for such inactivation of cdc18p/cdc6p because inhibition of mitotic cyclin/CDK activity leads to the accumulation of cdc18p/cdc6p and the re-replication of DNA before mitosis. (Jellapalli and Kelly, 1996). Thus, mitotic cyclin/CDK-dependent inactivation of cdc18p/cdc6p seems to be a necessary step in preventing genomic DNA from replicating more than once per cell cycle.

Association of Cyclin/CDKs With the Enzymes of DNA Replication

A direct interaction between cyclin/CDK complexes and the enzymes of DNA replication may also determine the ability of cells to enter into and/or progress through S phase. This is indicated from the observation that cyclin A/CDK2 is specifically co-localized with discrete sites of DNA replication in the nuclei of S phase cells (Cardoso et al., 1993). This would raise the possibility that S phase cyclin/CDKs may play a role in the assembly, activation, or maintenance of the multienzyme complex associated with DNA replication. Consistent with such a possibility is the observation that cyclin A/CDK2, but not cyclin B/cdc2, in HeLa cells is associated with a high molecular weight nuclear fraction consisting of DNA polymerase α and proliferating cell nuclear antigen (PCNA) during biochemical fractionation (Jaumot et al., 1994). Furthermore, $p21^{cip1}$, an inhibitor of most cyclin/CDKs, induced following DNA damage is shown to directly inhibit DNA replication by binding to and inhibiting PCNA associated with DNA polymerase δ (Waga et al., 1994; Li et al., 1994). From in vitro DNA replication studies, cyclin/CDK is also implicated in activating the origin-unwinding function of SV40 T antigen (Fotedar and Roberts, 1991; Roberts, 1993). Multisubunit single-stranded DNA-binding protein, referred to as *replication factor A* (RF-A), associated with DNA replication complex is also a cyclin/CDK substrate. RF-A binding to single-stranded DNA is required for T antigen–mediated unwinding of the SV40 duplex DNA containing origins of DNA replication (Wold and Kelly, 1988). However, cyclin/CDK phosphorylation of RF-A had no significant effect on SV40 in vitro DNA replication (Henricksen and Wold, 1994). In *Xenopus* egg extracts, assembly of DNA replication foci containing RF-A occur postmitotically on decondensing chromosomes. The p21 cyclin/CDK inhibitor did not prevent the formation of such an assembly of RF-A–containing complexes on decondensing chromosomes, but blocked the unwinding of double-stranded DNA required for DNA replication in them (Adachi and Laemmli, 1994). However, to date, no DNA unwinding protein associated with DNA replication in eukaryotes that requires cyclin/CDK for its activation has been identified.

Other Regulatory Events Associated With the Progression of Cells From G_1 Into S Phase

To identify specific role(s) of cyclin/CDKs in the onset of DNA replication, it is necessary that a full understanding of the molecular events contributing to the initiation and elongation of DNA replication be developed. In this context, it is also necessary to take into account the fact that the ability of cells to enter into S phase is normally determined by the extracellular stimuli involving growth factors and nutrients (Pardee, 1989). Mammalian cells require specific growth factors and cytokines for their transition from G_1 into S phase. The intracellular regulatory process(es) stemming from growth factor–receptor interactions on the membrane that lead to the induction of nuclear DNA replication and cell division are not fully understood. However, growth factors are known to activate membrane-bound phospholipase C, which converts phosphatidylinositol 4,5-bisphosphate (PIP_2) to diacylglycerol and inositol 1,4,5-trisphosphate (IP_3). IP_3 thus generated releases Ca^{2+} from intracellular stores, leading to the activation of several proteins and enzymes, including a calcium-receptor protein called *calmodulin* (CaM). The progression of cells from G_1 into S phase is dependent on Ca^{2+} (Means and Rasmussen, 1988; Means, 1994). Mammalian cells at the G_1/S boundary when deprived of Ca^{2+} seem to undergo dismantling of a pre-replicative structure and enter a state of quiescence (Whitfield et al., 1982). A similar situation occurs when cells at R point are deprived for specific growth factors, such as insulin-like growth factor-I, required for their entry into S phase (Campisi and Pardee, 1984), indicating that growth factor–induced entry of cells into S phase is being regulated by Ca^{2+}-mediated events. A role for the Ca^{2+} regulatory system in the control of the cell cycle is also indicated from the observation that the PSTAIRE, a conserved 16 amino acid sequence of CDKs from yeast and higher eukaryotes, triggers a specific increase in the concentration of intracellular free calcium by inducing its release from intracellular stores (Picard et al., 1990).

As a calcium receptor, CaM is implicated to play a pivotal role in progression of cells from G_1 into S phase (Reddy, 1994; Means, 1994). Although the specific mechanism of CaM action in regulating the transition of cells from G_1 into S phase remains elusive, its potential role in regulation of DNA replication is indicated from the observation that CaM-specific monoclonal antibodies inhibit DNA replication in permeabilized S phase cells (Reddy et al., 1992b). In Chinese hamster embryo fibroblast (Subramanyam et al., 1990) and hemopoietic progenitor (Reddy et al., 1992a, 1994; Reddy and Quesenberry, 1996) cells, expression and nuclear localization of a specific CaM-binding protein of 68 kD, called CaM-BP68, is associated with their specific growth factor/cytokine-dependent progression from G_1 into S phase. Furthermore, purified CaM-BP68 stimulates DNA replication in permeabilized density arrested hematopoietic progenitor cells (Reddy et al., 1994). CaM-BP68 in HeLa cells is shown to be tightly associated with the DNA polymerase α–primase complex involved in the initiation of DNA replication (Cao et al., 1995).

ENZYMOLOGY OF DNA SYNTHESIS

Since the formulation of the cell cycle model about 50 years ago, there have been tremendous advances in our understanding of the molecular events contributing to, and/or associated with, the progression of cells through G_1 and into S phase. However, one of the most formidable challenges during this period has been to identify regulatory mechanism(s) contributing to a simultaneous activation of over two dozen enzymes associated with DNA replication and its precursor synthesis abruptly at the beginning of S phase. At the biochemical level, it has been paradoxical to note that while most of the enzymes and substrates required for DNA synthesis are adequately represented in the extracts of G_1 phase cells, they are prevented from being catalytically active in intact cells until the cells are committed to enter into S phase (Reddy, 1982; Rode et al., 1980; Matherly et al., 1989). In addition, when key enzymes of deoxynucleotide metabolism are overexpressed in mutant cells, only a fraction of the total enzyme, comparable to that present in normal cells, is catalytically active (Ashman et al., 1981; Danenberg and Danenberg, 1989). For instance, when ribonucleotide reductase is overexpressed in Syrian hamster melanoma cells the levels of dNTPs remain unchanged relative to the levels observed in normal cells (Ashman et al., 1981). What biochemistry, therefore, prevents the enzymes of DNA synthesis from being catalytically active in intact cells during G_1 and other phases of the cell cycle when DNA is not being replicated?

Nuclear Translocation of the Enzymes Associated With DNA Synthesis

It is possible that a rapid induction of some of the key enzymes, such as thymidine kinase (Coppock et al., 1987; Dou et al., 1992) and ribonucleotide reductase (Bjorklund et al., 1990; Albert and Rozengurt, 1992), in late G_1 could trigger the activation of other enzymes in the pathway of deoxynucleotide metabolism, leading to the onset of DNA replication. However, such periodic changes in the level of selective enzymes does not explain how some of the overexpressed enzymes associated with this biosynthetic process are prevented from being fully active during DNA replication. While transcriptional activation of a number of genes associated with DNA synthesis is considered essential for the entry of cells into S phase, transport of these gene products across the nuclear membrane into the nuclei is functionally linked to the ability of cells to progress from G_1 into S phase. A number of enzyme activities associated with both the DNA replication and the DNA precursor synthesis are found to translocate from cytoplasm into the nucleus as the cells transit from G_1 into S phase. These enzymes include DNA polymerase α, DNA polymerase δ, DNA topoisomerase II, ribonucleotide reductase, thymidylate synthase, dihydrofolate reductase, nucleoside diphosphate kinase, and thymidine kinase (Reddy and Pardee, 1980; Hammond et al., 1989). The presence of ribonucleotide reductase in the nuclei was disputed based on a failure to detect either the activity of the enzyme in the nuclei isolated by detergent treatment of the cells (Leeds et al., 1985; Kucera and Paulus, 1986) or the immunoreactivity of the M1 and M2 subunits of the enzyme in the nuclear compartment of cells subjected to immunocytochemical procedures employing monoclonal antibodies (Engstrom et al., 1984; Engstrom and Rozell, 1988). However, the nuclei isolated by aqueous procedures from regenerating rat liver exhibited nuclear localization of ribonucleotide reductase (Youdale et al., 1984; Khatsernova et al., 1983), indicating that the detection of the enzyme activity in the nuclei is being influenced by the procedure employed for the isolation of the nuclei from the cells: the nuclei isolated by the cytocholasin B–enucleation method (Reddy and Pardee, 1980) and aqueous biochemical fractionation procedures (Youdale et al., 1984; Khatsernova et al., 1983) revealing nuclear localization and detergent-based isolation procedures (Leeds et al., 1985; Kucera and Paulus, 1986) exhibiting cytoplasmic localization of the enzyme. Similarly, an alternative immunocytochemical procedure employing antibodies that recognize C-terminal epitopes of M1 and M2 subunits revealed cytoplasmic as well as nuclear localization of ribonucleotide reductase in a number of cell types (Sikorska et al., 1990). Furthermore, the M1 subunit was found to be localized as a halo around isolated rat liver nuclei, and the changes in nuclear membrane–associated ribonucleotide reductase activity closely paralleled the changes in DNA synthesis (Sikorska et al., 1990). In keeping with the characteristic of most membrane-associated proteins, the M1 subunit of ribonucleotide reductase is glycosylated. Thus, the nuclear membrane association of ribonucleotide reductase may prevent certain monoclonal antibodies from binding to cognate epitopes on the enzyme subunits due to their interaction with other proteins in the nuclear membrane (Engstrom et al., 1984; Engstrom and Rozell, 1988). Nevertheless, considering that the sole purpose of deoxynucleotides generated by ribonucleotide reductase is to serve as substrates for DNA replication, it is likely that this and other enzymes of dNTP de novo synthesis are localized in a close proximity of DNA replication in S phase cells. Some of the least disruptive fractionation procedures employed to prepare nuclear fractions from actively proliferating cells is in agreement with such nuclear localization of the enzymes associated with both dNTP synthesis and DNA replication (Reddy and Pardee, 1980).

Multienzyme Complexes for DNA Synthesis

Physical Interactions Between the Enzymes of DNA Synthesis. Enzymes of DNA synthesis translocated into the nucleus are assembled into megacomplexes in proliferating mammalian cells. Several of the key enzymes of deoxynucleotide metabolism, including ribonucleotide reductase, thymidylate synthase, thymidylate kinase, and nucleoside diphosphate kinase, in nuclear extracts of regenerating rat liver or Novikoff tumor cells are reported to co-sediment with DNA polymerase α on sucrose density gradients (Baril et al., 1974). Such sedimentation of enzymes associated with dNTP synthesis and DNA replication is observed in a variety of mammalian cells, including Chinese hamster embryo fibroblast (CHEF/18) cells (Reddy and Pardee, 1980, 1982; Noguchi et al., 1983), mouse FM3A cells (Ayusawa et al., 1983), BHK fibroblast cells (Harvey and Pearson, 1988), and human lymphoblasts (Wickremasinghe et al., 1982, 1983; Wickremasinghe and Hoffbrand, 1983). Multienzyme complexes containing the enzymes of both dNTP synthesis and DNA replication were also observed in mammalian cells infected with herpes simplex virus-1 (Harvey and Pearson, 1988; Sclafini and Fangman, 1984; Jong et al., 1984) and adenovirus (Yamashita et al., 1977; Arens et al., 1977). In yeast the replication of 2 μm extrachromosomal plasmid DNA (Jazwinski and Edelman, 1982; 1984) and mitochondrial DNA (Murthy and Pashupathi, 1995) is also reported to be facilitated by multienzyme complexes of about 2 million Da and 40S, respectively. Although the complete subunit compositions of these large complexes remain to be identified, the enzymes that are known to be associated with the complexes include DNA polymerase I, DNA ligase, DNA topoisomerase, $3' \rightarrow 5'$ exonuclease, and ATPase.

Functional Significance of the Observed Interactions Between the Enzymes of DNA Synthesis. Physical aggregates of the enzymes isolated from nuclear fractions of proliferating cells are representative of functional complexes rather than random fortuitous pairing because such enzyme complexes could be recovered only from the proliferating (S phase), not from quiescent or G_1 phase cells (Reddy and Pardee, 1980; Noguchi et al., 1983). In synchronized CHEF cells there was a direct temporal correlation between the progression of cells from G_1 into S phase and the recovery of the complex fraction with DNA-synthesizing enzymes from their nuclear lysate (Reddy and Pardee, 1980; Reddy, 1982). In Syrian hamster melanoma cells overproducing ribonucleotide reductase (Ashman et al., 1981) or in Chinese hamster cells overproducing dihydrofolate reductase (Noguchi et al., 1983), only a constant fraction of the enzymes in proportion to cellular DNA content was found associated with the complexes, whereas the surplus enzyme was in a soluble form in the cytoplasm. Functional specificity of the enzymes associated with the complexes is also indicated by the observations that thymidylate synthase in wild-type murine FM3A cells co-sedimented with the complex fraction containing the DNA replicating enzymes, but heterologous thymidylate synthase expressed by human DNA transfected into thymidylate synthase–negative FM3A cells failed to co-sediment with the complex of DNA-synthesizing enzymes (Ayusawa et al., 1983). Thymidylate synthase encoded by a human gene, however, was functional in thymidylate synthase–negative immune cells to convert them into thymidine prototrophs when expressed at a high level. Thus there is a strong species specificity in sequestration of the dNTP-synthesizing enzymes into DNA synthesis complexes.

Kinetic Coupling Between the Enzymes of DNA Replication and dNTP Synthesis. Functional specificity of the multienzyme complexes containing the enzymes of dNTP synthesis and DNA replication is further indicated from the observation that they are capable of facilitating the incorporation of a distal precursor, such as ribonucleoside diphosphate (rNDP), into DNA. Such an incorporation in the presence of complexes isolated from proliferating calf thymus or CHEF cells was observed with a minimal accumulation of the intermediary deoxynucleotides deoxynucleoside diphosphates (dNDPs) and dNTPs that are generated from rNDPs in a sequential reaction by ribonucleotide reductase and nucleoside diphosphate kinase, respectively, prior to their incorporation into DNA (Noguchi et al., 1983). Similarly, multienzyme complexes isolated from human lymphoblastoid cells were able to channel thymidine, thymidine monophosphate (TMP), or deoxyuridine monophosphate (dUMP) into DNA without significant accumulation of deoxythymidine triphosphate (dTTP) (Wickremasinghe et al., 1982, 1983; Wickremasinghe and Hoffbrand, 1983). These studies with isolated complexes demonstrated kinetic coupling between the enzymes of dNTP synthesis and DNA replication, which allows the channeling of distal precursors into DNA through sequential reactions, limiting both the accumulation and the diffusion of intermediary deoxynucleotides prior to their incorporation into DNA.

This phenomenon of "channeling" was also observed in permeabilized S phase cells. The enzymes of dNTP synthesis and DNA replication in cells permeabilized with lysolecithin are accessible to exogenous precursors in a relatively physiological milieu (Castellot et al., 1978). It is observed that the incorporation of rNDPs into DNA in permeabilized cells was also accompanied by a negligible accumulation of intermediary deoxynucleotides, which were limited from diffusing and mixing freely with the outside pools prior to their incorporation into DNA (Reddy and Pardee, 1982). A similar, but low, level incorporation of ribonucleoside diphosphates into DNA of a permeabilized wild-type, but not thymidylate synthase–negative, mutant, FM3A (Ayusawa et al., 1983), and uninfected BHK cells (Harvey and Pearson, 1988) was also observed. This incorporation is highly sensitive to hydroxyurea, an inhibitor of ribonucleotide reductase, and the functional association of ribonucleotide reductase with the enzymes of DNA replication (Reddy and Pardee, 1982; Ayusawa et al., 1983). Although these studies with permeablized cells provided supporting evidence for the multienzyme complex–mediated channeling of distal precursors into DNA, the DNA that is being synthesized from rNDPs in such permeabilized cells was of low molecular weight, representing unligated Okazaki fragments with RNA primers (Reddy et al., 1986). More convincing evidence for channeling of distal precursors into DNA in permeabilized cells, therefore, requires the de-

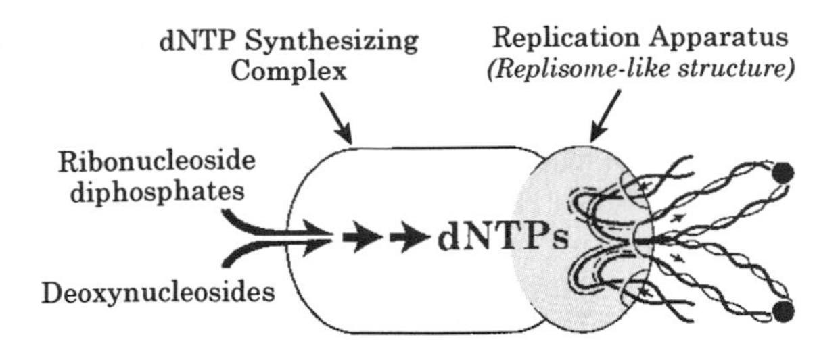

Figure 3.4. A hypothetical model depicting replitase complex in which dNTP-synthesizing complex is juxtaposed with replication apparatus.

velopment of experimental conditions that allow more stable and complete DNA replication from rNDPs. However, the concept of channeling and compartmentation of deoxynucleotides in mammalian cells is strongly supported by the kinetic studies in which the rate of incorporation of exogenous precursors into DNA in relation to the changes in endogenous nucleotide pools was monitored (Fridland, 1973; Scott and Forsdyke, 1980; Nicander and Reichard, 1983, 1985; Nguyen and Sadee, 1986; Leeds and Mathews, 1987; Reddy, 1989; Plucinski et al., 1990). A critical evaluation of much of the data supporting the concept of compartmentation of deoxynucleotide pools and their channeling into DNA during DNA replication in S phase cells is presented in detail elsewhere (Reddy and Fager, 1993).

Based on the physical and functional interactions between the enzymes of dNTP synthesis and DNA replication, and the channeling and compartmentation of deoxynucleotide pools associated with DNA replication in S phase cells, it is postulated that the DNA synthesis in mammalian cells, as in prokaryotes (Reddy et al., 1977; Reddy and Mathews, 1978; Allen et al., 1980; Chiu et al., 1982; Mathews et al., 1988), is being facilitated by allosteric interaction between the enzymes of DNA replication and dNTP synthesis. A macromolecular structure representing such interactions in mammalian cells is termed *replitase* (Reddy and Pardee, 1980) (Fig. 3.4). It is suggested that the formation of replitase as the cells transit from G_1 into S phase allows coordinated activation of the enzymes associated with both the DNA replication and DNA precursor synthesis (Reddy, 1982; Reddy and Pardee, 1983). Such interactions between the enzymes of DNA replication and DNA precursor synthesis could also maintain a high gradient of deoxynucleotides in the microvicinity of DNA replication without requiring any physical barriers for their compartmentation. Furthermore, the observation that the replitase complexes can only be isolated from the cells entering into S phase, but not from those in G_1, raises the possibility that the formation of such a complex may determine the ability of cells to enter into S phase.

Further structural and functional analyses of such complexes associated with DNA replication in S phase cells may lead to the identification of signaling molecules responsible for growth factor–dependent progression of cells from G_1 into S phase. In this regard it is intriguing to note that the CaM-BP68, a specific CaM-binding protein that is modulated in a direct response to growth factors and cytokines required in late G_1 for the induction of cells into S phase, is associated with the replitase (Subramanyam et al., 1990) by virtue of its tight binding to the DNA polymerase α–primase (Cao et al., 1995). This DNA polymerase α-primase is a subcomponent of a 21 S complex of enzymes for DNA synthesis isolated from HeLa cells and is capable of supporting SV40 DNA replication *in vitro* (Malkas et al., 1990; Li et al., 1993).

DNA Polymerases Associated With Nuclear DNA Replication

There are at least five distinct forms of DNA polymerases known to be present in mammalian cells. These enzymes are identified by the Greek letters α, β, γ, δ and ε (Wang, 1991). As the development of an in vitro DNA replication system using bacteriophage DNA has helped to define the DNA replication machinery in prokaryotes (Wickner et al., 1973; Alberts, 1985; Richardson, 1983; McMacken et al., 1988), the advent of a similar in vitro system employing the origins of simian virus 40 (SV40) DNA (Ariga and Sugano, 1983; Li and Kelly, 1984; Wobbe et al., 1985; Stillman and Gluzman, 1985) has contributed to the understanding of the functional roles of DNA polymerases and their associated proteins in DNA replication of eukaryotes. SV40 replication in permissive mammalian cells requires T antigen, the viral-coded 94 kD phosphoprotein that has ATPase activity and binds specifically to the DNA replication origin (ori) sequences in SV40 DNA through an ATP-dependent reaction (Dean et al., 1987; Borowiec et al., 1990). T antigen has helicase activity that allows unwinding of the DNA starting at the ori site on SV40 DNA (Stahl et al., 1986; Dodson et al., 1987). This unwinding of SV40 DNA at the ori sites allows cellular DNA polymerases and their accessory proteins to initiate and extend the DNA replication bidirectionally. Using this in vitro system, more than one form of DNA polymerase, involving DNA polymerases α, δ, and/or ε, are found to be required for complete replication of SV40 ori containing DNA (Waga and Stillman, 1994; Prelich et al., 1987a,b; Prelich and Stillman, 1988; Nathanel and Kaufman, 1990; Nathanel et al., 1992; Podust and Hubscher, 1993). In comparison with DNA polymerases α, δ and ε, DNA polymerases β and γ have no known role in nuclear DNA replication. DNA polymerase β is shown to be involved essentially in DNA repair by both genetic studies in yeast (Leem et al., 1994; Prasad et al., 1993b) and its biochemical characterization in mammalian cells (Singhal and Wilson, 1993; Singhal et al., 1995; Prasad et al., 1993a). While DNA polymerase β is believed to be involved in DNA repair, the likelihood of its function in nuclear DNA replication cannot be ruled out completely. DNA polymerase γ is perceived to be involved in mitochondrial DNA replication because of its exclusive mitochondrial localization (Fry and Loeb, 1986). Disruption of the MIP1 gene, which encodes DNA polymerase γ in budding yeast, leads to a total loss of mitochondrial DNA (Foury, 1989).

DNA Polymerase α. Although the specific involvement and the function of each of these polymerases in nuclear DNA replication of S phase cells remains elusive, DNA polymerase α is thought to be the major enzyme responsible for chromosomal DNA replication in a wide variety of eukaryotes ranging from yeasts to humans (Kaguni and Lehman 1988; Fry and Loeb, 1986). DNA polymerase α is a multimeric enzyme consisting of a 165–180 kD subunit with catalytic activity (Vishwanatha et al., 1986; Karawya et al., 1984; Kaguni et al., 1983), a 70 kD subunit without any clearly known enzymatic activity (Collins et al., 1993), and two (58 and 49 kD) subunits having DNA primase activity (Copeland and Wang, 1993). The catalytic subunit is found both glycosylated and phosphorylated (Hsi et al., 1990; Nasheuer et al., 1991: Copeland and Wang, 1991). These post-translational modifications may contribute to the observed variations in its apparent molecular weights ranging from 165 to 180 kD. Both the 165 kD catalytic subunit and the 78 kD subunit are phosphorylated in a cell cycle–dependent manner (Nasheuer et al., 1991).

Although the purified intact DNA polymerase α lacks readily detectable $3' \rightarrow 5'$ exonuclease activity, its 182 kD catalytic subunit when separated from the 73 kD subunit exhibited a cryptic $3' \rightarrow 5'$ exonuclease with a proofreading function (Skarnes et al., 1986; Cotterill et al., 1987; Reyland et al., 1988). Distinction between DNA polymerase α and other DNA polymerases, such as DNA polymerase δ, containing $3' \rightarrow 5'$ exonuclease activity, is drawn essentially from the differences in their sensitivities to specific inhibitors, such as butylphenyldeoxyguanosine triphosphate (BupdGTP) (Crute et al., 1986; Byrnes, 1985; Lee et al., 1985; Hammond et al., 1989), reactivity against antibodies that are specific to each of the polymerases (Wong et al., 1989), and the association of PCNA with other polymerases, but not with DNA polymerase α, as its accessory protein (Tan et al., 1986).

DNA polymerase α from yeast is designated as DNA polymerase I (pol I) and has the same subunit composition as that from human, mouse, and *Drosophila* (Sugino, 1995). The genes coding for all four subunits from yeast have been cloned and are found to be essential for yeast cell division (Campbell, 1993; Francesconi et al., 1993a). Using an SV40 in vitro DNA replication system, it is observed that the holoenzyme containing the DNA polymerase α–primase activities is capable of initiating DNA replication on both the leading and the lagging strands (Waga and Stillman, 1994). Initial studies with synchronized cells indicated that DNA polymerase α is cell cycle regulated with an increase in its activity at the beginning of S phase (Chiu and Baril, 1975; Spadari and Weissbach, 1974). However, it was subsequently shown that in normal cycling cells that are fractionated to enrich in different phases of cell cycle by elutriation, the DNA polymerase α and its transcripts remain relatively constant throughout the cell cycle, with only a nominal increase in S phase (Wong et al., 1988; Wahl et al., 1988). Although the expression of DNA polymerase α remained constant throughout the cell cycle, both the 165 kD catalytic subunit and the 70 kD subunit were phosphorylated in a cell cycle–regulated manner (Nasheuer et al., 1991). However, it is not known how this post-translation modification affects its activity during cell cycle.

DNA Polymerase δ. DNA polymerase δ is also implicated in eukaryotic DNA replication (Prelich and Stillman, 1988; Prelich et al., 1987a,b; Downey et al., 1990). It has two subunits, one a 124 kD catalytic subunit and the other a 48 kD subunit (Wang, 1991). PCNA associates with DNA polymerase δ as its accessory protein and stimulates the enzyme processivity for DNA synthesis (Tan et al., 1986; So and Downey, 1988; Prelich et al., 1987b; Downey et al., 1990). The 48 kD subunit of holoenzyme is shown to be required for PCNA-dependent stimulation of DNA polymerase δ processivity (Goulian et al., 1990; Chiang et al., 1993). The enzyme also contains tightly associated 3′ to 5′ exonuclease activity that has an editing function during DNA replication (Simon et al., 1991; Morrison et al., 1991). The content of DNA polymerase δ in the cells, like that of DNA polymerase α, at both the mRNA and protein levels remains constant throughout the cell cycle with a slight increase at the beginning of S phase (Zeng et al., 1994). Furthermore, unlike DNA polymerase α, DNA polymerase δ has not been reported to undergo any post-translational modification.

DNA polymerase δ is designated as DNA III in yeast (Sugino, 1995). Involvement of DNA polymerase δ in strand elongation during chromosomal DNA replication is demonstrated from temperature-sensitive mutants of *Sch. pombe* DNA pol III (Francesconi et al., 1993b). Although similar temperature-sensitive mutants of *Sac. cerevisiae* that are defective in the catalytic subunit of pol III exhibited a "cell cycle arrest" phenotype at nonpermissive temperatures, about 70% of the genomic DNA in these mutants seem to have replicated before the cell cycle arrest occurred (Conrad and Newlon, 1983). This indicates either that other DNA polymerases could partially compensate for the defect in the DNA pol III catalytic subunit or that the DNA pol III function is required toward the end of DNA replication during S phase in *Sac. cerevisiae.* However, in vitro studies with the SV40 DNA replication system have indicated that DNA polymerase δ is capable of elongating both the leading and the lagging strands of SV40 ori sequences containing DNA (Waga and Stillman, 1994). Based on the high frequency of recombination and sensitivity to ethylmethane sulfonate (MMS), of the cells containing mutations in the alleles of the gene encoding the catalytic subunit, DNA pol III is also implicated in recombinational repair in yeast (Aguilera and Klein, 1988; Gordenin et al., 1992). In situ studies, employing specific inhibitors of DNA polymerases, have also indicated DNA polymerase δ involvement in DNA repair of human cells (Dresler and Frattini, 1986; Nishida and Linn, 1988). Thus, the actual involvement of DNA polymerase δ in DNA replication and/or DNA repair remains to be established.

DNA Polymerase ε. DNA polymerase ε is yet another DNA polymerase whose involvement in DNA replication is implicated largely from studies with DNA pol II, a yeast homologue of mammalian DNA polymerase ε (Hamatake et al., 1990; Araki et al., 1992). While the exact subunit composition of DNA polymerase ε from mammalian cells is not yet clearly established, a highly purified enzyme from human cells contained a 255 kD catalytic subunit and a 55 kD subunit. By comparison, the enzyme from yeast contained five subunits with molecular weights ranging from 256 to 29 kD (256, 80, 34, 31, and 29 kD subunits) (Sugino, 1995). The gene from yeast (Morrison et al., 1990) and the cDNA from human cells (Syvaoja, 1990; Kesti et al., 1993) encoding the catalytic subunit of DNA polymerase ε have been cloned, and both were found to have open reading frames that encode 255 kD protein. DNA polymerase ε, like DNA polymerase δ, contains intrinsic 3′ to 5′ exonuclease activity required for editing function during DNA replication (Morrison et al., 1993; Morrison and Sugino, 1994), but lacks sensitivity to PCNA stimulation (Syvaoja and Linn, 1989; Morrison et al., 1990; Syvaoja, 1990; Weiser et al., 1991). However, it can form stable complexes with PCNA at primer terminus in the presence of replication factor-C (RF-C) and ATP (Burgers, 1991; Lee et al., 1991; Podust et al., 1992). Genetic studies in yeast have indicated that DNA polymerase ε is essential for completion of S phase (Araki et al., 1992), possibly through its involvement in leading-strand DNA synthesis (Morrison et al., 1990; Morrison and Sugino, 1992). However, in an in vitro SV40 DNA replication system DNA polymerase ε increased the length of Okazaki fragments synthesized in the presence of DNA polymerase α–primase and could not substitute for DNA polymerase δ in generating leading-strand products of sufficient length, indicating its possible role in Okazaki fragment synthesis (Lee et al., 1991). This enzyme was initially identified in mammalian cells as being involved in DNA repair (Chang, 1977; Nishida et al., 1988). Consistent with its role in DNA repair, DNA polymerase ε along with RF-C, PCNA, and DNA ligase were found to be essential components of an in vitro mammalian nucleotide excision repair system reconstituted with 30 purified proteins from human cells (Aboussekhra et al., 1995).

At this stage it seems as though there are more DNA polymerases designated to perform DNA replication than are required in vivo. Much of our understanding of the potential roles of DNA polymerases α, δ, and ε in nuclear DNA replication has derived from the observations made in reconstituted in vitro replication of SV40 DNA using purified proteins. However, it is not known how reflective these observations are of an in vivo condition inside the living cells, where each of these polymerases is likely to interact with a number of other cellular proteins and undergo post-translational modification in response to mitogenic stimulation. Even in the in vitro SV40 DNA replication system there are conflicting reports on the involvement of individual DNA polymerase(s) in leading and lagging-strand synthesis (Morrison et al., 1990; Burgers, 1991; Bambara and Jessee, 1991). Therefore, additional studies are required to establish the functional specificities of individual polymerase(s) at the replication fork in living cells.

ORIGINS OF DNA REPLICATION

As targets for regulating proteins and the enzymes of DNA replication, specific site(s) on eukaryotic chromosomes at which DNA replication is initiated have been the focus of intense investigation to understand the molecular event contributing to the onset of DNA replication. Pulse labeling and autoradiography experiments have indicated that the long DNA fibers, ranging in length from 500–1,800 μm (Cairns, 1966; Huberman and Riggs, 1966) to more than 2 cm (Sasaki and Norman, 1966) in mammalian chromosomes replicate in separate tandemly joined units of about 30 μm (Cairns, 1966; Huberman and Riggs, 1968). A unit of DNA replicated from a single initiation site is termed a *replicon* (Jacob et al., 1963). In a functional analogy to the operon model, in which gene expression is regulated either positively or negatively by the binding of an initiator protein to its promoter region, the replicon model postulates that the initiation of DNA replication is determined by the binding of trans-acting proteins ("initiators") to the cis-acting DNA sequences ("replicators") in a DNA template. In prokaryotes and lower eukaryotes such as *Sac. cerevisiae* and *Sch. pombe,* elements corresponding to both the initiator proteins and the replicator sequences have been identified (see below). Binding of the initiators to the replicators facilitates localized unwinding of the DNA, thereby allowing the replication machinery to initiate DNA replication. The regions or the segments of DNA in a chromosome at which DNA replication is initiated are the ori. Although specific DNA sequences and structures interacting with replication proteins have been identified to serve as the ori in prokaryotes (Bramhill and Kornberg, 1988), animal viruses (DePamphilis, 1987; Challberg and Kelly, 1989; Stillman, 1989), and budding yeast *Sac. cerevisiae* (Walker et al., 1991; Deshpande and Newlon, 1992; Marahrens and Stillman, 1992; Rivier and Rine, 1992), ori in multicellular eukaryotes (metazoans) remain elusive.

Origins of Replication in Yeast

Autonomously replicating sequence (ARS) elements with origin function in the budding yeast *Sac. cerevisiae* were identified by a selective screening process in which segments of yeast DNA sequences in plasmids were tested for their ability to promote extrachromosomal replication of plasmids (Stinchcomb et al., 1979). From base substitution and deletion analyses (Kearsey, 1984; Celniker et al., 1984; Maine et al., 1984), a short (11 bp) conserved core consensus sequence in ARS elements (Broach et al., 1982) is shown to be essential for autonomous replication of plasmids.

A two-dimensional gel electrophoresis method devised to obtain a detailed view of active replicons further confirmed the role of ARS elements in initiation of DNA replication (Brewer and Fangman, 1987, 1988; Huberman et al., 1988; Linskens and Huberman, 1988). These studies also revealed that, while ARS elements are essential for initiation of DNA replication, not all ARS elements exhibit origin function within a given cell. For example, ARS in tandemly repeated ribosomal DNA are used in approximately 20% of cell cycles on the average (Fangman and Brewer, 1991), whereas ARS 501 that is replicated late in S phase is activated in almost every cycle (Ferguson et al., 1991). Furthermore, not all replication origins within a given cell are activated at the same time during S phase. There are some origins that are activated in early S phase (e.g., ARS 1), and there are some that are activated in late S phase (e.g., ARS 501). These differences in the timing of their activation in S phase seem to be determined by the context of their location in the chromosome. For instance, moving ARS 501 from its normal telomeric location to a circular, but not to a linear, plasmid causes it to activate in early S phase, whereas placing a copy of ARS 1 near a telomere converts it to activate in late S phase (Ferguson and Fangman, 1992). It is also interesting to note from two-dimensional gel analysis that the deletion of an active replication origin in yeast does not prevent its surrounding DNA from replicating. The region around the deletion in a replicon is replicated by the forks extending from a functional origin in a neighboring replicon, but not by initiating at "nonorigin" sequences (Newlon et al., 1993). Thus, there is a flexibility in the ability of the fork generated from an ori region in one replicon to extend into an adjacent replicon lacking a functional origin.

Replication origins comparable to those found in *Sac. cerevisiae* are also present in *Sch. pombe*. However, the origin sequences in *Sch. pombe* are much longer, being on the order of 500–1,000 bp, and appear to be more diffuse and functionally less efficient than those in *Sac. cerevisiae* (Caddle and Calos, 1994; Wohlgemuth et al., 1994; Clyne and Kelly, 1995; Dubey et al., 1996).

Origins of Replication in Mammalian Cells

Although it is recognized that in mammalian cells, as in yeast, the initiation of DNA replication occurs mostly at intergenic regions in chromosomes, the sequences and the structural identities of replicators in mammalian cells remain elusive. In comparison to discrete origins of 100–200 bp in the *Sac. cerevisiae* chromosome, initiation of DNA replication in mammalian cells is known to occur in large zones, ranging in size from 0.5 to 55 kb. A major limitation in identifying the ori in mammalian cells has been the lack of a plasmid DNA transfection assay suitable for identifying specific mammalian DNA sequences that are able to autonomously replicate plasmids. When an Epstein-Barr virus (EBV) vector containing nuclear retention signal, but lacking origin function, was used to isolate human DNA fragments containing origin function, any large DNA sequences inserted into the vector and introduced into human cells by transfection were able to replicate (Krysan et al., 1989; Heinzel et al., 1991). These studies raised the possibility that the initiation of DNA replication occurs in any of the large number of sites that are distributed throughout the entire length of human DNA insert. However, none of the individual initiation sites with short DNA sequences could be subcloned from these large human DNA inserts. Thus, from these studies it seemed that it is the length of the DNA insert, rather than any specific sequence in it, that determines its ability to replicate autonomously. In accordance with this view, tandem repeats of the same sequence were able to replicate more efficiently than the single copy of the same sequence (Krysan et al., 1993).

Attempts to identify mammalian DNA sequences specific for autonomous replication of plasmids in these assays may have been complicated by the absence of chromatin structure in the DNA inserted into plasmids. It is possible that the organization of mammalian DNA in chromosomes may restrict initiation of its replication at random sites and allow only selective sites in it to serve as the ori. Although the lack of chromosomal organization in the DNA inserted into plasmids has made it difficult to isolate mammalian ARS by plasmid DNA transfection assay, a number of other approaches involving the analysis of nascent DNA and replication intermediates have revealed that the replication in mammalian cells, as in yeast, initiates at specific sites on chromosomes. Methods employed to analyze nascent DNA or replication intermediates in these approaches can be grouped into two general categories. In one, nascent DNA from synchronized cells is somehow radiolabeled and isolated and then annealed to specific DNA probes to determine the direction of their synthesis from an initiation site in the chromosomal domain of interest. In the second category, restriction fragments of replication intermediates isolated from the cells in S phase are subjected to two-dimensional gel electrophoresis to fractionate the DNA fragments based on their size and shape. This approach takes advantage of the fact that the replication intermediates with unique shapes, based on the presence of replication bubbles and forks in them, exhibit different mobility patterns during gel electrophoresis (Fangman and Brewer, 1991; Huberman, 1994). The genomic location of the replication intermediates with bubbles, representing the ori, are then mapped by hybridization of gel electrophoresis blots with sequence-specific radiolabeled DNA probes spanning the length of the chromosomal domain of interest.

Using different methods representing any of these two categories, a number of genomic regions containing preferential sites for initiation of DNA replication in mammalian cells have been identified. These include dihydrofolate reductase (DHFR) (Leu and Hamlin, 1989; Vaughn et al., 1990; Burhans et al., 1990; Dijkwel and Hamlin, 1992, 1995), carbamoyl phosphate synthetase–aspartate transcarbamylase–dihydrooratase (CAD) (Kelly et al., 1995), and rhodopsin (Gale et al., 1992) gene loci in hamster cells; histone gene repeats (Shinomiya and Ina, 1993), the DNA polymerase α gene (Shinomiya and Ina, 1994), and the chorion gene (Orr-Weaver, 1991) in *Drosophila* cultured cells; the DNA puff II/9A gene in the fungus fly *Sciara coprophila* (Liang et al., 1993); the adenosine deaminase (ADA) region of the mouse genome (Virta-Pearlman et al., 1993; Carroll et al., 1993); and the c-myc (Vassilev and Johnson, 1990), β-globin (Kitsberg et al., 1993), and rRNA (Little et al.,

1993; Yoon et al., 1995) genes in human cells. One of the most extensively studied among these is the DHFR locus.

Origin of Replication in the DHFR Domain of Chinese Hamster Cells

Chinese hamster ovary cells subjected to selective pressure to develop resistance against methotrexate yielded the CHOC 400 cell line. These cells contain 1,000 copies of the gene encoding the target enzyme of methotrexate, DHFR. The amplified domains of DHFR, referred to as *amplicons,* in these cells are 240 kb in length and are arranged as tandem repeats in the chromosome (Milbrandt et al., 1981; Looney and Hamlin, 1987). The presence of such a high copy number of a segment of DNA in the chromosome of this cell line has made it an attractive subject for most of the available mapping techniques to localize replication sites within the repeated sequence. The presence of the same initiation site in each of the 1,000 copies of chromosome domain makes its detection easy and more accurate than the one that is normally present in two copies per diploid genome.

All mapping methods applied to the amplified DHFR domain in CHOC 400 cells have indicated that the replication is initiated at a preferred site downstream of the DHFR gene. However, the length of the DNA containing the potential ori in this region varied depending on the method employed for its mapping. Initial attempts to map ori in the DHFR domain by analyzing early-radiolabeled restriction fragments in synchronized CHOC 400 cells entering into S phase have indicated that the replication begins within a 28 kb region downstream of the DHFR gene (Heintz and Hamlin, 1982). Using a sensitive method of quantitative analysis of the early-labeled restriction fragments, the replication in the DHFR domain was found to initiate at two preferred sites, referred to as "ori-β" and "ori-α" (Leu and Hamlin, 1989). These sites are located 22 kb apart in the intergenic region between the DHFR and 2BE2121 genes. A similar restriction fragment analysis of the DNA labeled with radioactive thymidine during the first 2 minutes of cell entry into S phase has narrowed the region in which replication is initiated to a 4.3 kb Xba1 restriction fragment surrounding ori-β (Burhans et al., 1986a). Treatment of the cells with AraC, an inhibitor of DNA replication, limited replication to this region, further indicating the presence of an initiation site within a 4.3 kb region of the DHFR domain (Burhans et al., 1986b). An extension of the early-labeled restriction fragment analysis method to a mouse cell line containing amplified ADA gene also localized the initiation site to a region of several kilobases in the amplified domain (Carroll et al., 1993).

In an effort to further narrow down the region of the DHFR domain in which replication is initiated, Handeli et al. (1989) employed a method to isolate leading-strand DNA emanating from initiation sites and anneal them to positive (+) and negative (−) strands of DNA probes representing various regions of the DHFR domain. In this approach, the initiation site is marked at the region where the hybridization of the leading strand switches from a (+) template strand to a (−) template strand. These analyses localized the initiation site to a several kilobase region that is approximately 15 kb downstream from the DHFR gene (Handeli et al., 1989; Burhans et al., 1991). By using an analogous lagging-strand assay in which radiolabeled short nascent DNA representing Okazaki fragments synthesized in permeabilized CHO cells were examined for hybridization to (+) and (−) template strands of DNA in the ori-β locus, a 0.45 kb region located 17 kb downstream from the DHFR gene was shown to contain initiation sites (Burhans et al., 1990). These observations pointing to the initiation of DNA replication from a single locus within the ori-β region did not seem to be an artifact of cell synchronization procedures employed in these studies or to be unique to the cells with artificially amplified DNA sequences. This is because an alternative method involving the polymerase chain reaction in exponentially growing CHO cells containing a single-copy DHFR gene have also revealed an initiation site within a 3 kb region of ori-β (Vassilev et al., 1990). Thus, a number of approaches involving the analysis of nascent DNA synthesized in CHOC 400 or CHO cells have provided evidence for the presence of a specific initiation site in the vicinity of the ori-β region of the DHFR domain.

However, in contrast to the observations with radiolabeled nascent DNA analysis, the two-dimensional gel electrophoresis analysis of replication intermediates consistently revealed a de-localized image of initiation sites in a 55 kb region encompassing ori-β and ori-α (Vaughn et al., 1990; Dijkwel and Hamlin, 1992). When a two-dimensional electrophoresis technique designed to measure the size of nascent DNA strands and to determine the direction of replication fork movement in a genomic region of interest (Nawotka and Huberman, 1988) was applied to DHFR domain of CHOC 400 cells in early S phase, short nascent DNA strands were found to have emanated from scattered sites throughout the intergenic region between DHFR and 2BE2121 genes (Dijkwel and Hamlin, 1992). While the replication of initiation sites in the intergenic region extended bidirectionally, those in the DHFR gene moved outward unidirectionally. Although these observations with a two-dimensional gel analysis method also indicate that the initiation of DNA replication occurs in a defined zone between DHFR and 2BE2121 genes, the initiation zone localized by this method is found to be much larger than that mapped by other methods.

This apparent discrepancy in the length of the region containing potential initiation sites identified in DHFR domain by the two-dimensional gel analysis method and the density or radioactive labeled nascent DNA analysis method is suggested to have resulted possibly from the differences in the replication products analyzed in the two different methods (Dijkwel and Hamlin, 1996). The replication intermediates analyzed by the two-dimensional gel analysis method are suggested to be mostly stable as they are at a steady-sate level at the time of their isolation, whereas the labeled nascent DNA analyzed by the other methods are mostly of a short half-life as they are isolated from the cells immediately after pulse labeling with radioactive nucleoside or nucleotide or with density label bromodeoxyuridine. Furthermore, the relatively higher sensitivities of two-dimensional gel analysis methods may allow the detection of trace amounts of individual initiation sites that perhaps are difficult to be detected by the other methods (Burhans and Huberman, 1994). Although trace amounts of replication bubbles, representing individual initiation sites, are seen throughout the 55 kb initiation zone of the DHFR domain, a quantitative analysis of replication intermediates containing bubbles indicated their abundance essentially in the 12 kb region surrounding ori-β (Dijkwel and Hamlin, 1992). A somewhat related observation was made when two-dimensional gel analysis methods were employed to map initiation sites in the *Drosophila* chorion gene region. This region in *Drosophila* follicle cells in amplified more than 60-fold during oogenesis. The two-dimensional gel analysis methods revealed that while a trace amount of initiation sites were distributed throughout the amplified region, a majority of them were localized in a narrow region close to the "amplification control element" (ACE) (Delidakis and Kafatos, 1989; Heck and Spradling, 1990; Orr-Weaver, 1991).

Origins of Replication in Metazoans are Genetically Determined

Although specific DNA sequences, analogous to those characterized as replicators in simple prokaryotes and yeast *Sac. cerevisiae,* have not been found in metazoans, origins of replication in higher eukaryotes, as in lower organisms, seem to be conserved and genetically determined. This is implicit in the observation that the replication is initiated at the same specific site in a genomic region when the locus containing that region is present in either two copies per cell or over 1,000 copies per cell as observed in the case of hamster DHFR (Handeli et al., 1989; Vassilev et al., 1990; Dijkwel and Hamlin, 1992, 1995) and mouse ADA (Virta-Pearlman et al., 1993) gene loci. This is further reinforced by the observation that the ori regions of the hamster DHFR domain (Handeli et al., 1989) and the *Drosophila* chorian gene (Orr-Weaver, 1991) retain the ability to initiate DNA replication when they are translocated to other chromosomal sites. Similarly, the activities of the ori region in Syrian hamster CAD gene (Kelly et al., 1995) or the Chinese hamster DHFR domain (Gilbert et al., 1993) are maintained when they are transfected into Chinese hamster cells or incubated with replication-competent protein extract of *Xenopus* oocytes, respectively. Most importantly, the deletion of an 8 kb region containing initiation sites from the human β-globin gene cluster abolishes its bidirectional replication; the β-globin gene locus with the deletion is replicated passively by the replication forks extending in only the 5′ to 3′ direction from active initiation sites in adjacent genomic regions (Kitsberg et al., 1993).

Initiation of replication in the human β-globin gene locus occurs in an 8 kb region located 50 kb downstream of the locus control region

(LCR). The sequences covering LCR are also required for initiation of replication as the deletion of LCR abolishes initiation in the entire locus (Aladjem et al., 1995). These observations indicate that the initiation of replication at specific sites in metazoan chromosomes is determined not only by the conserved sequences at which replication is initiated but also by their interaction with other sequence elements located at a distance in the chromosome. For an ori to be active, it seems necessary that its location in the chromosome would permit its interaction with other regulatory sequences or the factors associated with such sequences. Alternatively, it is possible that transcriptional activity in the genes located at a distance from the initiation sites may change chromosomal architecture in such a way that the initiation sites become accessible to the replication machinery. These possibilities are also reflected in the observation that the deletion of the promoter in the DHFR gene locus abrogates initiation of replication in ori-β regions located several kilobases downstream of the DHFR gene (Hamlin and Dijkwel, 1995). However, an assertion that transcriptional activity per se is responsible for the initiation of replication in the genomic region is confounded by the observation that the transcriptional activation of genes during embryo development represses initiation of replication within the transcribed regions. For instance, in early stages of *Xenopus* embryo development, replication in both the intragenic and intergenic regions of ribosomal RNA gene (rDNA) locus is initiated at 9–12 kb intervals. However, in late blastula stage, when the rDNA gene becomes transcriptionally active, initiation of replication within the transcribed region is repressed while that in nontranscribed intergenic regions continues to persist in ensuing divisions of embryo development (Hyrien et al., 1995). These observations suggest that chromatin remodelling, which occurs to facilitate gene transcription during embryo development, rather than the transcriptional activity itself may specify the sites at which replication is initiated.

Relationship Between Transcription and Replication in an Ori Region

DNA replication in eukaryotes is localized in both time and space, specific regions of chromosomal DNA replicating at specific intervals and at limited numbers of sites within the nucleus during S phase. As mentioned above, depending on the location of yeast ARS elements on the chromosome, they replicate either early or late in S phase (Ferguson et al., 1991). In metazoans the timing of initiation of replication in a genomic region seems to be dictated by the transcriptional activity in that region. Transcriptionally active and inactive regions in a metaphase chromosome can be distinguished by their differential staining with Giemsa. Transcriptionally active regions appearing as light bands after Giemsa staining were found to replicate early in S phase, whereas transcriptionally inactive regions staining as dark bands with Giemsa were found to replicate late in S phase (Yunis et al., 1977; Holmquist et al.,

1982). Furthermore, the same gene in different cell types may replicate either early or late in S phase depending on whether it is being expressed or not in a given cell type. For example, a genomic region containing a cystic fibrosis (CF) gene replicates early in S phase in the cells expressing the CF gene, whereas it replicates late in S phase in the cells that do not express the gene (Selig et al., 1992). Cell fusion studies have also revealed a tight coordination between transcriptional activity and the timing of replication in the β-globin gene locus (Dhar et al., 1989). When β-globin gene expression in mouse hepatoma cells was repressed following their fusion with mouse erythroleukemia (MEL) cells, replication of the β-globin gene locus shifted to a later time in S phase. Similarly, when the β-globin gene was activated in human fibroblasts by fusion with MEL cells, the replication of the entire locus shifted to an earlier time in S phase. Although it is evident from these observations that the transcription and replication occur coordinately in specific genomic regions, the causal relationship between these processes remains elusive.

Conceivable Models for the Functional Origins of Replication in Metazoans

The observations described above, when taken together, raise the possibility that while replication in metazoans is initiated at multiple sites in a broad region, most of the initiations become futile, and only those at selected sites within the ori region will be effective in allowing bidirectional replication of a replicon. Based on this possibility, and in an attempt to reconcile the differences in the lengths of ori regions of the DHFR domain identified by different origin mapping methods, several models have been proposed to explain how functional ori are manifested in intergenomic regions of multicellular eukaryotes. In one model, it is suggested that the DNA replication initiates at a number of sites in a broad region and extends unidirectionally toward a specific site from where the replication becomes bidirectional (Linskens and Huberman, 1990). In a second model, DNA primers are suggested to first synthesize at multiple sites in a large uncoiled region of duplex DNA before bidirectional replication is initiated at a fixed site (Benbow et al., 1992). In third model, it is proposed that the replication is initiated at multiple sites on naked DNA, but their elongation is suppressed by the organization of DNA into chromatin; only selected initiation sites in chromatin that are associated with nuclear structures are capable of further unwinding and promoting DNA replication (DePamphilis, 1993a–c; Burhans and Huberman, 1994). Validations of these models require that an understanding of the structural and functional organization of the DNA and its interaction with proteins and enzymes associated with DNA replication in the context of chromatin and nuclear architecture be fully developed. It is equally important that the specific role of individual DNA polymerases and their accessory proteins in the initiation and elongation of replication forks be defined.

NUCLEAR CONTEXT IN THE CONTROL OF DNA REPLICATION

Autoradiography analysis of pulse-labeled DNA spreads has revealed that the chromosomal DNA replicates in clusters of synchronously initiated replicons and that different clusters initiate at different times during S phase (Hand, 1978; Dubey and Raman, 1987). Replication in a cluster of replicons is initiated at a discrete site within the nuclei. This is revealed from the studies in which cells were pulse labeled with either bromodeoxyuridine (BrdU) or biotinylated deoxyuridine triphosphate (dUTP), and the sites containing BrdU or biotinylated dUTP incorporated DNA in the nuclei were visualized after staining with fluorescent antibodies against BrdU or streptavidin, respectively (Nakamura et al., 1986; Nakayasu and Berezney, 1989). Incorporation of BrdU or biotinylated dUTP into DNA caused a punctate staining in the nuclei with individual sites of incorporation seen as bright spots or "foci" under a fluorescence microscope. Each S phase cell contained about 100–300 such foci distributed throughout the nucleus except in the nucleoli. The number of replication forks present in each focus can range from 20–40 in mammalian culture cells (Nakamura et al., 1986; Hassan and Cook, 1993) to 300–1,000 in *Xenopus* eggs (Blow and Laskey, 1986; Mills et al., 1989). A relatively large number of replication forks in each replication focus of eggs may account for the rapid rate of replication observed in them. In somatic cells replication foci detected in early S phase grow in size with an increase in BrdU labeling time until all the replicons in them are matured. These foci are then replaced by new ones as the S phase progresses. Thus, there is a turnover of the replication foci as the foci containing mature replicons are replaced by the ones with newly initiated replicons. This continues throughout S phase until all chromosomal DNA in the nuclei is completely replicated (Nakamura et al., 1986; Nakayasu and Berezney, 1989). However, when demembranated *Xenopus* sperm nuclei were induced to replicate in the presence of egg extract, a single pattern with a fixed number, size and distribution of replication foci was observed through out the S phase (Mills et al., 1989). Interestingly, when human fibroblast nuclei were similarly incubated with *Xenopus* extract, a single pattern of replication foci, resembling that seen in sperm nuclei, was observed (Kill et al., 1991).

Nuclei incubated with egg extract undergo extensive reorganization of their chromatin and nuclear envelop. A common pattern of replication foci seen in the nuclei from two different sources after their incubation with egg extract indicates that the nuclear reorganization by the components in egg extract may determine the pattern of replication foci seen in these nuclei. Thus, it seems that the chromatin and nuclear structure, rather than specific DNA sequences, specify the sites at which DNA is replicated in metazoan nuclei.

Replication Foci Detected in S Phase Cells Are Attached to the Nuclear Structure

A patten of replication foci similar to that seen in intact cells labeled with BrdU is also seen in permeabilized cells that are allowed to incorporate biotinylated dUTP into their newly replicating DNA (Nakayasu and Berezney, 1989). The patten of replication foci observed in the nuclei of permeabilized cells was preserved even after their extraction with DNase 1 and 0.2 M ammonium sulfate. Furthermore, the nuclear matrix structure remaining after the extraction of cells with DNase 1 and 0.2 M ammonium sulfate retained the ability to incorporate biotinylated dUMP into DNA, and the nascent DNA synthesized on the templates attached to the nuclear matrix exhibited a pattern of replication foci similar to that in intact calls labeled with BrdU (Nakayasu and Berezney, 1989). These latter observations point to a physical association between template DNA, as well as the enzymes of DNA replication, and the nuclear matrix structure. A number of earlier studies have also shown that the nascent DNA synthesized in a variety of mammalian cells is firmly attached to the nuclear matrix (Berezney and Coffey, 1975; Dijkwel et al., 1979; McCready et al., 1980; Pardoll et al., 1980; Vogelstein et al., 1980; Mirkovitch et al., 1984; Jackson and Cook, 1986a). Furthermore, gel electrophoretic analyses have revealed that the nascent DNA and replication forks partition with nuclear matrix (Vaughn et al., 1990). Attachment also of replication origins to the nuclear matrix is indicated from the observations that the radiolabel incorporated into DNA at the onset of S phase stays in close proximity to the matrix during G_2 and the next S phase, whereas the radiolabel incorporated at a later time in S phase is chased into surrounding DNA loops away from the matrix (Aelen et al., 1983; Carri et al., 1986; Dijkwel et al., 1986).

Replicating DNA Is Associated With Nuclear Matrix

Segments of DNA physically associated with nuclear matrix are referred to as *matrix-attached regions* (MARs). Although no consensus sequence has yet been discerned for MARs, MARs activity is suggested to reside in certain sequence motifs (Nakagomi et al., 1994). MARs are often localized in the vicinity of ori (Amati and Gasser, 1988, 1990; Dijkwel and Hamlin, 1988). It is hypothesized that one potential role of MARs is to retain cis-regulatory sequences in a nuclear subcompartment that is accessible to trans-acting factors required for replication and transcription (Mirkovitch et al., 1984).

In yeast, two of the ARS (ARS1 and HMR-E origin) that are known to function as chromosomal ori have been shown to reside next to MARs (Amati and Gasser, 1988). MARs greatly enhance ARS activity when placed next to an ARS element in the same plasmid (Amati and Gasser,

1990). Interestingly, a large stretch of *Drosophila* chromosomal DNA containing MARs introduced into yeast exhibited ARS activity following its binding to nuclear scaffolds (Brun et al., 1990; Amati and Gasser, 1990). In CHOC 400 cells MARs were localized midway between ori-β and ori-γ of the DHFR gene locus, and the frequency of association of MARs with nuclear matrix in CHOC 400 and CHO cells seems to be in proportion to the initiation activity in the ori region of the DHFR domain (Dijkwel and Hamlin, 1988, 1992). The lack of a fuller understanding of the ori in metazoans has made it difficult to evaluate the role of MARs in initiation of DNA replication.

For a MAR to be functionally associated with the matrix, a specific protein or a complex of proteins that recognizes MAR sequence motifs is required to be present in the nuclear matrix structure. In vitro studies have lead to the identification of a number of proteins that interact with MARs (Dickinson et al., 1992; Romig et al., 1992; Nakagomi et al., 1994). However, any potential in vitro role of these proteins in anchoring MARs in genomic DNA to the matrix or their involvement in initiation of DNA replication is not known.

Localization of Enzymes and Proteins at the Sites of DNA Replication

Incorporation of biotinylated dUTP into DNA in isolated nuclear matrix structures (Nakayasu and Berezney, 1989) indicates that, in addition to the template DNA, these matrix preparations should contain enzymes of DNA replication at the sites were DNA is being replicated. Several enzymes, including DNA polymerase α and primase, are found associated with nuclear matrix preparations in a cell cycle– and DNA replication–dependent manner (Smith and Berezney, 1980; Mikhailov and Tsanev, 1983; Nishizawa et al., 1984; Wood and Collins, 1986; Collins and Chu, 1987; Jackson and Cook, 1986b; Tubo and Berezney, 1987a,b).

Immunofluorescent microscopic studies further revealed that the replication enzymes and proteins in the nuclei are associated with a discrete granular structure exhibiting a punctate, as opposed to a diffused, distribution in the nuclei of cultured cells. Replication enzymes and proteins exhibiting a punctate distribution in the nuclei include DNA polymerase α (Bensch et al., 1982; Nakamura et al., 1984, 1986; Yamamoto et al., 1984), DNA ligase I (Lasko et al., 1990), PCNA (Madsen and Celis, 1985; Bravo and MacDonald-Bravo, 1987; Kill et al., 1991), and single-stranded DNA-binding protein RF-A (Wilcock and Lane, 1991; Adachi and Laemmli, 1992). In addition, two essential S phase protein kinases, cyclin A and CDK2, were also found associated with the replication foci (Cardoso et al., 1993). DNA polymerase α in chick embryos is shown to be tightly associated with granular structures attached to the nuclear matrix (Yamamoto et al., 1984). DNA polymerase α and PCNA are observed in replication foci just prior to the initiation of replication (Hutchison and Kill, 1989).

Electron microscopic analyses of resinless sections of immunolabeled

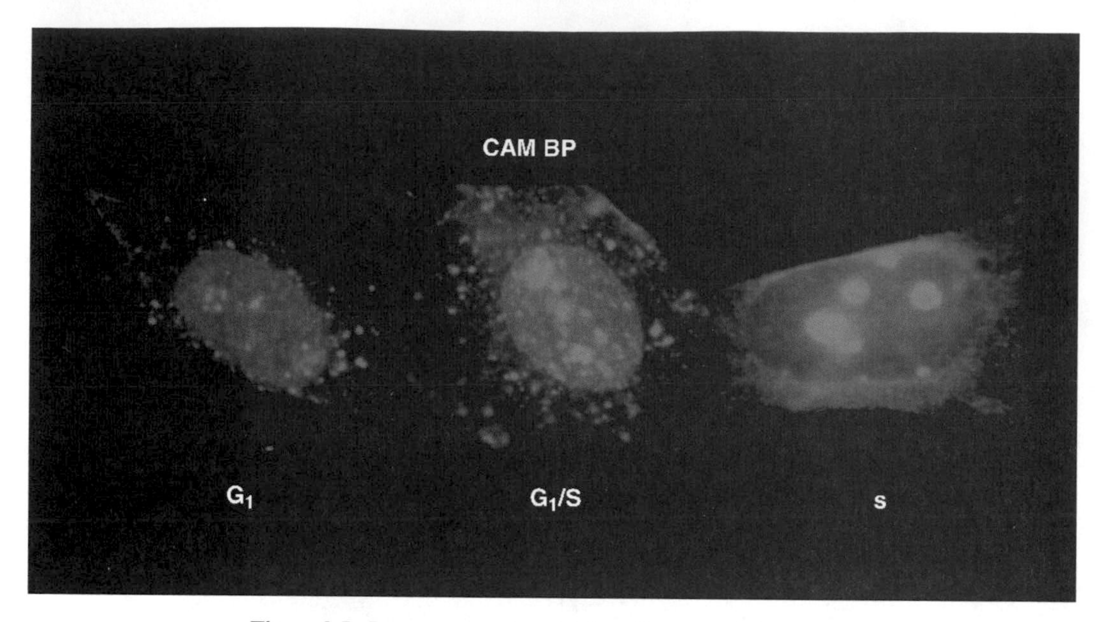

Figure 3.5. Immunocytochemical localization of CaM-BP68 in synchronized Chinese hamster embryo fibroblast (CHEF/18) cells. CHEF/18 cells grown on coverslips were synchronized by isoleucine deprivation as described by Reddy (1989). Zero hour (G_1), 6 hours (G_1/S), and 12 hours (S) following the release from isoleucine blocks, cells on the coverslips were extracted with buffer A (300 mM sucrose, 100 mM NaCl, 10 mM PIPES, 3 mM $MgCl_2$, 1 mM EGTA, pH 6.8, and 0.5% Triton X-100). This extraction procedure removes soluble cellular material, leaving cytoskeleton and nuclear architecture intact. Following extraction, cells were fixed on ice with 3.7% formaldehyde and subjected to immunostaining with rabbit polyclonal antibodies raised against purified CaM-BP68 and rhodamine-conjugated goat antirabbit IgG using a standard procedure. Cells were counterstained with DAPI, and photomicrography was performed with a Zeiss Axiofot microscope equipped for epifluorescence with a ×100 objective and single and multiple bandpass filter sets. Although a single stained cell is shown for each of the phases (G_1, G_1 phase; G_1/S, G_1/S boundary; and S, mid-to-late S phase), more than 85% of the cells on the coverslips representing each of the phases exhibited similar staining patterns under the synchronization conditions employed in these studies (E.G. Fey and G.P.V. Reddy, unpublished data). (See color plate.)

cells revealed that both DNA polymerase α and PCNA are associated with dense structures in which DNA is being replicated (Hozak et al., 1994). These structures, referred to as *replication factories,* are attached to nucleoskeleton. These factories appear at the end of G_1 phase and increase in size and decrease in number as the S phase progresses (Hozak et al., 1994). Immunofluorescence studies have revealed a similar nucleoskeletal localization pattern of a specific calmodulin-binding protein (CaM-BP68) (Fig. 3.5) (G.P.V. Reddy and E.G. Fry, unpublished data) that is tightly associated with the DNA polymerase α–primase complex (Cao et al., 1995) and is shown to be involved in DNA replication (Reddy et al., 1994).

Interestingly, DNA methyltransferase, which methylates deoxycytidine residues in mammalian genome, is also shown to associate with replication foci only during S phase (Leonhardt et al., 1992). This activity was reported earlier to be associated with the replitase complex isolated from the nuclei of S phase cells (Noguchi et al., 1983). In addition, it is observed that a 516 bp region in the DHFR ori-β locus and 127 bp in ori^{S14} (the ori region of the gene encoding ribosomal protein S14) is densely methylated during DNA replication (Tasheva and Roufa, 1994). Although a specific role of DNA methyltransferase in replication is not clear, these observations raise the possibility that methylation of DNA, as it is known to affect transcription, may also play a role in activating ori. In somatic cells B-type lamins are also localized to the replication foci (Moir et al., 1994). Immunodepletion of lamin B3 from *Xenopus* egg extract allows normal assembly of a nuclear envelop, but prevents the DNA from replicating (Newport et al., 1990; Meier et al., 1991; Jenkins et al., 1993). These observations are consistent with a direct role of B-type lamins in DNA replication (Hutchison et al., 1994).

Nuclear Structure Is Required for DNA Replication

As described above, it is evident that the nuclear structure forms the basis on which DNA is replicated in metazoans. This is analogous to the situation in bacteria and other microorganisms in which replicating DNA is attached to the cell membrane (Ryter, 1968; Firshein and Gillmorδ, 1970; Fuchs and Hanawalt, 1970). For DNA replication to initiate at specific sites it seems necessary that the DNA be first assembled into chromatin and organized into a nuclear structure.

Our understanding of the role of nuclear structure in DNA replication has derived largely from studies with *Xenopus* egg extract. *Xenopus* egg extract, recovered following low-speed centrifugation of crushed eggs, contains all major components required for both the assembly of nuclei and the initiation of DNA replication. These activities reside in the particulate material of the egg extract, as its removal by high-speed centrifugation makes the extract incapable of assembling nuclei or initiation DNA replication (Lohka and Masui, 1984; Newport, 1987; Sheehan et al., 1988). DNA, either in its naked form (purified) or in chromatin (when it is in demembranated *Xenopus* sperm nuclei or isolated somatic cell nuclei), incubated with egg extract undergoes replication after its assembly into interphase nuclei containing nuclear pores and envelop (Lohka and Masui, 1984; Vigers and Lohka, 1992). In this system only the fraction of naked DNA that is assembled into nuclei is replicated (Blow and Sleeman, 1990), indicating that the assembly of nuclei is a prerequisite for the initiation of DNA replication.

In addition to providing a structural support to the enzymes and proteins required for DNA replication, the nuclear envelop may also play a role in a selective accumulation of proteins required for the initiation of DNA replication inside the nucleus. This is indicated from the observation that the disruption of nuclear protein import prevents initiation of DNA replication in *Xenopus* egg extract (Cox, 1992). Furthermore, the nuclear structure seems to determine the specificity of sites at which DNA replication initiates in a mammalian chromosome. This is evident from the observation that the replication in the DHFR locus initiates preferentially at an ori region when intact Chinese hamster nuclei are incubated in *Xenopus* egg extract. However, initiation occurs throughout the DHFR locus, without any preference, at random sites when the nuclei with disrupted membrane or naked DNA containing the locus region is incubated in the same extract (Gilbert et al., 1995). It is further shown that replication initiates at preferred sites in the ori region of the DHFR locus only when the nuclei incubated in *Xenopus* egg extract are derived from late G_1, not from early G_1, phase cells (Wu and Gilbert, 1996). These observations raise an interesting possibility that the ori requires some component(s) of nuclear structure that are being recruited or activated during late G_1 phase.

Replicative Processes at Fixed Sites Within the Nuclei

One of the unresolved issues has been whether the DNA in the nucleus of a living cell is replicated by the involvement of a battery of soluble enzymes along a DNA template that is distributed evenly throughout the nucleus or by the spooling of DNA through a complex of enzymes localized at fixed sites in subnuclear compartments. Several observations described above pointing to the synchronous replication of a cluster of replicons at fixed sites in the nucleus (replication foci) and the presence of a number of enzymes and proteins required for DNA replication at the sites where DNA is being replicated tend to support the latter view. A hypothetical model in which DNA is replicated by its spooling through a complex of immobilized enzymes is presented in Figure 3.6.

According to this model, several adjacent replicons in a chromosome that replicate synchronously are arranged in loops by the binding of ori regions in each replicon to the replication apparatus, a replisome-like structure. As described above (see Fig. 3.4), the replication apparatus is a subcomponent of the replitase complex containing the enzymes of both DNA replication and DNA precursor synthesis. There could be as many replication apparatuses as the number of replication forks at each replication site (replication focus). A group of replication apparatuses supporting DNA replication in a cluster of replicons may represent a "replication factory" seen during resinless electron microscopic analysis of nuclei (Hozak et al., 1993). Once the replication is initiated, it extends bidirectionally in the 5′ to 3′ direction at replication forks that remain associated with the replication apparatus. As DNA from both sides of an ori region in a replicon spools through the complex of enzymes (as indicated by the arrows in Fig. 3.6), two adjacent replication forks continue to extend until the replication in a replicon comes close to completion. Two daughter strands generated by this process will loop out from around the site where the ori region of a replicon was initially bound to the replication apparatus. Depending on the number of replicons in

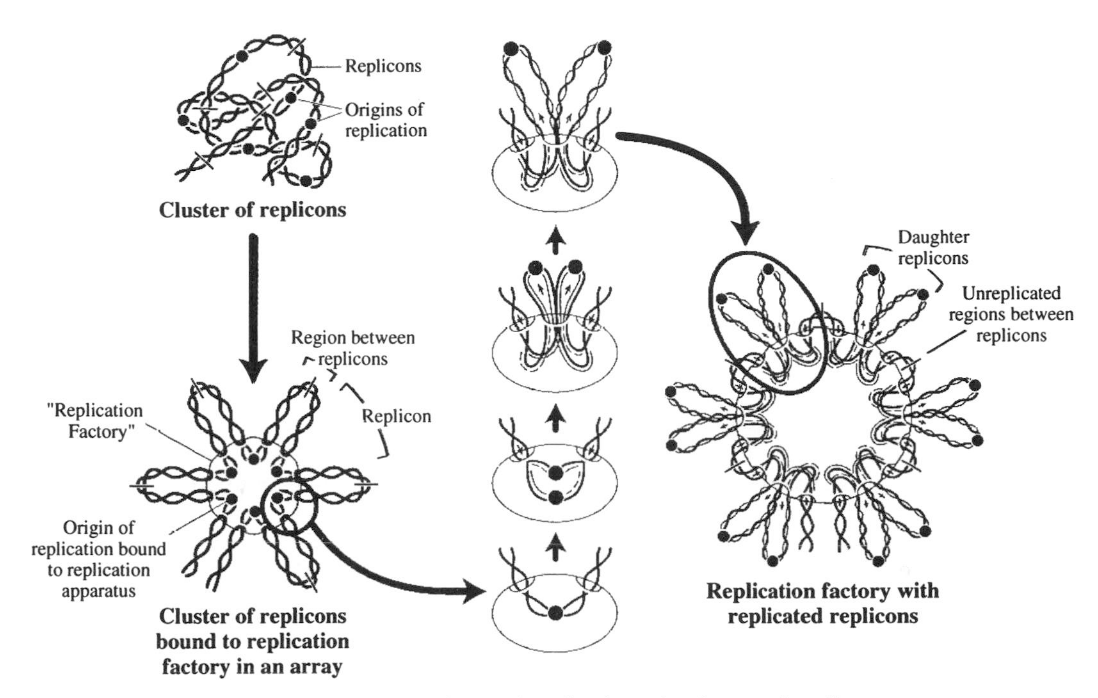

Figure 3.6. A model depicting organization and replication of a cluster of replicons at fixed sites containing replication factories attached to nuclear structure.

a cluster and the pattern in which the replication apparatuses are arranged on the nucleoskeleton, replication foci may exhibit a circular or a horseshoe shape as seen in BrdU-incorporated cells labeled with immunofluorescent antibodies (Nakamura et al., 1986). This model accommodates most of the biochemical (existence of replication enzymes in megacomplexes), biophysical (physical and functional organization of replicating DNA that is over 100 cm in length inside a nuclear compartment that is no more than 10 μm in diameter), and structural (association of enzymes of DNA replication with replication foci attached to nucleoskeleton) aspects of DNA replication observed in studies with mammalian cells.

However, the model as presented in Figure 3.6 falls short of explaining how the stretches of DNA between two replicons in a replication factory and that between replication factories are fully replicated. A possible solution to this puzzle may lie in the observation that several adjacent replication factories merge as the S phase progresses (Hozak et al., 1993). It is possible that further clustering of the replication apparatuses within individual replication factories, and of the replication factories themselves, may allow intra- and inter-replicon DNA to replicate fully by the end of S phase. Additional studies on the enzymes involved in replication of terminal DNA sequences in a replicon and their physical and functional interactions with replication apparatuses may shed further light on the mechanism by which DNA is fully replicated during S phase.

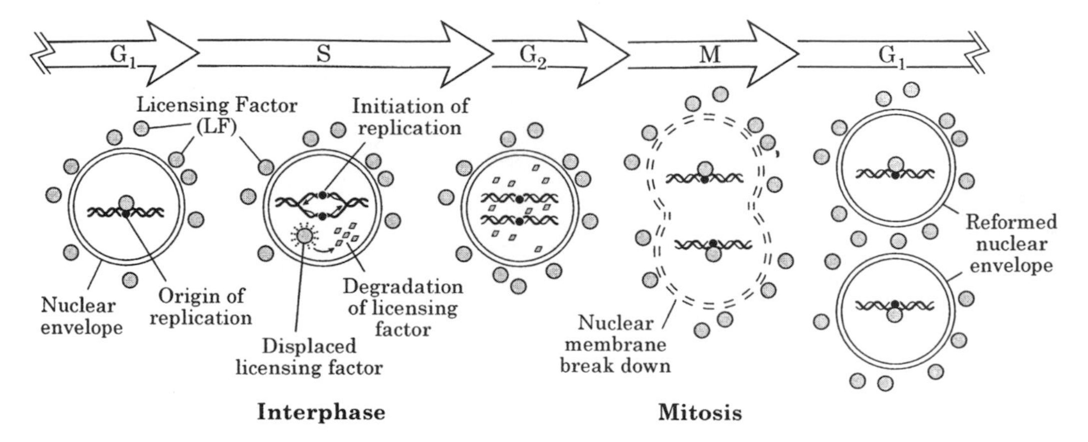

Figure 3.7. Proposed role of nuclear membrane and putative "licensing factor" in regulation of DNA replication during cell cycle.

Role of Nuclear Membrane in Limiting the Replication to Once per Cell Cycle

It is evident from cell fusion studies (Rao and Johnson, 1970) that the G_2 nucleus must go through mitosis before it is able to replicate again. Breakdown of nuclear membrane seems to be the primary event during mitosis that confers re-replicating ability to DNA in G_2 nuclei. This is revealed from the observation that G_2 nuclei incubated with fresh *Xenopus* egg extract are able to replicate again, without undergoing mitosis, only if their nuclear membrane is transiently permeabilized with nonionic detergents or lysolecithin (Blow and Laskey, 1988; Leno et al., 1992; Coverley et al., 1993). Collectively these observations indicate that, for the DNA in G_2 nuclei to replicate again, it must be exposed to a cytoplasmic replication-initiating factor that is incapable of entering into the nucleus unless its membrane breaks down at mitosis. The replication factor, whose access to the DNA requires breakdown of the nuclear membrane and whose activity is essential for the initiation of DNA replication in mammalian cells, is referred to as the *licensing factor* (Blow and Laskey, 1988).

According to the model (Fig. 3.7) proposed by Blow and Laskey (1988), binding of this factor to chromatin in reassembled mature nuclei allows initiation of replication at its binding sites. Once the replication is initiated during S phase, licensing factor bound to chromatin at initiation sites is inactivated or destroyed, making it incapable of binding to chromatin again. This will ensure that nuclear DNA does not replicate more than once during a single cell cycle. For nuclear DNA to replicate again, cytoplasmic licensing factor must gain access to the DNA, which requires breakdown of nuclear membrane at mitosis. Thus, the nuclear membrane seems to play an essential role in both providing a basic structure on which DNA is replicated inside the nucleus (described above) and ensuring that any segment of nuclear DNA does not replicate more than once per cell cycle.

FACTORS BINDING TO THE ORIGINS OF REPLICATION

What proteins and enzymes constitute the licensing factor? The basic criteria for the component(s) of this factor seem to be that they bind to DNA specifically at the sites where replication is being initiated and that their binding be transient, occurring just prior to the time when replication is initiated in S phase. Much of our understanding of such proteins involved in initiation of replication has once again originated from elegant studies in yeast. Knowing that the replication in *Sac. cerevisiae* initiates specifically at ARS elements (Stinchcomb et al., 1979; Brewer and Fangman, 1987, 1988), investigators designed biochemical and genetic studies to identify proteins that bind specifically to these elements. ARS elements consist of an essential core sequence A (Broach et al., 1983; Newlon, 1988) flanked on its 3′ end by two or three functionally conserved B (B1, B2, and B3) domains. Although B domains individually are not essential, they seem to influence the frequency with which an origin is used and are suggested to be required for ARS function (Marahrens and Stillman, 1992; Rao et al., 1994; Theis and Newlon, 1994; Huang and Kowalski, 1996). Sequences 5′ to the core (A) domain are also known to be necessary for ARS activity (Newlon and Theis, 1993). The B3 domain of ARS elements is suggested to enhance the initiation activity by its binding to a transcription factor called *ARS-binding factor 1* (Abf1) (Marahrens and Stillman, 1992). However, the B3 domain does not seem to be essential for the initiation of replication as it can be replaced by the sequence domains that bind to other transcription factors (Marahrens and Sillman, 1992). Several other proteins, including a single-stranded DNA-binding protein, are known to bind to ARS elements (Kuno et al., 1990; Hofmann and Gasser, 1991; Schmidt et al., 1991; Diffley and Stillman, 1988), but their function in initiation of replication remains to be characterized.

With ARS1, one of the most extensively studied yeast ARS elements, in a DNase protection assay, Bell and Stillman (1992) discovered a multiprotein complex in yeast nuclear extract that binds specifically to A and B domains. A purified multiprotein complex, referred to as an *origin recognition complex* (ORC), consists of six protein subunits of 120, 72, 62, 56, 53, and 50 kD molecular weights. The binding of ORC to the ARS1 element requires ATP, a characteristic of the proteins, including SV40 T antigen (Borowiec et al., 1990) and *Escherichia coli* dnaA protein (Kornberg and Baker, 1992), which are required for initiation of replication. In vivo studies involving nucleotide resolution genomic footprinting techniques have also revealed binding of proteins to A and B1 domains of ARS1 elements in a pattern similar to that seen in vitro with ORC (Diffley and Cocker, 1992). ORC binding to ARS1 is suggested to involve wrapping of DNA in A and B1 domains around the protein complex. Mutations in the A domain that reduce ARS activity in a plasmid stability assay are shown to cause a corresponding decrease in ORC binding to ARS1 (Bell and Stillman, 1992). Furthermore, mutations in the genes encoding ORC subunits are known to reduce the efficiency with which replication initiates in yeast (Loo et al., 1995; Liang et al., 1995). These observations indicate that ORC, by binding to ARS elements, plays an important role in the initiation of replication. Homologues of ORC subunits have also been identified in other species and in higher eukaryotes (Muzi-Falconi and Kelly, 1995; Gavin et al., 1995; Gossen et al., 1995; Leatherwood et al., 1996; Carpenter et al., 1996).

Role of ORC in Initiation of Replication

Although ORCs are known to be required for the initiation of replication, their binding to DNA did not fluctuate during the cell cycle; they are bound to DNA not just during the S phase, but throughout the cell cycle (Diffley and Cocker, 1992), indicating that the binding of ORC itself to DNA is necessary but not sufficient for initiation of replication. This is reinforced by the studies with *Xenopus* egg extract in which the replicon size and the time required for the completion of replication in sperm nuclei increased as the number of nuclei in a fixed volume of extract increased (Walter and Newport, 1997). The increase in replicon size is accompanied by the decrease in number of replicons per nuclei. These changes in the size and the number of replicons per nuclei were found to be independent of chromatin-bound ORC content of the nuclei. These observations raise the possibility that the initiation of replication is controlled by an unidentified factor that becomes limited as the number of nuclei increases in a fixed volume of extract (Walter and Newport, 1997). That factor(s) other than ORC might be rate limiting in the initiation of replication is also indicated by the observation that the footprinting pattern due to ORC binding to DNA changes as the cells exit mitosis and enter into early G_1 (Diffley et al., 1994). In these studies a nuclease hypersensitive site in the B1 domain of ARS1 was seen only during S, G_2, and M, but not in early G_1, stages of the cell cycle. Furthermore, the detection of this hypersensitive site in the B1 domain was dependent on the cdc7p protein kinase (Diffley et al., 1994). These observations raise the possibility that the binding of additional proteins to the preexisting ORC–DNA complexes, or post-translation modification of the protein subunits in ORC as the cells are transiting from G_1 into S phase, is required for the initiation of replication.

These findings are consistent with a model (Fig. 3.8) in which the replication complexes assembled at the ori exist in two alternating states. In one the complex acquires a pre-replicative state, referred to as the *pre-replication complex* (pre-RC), that is ready to be activated as the cells enter into S phase. As the replication is initiated, pre-RC at an ori becomes inactivated by its conversion to a postreplicative state consisting of ORC. For the postreplication complex (post-RC) to reacquire the pre-RC state, the ORC–DNA complex must interact with cytoplasmic factor(s) that bind to and/or modify ORC during G_1. Thus, in going from a post-RC to a pre-RC state, ORC seem to serve as a landing pad for the proteins that control the initiation of DNA replication and, therefore, the entry of cells into S phase.

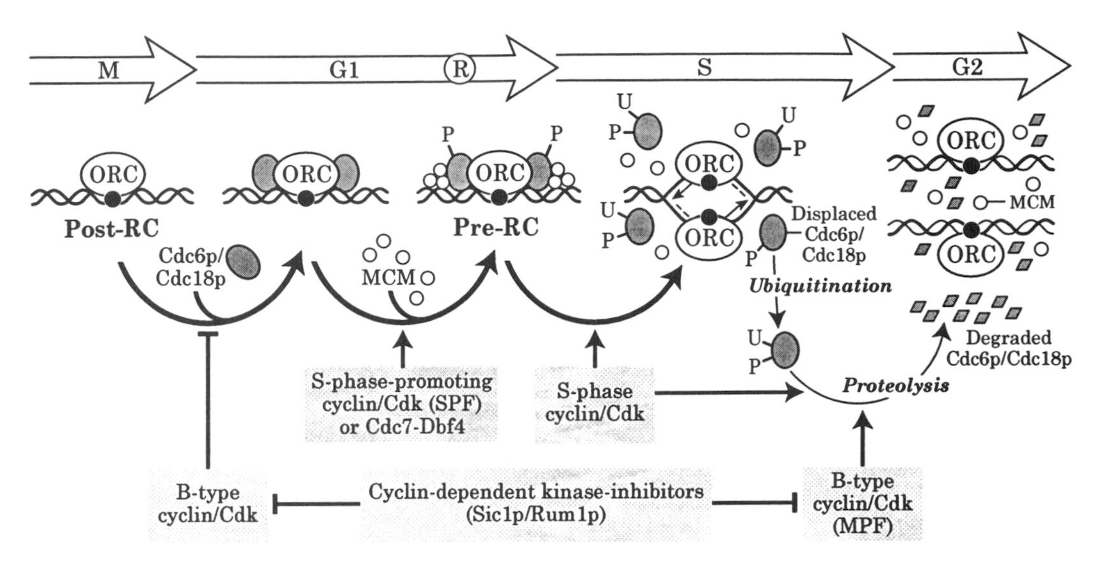

Figure 3.8. A summary of events leading to the entry of cells into S phase and regulation of DNA replication.

Interaction of Cell Cycle Regulatory Proteins With ORC

As described above, cdc6p in *Sac. cerevisiae* and its homologue cdc18p in *Sch. pombe* are known to be required for the initiation of DNA replication (Bueno and Russell, 1992; Liang et al., 1995; Piatti et al., 1995; Kelly et al., 1993; Muzi-Falconi et al., 1996). The level of these proteins in the cells oscillates with cell cycle, reaching a peak in late G_1, a property characteristic of regulatory proteins involved in transition of cells from G_1 into S phase (Piatti et al., 1995; Muzi-Falconi et al., 1996; Nishitani and Nurse, 1995). An early indication that cdc6p, and perhaps cdc18p, may interact directly with the ori is derived from the observation that the minichromosome loss seen in cdc6 mutants can be suppressed by the addition of extra ori to the minichromosomes (Hogan and Koshland, 1992). A direct role of cdc6p in initiation of replication is also evident from the observation that the formation of pre-RC in budding yeast requires cdc6p (Cocker et al., 1996; Dahmann et al., 1995). Genetic studies revealed functional interactions between cdc6p and the ORC subunits orc2p, orc5p, and orc6p (Li and Herskowitz, 1993; Loo et al., 1995). Biochemical studies have shown physical interactions between cdc6p and ORC in *Sac. cerevisiae* (Liang et al., 1995), and cdc18 and orp2p, the fission yeast homologue of budding yeast Orc2p, in *Sch. pombe* (Leatherwood et al., 1996). There is also a considerable sequence homology between cdc6p/cdc18p and one of the ORC subunits, orc1p/orp1p (Muzi-Falconi and Kelly, 1995; Gavin et al., 1995). These observations raise the possibility that orc1p/orp1p homology regions in cdc6p/cdc18p may play a role in its interaction with orc2p/orp2p and other subunits of ORC in pre-RC. cdc6p/cdc18p- and ORC-related proteins have also been identified in *Xenopus* extract (Coleman et al., 1996; Carpenter et al., 1996). As indicated from studies in yeast, the binding of cdc6p/cdc18p in *Xenopus* (Xcdc6) to chromatin requires *Xenopus* orc2p (Xorc2). Furthermore, Xcdc6 in *Xenopus* egg extract is shown to be required for initiation, but not for elongation, of replication forks. Thus, from these studies it is evident that cdc6p/cdc18p plays a critical role in initiation of replication by its binding to ORC.

Minichromosome Maintenance Proteins

Genetic screening of the budding yeast mutants defective in maintaining plasmids (minichromosomes) containing single ARS elements has also lead to the identification of genes encoding minichromosome maintenance (Mcm) proteins (Maine et al., 1984). A high frequency of minichromosome loss is related to the defect in initiation of replication in these Mcm mutants (Maiti and Sinha, 1992; Yan et al., 1993). In Mcm mutants the frequency of minichromosome loss seems to be determined by the ARS elements that it contains, minichromosomes containing certain ARS elements being lost at a much higher frequency than those that contain some other ARS elements (Sinha et al., 1986; Gibson et al., 1990; Yan et al., 1991; Chen et al., 1992). Furthermore, chromosome loss in certain Mcm mutants can be suppressed by the insertion of additional ARS elements in the minichromosome (Ray et al., 1994). Cells defective in these genes not only fail to maintain episomes containing ARS elements but also get arrested in cell cycle with partially replicated DNA at nonpermissive conditions. Possible involvement of these proteins in the initiation, but not in elongation, of replication is indicated by the observation that the temperature-sensitive Mcm mutants arrested in S phase by hydroxyurea treatment are able to complete S phase at the nonpermissive temperature (Hennessy et al., 1990). These observations taken together indicate a direct role of Mcm proteins in initiation of replication by their interaction with the origins of replication.

At least five essential Mcm genes (Mcm2, Mcm3, Mcm4/cdc54, Mcm5/cdc46, and Mcm7/cdc47) have been identified in *Sac. cerevisiae* (Chong et al., 1996). These gene products are required for viability as well as for the entry of cells into S phase (Gibson et al., 1990; Hennessy et al., 1991; Maiorano et al., 1996). Homologues of some of these Mcm proteins have also been identified in *Sch. pombe* (Coxon et al., 1992; Miyake et al., 1993; Forsburg and Nurse, 1994; Takahashi et al., 1994) and in metazoans, including mouse (Thommes et al., 1992), human (Hu et al., 1993; Todorov et al., 1994), *Drosophila* (Treisman et al., 1995), and *Xenopus* (Kubota et al., 1995; Madine et al., 1995a). Mouse Mcm3 protein was initially identified as P1 protein that copurified with DNA polymerase α (Thommes et al., 1992; Maiorano et al., 1993). Microinjection of antibodies against P1 protein into mouse cells (Thommes et al., 1992) and those against BM28 (human Mcm2) protein into human cells (Todorov et al., 1994) prevented their transition from G_1 into S phase. *Drosophila* embryo mutants defective in Mcm2 exhibited a delayed, or a lack of, DNA synthesis in the cells of imaginal discs and central nervous system (Treisman et al., 1995).

Xenopus egg extract treated with 6-dimethylaminopurine (6-DMAP), a protein kinase inhibitor, is incapable of supporting DNA replication. Using this extract in an assay to reconstitute replication capability, Chong et al. (1995) purified replication licensing factors, RLF-M and RLF-B, that are capable of restoring replication-supporting activity in the inactive extract. RLF-M purified to apparent homogeneity by this assay contained at least three polypeptides of 115, 106, and 92 kD molecular weights. While all subunits of the RLF-M complex cross-reacted with antibodies raised against conserved peptide sequences in the Mcm family of proteins, a 106 kD subunit was identified as Mcm3, and 115 and 92 kD subunits may possibly represent Mcm2 and Mcm5, respectively (Chong et al., 1995). *Xenopus* egg extract immunodepleted of Mcm3 is incapable of supporting chromatin replication (Chong et al., 1995; Kubota et al., 1995; Madine et al., 1995a). The replication activity in Mcm3-immunodepleted extract could be restored by the addition of RLF-M (Chong et al., 1995). However, RLF-M in its purified form was found to be relatively less effective than the crude form in restoring the activity, suggesting that factors other than those present in purified RLF-M complex are necessary for its activity. It is observed that addition of RLF-B to purified RLF-M fully restores the activity, but RLF-B alone is incapable of restoring the activity (Chong et al., 1995). Thus, RLF-M–dependent initiation of DNA replication in *Xenopus* egg extract is being mediated by RLF-B. The purified RLF-M in combination with RLF-B, therefore, seems to fit the picture of a licensing factor envisioned to be required for the initiation of replication in cells entering into S phase (Fig. 3.7). Consistent with this are the observations that Mcm3 dislodges from chromatin during DNA replication and that its reassociation with chromatin requires both RLF-B and the breakdown of nuclear membrane (Chong et al., 1995; Madine et al., 1995a,b; Kubota et al., 1995). Mcm proteins exhibit nuclear localization throughout the cell cycle (Thommes et al., 1992; Kimura et al., 1994; Todorov et al., 1994). Furthermore, Mcm3 and other proteins of the Mcm complex are able to enter freely into nuclei without requiring the breakdown of nuclear membrane (Madine et al., 1995a). Once inside the nuclei, however, their binding to the chromatin shows an absolute requirement for the breakdown of nuclear membrane (Madine et al., 1995b). These observations indicate the inability of RLF-B–like factor(s), required for the association of Mcm proteins with chromatin, to cross the nuclear membrane in metazoans.

Although RLF-B has not yet been characterized, cdc6 protein in *Xenopus* (Xcdc6) extract is shown to be similarly required for the binding of Mcm3 to chromatin (Coleman et al., 1996). In these studies, unlike Xorc2, which remains bound to chromatin, Xcdc6 is dislodged from chromatin immediately following the initiation of replication and found associated largely with nuclear envelop in postreplicative nuclei. However, XMcm3 is released not at the time of initiation per se, but as the replication forks begin to extend bidirectionally. This is indicated from the observation that aphidicobin, which blocks the elongation of replication forks but not the initiation of replication, greatly reduces the dissociation of XMcm3 from chromatin (Chong et al., 1995; Kubota et al., 1995; Madine et al., 1995a). XMcm3 released from the chromatin is found distributed throughout the nuclear compartment in a soluble form.

While Mcm proteins are known to be required for the onset of replication, their specific role in DNA replication remains elusive. From amino acid sequence analyses these proteins are believed to contain predicted consensus sequences for DNA-dependent ATPase activity (Koonin, 1993). Because similar sequence motifs are also present in DNA and RNA helicases, it is predicted that Mcm proteins may mediate ATP-dependent unwinding of double-stranded DNA at the ori. However, no such activity has yet been found in isolated Mcm proteins. Furthermore, it is reported that mouse P1 (Mcm3) (Kimura et al., 1994) and human BM28 (Mcm2) (Todorov et al., 1995) proteins undergo periodic changes in their phosphorylated states and in their intranuclear distribution during cell cycle. It is observed that Mcm proteins in G_1 phase are hyperphosphorylated, and, following the onset of S phase they gradually become underphosphorylated. It is not clear whether these apparent changes in phosphorylated states of Mcm proteins are a cause or a consequence of their binding to chromatin and contribute in any way to their function and/or stability within the nuclei.

A SUMMARY OF EVENTS LEADING TO THE ONSET OF S PHASE AND REGULATION OF DNA REPLICATION

Based on the observations described above, a sequential order of events leading to the assembly of replication initiation complexes at the ori has been suggested (Coleman et al., 1996; Rowles et al., 1996). As shown in Figure 3.8, during late mitosis and early G_1, Mcm proteins are loaded onto ORCs, which are bound to the ori. This loading of Mcm proteins onto ORCs requires prior association of cdc6p/cdc18p class proteins with ORC. The entry of cdc6p/cdc18p class proteins into the nuclei requires the breakdown of nuclear membrane, which normally occurs during mitosis. However, in yeast, where there is no detectable breakdown of nuclear membrane during mitosis, their entry into nuclei seems to be facilitated by a cell cycle–regulated transport or nuclear membrane permeability. Thus the entry of cdc6p/cdc18p class proteins into nuclei seems to be the rate-limiting step in the overall assembly of replication initiation complexes at the ori. This complex of ORC cdc6p/cdc18p class proteins and Mcm proteins assembled at the ori during mitosis and G_1 phase is referred to as a *pre-replication complex* (pre-RC).

At present, specific events leading to the activation of pre-RC to trigger initiation of DNA replication and the entry of cells into S phase are not entirely clear. However, observations described above strongly suggest the possibility that cyclins and CDKs, constituting S phase–promoting factor (SPF), phosphorylate the components of pre-RC triggering their activation at the G_1/S boundary. As described above, the activation of cyclin/CDKs at the G_1/S boundary in turn requires the destruction of CDK inhibitors. CDK inhibitors, such as Sic1p, are marked for ubiquitin-dependent proteolysis by G_1 cyclins and CDKs (Verma et al., 1997).

In budding yeast, the activity of another protein kinase encoded by the cdc7 gene is also shown to be required for the initiation of replication (Hereford and Hartwell, 1974; Kitada et al., 1992; Jackson et al., 1993). This kinase seems to trigger pre-RC activation through a cyclin/CDK-independent pathway. cdc7p kinase activity is shown to fluctuate with cell cycle, and its activation at the G_1/S boundary is dependent on its interaction with the regulatory protein Dbf4p (Jackson et al., 1993; Yoon and Campbell, 1991). Cells lacking either cdc7p or Dbf4p fail to initiate DNA replication even though they contain normal S phase–promoting cyclin/CDK activity and are able to transit through the START point in late G_1 phase. It is shown that cdc7p binds to ORC, and Dbf4p, like orc6p, interacts with the ori (Dowell et al., 1994). From these observations, it is possible that Dbf4p may target cdc7p kinase to ORC, allowing its interaction with, and phosphorylation of, Mcm proteins and other cell cycle regulatory proteins at the ori. While cdc7p/Dbf4p-mediated events are necessary for initiation of replication in budding yeast, proteins homologous to cdc7p and Dbf4p have not yet been identified in either fission yeast or higher eukaryotes.

It is conceptualized from these observations that the phosphorylation of pre-RC components, cdc6p/cdc18p, Mcm proteins, and/or ORC, by S phase–promoting cyclin/CDKs and cdc7p/Dbf4p may lead to the recruitment of the enzymes of DNA replication, including DNA polymerase α–primase, to the ori. Initiation of replication would then displace cdc6p/cdc18p, and subsequently Mcm proteins as the replication forks extend, from ORC. The phosphorylated state of the displaced cdc6p/cdc18p may also target them for ubiquitin-dependent proteolysis as described above. Such degradation of displaced cdc6p/cdc18p would ensure that no new pre-RCs are assembled to reinitiate DNA replication from the same origin during the remainder of S phase and prior to the passage of cells through mitosis. This is consistent with the observation that the mutations in genes encoding the components of ubiquitin-dependent protein degradation complex would lead to reinitiation of replication at the origins within a single cell cycle. Furthermore, overexpression of cdc18p in fission yeast leads to a continuous accumulation of DNA due to reinitiation of replication at each origin within the same S phase (Nishitani and Nurse, 1995; Muzi-Falconi et al., 1996). Thus, timely degradation of phosphorylated cdc6p/cdc18p displaced from the ori by ubiquitin-dependent proteolysis is essential for limiting the initiation of replication at an ori to once per cell cycle.

As described above, inactivation of mitotic CDK by various methods, including overexpression of rum1p in fission yeast (Moreno and Nurse, 1994) or Sic1p in budding yeast (Dahmann et al., 1995), also leads to re-replication of DNA without intervening mitosis. However, this re-replication due to CDK inactivation seems to result in the accumulation of DNA in full genome increments rather than in a continuous increase, which, as described above, occurs if reinitiation takes place at each origin within a single S phase. These observations indicate that, in the absence of mitotic CDK, cells lose the controls that limit their re-entry into S phase before mitosis, but retain the ability to prevent reinitiation of replication at each origin in a single S phase. Thus, active mitotic CDK plays an essential role in preventing the onset of S phase. In vivo experiments in budding yeast revealed a direct correlation between the inhibition of Clb/CDK and the assembly of pre-RC on chromatin (Dahmann et al., 1995). This observation may imply that high Clb/CDK activity in cells may somehow prevent nuclear accumulation of cdc6p/cdc18p class proteins required for the assembly of pre-RC. Therefore, mitotic CDK must be inactivated, which normally occurs at the end of mitosis, for a competent pre-RC to assemble during G_1 phase. Once again, an increase in S phase–promoting cyclin/CDK, which occurs in direct relation to the increase in cell size, would trigger the entry of cells into S phase by activating pre-RC to initiate DNA replication.

REFERENCES

Aboussekhra AM, Biggerstaff M, Shivji KMM, Vilpo JA, Monocollin V, Podust V, Protic M, Hubscher U, Egly J-M, Wood RD (1995): Mammalian DNA nucleotide excision repair reconstituted with purified protein components. Cell 80:859–868.

Adachi Y, Laemmli UK (1992): Identification of nuclear pre-replication centers poised for DNA synthesis in *Xenopus* egg extract: Immunolocalization study of replication protein A. J Cell Biol 119:1–15.

Adachi Y, Laemmli UK (1994): Study of the cell cycle–dependent assembly of DNA pre-replicated centers in *Xenopus* egg extracts. EMBO J 13:4153–4164.

Aelen JMA, Opstelten RJG, Wanka F (1983): Organization of DNA replication in *Physarum polycephalum.* Attachment of origins of replicons and replication forks to the nuclear matrix. Nucleic Acids Res 11:1181–1195.

Aguilera A, Klein HL (1988): Genetic control of intrachromosomal recombination in *Saccharomyces cerevisiae.* I. Isolation and genetic characterization of hyperrecombination mutations. Genetics 119:779–790.

Aladjem MI, Groudine M, Brody LL, Dieken ES, Fournier RE, Wahl GM, Epner EM (1995): Participation of the human β-globin locus control region in initiation of DNA replication. Science 270:815–819.

Albert DA, Rozengurt E (1992): Synergistic and coordinate expression of the gene encoding ribonucleotide reductase subunits in Swiss 3T3 cells: Effect of multiple signal-transduction pathways. Proc Natl Acad Sci USA 89:1597–1601.

Alberts BM (1985): Protein machines mediate the basic genetic processes. Trends Genet 1:26–30.

Allen R, Reddy GPV, Laser GC, Mathews CD (1980): T4 ribonucleotide reductase—Physical and kinetic linkage to other enzymes of deoxynucleotide biosynthesis. J Biol Chem 225:7583–7588.

Amati B, Gasser SM (1988): Chromosomal ARS and CEN elements bind specifically to the yeast nuclear scaffold. Cell 54:967–978.

Amati B, Gasser SM (1990): *Drosophila* scaffold attached regions bind nuclear scaffolds and can function as ARS elements in both budding and fission yeast. Mol Cell Biol 10:5442–5454.

Amon A, Irniger S, Nasmyth K (1994): Closing the cell cycle circle in yeast: G2 cyclin proteolysis initiated at mitosis persists until the activation of G1 cyclins in the next cycle. Cell 77:1037–1050.

Araki H, Ropp PA, Johnson AL, Johnston LH, Morrison A, Sugino A (1992): DNA polymerase II, the probable homolog of mammalian DNA polymerase ε, replicates chromosomal DNA in the yeast *Saccharomyces cerevisiae.* EMBO J 11:733–740.

Arens M, Yamashita T, Padmanabhan R, Tsuruo T, Green M (1977): Adenovirus deoxyribonucleic acid replication. Characterization of the enzyme activities of soluble replication system. J Biol Chem 252:7947–7953.

Ariga H, Sugino A (1983): Initiation of simian virus 40 DNA replication in vitro. J Virol 48:481–491.

Ashman CR, Reddy GPV, Davidson RL (1981): Bromodeoxyuridine mutagenesis, ribonucleotide reductase activity and deoxyribonucleotide pools in hydroxyurea-resistant mutants. Somat Cell Mol Genet 7:751–768.

Ayusawa D, Shimizu K, Koyama H, Takeishi K, Seno T (1983): Unusual aspects of human thymidylate synthase in mouse cells introduced by DNA-mediated gene transfer. J Biol Chem 258:48–53.

Bai C, Sen P, Hofmann K, Ma L, Goebl M, Harper JW, Elledge SJ (1996): SKP1 connects cell cycle regulators to the ubiquitin proteolysis machinery through a novel motif, the F-box. Cell 86:263–274.

Baldin V, Lukas J, Marcote MJ, Pagano M, Draetta G (1993): Cyclin D1 is a nuclear protein required for cell cycle progression in G1. Genes Dev 7:812–821.

Bambara RA, Jessee CB (1991): Properties of DNA polymerase δ and $\acute{\varepsilon}$, and their roles in eukaryotic DNA replication. Biochim Biophys Acta 1088:11–24.

Bandara LR, Adamczewski JP, Hunt T, Thangue NB (1991): Cyclin A and the retinoblastoma gene product complex with a common transcription factor. Nature 352:249–251.

Bandara LR, Adamczewski JP, Zamanian M, Poon RY, Hunt T, Thangue NB (1992): Cyclin A recruits $p33^{cdk2}$ to the cellular transcription factor DRTF1. J Cell Sci (suppl) 16:77–85.

Baril EF, Baril B, Elford H, Luftig R (1974): DNA polymerases and a possible multienzyme complex for DNA biosynthesis in eukaryotes. In Kolber AR and Kohiyama M (eds): Mechanisms and DNA Regulation of Replication. New York, Plenum Press, 275–291.

Beach D, Durkacz B, Nurse P (1982): Functionally homologous cell cycle control genes in budding and fission yeast. Nature 300:706–709.

Beijersbergen RL, Carlee L, Kerkhoven RM, Bernards R (1995): Regulation of the retinoblastoma protein–related p107 by G_1 cyclin complexes. Genes Dev 9:1340–1353.

Bell SP, Stillman B (1992): ATP-dependent recognition of eukaryotic origins of DNA replication by a multiprotein complex. Nature 357:128–134.

Benbow RM, Zhao J, Larson DD (1992): On the nature of origins of DNA replication in eukaryotes. BioEssays 14:661–670.

Bensch KG, Tanaka S, Hu S-Z, Wang TS-F, Korn D (1982): Intracellular localization of human DNA polymerase α with monoclonal antibodies. J Biol Chem 257:8391–8396.

Berezney R, Coffey DS (1975): Nuclear protein matrix: Association with newly synthesized DNA. Science 189:291–293.

Bjorklund S, Skog S, Tribukait B, Thelander L (1990): S-phase–specific expression of mammalian ribonucleotide reductase R1 and R2 subunit mRNAs. Biochemistry 29:5452–5458.

Blow JJ, Laskey RA (1986): Initiation of DNA replication in nuclei and purified DNA by a cell-free extract of *Xenopus* eggs. Cell 47:577–587.

Blow JJ, Laskey RA (1988): A role for the nuclear envelop in controlling DNA replication within the cell cycle. Nature 332:546–548.

Blow JJ, Nurse P (1990): A cdc2-like protein is involved in the initiation of DNA replication in *Xenopus* egg extracts. Cell 62:855–862.

Blow JJ, Sleeman AM (1990): Replication of purified DNA in *Xenopus* egg extracts is dependent on nuclear assembly. J Cell Sci 95:383–391.

Bodrug SE, Warner BJ, Bath ML, Lindeman GJ, Harris AW, Adams JM (1994): Cyclin D1 transgene impedes lymphocyte maturation and collaborates in lymphoma genesis with the myc gene. EMBO J 13:2124–2130.

Booher RN, A1fa CE, Hyams JS, Beach DH (1989): The fission yeast cdc2/cdc13/suc1 protein kinase: Regulation of catalytic activity and nuclear localization. Cell 58:485–497.

Borowiec JA, Dean FB, Bullock PA, Hurwitz J (1990): Binding and unwinding—How T antigen engages the SV40 origin of DNA replication. Cell 60:181–184.

Bramhill D, Kornberg A (1988): A model for initiation at origins of DNA replication. Cell 54:915–918.

Bravo R, MacDonald-Bravo H (1987): Existence of two populations of cyclin/proliferating cell nuclear antigen during the cell cycle: Association with DNA replication sites. J Cell Biol 105:1549–1554.

Brewer BJ, Fangman WL (1987): The localization of replication origins on ARS plasmids in *S. cerevisiae.* Cell 51:463–471.

Brewer BJ, Fangman WL (1988): A replication fork barrier at the 3′ end of yeast ribosomal RNA genes. Cell 55:637–643.

Broach J, Li Y-Y, Feldman J, Jayaram M, Abraham J, Nasmyth KA, Hicks JB (1982): Localization and sequence analysis of yeast origins of DNA replication. Cold Spring Harbor Symp Quant Biol 47:1165–1173.

Brun C, Dang Q, Miassod R (1990): Studies on a 800 kb stretch of the *Drosophila* X chromosome: Comapping of a sub-class of scaffold-attached regions with sequences able to replicate autonomously in yeast. Mol Cell Biol 10:5455–5463.

Bueno A, Russell P (1992): Dual functions of CDC6: A yeast protein required for DNA replication also inhibits nuclear division. EMBO J 11:2167–2176.

Burgers PM (1991): *Saccharomyces cerevisiae* replication factor C II. Formation and activity of complexes with the proliferating cell nuclear antigen and with DNA polymerase δ and ε. J Biol Chem 266:22698–22706.

Burhans WC, Huberman JA (1994): DNA replication origins in animal cells: A question of context. Science 263:639–640.

Burhans WC, Selegue JE, Heintz NH (1986a): Replication intermediates formed during initiation of DNA synthesis in methotrexate-resistant CHOC 400 cells are enriched for sequences from a specific, amplified restriction fragment. Biochemistry 25:441–449.

Burhans WC, Selegue JE, Heintz NH (1986b): Isolation of the origin of replication associated with Chinese hamster ovary dihydrofolate reductase domain. Proc Natl Acad Sci USA 83:7790–7794.

Burhans WC, Vassilev LT, Caddle MS, Heintz NH, DePamphilis ML (1990): Identification of an origin of bidirectional DNA replication in mammalian chromosomes. Cell 62:955–965.

Burhans WC, Vassilev LT, Wu J, Sogo JM, Nallaseth FS, DePamphilis ML (1991): Emetine allows identification of origins of mammalian DNA replication by unbalanced DNA synthesis, not through conservative nucleosome segregation. EMBO J 10:4351–4360.

Byrnes JJ (1985): Differential inhibitors of DNA polymerases alpha and delta. Biochem Biophys Res Commun 132:628–634.

Caddle MS, Calos MP (1994): Specific initiation at an origin of replication from *Schizosaccharomyces pombe.* Mol Cell Biol 14:1796–1805.

Cairns J (1996): Autoradiography of HeLa cell DNA. J Mol Biol 15:372–373.

Campbell JL (1993): Yeast DNA replication. J Biol Chem 268:25261–25264.

Campisi J, Pardee AB (1984): Post-transcriptional control of the onset of DNA synthesis by an insulin-like growth factor. Mol Cell Biol 4:1807–1814.

Cao L, Faha B, Dembski M, Tsai LH, Harlow E, Dyson N (1992): Independent binding of the retinoblastoma protein and p107 to the transcription factor E2F. Nature 355:176–179.

Cao PQ, McGrath CA, Baril EF, Quesenberry PJ, Reddy GPV (1995): The 68 kDa calmodulin-binding protein is tightly associated with multiprotein DNA polymerase α–primase complex in HeLa cells. Biochemistry 34:3878–3883.

Cardoso Mc, Leonhardt H, Nadal-Ginard B (1993): Reversal of terminal differentiation and control of DNA replication: Cyclin A and cdk2 specifically localize at subnuclear sites of DNA replication. Cell 74:979–992.

Carpenter PB, Mueller PR, Dunphy WG (1996): Role for a *Xenopus* Orc2-related protein in controlling DNA replication. Nature 379:357–360.

Carri MT, Micheli G, Graziano E, Pace T, Buongiorno-Nardelli M (1986): The relationship between chromosomal origins of replication and the nuclear matrix during the cell cycle. Exp Cell Res 164:426–436.

Carroll SM, DeRose ML, Kolman JJ, Nonet GH, Kelly RE, Wahl GM (1993): Localization of a bidirectional DNA replication origin in the wild type and in episomally amplified murine ADA loci. Mol Cell Biol 13:2971–2981.

Castellot JJ Jr, Miller MR, Pardee AB (1978): Animal cells reversibly permeable to small molecules. Proc Natl Acad Sci USA 75:351–355.

Celniker SE, Sweder K, Srienc F, Bailey JE, Campbell JL (1984): Deletion mutations affecting autonomously replicating sequence ARS1 of *Saccharomyces cerevisiae.* Mol Cell Biol 4:2455–2466

Challberg MD, Kelly TJ (1989): Animal virus DNA replication. Annu Rev Biochem 58:671–717.

Chang LMS (1977): DNA polymerases from bakers' yeast. J Biol Chem 252:1873–1880.

Chen Y, Hennessy KM, Bostein D, Tye BK (1992): CDC46/MCM5, a yeast protein whose subcellular localization is cell cycle–regulated, is involved in

DNA replication at autonomously replicating sequences. Proc Natl Acad Sci USA 89:10459–10463.

Chiang CS, Mitsis PG, Lehman IR (1993): DNA polymerase δ from embryos of *Drosophila melanogaster.* Proc Nat Acad Sci USA 90:9105–9109.

Chiu C-S, Cook KS, Greenberg GR (1982): Characteristics of a bacteriophage T4-induced complex synthesizing deoxynucleotides. J Biol Chem 257:15087–15097.

Chiu RC, Baril EF (1975): Nuclear DNA polymerases and the HeLa cell cycle. J Biol Chem 250:7951–7957.

Chong PJ, Mahbubani HM, Khoo CY, Blow JJ (1995): Purification of an Mcm-containing complex as a component of the DNA replication licensing factor. Nature 375:418–421.

Chong PJ, Thommes P, Blow JJ (1996): The role of MCM/P1 protein in the licensing of DNA replication. Trends Biol Sci 21:102–106.

Ciechanover A (1994): The ubiquitin–proteosome proteolytic pathway. Cell 79:13–21.

Clyne RK, Kelly TJ (1995): Genetic analysis of an ARS element from the fission yeast *Schizosaccharomyces pombe.* EMBO J 14:6348–6357.

Cocker JH, Piatti S, Santocanale C, Nasmyth K, Diffley JFX (1996): An essential role for the cdc6 protein in forming the prereplicative complexes in budding yeast. Nature 379:180–182.

Coleman TR, Carpenter PB, Dunphy WG (1996): The *Xenopus* Cdc6 protein is essential for the initiation of a single round of DNA replication in cell-free extracts. Cell 87:53–63.

Collins JM, Chu AK (1987): Binding of the DNA polymerase alpha–DNA primase complex to the nuclear matrix in HeLa cells. Biochemistry 26:5600–5607.

Collins KL, Russo AA, Tseng BY, Kelly TJ (1993): The role of the 70 kDa subunit of human DNA polymerase α in DNA replication. EMBO J 12:4555–4566.

Conrad MN, Newlon CS (1983): *Saccharomyces cerevisiae* cdc2 mutants fail to replicate approximately one-third of their nuclear genome. Mol Cell Biol 3:1000–1012.

Copeland WC, Wang TS-F (1991): Catalytic subunit of human DNA polymerase α overproduced from baculovirus-infected insect cells. Structural and enzymological characterization. J Biol Chem 266:22739–22748.

Copeland WC, Wang TS-F (1993): Enzymatic characterization of the individual mammalian primase subunits reveals a biphasic mechanism for initiation of DNA replication. J Biol Chem 268:26179–26189.

Coppock DL, Pardee AB (1987): Control of thymidine kinase mRNA during the cell cycle. Mol Cell Biol 7:2925–2932.

Correa-Bordes J, Nurse P (1996): $p25^{ruml}$ orders S phase and mitosis by acting as an inhibitor of the $p34^{cdc2}$ mitotic kinase. Cell 83:1001–1009.

Cotterill SM, Reyland ME, Loeb LA, Lehman IR (1987): A cryptic proofreading $3' \rightarrow 5'$ exonuclease associated with polymerase subunit of the DNA polymerase–primase from *Drosophila melanogaster.* Proc Natl Acad Sci USA 84:5635–5639.

Coverley D, Downes CS, Romanowski P, Laskey RA (1993): Reversible effects of nuclear membrane permeabilization on DNA replication: Evidence for a positive licensing factor. J Cell Biol 122:985–992.

Cox LS (1992): DNA replication in cell-free extracts from *Xenopus* eggs is prevented by disrupting nuclear envelop function. J Cell Sci 101:43–53.

Coxon A, Maundrell K, Kearsey SE (1992): Fission yeast cdc21$^+$ belongs to a family of proteins involved in an early step of chromosome replication. Nucleic Acids Res 20:5571–5577.

Crute JJ, Wahl AF, Bambara RA (1986): Purification and characterization of two new high molecular weight forms of DNA polymerase. Biochemistry 25:26–36.

Dahmann C, Diffley JFX, Nasmyth K (1995): S phase–promoting cyclin-dependent kinases prevent re-replication by inhibiting the transition of replication origins to a pre-replicative state. Curr Biol 5:1257–1269.

Dalton S (1992): Cell cycle regulation of the human cdc2 gene. EMBO J 11:1797–1804.

Danenberg KD, Danenberg PV (1989): Activity of thymidylate synthase and its inhibition by 5-fluorouracil in highly enzyme-overproducing cells resistant to 10-propargyl-5, 8 dideazafolate. Mol Pharmacol 36:219–223.

Dean FB, Dodson M, Echols H, Hurwitz J (1987): ATP-dependent formation of a specialized nucleoprotein structure by simian virus 40 (SV40) large antigen at SV40 replication origin. Proc Natl Acad Sci USA 84:8981–8985.

Delidakis C, Kafatos FC (1989): Amplification enhancers and replication origins in the autosomal chorion gene cluster of *Drosophila.* EMBO J 8:891–901.

Deng C, Zang P, Harper JW, Elledge SJ, Leder P (1995): Mice lacking p21$^{cip\ 1/waf1}$ undergo normal development, but are defective in G_1 check point control. Cell 82:675–684.

DePamphilis ML (1987): Replication of simian virus 40 and polyoma virus chromosomes. In Aloni Y (ed): Molecular Aspects of Papovaviruses. Boston: Martinus Nijhoff Publishing, pp 1–40.

DePamphilis ML (1993a): Eukaryotic DNA replication: Anatomy of an origin. Annu Rev Biochem 62:29–63.

DePamphilis ML (1993b): Origins of DNA replication that function in eukaryotic cells. Curr Opin Cell Biol 5:434–441.

DePamphilis ML (1993c): Origins of DNA replication in metazoan chromosomes. J Biol Chem 268:1–4.

Deshaies RJ, Chau V, Kirschner M (1995): Ubiquitination of the G_1 cyclin Cln2p by a Cdc34p-dependent pathway. EMBO J 14:303–312.

Deshpande A, Newlon CS (1992): The ARS consensus sequence is required for chromosomal origin function in *Saccharomyces cerevisiae.* Mol Cell Biol 12:4305–4313.

Devoto SH, Mudryj M, Pines J, Hunter T, Nevins JR (1992): A cyclin A–protein kinase complex possesses sequence-specific DNA binding activity. p33cdk2 is a component of the E2F–cyclin A complex. Cell 68:167–176.

Dhar V, Skoultchi AI, Schildkraut CL (1989): Activation and repression of a β-globin gene in cell hybrids is accompanied by a shift in the temporal replication. Mol Cell Biol 9:3524–3532.

Dickinson LA, Joh T, Kohwi Y, Kohwi-Shigematsu T. (1992): A tissue specific MAR/SAR DNA-binding protein with unusual binding site recognition. Cell 70:631–645.

Diffley JFX, Cocker JH (1992): Protein–DNA interactions at a yeast replication origin. Nature 357:169–172.

Diffley JFX, Cocker JH, Dowell SJ, Rowley A (1994): Two steps in the assembly of complexes at yeast replication origins in vivo. Cell 78:303–316.

Diffley JFX, Stillman B (1988): Purification of a yeast protein that binds to origins of DNA replication and a transcriptional silencer. Proc Natl Acad Sci USA 85:2120–2124.

Dijkwel PA, Hamlin JL (1988): Matrix attachment regions are positioned near replication initiation sites, genes, and an interamplicon junction in the amplified dihydrofolate reductase domain of Chinese hamster ovary cells. Mol Cell Biol 12:5398–5409.

Dijkwel PA, Hamlin JL (1992): Initiation of DNA replication in the dihydrofolate reductase locus is confined to the early S period in CHO cells synchronized with plant amino acid mimosine. Mol Cell Biol 12:3715–3722.

Dijkwel PA, Hamlin JL (1995): The Chinese hamster dihydrofolate reductase origin consists of multiple potential nascent-strand start sites. Mol Cell Biol 15:3023–3031.

Dijkwel PA, Hamlin JL (1996): Sequence and context effects on origin function in mammalian cells. J Cell Biochem 62:210–222.

Dijkwel PA, Mullenders LHF, Wanka F (1979): Analysis of the attachment of replicating DNA to a nuclear matrix in mammalian interphase nuclei. Nucleic Acids Res 6:219–230.

Dijkwel PA, Wenink PW, Poddighe J (1986): Permanent attachment of replication origins to the nuclear matrix in BHK cells. Nucleic Acids Res 14:3241–3249.

Dirick L, Bohm T, Nasmyth K (1995): Roles and regulation of cln–cdc28 kinases at the start of the cell cycle in *Saccharomyces cerevisiae.* EMBO J 14:4803–4813.

Dodson M, Dean FB, Bullock P, Echols H, Hurwitz J (1987): Unwinding of duplex DNA from the SV40 origin of replication by T antigen. Science 238:964–967.

Donovan JD, Toyn JH, Johnson AL, Johnston LH (1994): P40SDB25, a putative cdk inhibitor, has a role in the M/G_1 transition in *Saccharomyces cerevisiae.* Genes Dev 8:1640–1653.

Dou Q-F, Levin AH, Zhao S, Pardee AB (1993): Cyclin E and cyclin A as candidates for the restriction point protein. Cancer Res 53:1493–1497.

Dou Q-F, Markell PJ, Pardee AB (1992): Thymidine kinase transcription is regulated at G_1/S phase by a complex that contains retinoblastoma-like protein and a cdc2 kinase. Proc Natl Acad Sci USA 89:3256–3260.

Dowell SJ, Romanowski P, Diffley JFX (1994): Interaction of Dbf4, the Cdc7 protein kinase regulatory subunit, with yeast replication origins in vivo. Science 265:1243–1246.

Downey KM, Tan CK, So AG (1990): DNA polymerase δ: A second eukaryotic DNA replicase. BioEssays 12:231–236.

Dresler SL, Frattini MG (1986): DNA replication and UV-induced repair synthesis in human fibroblasts are much less sensitive than DNA polymerase α to inhibition by butylphenyl-deoxyguanosine triphosphate. Nucleic Acids 17:7093–7102.

Dubey DD, Kim S-M, Todorov IT, Huberman JA (1996): Large, complex modular structure of a fission yeast DNA replication origin. Curr Biol 6:467–473.

Dubey DD, Raman R (1987): Factors influencing replicon organization in tissues having different S phase duration in the mole rat, *Dandicota bengalensis.* Chromosoma 95:285–289.

Dulic V, Lees E, Reed SI (1992): Association of human cyclin E with a periodic G_1-S phase protein kinase. Science 257:1958–1961.

Ecker RE, Smith LD (1971): The nature and fate of *Rana pipiens* proteins synthesized during maturation and early cleavage. Dev Biol 24:559–576.

El-Deiry WS, Tokino T, Velculescu VE, Levy DB, Parsons R, Trent JM, Lin D, Mercer WE, Kinzler KW, Vogelstein B (1993): WAF1 a potential mediator of p53 tumor suppression. Cell 75:817–825.

Engstrom Y, Rozell B (1988): Immunocytochemical evidence for the cytoplasmic localization and differential expression during the cell cycle of the M1 and M2 subunits of mammalian ribonucleotide reductase. EMBO J 7:1615–1620.

Engstrom Y, Rozell B, Hansson H-A, Stemme S, Thelander L (1984): Localization of ribonucleotide reductase in mammalian cells. EMBO J 3:863–867.

Fang F, Newport JW (1991): Evidence that the G_1-S and G_2-M transitions are controlled by different cdc2 proteins in higher eukaryotes. Cell 66:731–742.

Fangman WL, Brewer BJ (1991): Activation of replication origin within yeast chromosomes. Annu Rev Cell Biol 7:375–402.

Ferguson BM, Brewer BJ, Reynolds AE, Fangman WL (1991): A yeast origin of replication is activated late in S phase. Cell 65:507–515.

Ferguson BM, Fangman WL (1992): A position effect on the time of replication origin activation in yeast. Cell 68:333–339.

Firpo EF, Koff A, Solomon MJ, Roberts JM (1994): Inactivation of Cdk2 inhibitor during interleukin 2-induced proliferation of human T lymphocytes. Mol Cell Biol 14:4889–4901.

Firschein W, Gillmor RG (1970): DNA–membrane complex: Macromolecular content and stimulation of enzymatic activity by polyadenylic acid. Science 169:66–68.

Fisher DL, Nurse P (1996): A single fission yeast mitotic cyclin B p34cdc2 kinase promotes both S phase and mitosis in the absence of G_1 cyclins. EMBO J 15:850–860.

Fitch I, Dahmann C, Surana U, Amon A, Nasmyth K, Goetsch L, Byers B, Futcher B (1992): Characterization of four B-type cyclin genes of the budding yeast *Saccharomyces cerevisiae.* Mol Biol Cell 3:805–818.

Forsburg SL, Nurse P (1994): The fission yeast cdc19$^+$ gene encodes a member of the MCM family of replication proteins. J Cell Sci 107:2779–2788.

Fotedar R, Roberts JM (1991): Association of p34^{cdc2} with replicating DNA. Cold Spring Harbor Symp Quant Biol 56:325–333.

Foury F (1989): Cloning and sequencing of the nuclear gene MIP1 encoding the catalytic subunit of the yeast mitochondrial DNA polymerase. J Biol Chem 264:20552–20560.

Francesconi S, Copeland WC, Wang TS-F (1993a): In vivo species specificity of DNA polymerase α. Mol Gen Genet 241:457–466.

Francesconi S, Park H, Wang TS-F (1993b): Fission yeast with DNA polymerase δ temperature-sensitive alleles exhibits cell division cycle phenotype. Nucleic Acids Res 21:3821–3828.

Fridland A (1973): DNA precursors in eukaryotic cells. Nature New Biol 243:105–107.

Fry M, Loeb LA (1986): Animal Cell DNA Polymerases. Boca Raton, FL. CRC Press, Inc.

Fuchs E, Hanawalt P (1970): Isolation and characterization of the DNA replication complex from *Escherichia coli.* J Mol Biol 52:301–322.

Gale JM, Tobey RA, D'Anna JA (1992): Localization and DNA sequence of a replication origin in the rhodopsin gene locus of Chinese hamster cells. J Mol Biol 224:343–352.

Gavin KA, Hidaka M, Stillman B (1995): Conserved initiator proteins in eukaryotes. Science 270:1667–1671.

Geng Y, Eaton EN, Picon M, Roberts JM, Lundberg AS, Gifford A, Sardet C, Weinberg RA (1996): Regulation of cyclin E transcription by E2Fs and retinoblastoma protein. Oncogene 12:1173–1180.

Gibson SI, Surosky RT, Tye BK (1990): The phenotype of minichromosome mutant MCM3 is characteristic of mutants defective in DNA replication. Mol Cell Biol 10:5707–5720.

Gilbert DM, Miyazawa H, DePamphilis ML (1995): Site-specific initiation of DNA replication in *Xenopus* egg extract requires nuclear structure. Mol Cell Biol 15:2942–2954.

Gilbert DM, Miyazawa H, Nallaseth FS, Ortega JM, Blow JJ, DePamphilis ML (1993): Site-specific initiation of DNA replication in metazoan chromosomes and the role of nuclear organization. Cold Spring Harbor Symp Quant Biol 58:475–485.

Girard F, Stausfeld U, Fernandez A, Lamb NJ (1991): Cyclin A is required for the onset of DNA replication in mammalian fibroblasts. Cell 67:1169–1179.

Glotzer M, Murray AW, Kirschner MW (1991): Cyclin is degraded by the ubiquitin pathway. Nature 349:132–138.

Goebl MG, Yochem J, Jentsch S, McGrath JP, Varshavsky A, Byers B (1988): The yeast cell cycle gene cdc34 encodes ubiquitin-conjugating enzyme. Science 241:1331–1335.

Gordenin DA, Malkova AL, Peterzen A, Kulikov VN, Pavlov YI, Perkins E, Resnick MA (1992): Transposon Tn5 excision in yeast: Influence of DNA polymerases alpha, delta, and epsilon and repair genes. Proc Natl Acad Sci USA 89:3785–3789.

Gossen M, Pak DTS, Hansen SK, Acharya JK, Botchan MR (1995): A *Drosophila* homolog of the yeast origin recognition complex. Science 270:1674–1677.

Goulian M, Herrmann SM, Sackett JW, Grimm SL (1990): Two forms of DNA polymerase δ from mouse cells. Purification and properties. J Biol Chem 265:16402–16411.

Gu Y, Turck CW, Morgan D (1993): Inhibition of cdk2 activity in vivo by an associated 20k regulatory subunit. Nature 366:707–710.

Guan K-L, Jenkins CW, Li Y, Nichols MA, Wu X, O'Keefe CL, Matera AG, Xiong Y (1994): Growth suppression by p18, a p16INK4/MTS1- and p14INK4B/MTS2-related cdk6 inhibitor, correlates with wild-type pRB function. Genes Dev 8:2939–2952.

Gyuris JEG, Goelemis E, Chertkov H, Brent R (1993): Cdi1, a human G_1 and S phase protein phosphatase that associates with Cdk2. Cell 75:791–804.

Hamaguchi JR, Tobey RA, Pines J, Crissman HA, Hunter T, Bradbury EM (1992): Requirement for $p34^{cdc2}$ kinase is restricted to mitosis in the mammalian cdc2 mutant FT210. J Cell Biol 117:1041–1053.

Hamatake RK, Hasgawa H, Clark AB, Bebenek K, Kunkel TA, Sugino A (1990): Purification and characterization of DNA polymerase II from the yeast *Saccharomyces cerevisiae.* Identification of the catalytic core and a possible holoenzyme form of the enzyme. J Biol Chem 265:4072–4083.

Hamlin JL, Dijkwel PA (1995): On the nature of replication origins in higher eukaryotes. Curr Opin Genet Dev 5:153–161.

Hammond RA, Miller MR, Gray MS, Reddy GPV (1989): Association of $3' \rightarrow 5'$ exodeoxyribonuclease activity with DNA replitase complex from S phase Chinese hamster embryo fibroblast cells. Exp Cell Res 183:284–293.

Hand R (1978): Eukaryotic DNA: Organization of the genome for replication. Cell 15:317–325.

Handeli S, Klar A, Meuth M, Cedar H (1989): Mapping replication units in animal cells. Cell 57:909–920.

Hannon GJ, Beach D (1994): P15INK4B is a potential effector of TGF-β–induced cell cycle arrest. Nature 371:257–261.

Harper JW, Adami GR, Wei N, Keyomarsi K, Elledge S (1993): The p21 cdk-interacting protein Cip1 is a potent inhibitor of G_1 cyclin-dependent kinases. Cell 75:805–816.

Hartwell LH (1971): Genetic control of the cell division cycle in yeast II. Genes controlling DNA replication and its initiation. J Mol Biol 59:183–194.

Hartwell LH, Culotti J, Reid B (1970): Genetic control of the cell-division cycle in yeast, I. Detection of mutants. Proc Natl Acad Sci USA 66:352–359.

Harvey G, Pearson CK (1988): Search for multienzyme complexes for DNA precursor pathways in uninfected mammalian cells and in cells infected with herpes simplex virus type-1. J Cell Physiol 134:25–36.

Hassan AB, Cook PR (1993): Visualization of replication sites in unfixed human cells. J Cell Sci 105:541–550.

Hayles J, Fisher D, Wollard A, Nurse P (1994): Temporal order of S phase and mitosis in fission yeast is determined by the state of the p34cdc2–mitotic B cyclin complex. Cell 78:813–822.

Heck MMS, Spradling AC (1990): Multiple replication origins are used during *Drosophila* chorion gene amplification. J Cell Biol 4:903–914.

Heintz NH, Hamlin JL (1982): An amplified chromosomal sequence that includes the gene for dihydrofolate reductase initiates replication within specific restriction fragments. Proc Natl Acad Sci USA 79:4083–4087.

Heinzel SS, Krysan PJ, Tran CT, Calos MP (1991): Autonomous DNA replication in human cells is affected by the size and the source of the DNA. Mol Cell Biol 11:2263–2272.

Hennessy KM, Clark CD, Botstein D (1990): Subcellular localization of yeast CDC46 varies with the cell cycle. Genes Dev 4:2252–2263.

Hennessy KM, Lee A, Chen E, Botstein D (1991): A group of interacting yeast DNA replication genes. Genes Dev 5:958–969.

Henricksen LA, Wold MS (1994): Replication protein A mutants lacking phosphorylation sites for $p34^{cdc2}$ kinase support DNA replication. J Biol Chem 269:24203–24208.

Hereford LM, Hartwell LH (1971): Defective DNA synthesis in permeabilized yeast mutants. Nature New Biol 234:171–172.

Hereford LM, Hartwell LH (1974): Sequential gene function in the initiation of *Saccharomyces cerevisiae* DNA synthesis. J Mol Biol 84:445–461.

Hilt W, Wolf DH (1996): Proteosomes: Destruction as a programme. Trends Biochem Sci 21:96–102.

Hinds PW, Mittnacht S, Dulic V, Arnold A, Reed SI, Weinberg RA (1992): Regulation of retinoblastoma protein functions by ectopic expression of human cyclins. Cell 70:993–1006.

Hofmann JF-X, Gasser SM (1991): Identification and purification of a protein that binds the yeast ARS consensus sequence. Cell 64:951–960.

Hogan E, Koshland D (1992): Addition of extra origins of replication to a minichromosome suppresses its mitotic loss in Cdc6 and Cdc14 mutants of *Saccharomyces cerevisiae.* Proc Natl Acad Sci USA 89:3098–3102.

Holmquist G, Gray M, Porter T, Jordan J (1982): Characterization of Giemsa dark- and light-band DNA. Cell 31:121–129.

Hozak P, Hassan AB, Jackson DA, Cook PR (1993): Visualization of replication factories attached to a nucleoskeleton. Cell 73:361–373.

Hozak P, Jackson DA, Cook PR (1994): Replication factories and nuclear bodies: The ultra structural characterization of replication sites during the cell cycle. J Cell Sci 107:2191–2202.

Hsi KL, Copeland WC, Wang TS-F (1990): Human DNA polymerase α catalytic polypeptide binds Con A and RCA and contains a specific labile site in the N-terminus. Nucleic Acids Res 18:6231–6237.

Hu B, Burkhart R, Schulte D, Musahl C, Knippers R (1993): The P1 family: A new class of mammalian proteins related to the yeast Mcm replication proteins. Nucleic Acids Res 21:5289–5293.

Huang R-Y, Kowalski D (1996): Multiple DNA elements in ARS 305 determine replication origin activity in a yeast chromosome. Nucleic Acids Res 24:816–823.

Huberman JA (1994): Analysis of DNA replication origins and directions by two-dimensional gel electrophoresis. In Fantes P, Brooks RF (eds): The Cell Cycle: A Practical Approach. New York: Oxford University Press, pp 213–234.

Huberman JA, Riggs AD (1966): Autoradiography of chromosomal DNA fibers from Chinese hamster cells. Proc Natl Acad Sci USA 55:599–606.

Huberman JH, Riggs AD (1968): On the mechanism of DNA replication in mammalian chromosomes. J Mol Biol 32:327–341.

Huberman JA, Zhu J, Davis LR, Newlon CS (1988): Close association of a DNA replication origin and an ARS element on chromosome III of the yeast, *Saccharomyces cerevisiae.* Nucleic Acids Res 14:6373–6384.

Hunter T (1993): Braking the cycle. Cell 75:839–841.

Hunter T, Pines J (1994): Cyclins and cancer. II. Cyclin D and CDK inhibitors come of age. Cell 79:573–582.

Hutchison CJ, Bridger JM, Cox LS, Kill IR (1994): Weaving a pattern from disparate threads: Lamin function in nuclear assembly and replication. J Cell Sci 107:3259–3269.

Hutchison CJ, Cox R, Drepaul RS, Gomperts M, Ford CC (1987): Periodic DNA synthesis in cell-free extracts of *Xenopus* eggs. EMBO J 6:2003–2010.

Hutchison CJ, Kill IR (1989): Changes in the nuclear distribution of DNA polymerase alpha and PCNA/cyclin during the progress of the cell cycle, in a cell-free extract of *Xenopus* eggs. J Cell Sci 93:605–613.

Hyrien O, Maric C, Mechali M (1995): Transition in specification of embryonic metazoan DNA replication origins. Science 270:994–997.

Jackson AL, Pahl PMB, Harrison K, Rosamond J, Sclafani RA (1993): Cell cycle regulation of the yeast Cdc7 protein kinase by association with the Dbf4 protein. Mol Cell Biol 13:2899–2908.

Jackson DA, Cook PR (1986a): Replication occurs at a nucleoskeleton. EMBO J 5:1403–1410.

Jackson DA, Cook PR (1986b): A cell cycle–dependent DNA polymerase activity that replicates intact DNA in chromatin. J Mol Biol 192:65–67.

Jackson PK, Chevalier S, Phillipe M, Kirschner MW (1995): Early events in DNA replication require cyclin E and are blocked by p21 Cip1. J Cell Biol 130:755–769.

Jacob F, Brenner S, Cuzin F (1963): On the regulation of DNA replication in bacteria. Cold Spring Harbor Symp Quant Biol 28:329–348.

Jallepalli PV, Kelly TJ (1996): Rum 1 and cdc18 link inhibition of cyclin-dependent kinase to the initiation of DNA replication in *Schizosaccharomyces pombe.* Genes Dev 10:541–552.

Jaumot M, Grana X, Giordano A, Reddy GPV, Agell N, Bachs O (1994): Cyclin/cdk2 complexes in the nucleus of HeLa cells. Biochem Biophys Res Commun 203:1527–1534.

Jazwinski SM, Edelman GM (1982): Protein complexes from active replicative fractions associate in vitro with the replication origins of yeast 2-μm DNA plasmid. Proc Natl Acad Sci USA 79:3428–3432.

Jazwinski SM, Edelman GM (1984): Evidence for participation of a multi-protein complex in yeast DNA replication in vitro. J Biol Chem 259:6852–6857.

Jenkins H, Holman T, Lyon C, Lane B, Stick R, Hutchison C (1993): Nuclei that lack a lamina accumulate karyophilic proteins and assemble a nuclear matrix. J Cell Sci 106:275–285.

Johnson DG, Ohtani K, Nevins JR (1994): Autoregulatory control of E2F-1 expression in response to positive and negative regulators of cell cycle progression. Genes Dev 8:1514–1525.

Jong AYS, Kuo CL, Cambell JL (1984): The CDC8 gene of yeast encodes thymidylate kinase. J Biol Chem 259:11052–11059.

Kaguni LS, Lehman IR (1988): Eukaryotic DNA polymerase–primase: Structure, mechanism and function. Biochim Biophys Acta 950:87–101.

Kaguni LS, Rossignol JM, Conaway RC, Lehman IR (1983): Isolation of an intact DNA polymerase–primase from embryos of *Drosophila melanogaster.* Proc Natl Acad Sci USA 80:2221–2225.

Karawya E, Swack J, Albert W, Fedorko J, Minna J, Wilson SH (1984): Identification of a higher molecular weight DNA polymerase α catalytic polypeptide in monkey cells by monoclonal antibody. Proc Natl Acad Sci USA 81:7777–7781.

Karlseder J, Rotheneder H, Wintersberger E (1996): Interaction of Sp1 with the growth and cell cycle-regulated transcription factor E2F. Mol Cell Biol 16:1659–1667.

Kato JY, Matsuoka M, Polyak K, Massague J, Sherr CJ (1994): Cyclic AMP-induced G1 phase arrest mediated by an inhibitor (p27 Kip1) of cyclin-dependent kinase 4 activation. Cell 79:487–496.

Kearsey S (1984): Structural requirements for the function of a yeast chromosomal replicator. Cell 37:299–307.

Kelly RE, DeRose ML, Draper BW, Wahl GM (1995): Identification of an origin of bidirectional DNA replication in the ubiquitously expressed mammalian CAD gene. Mol Cell Biol 15:4136–4148.

Kelly TJ, Martin GS, Forsburg SL, Stephen RJ, Russo A, Nurse P (1993): The fission yeast $cdc18^+$ gene product couples S phase to START and mitosis. Cell 74:371–382.

Kesti T, Frantti H, Syvaoja JE (1993): Molecular cloning of the cDNA for the catalytic subunit of human DNA polymerase ε. J Biol Chem 268:10238–10245.

Khatsernova BY, Silaeva SA, Sheremetevskaya TN (1983): Interrelationship between activity of nuclear ribonucleotide reductase and DNA synthesis of hepatocyte nuclei during liver tissue regeneration. Vopr Med Khim 29:61–64.

Kill IR, Bridger JM, Campbell KHS, Maldonado-Codina G, Hutchison CJ (1991): The timing of the formation and usage of replicase clusters in S phase nuclei of human diploid fibrolasts. J Cell Sci 100:869–879.

Kimura H, Nozak N, Sugimoto K (1994): DNA polymerase alpha associated protein P1, a murine homolog of yeast MCM3, changes in intranuclear distribution during the DNA synthetic period. EMBO J 13:4311–4320.

Kitada K, Johnston LH, Sugino T, Sugino A (1992): Temperature sensitive Cdc7 mutations of *Saccharomyces cerevisiae* are suppressed by the DBF4 gene, which is required for the G^1/S cell cycle transition. Genetics 131:21–29.

Kitsberg D, Selig S, Keshet I, Cedar H (1993): Replication structure of the human β-globin gene domain. Nature 366:588–590.

Knoblich JA, Lehner CF (1993): Synergistic action of *Drosophila* cyclins A and B during the G_2-M transition. EMBO J 12:65–74.

Knoblich JA, Sauer K, Jones L, Richardson H, Saint R, Lehner CF (1994): Cyclin E controls S phase progression and its down-regulation during *Drosophila* embryogenesis is required for the arrest of cell proliferation. Cell 77:107–120.

Koff A, Giordano A, Desai D, Yamashita K, Harper JW, Elledge S, Nishimoto T, Morgan DO, Fraza BR, Roberts JM (1992): Formation and activation of a cyclin E–cdk2 complex during the G1 phase of the human cell cycle. Science 257:1689–1694.

Koff A, Ohtsuki M, Polyak K, Roberts JM, Massague J (1993): Negative regulation of G1 in mammalian cells: Inhibition of cyclin E–dependent kinase by TGF-β. Science 260:536–539.

Koh J, Enders GH, Dynlacht BD, Harlow E (1995): Tumor-derived p16 alleles encoding proteins defective in cell cycle inhibition. Nature 375:506–510.

Koonin EV (1993): A common set of conserved motifs in a vast variety of putatitive nucleic acid–dependent ATPases including MCM proteins involved in the initiation of eukaryotic DNA replication. Nucleic Acids Res 21:2541–2547.

Kornberg A, Baker TA (1992): DNA Replication, 2nd ed. New York: WH Freeman.

Krek W, Ewen ME, Shirodkar S, Arany Z, Kaelin WG Jr, Livingston DM (1994): Negative regulation of the growth promoting transcription factor E2F-1 by a stably bound cyclin A–dependent protein kinase. Cell 78:161–172.

Krysan PJ, Haase SB, Calos MP (1989): Isolation of human sequences that replicate autonomously in human cells. Mol Cell Biol 9:1026–1033.

Krysan PJ, Smith JG, Calos MP (1993): Autonomous replication in human cells of multimers of specific human and bacterial sequences. Mol Cell Biol 13:2688–2696.

Kubota Y, Mimura S, Nishimoto SI, Takisawa H, Nojima H (1995): Identification of the yeast Mcm3-related protein as a component of *Xenopus* DNA replication licensing factor. Cell 81:601–609.

Kucera R, Paulus H (1986): Localization of the deoxyribonucleotide biosynthetic enzymes ribonucleotide reductase and thymidylate synthase in mouse L cells. Exp Cell Res 167:417–428.

Kuhne C, Linder P (1993): A new pair of B-type cyclins from Saccharomyces cerevisiae that function early in the cell cycle. EMBO J 12:3437–3447.

Kuno K, Murakami S, Kuno S (1990): Single-strand–binding factor(s) which interact with ARS1 of *Saccharomyces cerevisiae.* Gene 95:73–77.

Lam E W-F, Watson RJ (1993): An E2F-binding site mediates cell-cycle regulated repression of mouse B–myb transcription. EMBO J 7:2705–2713.

Lasko DD, Tomkinson AE, Lindahl T (1990): Biosynthesis and intracellular localization of DNA ligase I. J Biol Chem 265:12618–12622.

La Thange NB (1994): DRTF1/E2F: An expanding family of heterodimeric transcription factors implicated in cell-cycle control. Trends Biochem Sci 19:108–114.

Leatherwood J, Lopez-Girona A, Russel P (1996): Interaction of Cdc2 and Cdc18 with a fission yeast ORC2-like protein. Nature 379:360–363.

Lee MG, Nurse P (1987): Complimentation used to clone a human homologue of the fission yeast cell cycle control gene cdc2. Nature 327:31–35.

Lee MYWT, Toomey NL, Wright GE (1985): Differential inhibition of human placental DNA polymerases δ and α by BuPdGTP and BuAdATP. Nucleic Acids Res 13:8623–8630.

Lee SH, Pan ZQ, Kwong AD, Burgers PM, Hurwitz J (1991): Synthesis of DNA by DNA polymerase ε in vitro. J Biol Chem 266:22707–22717.

Leeds JM, Mathews CK (1987): Cell cycle–dependent effects on deoxynucleotide and DNA labeling by nucleoside precursors in mammalian cells. Mol Cell Biol 7:532–534.

Leeds JM, Slabaugh MB, Mathews CK (1985): DNA precursor pools and ribonucleotide reductase activity: Distribution between the nucleus and cytoplasm of mammalian cells. Mol Cell Biol 5:3443–3450.

Leem SH, Ropp PA, Sugino A (1994): The yeast *Saccharomyces cerevisiae* DNA polymerase IV: Possible involvement in double strand break DNA repair. Nucleic Acids Res 22:3011–3017.

Lees E, Faha B, Dulic V, Reed SI, Harlow E (1992): Cyclin E/cdk2 and cyclin A/cdk2 kinases associate with p107 and E2F in a temporally distinct manner. Genes Dev 6:1874–1885.

Lehner CF, O'Farrell PH (1989): Expression and function of *Drosophila* cyclin A during embryonic cell cycle progression. Cell 56:957–968.

Lehner CF, O'Farrell PH (1990): The roles of *Drosophila* cyclins A and B in mitotic control. Cell 61:535–547.

Leno GH, Downes CS, Laskey RA (1992): The nuclear membrane prevents replication of human G2 nuclei but not G1 nuclei in *Xenopus* egg extract. Cell 69:151–158.

Leonhardt H, Page AW, Weier H-U, Bestor TH (1992): A targeting sequence directs DNA methyltransferase to sites of DNA replication in mammalian nuclei. Cell 71:865–873.

Leu TH, Hamlin JL (1989): High resolution mapping of replication fork movement through the amplified dihydrofolate reductase domain in CHO cells by in-gel renaturation analysis. Mol Cell Biol 9:523–531.

Li C, Cao L-G, Wang Y-L, Baril EF (1993): Further purification and characterization of a multi-enzyme complex for DNA synthesis in human cells. J Cell Biochem 53:405–419.

Li JJ, Herskowitz I (1993): Isolation of ORC6, a component of the yeast origin recognition complex by a one-hybrid system. Science 262:1870–1874.

Li JJ, Kelly TJ (1984): Simian virus 40 DNA replication in-vitro. Proc Natl Acad Sci USA 81:6973–6977.

Li R, Waga S, Hannon GJ, Beach D, Stillman B (1994): Differential effects by the p21 CDK inhibitor on PCNA dependent DNA replication and DNA repair. Nature 371:534–537.

Liang C, Spitzer JD, Smith HS, Gerbi SA (1993): Replication initiates at a confined region during DNA amplification in *Sciara* DNA puff II/9A. Genes Dev 7:1072–1084.

Liang C, Weinreich M, Stillman B (1995): ORC and cdc6p interact and determine the frequency of initiation of DNA replication in the genome. Cell 81:667–676.

Linskens MH, Huberman JA (1988): Organization of replication of ribosomal DNA in *Saccharomyces cerevisiae.* Mol Cell Biol 8:4927–4935.

Linskens MHK, Huberman JA (1990): The two faces of higher eukaryotic DNA replication origins. Cell 62:845–847.

Little RD, Platt THK, Schildkraut CL (1993): Initiation and termination of DNA replication in human and RNA genes. Mol Cell Biol 13:6600–6610.

Lohka MJ, Hayes MK, Maller JL (1988): Purification of maturation-promoting factor, an intracellular regulator of early mitotic events. Proc Natl Acad Sci USA 85:3009–3013.

Lohka MJ, Masui Y (1984): Roles of cytosol and cytoplasmic particles in nuclear envelop assembly and sperm pronuclear formation in cell-free preparation amphibian eggs. J Cell Biol 98:1222–1230.

Loo S, Fox CA, Rine J, Kobayashi R, Stillman B, Bell S (1995): The origin recognition complex in silencing, cell cycle progression, and DNA replication. Mol Biol Cell 6:741–756.

Looney JE, Hamlin JL (1987): Isolation of the amplified dihydrofolate reductase domain from methotrexate-resistant Chinese hamster ovary cells. Mol Cell Biol 7:569–577.

Lovec H, Grzeschiczek A, Kowalski MB, Moroy T (1994): Cyclin D1/bcl-1 cooperates with myc genes in the generation of B-cell lymphoma in transgenic mice. EMBO J 13:3487–3495.

Luca FC, Shibuya EK, Dohrmann CE, Ruderman JV (1991): Both cyclin AΔ60 and BΔ97 are stable and arrest cells in M-phase, but only cyclin BΔ97 turns on cyclin destruction. EMBO J 10:4311–4320.

Lukas J, Muller H, Bartkova J, Spitkovsky D, Kjerulff AA, Durr PJ, Strauss M, Bartek J (1994): DNA tumor virus oncoproteins and retinoblastoma gene mutations share the ability to relieve the cells requirement for cyclin D1 function in G1. J Cell Biol 125:625–638.

Lukas J, Parry D, Aagaard L, Mann DJ, Bartkova J, Strauss M, Peters G, Bartek

J (1995): Retinoblastroma–protein-dependent cell cycle inhibition by the tumor suppressor p16. Nature 375:503–506.

Madine MA, Khoo C-Y, Mills AD, Laskey RA (1995a): MCM3 complex required for cell cycle regulation of DNA replication in vertebrate cells. Nature 375:421–424.

Madine MA, Khoo C-Y, Mills AD, Laskey RA (1995b): Nuclear envelop prevents replication by restricting binding of MCM3 to chromatin. Curr Biol 5:1270–1279.

Madsen P, Celis JE (1985): S phase patterns of cyclin (PCNA) antigen staining resemble topographical patterns of DNA synthesis. A role for cyclin in DNA replication? FEBS Lett 193:5–11.

Maine GT, Sinha P, Tye BK (1984): Mutants of *S. cerevisiae* defective in maintenance of minichromosomes. Genetics 106:365–385.

Maiorano D, Blom van Assendelft G, Hu B, Burkhart R, Schulte D, Musahl C, Knippers R (1993): The P1 family: A new class of mammalian proteins related to the yeast Mcm replication proteins. Nucleic Acids Res 21:5289–5293.

Maiorano D, Blom van Assendelft G, Kearsey SE (1996): Behaviour of Cdc21 in the fission yeast cell cycle: Requirement for S phase entry and coordination of mitosis with DNA replication. EMBO J 15:861–872.

Maiti AK, Sinha P (1992): The MCM2 mutation of yeast affects replication, rather than segregation or amplification of the two-micron plasmid. J Mol Biol 224:545–558.

Malkas LH, Hickey RJ, Baril EF (1990): Multienzyme complexes for DNA synthesis in eukaryotes: P-4 revisited. In Strauss P, Wilson S (eds): The Eukaryotic Nucleus. Molecular Biochemistry and Macromolecular Assemblies, vol 1. Caldwell, NJ: Teleford Press, pp 45–68.

Marahrens Y, Stillman B (1992): A yeast chromosomal origin of DNA replication defined by multiple functional elements. Science 255:817–823.

Martin-Castellanos C, Labib K, Moreno S (1996): B-type cyclins regulate G_1 progression in fission yeast in opposition to the $p25^{rum1}$ cdk inhibitor. EMBO J 15:839–849.

Masui Y, Markert CL (1971): Cytoplasmic control of nuclear behavior during meiotic maturation of frog oocytes. J Exp Zoo 177:129–145.

Matherly LH, Schuetz JD, Westin E, Goldman ID (1989): A method for synchronization of cultured cells with aphidicolin: Application to the large-scale synchronization of L1210 cells and the study of the cell regulation of thymidylate synthase and dihydrofolate reductase. Anal Biochem 182:338–345.

Mathews CK, Moen LK, Wang Y, Sargent RG (1988): Intracellular organization of DNA precursor biosynthetic enzymes. Trends Biochem Sci 13:394–397.

Mathias N, Johnson S, Winey M, Adams AEM, Goetsch L, Pringle JR, Byers B, Goebl MG (1996): Cdc53 acts in concert with cdc4p and cdc34p to control the G1- to S-phase transition and identifies a conserved family of proteins. Mol Cell Biol 16:6634–6643.

McCready SJ, Godwin J, Mason DW, Brazell IA, Cook PR (1980): DNA is replicated at the nuclear cage. J Cell Sci 46:365–386.

McMaken R, Mensa-Wilmot K, Alfano C, Seaby R, Carroll K, Gomes B, Stephens K (1988): Reconstitution of purified protein systems for the replication and regulation of bacteriophage λ DNA replication. Cancer Cells 6:25–34.

Means AL, Slansky JE, McMahon SL, Knuth MW, Farnham PJ (1992): The

HIP1 binding stie is required for growth regulation of the dihydrofolate reductase gene promoter. Mol Cell Biol 12:1054–1063.

Means AR (1994): Calcium, calmodulin and cell cycle regulation. FEBS Lett 347:1–4.

Means AR, Rasmussen CD (1988): Calcium, calmodulin, and cell proliferation. Cell Calcium 9:313–319.

Meier J, Campbell KHS, Ford CC, Stick R, Hutchison CJ (1991): The role of lamin Liic in nuclear assembly and DNA replication, in cell-free extracts of *Xenopus* egg. J Cell Sci 98:271–279.

Mikhailov VS, Tsanev R (1983): Intranuclear localization of DNA polymerase alpha and beta in regenerating rat liver. Int J Biochem 15:855–859.

Milbrandt JD, Heintz NH, White WC, Rothman SM, Hamlin JL (1981): Methotrexate-resistant Chinese hamster cells have amplified a 135-kilobase-pair region that includes the dihydrofolate reductase gene. Proc Natl Acad Sci USA 78:6043–6047.

Mills AD, Blow JJ, White JG, Amos WB, Wilcock D, Laskey RA (1989): Replication occurs at discrete foci spaced throughout nuclei replicating in vitro. J Cell Sci 94:471–477.

Minshull J, Blow JJ, Hunt T (1989): Translation of cyclin mRNA is necessary for extracts of activated *Xenopus* eggs to enter mitosis. Cell 56:947–956.

Mirkovitch J, Mirault ME, Laemmli UK (1984): Organization of the higher-order chromatin loop: Specific DNA attachment sites on nuclear scaffold. Cell 39:223–232.

Miyake S, Okishio N, Samejima I, Hiraoka Y, Toda T, Saitoh I, Yanagida M (1993): Fission yeast genes nda1$^+$ and nda4$^+$, mutations of which lead to S phase block, chromatin alteration and Ca^{2+} suppression, are members of CDC46/MCM2 family. Mol Biol Cell 4:1003–1015.

Moir RD, Montag-Lowy M, Goldman RD (1994): Dynamic properties of nuclear lamins: Lamin B is associated with sites of DNA replication. J Cell Biol 125:1201–1212.

Mondesert O, McGowan CH, Russell P (1996): Cig2, a B-type cyclin, promotes the onset of S in *Schizosaccharomyces pombe.* Mol Cell Biol 16:1527–1533.

Moreno P, Hayles J, Nurse P (1989): Regulation of p34^{cdc2} proteins kinase during mitosis. Cell 58:361–372.

Moreno S, Nurse P (1994): Regulation of progression through the G_1 phase of the cell cycle by the rum1$^+$ gene. Nature 367:236–242.

Morrison A, Araki H, Clark AB, Hamatake RK, Sugino A (1990): A third essential DNA polymerase in *S. cerevisiae*. Cell 62:1143–1151.

Morrison A, Bell J, Kunkel TA, Sugino A (1991): Eukaryotic DNA polymerase amino acid sequence required for $3' \rightarrow 5'$ exonuclease activity. Proc Natl Acad Sci USA 88:9473–9477.

Morrison A, Johnson AL, Johnston LH, Sugino A (1993): Pathway correcting DNA replication errors in *Saccharomyces cerevisiae.* EMBO J 12:1467–1473.

Morrison A, Sugino A (1992): Roles of POL3, POL2 and PMS1 genes in maintaining accurate DNA replication. Chromosoma (suppl) 102:S147–S149.

Morrison A, Sugino A (1994): The $3' \rightarrow 5'$ exonuclease of both DNA polymerases delta and epsilon participate in correcting errors of DNA replication in *Saccharomyces cerevisiae.* Mol Gen Genet 242:289–296.

Mudryj M, Devoto SH, Hiebert SW, Hunter T, Pines J, Nevins JR (1991): Cell cycle regulation of the E2F transcription factor involves an interaction with cyclin A. Cell 65:1243–1253.

Mudryj M, Hiebert SW, Nevins JR (1990): A role for the adenovirus inducible E2F transcription factor in a proliferation dependent signal transduction pathway. EMBO J 9:2179–2184.

Murthy V, Pashupathi K (1995): Isolation and characterization of a multienzyme complex containing DNA replicative enzymes from mitochondria of *S. cerevisiae*. Multienzyme complex from yeast mitochondria. Mol Biol Rep 20:135–141.

Muzi-Falconi M, Brown GW, Kelly TJ (1996): Cdc18^{+} regulates initiation of DNA replication in *Schizosaccharomyces pombe*. Proc Natl Acad Sci USA 93:1666–1670.

Muzi-Falconi M, Kelly TJ (1995): Orp1, a member of the cdc18/cdc6 family of S phase regulators, is homologous to a component of the origin recognition complex. Proc Natl Acad Sci USA 92:12475–12479.

Nakagomi K, Kohwi Y, Dickinson LA, Kohwi-Shigematsu T (1994): A novel DNA binding motif in the nuclear matrix attachment DNA-binding proteins SATB1. Mol Cell Biol 14:1852–1860.

Nakamura H, Morita T, Masaki S, Yoshida S (1984): Intracellular localization and metabolism of DNA polymerase α in human cells visualized with monoclonal antibody. Exp Cell Res 151:123–133.

Nakamura H, Morita T, Sato C (1986): Structural organizations of replicon domains during DNA synthetic phase in the mammalian nucleus. Exp Cell Res 165:291–297.

Nakayasu H, Berezney R (1989): Mapping replicational sites in the eukaryotic cell nucleus. J Cell Biol 108:1–11.

Nasheuer H-P, Moore A, Wahl AF, Wang TS-F (1991): Cell cycle–dependent phosphorylation of human DNA polymerase α. J Biol Chem 266:7893–7903.

Nasmyth K, Hunt T (1993): Cell cycle. Dams and sluices. Nature 366:634–635.

Nasmyth K, Nurse P (1981): Cell division cycle mutants altered in DNA replication and mitosis in the fission yeast *Schizosaccharomyces pombe*. Mol Gen Genet 182:119–124.

Nasmyth K, Nurse P, Fraser RS (1979): The effect of cell mass on the cell cycle timing and duration of S phase in fission yeast. J. Cell Sci 39:215–233.

Nathanel T, Kaufmann G (1990): Two DNA polymerases may be required for synthesis of the lagging DNA strand of simian virus 40. J Virol 64:5912–5918.

Nathanel T, Zlotkin T, Kaufmann G (1992): Assembly of simian virus 40 Okazaki pieces from DNA primers is reversibly arrested by ATP depletion. J Virol 66:6634–6640.

Nawotka KA, Huberman JA (1988): Two-dimensional gel electrophoretic method for mapping DNA replicons. Mol Cell Biol 8:1408–1413.

Newlon CS (1988): Yeast chromosome replication and segregation. Microbiol Rev 52:568–601.

Newlon CS, Collins I, Dershowitz AM, Greenfeder SA, Ong LY, Theis JF (1993): Analysis of replication origin function on chromosome III of *Saccharomyces cerevisiae*. Cold Spring Harbor Symp Quant Biol 58:415–423.

Newlon CS, Theis JF (1993): The structure and function of yeast ARS elements. Curr Opin Genet Dev 3:752–758.

Newport J (1987): Nuclear reconstitution in vitro: Stages of assembly around protein-free DNA. Cell 48:205–217.

Newport J, Wilson KL, Dunphy WG (1990): A lamin-independent pathway for nuclear envelop assembly. J Cell Biol 111:2247–2259.

Nguyen BT, Sadee W (1986): Compartmentation of guanine nucleotide precursors for DNA synthesis. Biochem J 234:263–269.

Nicander B, Reichard P (1983): Dynamics of pyrimidine deoxynucleoside triphosphate pools in relationship to DNA synthesis in 3T6 mouse fibroblasts. Proc Natl Acad Sci USA 80:1347–1354.

Nicander B, Reichard P (1985): Relationship between synthesis of deoxynucleotides and DNA replication in 3T6 fibroblasts. J Biol Chem 260:5376–5381.

Nishida C, Linn S (1988): DNA repair synthesis in permeabilized human fibroblasts mediated by DNA polymerase δ and application for purification of xeroderma pigmentosum factors. Cancer Cells 6:411–415.

Nishida C, Reinhard P, Linn S (1988): DNA repair synthesis in human fibroblast requires DNA polymerase δ. J Biol Chem 263:501–510.

Nishitani H, Nurse P (1995): $P85^{cdc18}$ has a major role controlling the initiation of DNA replication in fission yeast. Cell 83:397–405.

Nishizawa M, Tanabe K, Takahashi T (1984): DNA polymerase alpha and DNA topoisomerases solubilize from nuclear matrices of regenerating rat liver. Biochem Biophys Res Commun 124:917–924.

Noda A, Ning Y, Venables SF, Pereira-Smith OM, Smith JR (1994): Cloning of senescent cell-derived inhibitors of DNA synthesis using an expression screen. Exp Cell Res 211:90–98.

Noguchi H, Reddy GPV, Pardee AB (1983): Rapid incorporation of label from ribonucleoside diphosphates into DNA by a cell free high molecular weight fraction from animal cell nuclei. Cell 32:443–451.

Nugroho TT, Mendenhall MD (1994): An inhibitor of yeast cyclin-dependent protein kinase plays an important role in ensuring the genomic integrity of daughter cells. Mol Cell Biol 14:3320–3328.

Nurse P (1975): Genetic control of cell size at cell division in yeast. Nature 256:457–461.

Nurse P, Thuriaux P (1977): Controls over the timing of DNA replication during the cell cycle of fission yeast. Exp Cell Res 107:365–375.

Nurse P, Thuriaux P (1980): Regulatory genes controlling mitosis in the fission yeast *Schizosaccharomyces pombe.* Genetics 96:627–637.

Nurse P, Thuriaux P, Nasmyth K (1976): Genetic control of the cell division cycle in the fission yeast *Schizosaccharomyces pombe.* Mol Gen Genet 146:167–178.

Ogris E, Rotheneder H, Mudrak I, Pichler A, Winterberger (1993): A binding site for transcription factor E2F is a target for transactivation of murine thymidine kinase by polyoma virus large T antigen and plays an important role in growth regulation of the gene. J Virol 67:1765–1771.

Ohtsubo M, Roberts JM (1993): Cyclin-dependent regulation of G_1 in mammalian fibroblasts. Science 259:1908–1912.

Orr-Weaver TL (1991): *Drosophila* chorion genes: Cracking the eggshell's secrets. BioEssays 13:97–105.

Pagano M, Pepperkok R, Verde F, Ansorge W, Draetta G (1992): Cyclin A is required at two points in the human cell cycle. EMBO J 11:961–971.

Pardee AB (1974): A restriction point for control of normal animal cell proliferation. Proc Natl Acad Sci USA 7:1286–1290.

Pardee AB (1989): G_1 events and regulation of cell proliferation. Science 246:603–608.

Pardoll DM, Vogelstein B, Coffey DS (1980): A fixed site of DNA replication in eukaryotic cells. Cell 19:527–536.

Paris J, LeGuellec R, Couturier A, LeGuellec K, Omilli F, Camonis J, MacNeill S, Philppe M (1991): Cloning by differential screening of a *Xenopus* cDNA coding for a protein highly homologous to cdc2. Proc Natl Acad Sci USA 88:1039–1043.

Parker SB, Eichele G, Zhang P, Rawls A, Sands AT, Bradley A, Olson EN, Harper JW, Elledg SJ (1995): P53-independent expression of p21Cip1 in muscle and other terminally differentiating cells. Science 267:1024–1027.

Pearson BE, Nasheuer HP, Wang TS-F (1991): Human DNA polymerase alpha gene: Sequences controlling expression in cycling and serum-stimulated cells. Mol Cell Biol 11:2081–2095.

Piatti S, Lengauer C, Nasmyth K (1995): Cdc6 is an unstable protein whose de novo synthesis in G_1 is important for the onset of S phase and for preventing a "reductional" anaphase in the budding yeast *Saccharomyces cerevisiae.* EMBO J 14:3788–3799.

Picard A, Cavadora J-C, Lory P, Bernengo J-C, Ojeda C, Doree M (1990): Microinjection of a conserved peptide sequence of $p34^{cdc2}$ induces Ca^{++} transient in oocytes. Science 247:327–329.

Pines J, Hunter T (1990): Human cyclin A is adenovirus E1A–associated protein P60 and behaves differently from cyclin B. Nature 346:760–763.

Plucinski TM, Fager RS, Reddy GPV (1990): Allosteric interaction of components of the replitase complex is responsible for enzyme cross-inhibition. Mol Pharmacol 38:114–120.

Podust VN, Georgaki A, Strack B, Hubscher U (1992): Calf thymus RF-C as an essential component for DNA polymerase δ and ε holoenzymes function. Nucleic Acids Res 20:4159–4165.

Podust VN, Hubscher U (1993): Lagging strand DNA synthesis by calf thymus DNA polymerase α, β, δ and ε in the presence of auxiliary proteins. Nucleic Acids Res 21:841–846.

Polyak K, Lee MH, Erdjument-Bromage H, Koff A, Roberts JM, Tempst P, Massague J (1994): Cloning of $P27^{Kip1}$, a cyclin-dependent kinase inhibitor and a potential mediator of extracellular antimitogenic signals. Cell 78:59–66.

Prasad R, Kumar A, Widen SG, Casas FJ, Wilson SH (1993a): Identification of residues in the single stranded DNA-binding site of the 8-kDa domain of rat DNA polymerase β by UV cross-linking. J Biol Chem 268:22746–22755.

Prasad R, Widen SG, Singhal RK, Wilkins J, Prakash L, Wilson SH (1993b): Yeast open reading frame YCR14C encodes a DNA β-polymerase-like enzyme. Nucleic Acids Res 21:5301–5307.

Prelich G, Kostura M, Marshak DR, Mathews MB, Stillman B (1987a): The cell cycle regulated proliferating cell nuclear antigen is required for SV40 DNA replication in vitro. Nature (Lond) 326:471–475.

Prelich G, Stillman B (1988): Coordinated leading and lagging strand synthesis during SV40 DNA replication in vitro requires PCNA. Cell 53:117–126.

Prelich G, Tan C-K, Mikostura M, Mathews MB, So AG, Downey KM, Stillman B (1987b): Functional identity of proliferating cell nuclear antigen and a DNA polymerase-δ auxiliary protein. Nature (Lond) 326:517–520.

Quelle DE, Ashmum RA, Shurtleff SA, Kato JY, Bar Sagi D, Roussel MF, Sherr CJ (1993): Overexpression of mouse D-type cyclins accelerates G1 phase in rodent fibroblasts. Genes Dev 7:1559–1571.

Rao H, Marahrens Y, Stillman B (1994): Functional conservation of multiple elements in yeast chromosomal replicators. Mol Cell Biol 14:7643–7651.

Rao PN, Johnson RT (1970): Mammalian cell fusion: Studies on the regulation of DNA synthesis and mitosis. Nature 225:159–164.

Ray A, Roy N, Maitra M, Sinha P (1994): A 61-kb ring chromosome shows an ARS-dependent increase in its mitotic stability in the MCM2 mutant of yeast. Curr Genet 26:403–409.

Reddy GPV (1982): Catalytic function of thymidylate synthase is confined to S phase due to its association with replitase. Biochem Biophys Res Commun 109:908–915.

Reddy GPV (1989): Compartmentation of deoxypyrimidine nucleotides for nuclear DNA replication in S phase mammalian cells. J Mol Rec 2:75–83.

Reddy GPV (1994): Cell cycle: Regulatory events in $G_1 \rightarrow$ S transition of mammalian cells. J Cell Biochem 54:379–386.

Reddy GPV, Deacon D, Reed WC, Quesenberry PJ (1992a): Growth factor–dependent proliferative stimulation of hemopoietic cells is associated with the nuclear localization of 68 kDa calmodulin-binding protein. Blood 79:1946–1955.

Reddy GPV, Fager RS (1993): Replitase: A complex integrating dNTP synthesis and DNA replication. Crit Rev Eukaryot Gene Expression 3:255–277.

Reddy GPV, Klinge EM, Pardee AB (1986): Ribonucleotides are channeled into a mixed DNA–RNA polymer by permeabilized hamster cells. Biochem Biophys Res Commun 135:340–346.

Reddy GPV, Mathews CK (1978): Functional compartmentation of DNA precursors. J Biol Chem 253:3461–3467.

Reddy GPV, Pardee AB (1980): A multienzyme complex for metabolic channeling in mammalian DNA replication. Proc Natl Acad Sci USA 77:3312–3316.

Reddy GPV, Pardee AB (1982): Coupled ribonucleoside diphosphate reduction, channeling, and incorporation into DNA of mammalian cells. J Biol Chem 257:12527–12531.

Reddy GPV, Pardee AB (1983): Inhibitor evidence for allosteric interaction in the replitase multienzyme complex. Nature 303:86–88.

Reddy GPV, Quesenberry PJ (1996): Stem cell factor enhances interleukin-3–dependent induction of 68 kDa calmodulin-binding protein and thymidine kinase activity in NFS-60 cells. Blood 87:3195–3202.

Reddy GPV, Reed WC, Deacon DH, Quesenberry PJ (1994): Growth factor modulated calmodulin-binding protein stimulates nuclear DNA synthesis in hematopoietic progenitor cells. Biochemistry 33:6605–6610.

Reddy GPV, Reed WC, Sheehan E, Sacks DB (1992b): Calmodulin-specific antibodies inhibit DNA replication in mammalian cells. Biochemistry 31:10426–10430.

Reddy GPV, Singh A, Stafford ME, Mathews CK (1977): Enzyme associations in T4 phage DNA precursor synthesis. Proc Natl Acad Sci USA 74:3152–3156.

Reed SI, Dulic V, Lew DJ, Richardson HE, Wittenberg C (1992): G_1 control in yeast and animal cells. Ciba Found Symp 170:7–19.

Resnitzky D, Gossen M, Bujard H, Reed SI (1994): Acceleration of G_1/S phase transition by expression of cyclin D1 and E with an inducible system. Mol Cell Biol 14:1669–1679.

Resnitzky D, Hengst L, Reed SI (1995): Cyclin A–associated kinase activity is rate limiting for entrance into S phase and is negatively regulated in G_1 by p27Kip1. Mol Cell Biol 15:4347–4352.

Resnitzky D, Reed SI (1995): Different roles for cyclins D1 and E in regulation of the G1-to-S transition. Mol Cell Biol 15:3463–3469.

Reyland ME, Lehman IR, Loeb LA (1988): Specificity of proofreading by the 3′ to 5′ exonuclease of the DNA polymerase–primase of *Drosophila* melanogaster. J Biol Chem 263:6518–6524.

Riabowol K, Draetta G, Brizuela L, Vandre D, Beach D (1989): The cdc2 kinase is a nuclear protein that is essential for mitosis in mammalian cells. Cell 57:393–401.

Richardson CC (1983): Bacteriophage T7 minimal requirement for the replication of a complex DNA molecule. Cell 33:315–317.

Richardson H, Lew DJ, Henze M, Sugimoto K, Reed SI (1992): Cyclin-B homologs in *Saccharomyces cerevisiae* function in S phase and in G_2. Genes Dev 6:2021–2034.

Rivier DH, Rine J (1992): An origin of DNA replication and a transcription silencer require a common element. Science 256:659–663.

Roberts JM (1993): Turning DNA replication on and off. Curr Opin Cell Biol 5:201–206.

Rode W, Scanlon KJ, Moroson BA, Bertino JR (1980): Regulation of thymidylate synthase in mouse leukemia cells (L1210). J Biol Chem 255:1305–1311.

Romig H, Fackelmayer FO, Renz A, Ramsperger U, Richter A (1992): Characterization of SAF-A, a novel nuclear DNA-binding protein from HeLa cells with high affinity for nuclear matrix/scaffold attachment DNA elements. EMBO J 11:3431–3440.

Rowles A, Chong JPJ, Brown L, Howell M, Evan GI, Blow JJ (1996): Interaction between the origin recognition complex and the replication licensing system in *Xenopus.* Cell 87:287–296.

Rusch HP, Sachsenmaier W, Behrens K, Gruter V (1966): Synchronization of mitosis by the fusion of the plasmodia of *Physarum polycephalum.* J Cell Biol 31:204–209.

Ryter A (1968): Association of the nucleus and the membrane of bacteria: A morphological study. Bacteriol Rev 32:39–54.

Sasaki MS, Norman A (1966): DNA fibers from human lymphocyte nuclei. Exp Cell Res 44:642–645.

Sauer K, Knoblich JA, Richardson H, Lehner CF (1995): Distinct modes of cyclin E/cdc2 kinase regulation and S-phase control in mitotic and endoreduplication cycle of *Drosophila* embryogenesis. Genes Dev 9:1327–1339.

Schmidt AMA, Herterich SU, Krauss G (1991): A single-stranded DNA binding protein from *S. cerevisiae* specifically recognizes T-rich strand of the core sequence of ARS elements and discriminates against mutant sequences. EMBO J 10:981–985.

Schneider BL, Yang Q-H, Futcher AB (1996): Linkage of replication to START by the cdk inhibitor Sic1. Science 272:560–562.

Schwob E, Bohm T, Mendenhall MD, Nasmyth K (1994): The B type cyclin kinase inhibitor p40S1C1 controls the G1 to S transition in *S. cerevisiae.* Cell 79:233–244.

Schwob E, Nasmyth K (1993): CLB5 and CLB6, a new pair of B cyclins involved in DNA replication in *Saccharomyces cerevisiae.* Genes Dev 7:1160–1175.

Sclafini RA, Fangman WL (1984): Yeast gene CDC8 encodes thymidylate kinase and is complemented by herpes thymidine kinase gene TK. Proc Natl Acad Sci USA 81:5821–5824.

Scott FW, Forsdyke DR (1980): Isotope dilution analysis of the effects of deoxyguanosine and deoxyadenosine on the incorporation of thymidine and deoxycytidine by hydroxyurea-treated thymus cells. Biochem J 190:721–730.

Selig S, Okumura K, Ward DC, Cedar H (1992): Delineation of DNA replication time zones by fluorescence in situ hybridization. EMBO J 11:1217–1225.

Serrano M, Hannon GJ, Beach D (1993): A new regulatory motif in cell cycle control causing specific inhibition of cyclin D/cdk4. Nature 366:704–707.

Seufert W, Futcher B, Jentsch S (1995): Role of ubiquitin-conjugating enzyme in degradation of S- and M-phase cyclins. Nature 373:78–81.

Sheehan MA, Mills AD, Sleeman AM, Laskey RA, Blow JJ (1988): Steps in the assembly of replication-competent nuclei in a cell-free system from *Xenopus* eggs. J Cell Biol 106:1–12.

Sherr CJ (1993): Mammalian G_1 cyclins. Cell 7:1059–1065.

Sherr CJ (1994): G_1 phase progression: Cycling on cue. Cell 79:551–555.

Sherr CJ (1996): Cancer cell cycle. Science 274:1672–1677.

Sherr CJ, Roberts JM (1995): Inhibitors of mammalian G_1 cyclin-dependent kinases. Genes Dev 9:1149–1163.

Shinomiya T, Ina S (1993): DNA replication of histone gene repeats in *Drosophila melanogaster* tissue culture cells: Multiple initiation sites and replication pause sites. Mol Cell Biol 13:4098–4106.

Shinomiya T, Ina S (1994): Mapping an initiation origin of DNA replication at a single-copy chromosomal locus in *Drosophila melanogaster* cells by two-dimensional gel methods and PCR-mediated nascent-strand analysis: Multiple replication origins in a broad zone. Mol Cell Biol 14:7394–7403.

Sicinski P, Donaher JL, Parker SB, Li T, Fazeli A, Gardner H, Haslam SZ, Bronson RT, Elledge SJ, Weinberg RA (1995): Cyclin D1 provides a link between development and oncogenesis in the retina and breast. Cell 82:621–630.

Sikorska M, Brewer LM, Youdale T, Richards R, Whitfield JF, Houghton RA, Walker PR (1990): Evidence that mammalian ribonucleotide reductase is a nuclear membrane associated glycoprotein. Biochem Cell Biol 68:880–888.

Simanis V, Hayles J, Nurse P (1987): Control over the onset of DNA synthesis in fission yeast. Philos Trans R Soc Lond Ser B 317:507–516.

Simon M, Giot L, Faye G (1991): The 3′ to 5′ exonuclease activity located in the DNA polymerase-delta subunit of *Saccharomyces cerevisiae* is required for accurate replication. EMBO J 10:2165–2170.

Singhal RK, Prasad R, Wilson SH (1995): DNA polymerase β conducts the gap-filling step in uracil-initiated base excision repair in a bovine testis nuclear extract. J Biol Chem 268:15906–15911.

Singhal RK, Wilson SH (1993): Short gap-filling synthesis by DNA polymerase β is processive. J Biol Chem 268:15906–15911.

Sinha P, Chang V, Tye BK (1986): A mutant that affects the function of autonomously replicating sequences in yeast. J Mol Biol 192:805–814.

Skarnes W, Bonin P, Baril EF (1986): Exonuclease activity associated with a multiprotein form of HeLa cell DNA polymerase α. Purification and properties of exonuclease. J Biol Chem 261:6629–6636.

Smith HC, Berezney R (1980): DNA polymerase alpha is tightly bound to the nuclear matrix of actively replicating liver. Biochem Biophys Res Commun 97:1541–1547.

So AG, Downey KM (1988): Mammalian DNA polymerase α and δ: Current status in DNA replication. Biochemistry 27:4591–4595.

Spadari S, Weissbach A (1974): Interrelation between DNA synthesis and various DNA polymerase activities in synchronized HeLa cells. J Mol Biol 86:11–20.

Stahl H, Droge P, Knippers R (1986): DNA helicase activity of SV40 large tumor antigen. EMBO J 5:1939–1944.

Stern B, Nurse P (1996): A quantitative model for the cdc2 control of S phase and mitosis in fission yeast. Trends Genet 12:345–350.

Stillman B (1989): Initiation of eukaryotic DNA replication in vitro. Annu Rev Cell Biol 5:197–245.

Stillman B, Gluzman Y (1985): Replication and super-coiling of simian virus 40 DNA in cell extracts from human cells. Mol Cell Biol 5:2051–2060.

Stinchcomb DT, Struhl K, Davis RW (1979): Isolation and characterization of a yeast chromosomal replicator. Nature (Lond) 282:39–43.

Strausfeld UP, Howell M, Descombes P, Chavalier S, Rempel RE, Adamczewski J, Maller JL, Hunt T, Blow JJ (1966): Both cyclin A and cyclin E have S-phase promoting (SPF) activity in *Xenopus* egg extracts. J Cell Sci 109:1555–1563.

Strausfeld UP, Howell M, Rempel R, Maller JL, Hunt T, Blow JJ (1994): Cip1 blocks the initiation of DNA replication in *Xenopus* extracts by inhibition of cyclin-dependent kinases. Curr Biol 4:876–883.

Subramanyam C, Honn SC, Reed WC, Reddy GPV (1990): Nuclear localization of 68 kDa calmodulin-binding protein is associated with the onset of DNA replication. J Cell Physiol 144:423–428.

Sugino A (1995): Yeast DNA polymerases and their role at the replication fork. Trends Pharmacol Sci 20:319–323.

Surana U, Robitsch H, Price C, Schuster T, Fitch I, Futcher AB, Nasmyth K (1991): The role of cdc28 and cyclins during mitosis in the budding yeast *S. cerevisiae.* Cell 65:145–161.

Syvaoja JE (1990): DNA polymerase ε: The latest member in the family of mammalian DNA polymerases. BioEssays 12:533–536.

Syvaoja JE, Linn S (1989): Characterization of a large form of DNA polymerase δ from Hela cells that is insensitive to proliferating cell nuclear antigen. J Biol Chem 264:2489–2497.

Takahashi K, Yamada H, Yanagida M (1994): Fission yeast minichromosome loss mutants mis cause lethal aneuploidy and replication abnormality. Mol Biol Cell 5:1145–1158.

Tan CK, Castillo C, So AG, Downey KM (1986): An auxiliary protein for DNA polymerase-delta from fetal calf thymus. J Biol Chem 261:12310–12316.

Tasheva ES, Roufa DJ (1994): Densely methylated DNA islands in mammalian chromosomal replication origins. Mol Cell Biol 14:5636–5644.

Theis JF, Newlon CS (1994): Domain B of ARS307 contains two functional elements and contributes to chromosomal replication origin function. Mol Cell Biol 14:7652–7659.

Thommes P, Fett R, Schray B, Burkhart R, Barnes M, Kennedy C, Brown NC, Knippers R (1992): Properties of nuclear P1 proteins, a mammalian homologue of the yeast MCM3 replication protein. Nucleic Acids Res 20:1069–1074.

Todorov IT, Attaran A, Kearsey SA (1995): BM28, a human member of the MCM 2-3-5 family, is displaced from chromatin during DNA replication. J Cell Biol 129:1433–1445.

Todorov IT, Pepperkok R, Philipova RN, Kearsey SE, Ansorge W, Werner D (1994): H human nuclear protein with sequence homology to a family of early S phase proteins required for entry into S phase and cell division. J Cell Sci 107:253–265.

Treisman JE, Follette PJ, O'Farrell PH, Rubin GM (1995): Cell proliferation and DNA replication defects in a *Drosophila* MCM2 mutant. Genes Dev 9:1709–1715.

Tsai LH, Harlow E, Meyerson M (1991): Isolation of the human cdk2 gene that encodes the cyclin A– and adenovirus E1A–associated p33 kinase. Nature 353:174–177.

Tubo RA, Berezney R (1987a): Pre-replicative association of multiple replicative enzyme activities with the nuclear matrix during rat liver regeneration. J Biol Chem 262:1148–1154.

Tubo RA, Berezney R (1987b): Nuclear matrix–bound DNA primase. Elucidation of an RNA priming system in nuclear matrix isolated from regenerating rat liver. J Biol Chem 262:6639–6642.

Tyers M (1996): The cyclin-dependent kinase inhibitor p40SIC1 imposes the requirement for Cln G_1 cyclin function at START. Proc Natl Acad Sci USA 93:7772–7776.

Tyson JJ, Novak B, Odel GM, Chen K, Thron CD (1996): Chemical kinetic theory: Understanding cell cycle regulation. Trends Biochem Sci 21:89–96.

Vassilev LT, Burhans WC, DePamphilis ML (1990): Mapping an origin of DNA replication at a single copy locus in exponentially proliferating mammalian cells. Mol Cell Biol 10:4685–4689.

Vassilev LT, Johnson EM (1990): An initiation zone of chromosomal DNA replication located upstream of the c-myc gene in proliferating HeLa cells. Mol Cell Biol 10:4899–4904.

Vaughn JP, Dijkwel PA, Hamlin JL (1990): Replication initiates in a broad zone in the amplified CHO dihydrofolate reductase domain. Cell 61:1075–1087.

Verma R, Annan RS, Huddleston MJ, Carr SA, Reynard G, Deshaies RJ (1997): Phosphorylation of Sic1p by G1 CdK required for its degradation and entry into S phase. Science 278:455–460.

Vigers GP, Lohka MJ (1992): Regulation of nuclear envelop precursor function during cell division. J Cell Sci 102:273–284.

Virta-Pearlman VJ, Gunaratne PH, Chinault AC (1993): Analysis of a replication initiation sequence from the adenosine deaminase region of the mouse genome. Mol Cell Biol 13:5931–5942.

Vishwanatha JK, Coughlin SA, Wesolowski-Owen M, Baril EF (1986): A multiprotein form of DNA polymerase α from Hela cells. Resolution of its associated catalytic activities. J Biol Chem 261:6619–6628.

Vogelstein B, Pardol DM, Coffey DS (1980): Supercoiled loops and eukaryotic DNA replication. Cell 22:79–85.

Waga S, Hannon GJ, Beach D, Stillman B (1994): The p21 inhibitor of cyclin-dependent kinases controls DNA replication by interaction with PCNA. Nature 369:574–578.

Waga S, Stillman B (1994): Anatomy of a DNA replication fork revealed by reconstitution of SV40 DNA replication in vitro. Nature 369:207–212.

Wahl AF, Geis AM, Spain BH, Wong SW, Korn D, Wang TS-F (1988): Gene expression of human DNA polymerase α during cell proliferation and the cell cycle. Mol Cell Biol 8:5016–5025.

Walker SS, Malik AK, Eisenberg S (1991): Analysis of the interactions of functional domains of a nuclear origin of replication from *Saccharomyces cerevisiae.* Nucleic Acids Res 19:6255–6262.

Walter J, Newport JW (1997): Regulation of replicon size in *Xenopus* egg extracts. Science 275:993–995.

Wang TS-F (1991): Eukaryotic DNA polymerases. Annu Rev Biochem 60:513–552.

Weinberg RA (1995): The retinoblastoma protein and cell cycle control. Cell 81:323–330.

Weiser T, Gassmann M, Thommes P, Ferrari E, Hafkemeyer P, Hubscher U (1991): Biochemical and functional comparison of DNA polymerase α, δ and ε from calf thymus. J Biol Chem 266:10420–10428.

Whitfield JF, MacManus JP, Boynton AL, Durkin J, Jones A (1982): Futures of calcium, calmodulin-binding proteins in the quest for an understanding of cell proliferation and cancer. In Corradino RA (ed): Functional Regulation at the Cellular and Molecular Levels. New York: Elsevier/North Holland, Inc., pp 61–87.

Wickner W, Schekman R, Geider K, Kornberg A (1973): A new form of DNA polymerase III and a copolymerase replicate a long single stranded primer template. Proc Natl Acad Sci USA 70:1764–1767.

Wickremasinghe RG, Hoffbrand AV (1983): Inhibition by aphidicolin and dideoxythymidine triphosphate of a multienzyme complex of DNA synthesis from human cells. FEBS Lett 159:175–179.

Wickremasinghe RG, Yaxley JC, Hoffbrand AV (1982): Solubilization and partial characterization of a multienzyme complex of DNA synthesis from human lymphoblastoid cells. Eur J Biochem 126:589–596.

Wickremasinghe RG, Yaxley JC, Hoffbrand AV (1983): Gel filtration of a complex of DNA polymerase and DNA precursor–synthesizing enzymes from a human lymphoblastoid cell line. Biochim Biophys Acta 740:243–248.

Wilcock D, Lane DP (1991): Localization of p53, retinoblastoma and host replication proteins at sites of viral replication in herpes-infected cells. Nature 349:429–431.

Wobbe CR, Dean F, Weissback L, Hurwitz J (1985): In vitro replication of duplex circular DNA containing the simian virus 40 DNA origin site. Proc Natl Acad Sci USA 82:5710–5714.

Wohlgemuth JG, Bulboaca GH, Moghadam M, Caddle MS, Calos MP (1994): Physical mapping of origins of replication in the fission yeast *Schizosaccharomyces pombe.* Mol Biol Cell 5:839–849.

Wold MS, Kelly TJ (1988): Purification and characterization of replication protein A, a cellular protein required for in vitro replication of simian virus DNA. Proc Natl Acad Sci USA 84:3643–3647.

Wong SW, Wahl AF, Yuan PM, Arai N, Pearson BE, Arai K-I, Korn D, Hunkapiller MW, Wang TS-F (1988): Human DNA polymerase α gene expression is cell proliferation dependent and its primary structure is similar to both prokaryotic and eukaryotic replicative DNA polymerases. EMBO J 7:37–47.

Wood SH, Collins JM (1986): Preferential binding of DNA primase to the nuclear matrix in HeLa cells. J Biol Chem 261:7119–7122.

Wu J-R, Gilbert DM (1996): A distinct G_1 step required to specify the Chinese hamster DHFR replication origin. Science 271:1270–1272.

Xiong Y, Zhang H, Beach D (1993): Subunit rearrangement of the cyclin-dependent kinases is associated with cellular transformation. Genes Dev 7:1572–1583.

Yaglom J, Linskens MH, Sadis S, Rubin DM, Futcher B, Finley D. (1995): p34cdc28-mediated control of Cln3 cyclin degradation. Mol Cell Biol 15:731–741.

Yamamoto S, Takahashi T, Matsukage A (1984): Tight association of DNA polymerase alpha with granular structures in the nuclear matrix of chick embryo cells: Immunocytochemical detection with monoclonal antibody against DNA polymerase alpha. Cell Struct Funct 9:83–90.

Yamashita T, Arens M, Green M (1977): Adenovirus deoxyribonucleic and replication. Isolation of a soluble replication system and analysis of the in vitro DNA product. J Biol Chem 252:7940–7946.

Yan H, Gibson SI, Tye BK (1991): MCM2 and MCM3, two proteins important for ARS activity, are related in structure and function. Genes Dev 5:944–957.

Yan H, Harchant AM, Tye BK (1993): Cell cycle–regulated nuclear localization of MCM2 and MCM3, which are required for the initiation of DNA synthesis at chromosomal replication origins in yeast. Gene Dev 7:2149–2160.

Yoon H-J, Campbell JL (1991): The Cdc7 protein of *Saccharomyces cerevisiae* is a phosphoprotein that contains protein kinase activity. Proc Natl Acad Sci USA 88:3574–3578.

Yoon Y, Sanchez JA, Brun C, Huberman JA (1995): Mapping of replication initiation sites in human ribosomal DNA by nascent-strand abundance analysis. Mol Cell Biol 15:2482–2489.

Youdale T, Frappier L, Whitfield JF, Rixon RH (1984): Changes in cytoplasmic and nuclear activities of the ribonucleotide reductase EC-1.17.4.1 holoenzyme and its subunits in regenerating liver cells in normal and thyroparathyroidectomized rats. Can J Biochem Cell Biol 62:914–920.

Yunis JJ, Kuo MT, Saunders GF (1977): Localization of sequences specifying messenger RNA to light-staining G-bands of human chromosomes. Chromosoma 61:335–344.

Zeng XR, Hao H, Jiang Y, Lee MY (1994): Regulation of human DNA polymerase δ during the cell cycle. J Biol Chem 269:24027–24033.

Zindy F, Lamas E, Chenivesse X, Sobczak J, Wang J, Fesquet D, Henglein B, Brechot C (1992): Cyclin A is required in S phase in normal epithelial cells. Biochem Biophys Res Commun 182:1144–1154.

CHAPTER 4

MITOSIS: THE CULMINATION OF THE CELL CYCLE

GREENFIELD SLUDER, EDWARD H. HINCHCLIFFE, and CONLY L. RIEDER
Department of Cell Biology, University of Massachusetts Medical Center, Worcester, MA 01655 (G.L., E.H.H.); Laboratory of Cell Regulation, Division of Molecular Medicine, Wadsworth Center, Albany, NY 12201-0509 and Department of Biomedical Sciences, State University of New York, Albany, NY 12222 (C.L.R.)

INTRODUCTION

> . . . Double or nothing. With few exceptions a living cell either reproduces or dies; the principle is so simple that no one has bothered to call it a principle. A cell is born in the division of a parent cell. It then doubles in every respect: in every part, in every kind of molecule, even in the amount of water it contains. Thereafter it divides with such equal justice that each new daughter cell is an identical copy of the parent. This doubling and halving, the cycle of growth and division, is known generally as the cell cycle. . . .

In this short paragraph Daniel Mazia (1974) elegantly summarized the purpose of the cell cycle: the formation of two genetically identical daughter cells. The culmination of the cell cycle and the reason for all the events of growth and duplication is known as *mitosis.* In this chapter we briefly review the events of mitosis in animal cells and discuss some of the regulatory mechanisms that ensure the equal segregation of the genome.

PHASES OF MITOSIS

Traditionally, mitosis has been separated into five phases: prophase, prometaphase, metaphase, anaphase, and telophase (Mazia, 1961). Over

The Molecular Basis of Cell Cycle and Growth Control, Edited by G.S. Stein, R. Baserga, A. Giordano, and D.T. Denhardt
ISBN 0-471-15706-6, pages 155–182.

the years these stages, which are based on the position of the chromosomes relative to the spindle poles, have served as convenient labels to indicate how far the cell has progressed through mitosis and define in a shorthand fashion what events are occurring at a particular time. However, it is important to keep in mind that the definitions of these stages are based on morphology as seen by the light microscope, which does not reveal the underlying biochemical events at work. Today we know that many of these morphological events begin well before they become visible in the light microscope and thus fall into more than one of the traditional stages. In assigning stages to mitosis one is slicing the flow of events into separate pieces based on the particular criteria used, be they morphological or biochemical. Thus, in the light of new advances in our understanding of mitotic events, some of the classic terms are losing their precise meaning when used as undefined labels.

PROPHASE

Traditionally, mitosis is defined to start in prophase (Figs. 4.1A, 4.2A), which begins when the light microscopist can first detect the presence of condensing chromosomes in the nucleus and ends with nuclear envelope breakdown (NEB). As prophase progresses, the chromosomes become progressively more condensed, the nucleoli dissipate, and the extensive interphase cytoplasmic microtubule array becomes reorganized into two focal arrays known as *asters* (Fig. 4.2A). Ultimately these astral arrays, which are generated from the replicated centrosomes, separate to form the spindle poles and to supply the microtubules used to construct the mitotic apparatus. However, the extent to which the asters are formed and separated by the time of NEB varies greatly between cells, even for neighboring cells in the same culture (reviewed in Rieder, 1990). In some cases the duplicated centrosomes remain close together with little evidence of astral microtubule assembly. In others, both asters are well developed and have separated to opposite sides of the nucleus before the end of prophase (Fig. 4.2A).

The force-producing mechanism for spindle pole separation has been the subject of much debate. Some favor the proposal that forces exerted between the two overlapping and antiparallel astral microtubule arrays push the poles apart, which is clearly the mechanism for spindle pole separation in yeast and diatoms (reviewed in Hogan and Cande, 1990). However, in vertebrate somatic cells, the asters continue to separate with normal kinetics even when their arrays of microtubules no longer overlap (Waters et al., 1993). In such cells the force-generating mechanism for centrosome separation is intrinsic to each aster, and the centrosomes pull themselves apart, perhaps as their associated arrays of astral microtubules interact with minus end–directed motor molecules (e.g., cytoplasmic dynein) that are anchored in the cytoplasm (reviewed in Ault and Rieder, 1994). It is also possible that the same cells use

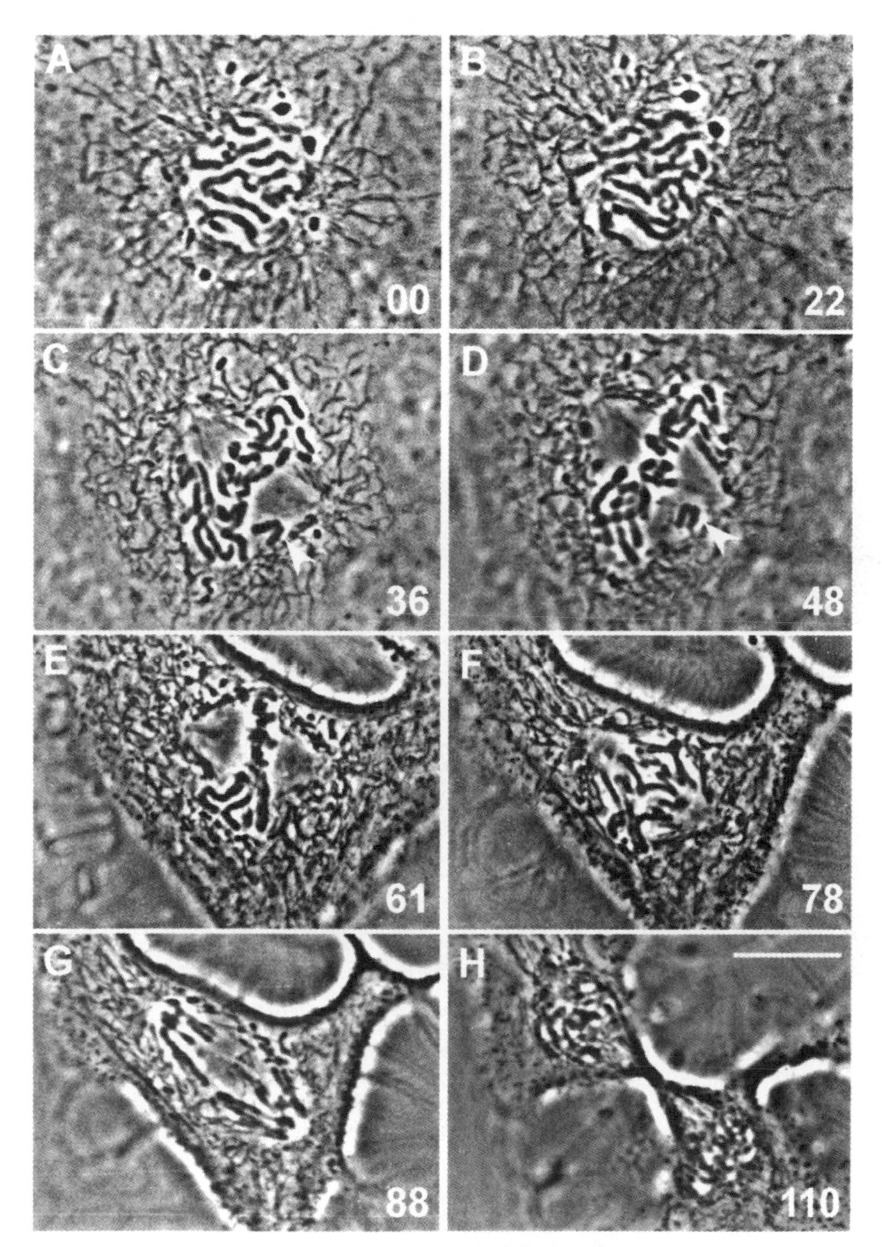

Figure 4.1. Sequential phase-contrast images of a living PtK1 cell in the process of mitosis. **(A)** By *late prophase,* the chromosomes are condensed within the nucleus, and the nucleolar organizers have dissipated. **(B)** *Prometaphase* is initiated when the nuclear envelope breaks down to allow the chromosomes to interact with the centrosomes to form the spindle. **(C)** By *mid-prometaphase,* all of the chromosomes have acquired a bipolar alignment, but one (white arrowhead) is still monooriented. **(D)** By *late prometaphase,* this last monooriented chromosome (white arrowhead) has become bioriented and is congressing to the spindle equator. **(E)** At *metaphase,* all of the chromosomes are positioned on the spindle equator, at approximately equal distances between the two poles. **(F)** As *anaphase* begins, the chromatids separate and move toward their respective poles. **(G)** During *late anaphase,* the two spindle poles move farther apart in a process known as *anaphase B,* which additionally separates the two genomes. **(H)** During *telophase,* the cytokinesis pinches the cell into two daughter cells, and a nuclear envelope re-forms around the two groups of chromosomes. Times in minutes are given in the lower right corners of each part. Bar in H = 15 μm.

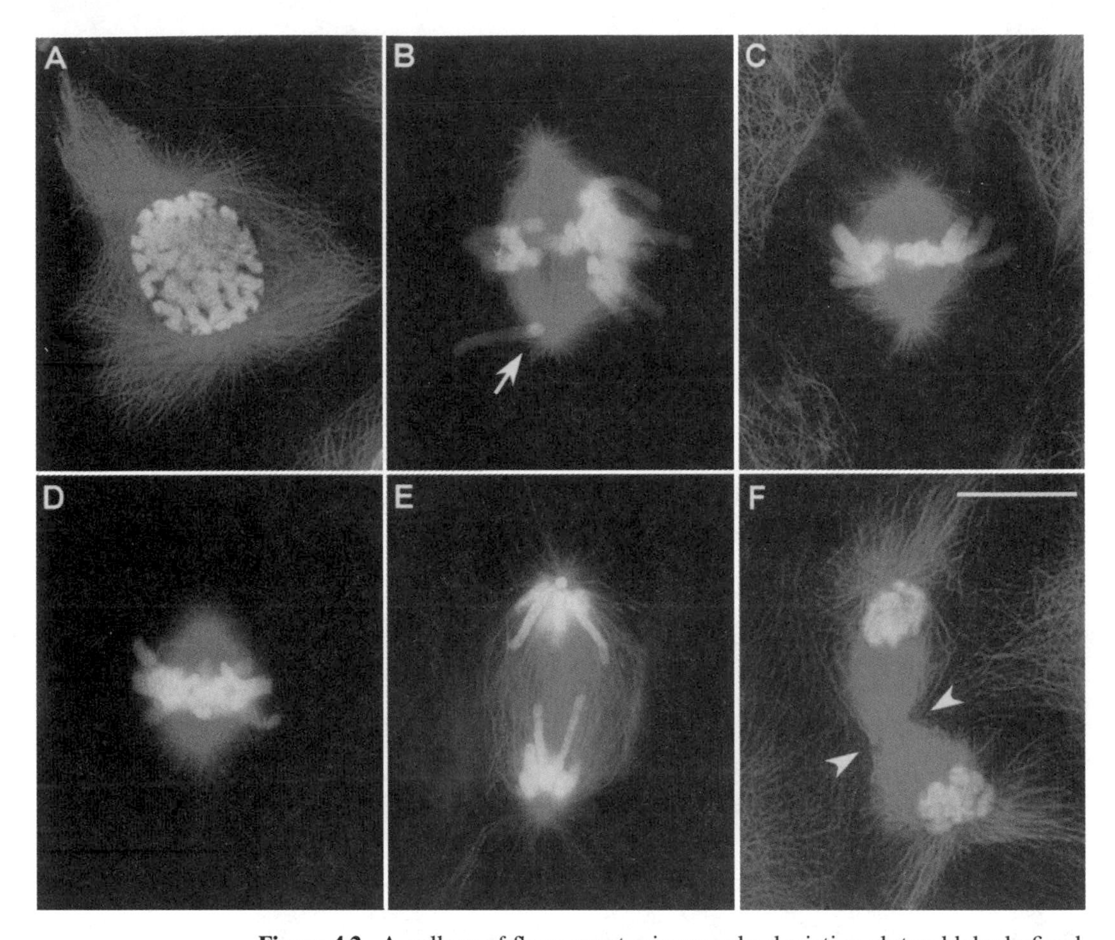

Figure 4.2. A gallery of fluorescent micrographs depicting glutaraldehyde-fixed and lysed PtK1 cells in various stages of mitosis. In these examples the microtubules (red) were stained using indirect immunofluorescent methods, while the chromosomes (green) were stained with the DNA probe Hoechst 33342. **(A)** A *late prophase* cell in which the two centrosomes and their associated radial arrays of astral microtubules have separated to opposite sides of the nucleus. Note that there are still many cytoplasmic microtubules in this cell that are not associated with the asters. **(B)** A *mid-prometaphase* cell that contains one monooriented chromosome (white arrow) and several congressing chromosomes. By this time, all of the cytoplasmic microtubules have disassembled, and most of the astral microtubules have been incorporated into the spindle. **(C)** A *metaphase* cell in which all of the chromosomes are aligned on the spindle equator. **(D)** A cell just entering *anaphase* in which the chromatids are disjoining. Note the compact nature of the spindle (cf. C and D). **(E)** A *late anaphase* cell in which the two groups of chromosomes are already at the spindle poles, which themselves are moving farther apart (anaphase B). **(F)** A *telophase* cell in which the two groups of well-separated chromosomes are reforming nuclei and in which cytokinesis (between the white arrowheads) is almost complete. The prominent bundle of microtubules between the two nuclei participates in cytokinesis and is known as the *midbody*. Bar in F = 15 μm. (Courtesy of Dr. Alexey Khodjakov.) (See color plate.)

multiple mechanisms to separate their asters (reviewed in Whitehead et al., 1996).

Although the initiation of prophase is generally defined as the first visible signs of chromosome condensation, it is not clear that this phase has a sharply defined beginning. There is evidence that the chromosome condensation cycle is a continuum (Mazia, 1961; Pederson, 1972; Pederson and Robbins, 1972), beginning in S phase and reaching its greatest extent in early anaphase (Bajer, 1959, 1965). After mitosis the chromosomes become progressively more decondensed until S phase (Pederson and Robbins, 1972). For many years it was thought that chromosome condensation was mediated, at least in part, by the phosphorylation of histone H1 by the active $p34^{cdc2}$ kinase (CDK1) (reviewed in Bradbury, 1992). However, recent data have shown that chromosome condensation occurs normally in *Xenopus* extracts even after histone H1 has been depleted and it does not appear to require CDK1 activity (Ohsumi et al., 1993). Instead, the initiation of chromosome condensation is better correlated with the phosphorylation of histone H3 (reviewed in Hendzel et al., 1997), which occurs in late G_2, and it also involves chromosome condensation protein complexes that are crucial for forming and maintaining higher order chromosome structure (Hirano and Mitchison, 1994; Hirano et al., 1997; Kimura and Hirano, 1997).

During G_2 and into prophase, cyclin B2 is synthesized and accumulates progressively in the cytoplasm where it rapidly associates with CDK1 whose activity is believed to drive the cell into mitosis. However, once this CDK1/cyclin B2 complex forms, its activity is inhibited by the phosphorylations on Thr-14 and Tyr-15 of CDK1 by the Myt1 kinase located in the cytoplasm (Dunphy and Newport, 1989; Solomon et al., 1990; Li et al., 1997). Then, at some point in prophase, the CDK1/cyclin B2 complex translocates into the nucleus (Pines and Hunter, 1991; Gallant and Nigg, 1992). This nuclear import is thought to occur when the cytoplasmic retention signal (CRS) associated with the cyclin B2 subunit becomes phosphorylated (e.g., Li et al., 1997); whether import occurs at the initiation of prophase, or later, remains to be determined. Regardless, once within the nucleus the CDK1/cyclin B complex is thought to be maintained in an inactive state through the phosphorylation of Try-15 by the nuclear wee1 kinase (e.g., Heald et al., 1993; Parker et al., 1995). Then, in response to an unknown signal, the CDK1/cyclin B within the nucleus is activated by the influx of cdc25, whose activity is also regulated by phosphorylation (Peng et al., 1997).

In recent years, as an understanding of the cyclin dependent kinases that control the cell cycle has increased, the beginning of mitosis has been increasingly defined as NEB. Using NEB as the marker for the start of mitosis has the advantage in that it is a discrete irreversible event that can be used for timing studies, and it integrates morphological changes in the cell with distinct biochemical events. NEB is controlled, in part, by the activity of CDK1-cyclin B2, which allows the nuclear envelope to vesiculate by hyperphosphorylating its associated lamin proteins (reviewed in Gerace and Foisner, 1994; Fields and Thompson,

1995). NEB is not dependent on mechanical disruption by the action of microtubules because it occurs at the normal time when microtubule assembly is completely inhibited (Sluder, 1979; Rieder and Palazzo, 1992).

In addition to the nuclear events that lead to chromosome condensation and ultimately NEB, some cytoplasmic components, such as the duplicated centrosomes, also undergo extensive biochemical modifications during prophase. Immunological evidence reveals that some CDK1 molecules become complexed with cyclin B1, which then accumulates at the centrosome during G_2 (Bailly et al., 1992). This centrosome-associated CDK1/cyclin B1 is then activated near the G_2/M boundary by cdc25B, the concentration of which increases during prophase (Gabrielli et al., 1996). Possibly the activation of centrosome-associated CDK1/cyclin B1 leads to the hyperphosphorylation of those centrosomal proteins that are involved in converting the replicated interphase centrosome into two independent radial arrays of dynamically unstable microtubules (Vandre et al., 1984) (see Fig. 4.2A).

PROMETAPHASE

The breakdown of the nuclear envelope (NEB) occurs over a 1–2 minute interval, and it signals the start of prometaphase, the stage when the spindle forms (Figs. 4.1B–D, 4.2B). Three essential events must be accomplished during this phase if the division is to be normal: The cell must establish a bipolar spindle axis; the daughter chromatids of each replicated chromosome must become connected to opposing spindle poles (i.e., bioriented); and the chromosomes must become aligned at or near the spindle equator.

For animal cells spindle bipolarity is determined by the two radial arrays of centrosomal microtubules (asters) as they separate. As noted above, this may occur prior to NEB, or it may occur after NEB. Regardless, all of the microtubules used to construct the spindle are derived from the centrosomes (Sluder and Rieder, 1985; reviewed in Brinkley, 1985). This is clearly demonstrated by the fact that cells with only one centrosome inevitably assemble a monopolar spindle (Mazia et al., 1960; Bajer, 1982; Sluder and Begg, 1985) and that those with more than two centrosomes typically form multipolar spindles (Heneen, 1975; Sluder et al., 1997). Furthermore, at least in sea urchin zygotes, chromosomes alone in a mitotic cytoplasm will not establish a bipolar spindle; in fact, no microtubules are formed in their vicinity (Sluder and Rieder, 1985).

It is important to note, however, that spindle assembly in some meiotic cells can involve different mechanisms (see Rieder et al., 1993). Although all male meiotic cells (spermatocytes) have centrosomes, spindle assembly in female meiotic cells (oocytes) of some organisms can occur without centrosomes. In such cells large numbers of free microtubules assemble in the immediate vicinity of the chromosomes after NEB, perhaps in response to local and as yet undefined changes in cytoplasmic conditions

(Karsenti et al., 1984a,b; Heald et al., 1996). Initially these microtubules are randomly arrayed, but with time they are bundled into a bipolar fusiform array by cross-linking proteins (e.g., NuMA) that act in concert with minus end–directed microtubule-dependent motor molecules such as cytoplasmic dynein (Hyman and Karsenti, 1996; Heald et al., 1997; Gaglio et al., 1997). We emphasize, however, that this centrosome-independent spindle assembly pathway is not seen during mitosis and that it is therefore relevant only to some meiotic systems. In mitotic cells NuMA and the minus end–directed microtubule motor molecules serve to maintain the focused anchorage of spindle microtubules that have detached from the centrosome (see Keating et al., 1997) and keep the centrosome attached to the end of the spindle (Heald et al., 1997).

Kinetochores are paired complex macromolecular assemblies that form on opposite sides of the primary constriction, or centromeric region, of each chromosome (reviewed in Rieder, 1982; Yen and Schaar, 1996). A chromosome becomes attached to the forming spindle when its "sister" kinetochores become associated with astral microtubules growing from the spindle poles (Rieder and Alexander, 1990). This attachment process is remarkably dynamic and depends on the properties of the microtubule ends as well as the ability of kinetochores to interact with these astral microtubules. At NEB astral microtubules grow into the volume previously occupied by the nucleus that now contains the condensed chromosomes. Although there is net growth of these microtubules, the growing tip of each is dynamically unstable (reviewed in Cassimeris et al., 1987); for each microtubule, there is a variable period of growth followed by rapid shortening, either back to the centrosome or, more likely, to some intermediate length. Thereafter the tip grows again. The result is that the volume occupied by the chromosomes is constantly being probed by the tips of growing astral microtubules.

Chromosome attachment to the spindle is accomplished by the ability of kinetochores to capture the ends or walls of astral microtubules, thereby forming the kinetochore fibers of the spindle (reviewed in Rieder and Alexander, 1990; Mitchison, 1990; Skibbens et al., 1993). These fibers, in turn, serve as the scaffold on which the poleward (P) forces are produced to move the chromosomes. Normally, due to the stochastic nature of kinetochore fiber formation, the attachment of sister kinetochores to the spindle is asynchronous. As a rule, the first kinetochore to attach is the one located closest to and facing a spindle pole at NEB (reviewed in Rieder, 1990). This attachment "monoorients" the chromosome and allows the kinetochore to move toward that pole (see Rieder and Alexander, 1990; Khodjakov et al., 1996; Figs. 4.1C, 4.2B). Once near the pole monooriented chromosomes begin to undergo continuous oscillatory movements toward and away from the pole, which reflect the directionally unstable nature of the attached kinetochore (reviewed in Khodjakov and Rieder, 1996). When moving toward the pole (P motion), the kinetochore is translocated by forces produced primarily at, or acting on, the kinetochore in concert with the coordinated disassembly of the ends of microtubules at the kinetochore. During away-from-the pole (AP) motion the kinetochore appears to be in a "neutral," nonforce-producing state that allows it to be pushed AP, while its associated microtubules elongate by the action of a polar ejection force that acts along the length of the chromosome. Some think that this polar ejection force, or "polar wind," is produced by the growth of astral microtubules (Rieder et al., 1986; Khodjakov and Rieder, 1996; Ault et al., 1991; Cassimeris et al., 1994; reviewed in Rieder and Salmon 1994). However, others have suggested that it is due to microtubule plus end motors that are associated with the chromosomes arms (e.g., chromokinesins) (Wang and Adler, 1995; Tokai et al., 1996). Regardless of the mechanism, as the chromosome moves AP the microtubules associated with the following kinetochore must lengthen by the addition of tubulin subunits at that kinetochore.

The bipolar attachment of a monooriented chromosome occurs as the previously unattached kinetochore captures and stabilizes microtubules from the more distant aster (McEwen et al., 1997). Again, this is a stochastic process that relies on the chance encounter of a microtubule wall or tip with the kinetochore, and it may be facilitated by the constant positional changes of the monooriented chromosome (Rieder, 1990). When the unattached kinetochore eventually captures one or more microtubules, the now "bioriented" chromosome rapidly initiates movement to the spindle equator. During this "congression" process both sister kinetochores remain directionally unstable and continue to show transient periods of P and AP motions. However, net changes in chromosome position occur because, once bioriented, the motilities of sister kinetochores become coordinated to allow for changes in chromosome position, and this coordination is thought to be mediated by a tension-sensing mechanism that acts across the centromere (e.g., Skibbens et al., 1995).

Over a variable period of time all of the chromosomes become attached to the spindle in a bipolar fashion and move to the midpoint or equator of the spindle. The aggregate of chromosomes positioned near or on the spindle equator forms the "metaphase plate." The establishment of this equilibrium position for any given chromosome is thought to be due to a balance between P pulling forces on the sister kinetochores, which is not necessarily "on" or equal at any given time, and the action of polar ejection forces whose strength in each half spindle drop off from the pole to the equator as the density of the growing astral microtubules falls off (reviewed in Rieder and Salmon, 1994; Khodjakov and Rieder, 1996; McEwen et al., 1997). Thus, as a chromosome moves away from the metaphase plate toward a spindle pole it encounters a progressively stronger force pushing it away from that pole. The P-moving leading kinetochore, now under greater tension, has a higher probability of becoming directionally unstable and changing from P movement to AP movement. As a consequence, the chromosome moves back toward the metaphase plate. Although photographs of living or fixed cells might suggest that chromosomes at the metaphase plate cease moving, time-lapse cinematography reveals that individually they constantly oscillate back and forth across the metaphase plate, rarely making large excursions. In addition, the size of the chromosomes determines whether or not the chromosomes are evenly distributed through the metaphase plate. During spindle formation there is a tendency for the larger chromosomes to be excluded from the spindle and to be positioned at the periphery—with the kinetochores just within the spindle and the chromosome arms projecting into the cytoplasm. When viewed with the microscope along the axis of the spindle, cells with predominantly large chromosomes (e.g., newt lung cells and rat kangaroo cells) have a metaphase plate that looks like a ring of chromosomes. On the other hand, cells with very small chromosomes (such as HeLa, LLC-PK, and CHO cells) have a metaphase plate that is solidly packed with chromosomes.

METAPHASE

When all the chromosomes are bioriented and positioned near the spindle equator, the cell is considered to be in the metaphase stage of mitosis (Figs. 4.1E, 4.2C). This stage has been traditionally defined solely by morphological criteria, namely, the alignment of all chromosomes on the metaphase plate. Also, during metaphase the distance between the spindle poles decreases as the spindle becomes progressively more compacted (cf. Fig. 4.2C,D). By morphological criteria metaphase represents the culmination of spindle assembly events occurring during prometaphase. As a consequence, the classic cytological terms *metaphase arrest* and *metaphase block* have lost any real meaning when applied to cells treated with agents that prevent spindle microtubule assembly; such cells are arrested in prometaphase of mitosis and are not necessarily poised to initiate anaphase onset (reviewed in Rieder and Palazzo, 1992). As we continue to learn more about the molecular events that trigger the metaphase–anaphase transition and the pathways that control them, the definition of metaphase could change away from being strictly morphological to include biochemical criteria. For example, the start of metaphase might be defined as the time when the checkpoint monitoring chromosome attachment (described later) is relieved and the cell becomes committed to later initiate anaphase.

ANAPHASE

The anaphase stage of mitosis starts when the sister chromatids comprising each chromosome disjoin to form two independent chromosomes, each of which immediately begins moving toward its attached spindle pole at 1–2 μm/min. Disjunction of sister chromatids does not depend on pulling forces generated by the spindle; when microtubule assembly is completely prevented, chromatid disjunction still occurs as the cell eventually undergoes the metaphase–anaphase transition (Eigsti and Dustin, 1955; Sluder, 1979; reviewed in Bajer and Mole-Bajer, 1972;

Palazzo and Rieder, 1992). Also, the P motion of the newly disjoined anaphase chromosomes does not appear to arise from the sudden activation of P force producers that begin to act on the kinetochore only during anaphase. The directionally unstable sister kinetochores on a metaphase chromosome undergo constant tension-related switches between P and AP activity states, and disjunction of the chromatids suddenly relieves the tension on both sister kinetochores, which then allows them to switch into a P state of motion. When one kinetochore on a metaphase chromosome is destroyed by laser microsurgery, the chromosome moves toward the other spindle pole with the same kinetics of an anaphase chromosome (reviewed in Rieder et al., 1995). Anaphase ends when chromosome P motion is completed. At some point in mid-to-late anaphase the process of cytokinesis (Figs. 4.1H, 4.2F), which pinches the cytoplasm in two between the separating groups of chromosomes, is also initiated (reviewed in Rappaport, 1969; White and Borisy, 1983; Salmon, 1989; Oegema and Mitchison, 1997).

The term *metaphase–anaphase transition* is widely used today to represent entry into the anaphase portion of the cell cycle. However, this term implies far more than chromatid disjunction and subsequent P chromosome motion; it refers to a fundamentally important transition in the cell cycle that commits the cell to finish mitosis and enter the next cell cycle. Until recently chromatid disjunction and exit from mitosis were thought to be triggered by the same mechanistic pathway (the sudden inactivation of CDK1/cyclin B2). Now, however, we know that chromatid disjunction can occur even when the cell is prevented from exiting mitosis by the expression of a nondegradable form of cyclin B (which keeps CDK1 activity high) (Holloway et al., 1993; Wheatley et al., 1997; Hinchcliffe et al., 1998). This indicates that chromatid disjunction and exit from mitosis are mediated by separate but normally coordinated pathway (reviewed in Holloway, 1995; Straight et al., 1996).

At the metaphase–anaphase transition the anaphase-promoting complex (APC) becomes activated. These large macromolecular assemblies, originally called *cyclosomes,* promote the polyubiquitination of numerous specific proteins. In turn, this then targets them for degradation by proteosomes (reviewed in King et al., 1996). Immunofluorescence analyses of lysed HeLa cells suggest that the APCs are associated primarily with the spindle (Tugendreich et al., 1995). Targets of the APC and subsequent proteolysis include the proteins that hold the chromatids together and cyclin B2 (Glotzer et al., 1991; Holloway et al., 1993; Murray, 1995). These proteolytic events ensure that the metaphase–anaphase transition is irreversible for both chromatid disjunction and exit from mitosis. The loss of CDK1 activity not only starts the return of the cell to interphase but may also lead to the dephosphorylation, and consequent activation, of those myosin molecules that participate in cytokinesis (Satterwhite et al., 1992; Wheatley et al., 1997). Although the assembly of the actin-based cytokinetic apparatus is initiated at or shortly after the metaphase–anaphase transition, the actual furrowing process is not apparent until later (Fig. 4.1F–H).

During anaphase each chromosome moves to its respective pole (anaphase A: Fig. 4.1F), and the poles themselves move further apart (anaphase B: Figs. 4.1G, 4.2E). These two motions act additively to increase the distance between the two separating genomes. Although anaphase A and B movements usually start simultaneously upon chromatid disjunction (as in vertebrates), in some organisms they begin at different times (reviewed in Mazia, 1961), suggesting that in some cases they can be independently regulated. Anaphase A involves two coordinated events: the movement of chromosomes through the cytoplasm by P forces that act at the kinetochore and the shortening of the microtubules attached to the kinetochore via tubulin subunit loss at the kinetochore. The mechanism(s) by which the motive force is generated for chromosome P motion remain a subject of lively debate and may differ, to various extents, between organisms (reviewed in McIntosh and Pfarr, 1991; Sawin and Endow, 1993; Rieder and Salmon, 1994). Mechanisms that are capable of providing the force for P chromosome movement include (1) minus end–directed motor molecules associated with the kinetochore that act on the microtubules associated with, and disassembling at, the kinetochore (Rieder and Alexander, 1990; McIntosh and Pfarr, 1991; Thrower et al., 1996; Brown et al., 1996) and the (2) disassembly of microtubule subunits at the kinetochore while the kinetochore "hangs on" to the shortening end and/or the contraction of a nonmicrotubule matrix within the kinetochore fiber (Koshland et al., 1988; Steffen and Linck, 1992; reviewed in Inoue and Salmon, 1995). In addition, the slow depolymerization of kinetochore microtubule minus ends at the spindle pole, occurring throughout mitosis, contributes a force-producing component for chromosome P motion, particularly in late anaphase (Mitchison and Salmon, 1992; Sawin and Mitchison, 1994; Waters et al., 1996a). Because very little force is required to move even large chromosomes through the cytoplasm at the slow (approximately 1–2 μm/min) speeds normally seen in anaphase (reviewed in Nicklas, 1988), any of these mechanisms could, in principle, provide the requisite forces for anaphase A.

The motive force for separating the spindle poles during anaphase B could come from two processes acting singly or in combination. In yeast and diatoms microtubule plus end–directed motors (e.g., members of the kinesin superfamily), which are anchored to a matrix in the spindle midzone and cross-link adjacent antiparallel microtubules, push the poles apart by working against the overlapping pole-to-pole microtubules (Hogan and Cande, 1990; Hogan et al., 1993; reviewed in Ault and Rieder, 1994). However, recent work suggests that the separation of the centrosomes during anaphase B in vertebrate somatic cells occurs from a pulling mechanism intrinsic to each pole (Waters et al., 1993; Wheatley et al., 1997). Indeed, in these cells the overlapping antiparallel microtubules that connect the two centrosomes during metaphase detach from the centrosomes during anaphase—so that the centrosomes are no longer connected, as in yeast and diatoms (Mastronarde et al., 1993). The putative pulling forces that act on these centrosomes, to effect anaphase B, are presumably produced by the interaction of astral microtubules with minus end–directed microtubule motors anchored to the cell cortex or cytoplasmic structures such as the endoplasmic reticulum (see Vaisberg et al., 1993; Shaw et al., 1997).

TELOPHASE

Shortly after the completion of anaphase B, a nuclear envelope reforms around both of the separated genomes. In those cell in which all of the chromosomes come into close contact during late anaphase (e.g., vertebrates) a single nuclear envelope simply forms around the single mass of chromosomes. However, in other systems (e.g., sea urchin zygotes), in which the chromosomes are still separated and not touching at the end of anaphase, a nuclear envelope forms around each individual chromosome. These karyomeres then aggregate and fuse into a single nucleus. During this final "telophase" stage of mitosis (Figs. 4.1H, 4.2F), the cell also begins to cleave between the separated nuclei in a process known as *cytokinesis.* The cleavage apparatus is composed of a circumferential band of actin and myosin that coordinately contracts and disassembles so that the constricting furrow is not sterically constrained from completing cell division by a mass of actomyosin (see Fishkind and Wang, 1993; Oegema and Mitchison, 1997). Typically cytokinesis is not completed during telophase because after this stage the two daughter cells are still connected by a stem body that is composed of tightly bundled antiparallel microtubules embedded in a densely staining matrix material. For nonconfluent cells in culture, cytokinesis is completed as the daughter cells move apart, rupturing the connection between them with one daughter cell inheriting the stem body. During this time the centrosome inherited by each cell reassumes its role in the cell as the nucleation center and organizer of the cytoplasmic microtubule complex.

The events that occur during telophase are triggered by the inactivation of CDK1/cyclin B and are clearly separable from those occurring during anaphase. Indeed, when cells are induced to express a nondegradable form of cyclin B, the chromosomes complete both anaphase A and anaphase B, but the chromosomes do not decondense, the nuclear envelope fails to reform, the centrosome remains in the "mitotic" state, and cytokinesis does not occur (Holloway et al., 1993; Wheatley et al., 1997; Hinchcliffe et al., 1998).

CHECKPOINT CONTROLS FOR THE METAPHASE–ANAPHASE TRANSITION

Checkpoint Monitoring Kinetochore Attachment to the Spindle

The purpose of mitosis is the generation of genetically identical daughter cells. This requires the establishment of a strictly bipolar spindle axis and the attachment of all sister chromatids to opposite spindle poles before the cell initiates the metaphase–anaphase transition. The problem the cell faces is that incomplete chromosome attachment (i.e., monoorientation) to the spindle is a normal part of the mitotic process (concepts discussed in Nicklas, 1989; Rieder et al., 1994; Nicklas and Ward, 1994). As discussed earlier, a chromosome first attaches to the spindle when one of its kinetochores captures microtubules from one aster leading to monoorientation to that spindle pole. Because astral microtubule density drops with distance from the more distant pole and relatively few dynamically unstable microtubules grow sufficiently long to span the interpolar distance, the distal kinetochore on the monooriented chromosome remains unattached for a highly variable period of time. A recent study of normal PtK cells revealed that the time from nuclear envelope breakdown to the bipolar attachment of all 12 chromosomes ranged from approximately 7 minutes to 1 hour, with one case taking almost 3 hours (Rieder et al., 1994). Given this enormous variability in the amount of time required for completing the bipolar attachment of all chromosomes, the cell would risk unequal chromosome distribution if the time of the metaphase–anaphase transition were determined by an invariant timing mechanism.

This essential coordination between chromosome attachment to the spindle and anaphase onset is not left to chance; in almost all higher eukaryotic cells the metaphase–anaphase transition is subject to a "checkpoint" control that delays anaphase onset in response to perturbations in spindle microtubule assembly and chromosome attachment to the spindle (Sluder, 1979; Sluder and Begg, 1983; Hoyt et al., 1991; Li and Murray, 1991; reviewed in Rieder and Palazzo, 1992; Wells, 1996). Checkpoints are signal transduction pathways that block progression of the cell cycle until the event being monitored is completed (concepts reviewed in Hartwell and Wienert, 1989; Murray, 1992). The checkpoint for the metaphase–anaphase transition consists of a detector that monitors bipolar chromosome attachment and a signal transduction pathway that targets some element of the machinery that triggers the sequence of molecular events of the metaphase–anaphase transition (see Hardwick and Murray, 1995).

The first indications that some sort of spindle-dependent function profoundly influenced the temporal progression of mitosis can, in retrospect, be traced back to the first use of colchicine in cytology in the late 1800s and the early 1900s (reviewed in Eigsti and Dustin, 1955). Early workers found that the disassembly of the achromatic figure (spindle) with colchicine would arrest or greatly prolong the duration of "metaphase" as seen by condensed chromosomes. The implications of these observations for mechanisms of cell cycle control were understandably not appreciated at the time, and the explicit statement of the idea that the time of anaphase onset may be dependent on the attachment of all chromosomes to the spindle was first outlined by Mazia (1961). This was based on observations of Bajer and Mole-Bajer (1956; see also Zirkle, 1970a,b) that mitotic cells appeared to "wait" until the last chromosome moved to the metaphase plate before entering anaphase. Then, in the early 1990s, studies with yeast and mammalian cells provided evidence that improper centromeric assembly could lead to a delay in the metaphase–anaphase transition. For example, Spencer and Hieter (1992) found that mutational or deletional manipulation of the centromeric DNA of a supernumerary chromosome in yeast led to a prolongation of mitosis. Also, injection of affinity-purified autoimmune sera containing antibodies against the CENP family of centromeric proteins into G_2 HeLa cells allows seemingly normal spindle assembly and chromosome attachment but blocks the cell cycle in mitosis (Bernat et al., 1990; Earnshaw et al., 1991). Similarly, injection of cultured cells with monoclonal antibodies to CENP-E, a kinesin-like protein at the kinetochore, blocks or substantially delays entry into anaphase (Yen et al., 1991, 1992). Finally, for spermatocytes of some mantid species anaphase does not occur, and the cells eventually degenerate, when the sex chromosomes fail to normally pair (reviewed in Nicklas and Arana, 1992).

The first direct demonstration for the existence of a checkpoint for the metaphase–anaphase transition that monitors chromosome attachment to the spindle in mammalian somatic cells came from the detailed analysis of the temporal relationship between chromosome attachment and the duration of mitosis, defined as the time from nuclear envelope breakdown to anaphase onset (Rieder et al., 1994). Video time-lapse recordings of 126 normal PtK cells revealed that the duration of mitosis was variable, ranging from 23 to 198 minutes. Importantly, this variability was found to be entirely due to the range of times required for the last monooriented chromosome to establish connections to both poles of the spindle. Once the last chromosome attached to the spindle and started to congress to the metaphase plate, anaphase onset occurred on average 23 minutes later, regardless of how long the cell had a monooriented chromosome. In these cells even a single monooriented chromosome delayed the onset of anaphase for up to 3 hours. Put differently, only one unattached kinetochore out of a total of 24 for PtK1 cells is sufficient to profoundly delay progression through mitosis.

How the cell detects the presence of even a single unattached kinetochore against a background of many properly attached chromosomes is still not fully understood. However, significant progress has been made in the past few years in characterizing the functional properties of the feedback system and in the identification of some of the proteins involved. In principle there are two possible mechanisms by which the cell can delay anaphase until all of its chromosomes are properly bioriented. In the first, which is a positive feedback pathway, attached kinetochores produce a promoter of anaphase onset, and anaphase onset would not occur until every kinetochore was involved in signal production. In the second possibility, which is a negative feedback pathway, unattached kinetochores produce an inhibitor of the metaphase–anaphase transition (McIntosh, 1991). The inhibitor must, to some extent, be diffusible to prevent the initiation of the metaphase–anaphase transition in the entire cell. According to this hypothesis, inhibitor production presumably ceases once the last kinetochore attaches to the spindle and its concentration ultimately reaches below the threshold that permits the metaphase–anaphase transition to occur (discussed by McIntosh, 1991; Earnshaw et al., 1991). To differentiate between these possibilities, Rieder et al. (1995) used a laser microbeam to selectively destroy the unattached kinetochore on the last monooriented chromosome in PtK1 cells. Normally these cells enter anaphase on average 23 minutes after the last kinetochore attaches. When the unattached kinetochore on the last monooriented chromosome is destroyed, cells initiated anaphase on average 17 minutes later, even though the irradiated chromosome remains monooriented at one of the spindle poles. This result reveals that unattached kinetochores produce an inhibitor of the metaphase–anaphase transition and that monoorientation per se is not the event monitored by the checkpoint pathway.

The aspect of kinetochore attachment that causes inhibitor production to cease is not presently known with certainty. The most obvious possibilities include the extent to which the kinetochore is saturated with attached microtubules (Rieder et al., 1994) and, alternatively, the production of tension across the centromere due to poleward-directed forces exerted by motor molecules in the sister kinetochores (McIntosh, 1991; Li and Nicklas, 1995, 1997). Presumably tension, which normally exists between attached sister kinetochores, could be detected by changes in the activity of putative stretch-sensitive enzymes producing the inhibitor (McIntosh, 1991) or by mechanical distortion of the kinetochore that could separate subunits of a multisubunit complex essential for the signal transduction. Experimentally distinguishing between these two possibilities has been difficult because the phenomena are interrelated; the accumulation of microtubules at the kinetochore promotes poleward forces, and tension at the kinetochore stabilizes microtubule attachment (Nicklas and Ward, 1994). Nevertheless, tension across the chromatid pair is the currently favored model due to two lines of evidence. First, the checkpoint inhibition at the metaphase–anaphase transition is maintained when low doses of the microtubule-stabilizing drug taxol are applied to cells in which all the chromosomes have established bipolar connections to the spindle (Rieder et al., 1994). Taxol inhibits microtubule tip dynamics and leads to a relaxation of tension between kinetochores on chromosomes (Waters et al., 1996b), but *on average,* does not diminish the number of microtubules attached to individual kinetochores beyond the normal range (McEwen et al., 1997). Second, micromanipulation studies on mantid spermatocytes have linked the experimental application of tension on a monooriented chromosome with the ability of the cell to initiate anaphase (Li and Nicklas, 1997). In these meiotic cells univalent chromosomes form naturally in meiosis I when partners fail to pair. The two kinetochores of such univalents lie on the same side of the chromosome and function as a single unit. Although attached by some microtubules to one of the spindle poles, these kinetochores inhibit anaphase onset for many hours. However, when a microneedle is used to apply tension away from the spindle pole, the cell initiates anaphase.

Here it is not known to what extent tension at a kinetochore promotes the stability of microtubule attachment—a consideration suggested by the finding that tension promotes stable kinetochore attachment to the spindle in insect spermatocytes (Nicklas and Staehly, 1967; Nicklas, 1967; Nicklas and Koch, 1969). Also, in vertebrate somatic cells the attached kinetochore of a monooriented chromosome, on which the unattached kinetochore has been laser ablated, does not inhibit the onset of anaphase (Rieder et al., 1995). Perhaps the tension experienced by the attached kinetochore during mitosis in somatic cells is higher than that during meiosis in spermatocytes and thus does not inhibit the metaphase–anaphase transition. In any case, if tension is the event monitored, it must be tension at the individual kinetochore, not tension across the centromere as was earlier proposed (McIntosh, 1991). Laser ablation of just one kinetochore of a bioriented chromosome in PtK1 cells leaving the centromere intact relaxes tension across the chromatid pair, but this chromosome will not delay anaphase onset after the last naturally monooriented chromosome establishes bipolar connections to the spindle (Rieder et al., 1995).

A biochemical correlate has recently been found for an activity produced by unattached kinetochores. Unattached kinetochores contain an unknown phosphoprotein that reacts to the 3F3 antibody, a monoclonal originally raised against whole thiophosphorylated frog egg extracts (reviewed in Gorbsky, 1995). Unattached kinetochores of PtK1 cells stain brightly, while the kinetochores of chromosomes moving to the metaphase plate label weakly, often with one kinetochore staining more brightly than the other. Once chromosomes become aligned on the metaphase plate the kinetochores do not label (Gorbsky and Rickets, 1993). In grasshopper and mantid spermatocytes, all kinetochores react with the antibody to some extent, but only those that are unattached or not under tension show bright immunofluorescent staining (Gorbsky and Ricketts, 1993; Nicklas et al., 1995). Importantly, the experimental application of tension of monooriented univalents correlates with a decided reduction in staining brightness of kinetochores in these spermatocytes (Li and Nicklas, 1997). Whether the kinetochore protein(s) recognized by this antibody play a role in the checkpoint pathway is still not certain in that all kinetochores are stained with similar intensity throughout mitosis in newt lung cells (Waters et al., 1996b). However, we note the interesting finding that microinjection of the antibody into dividing PtK cells arrests the cells in mitosis for a time that is proportional to the amount of antibody injected (Campbell and Gorbsky, 1995). In such cells the chromosomes move to the metaphase plate while the antibody remains bound to the kinetochores, and, when the cells eventually initiate anaphase, the injected antibody is lost from the kinetochores. These results were interpreted as evidence that the antibody protects the phosphoepitopes at the kinetochore, thereby artificially maintaining the activation of the checkpoint.

Although the checkpoint for the metaphase–anaphase transition in some cell types, such as HeLa, leads to a permanent arrest in mitosis until the cells undergo apoptosis, it is important to note that many other kinds of cells will eventually show synchronous chromosome disjunction and progress into the next G_1 regardless of whether unattached kinetochores are present (reviewed in Rieder and Palazzo, 1992). Little information is available concerning how cells "escape" from the checkpoint or why the checkpoint in some cells is more leaky than others. Perhaps the inhibitory influence produced by unattached kinetochores acts to greatly slow but not stop the progression of molecular reactions that trigger the execution of the metaphase–anaphase transition. If so, differences between cell types in the extent to which the checkpoint slows these reactions could explain the variation in the behavior observed. Alternatively, cells may have evolved specific compensatory mechanisms that in time allow the completion of mitosis, albeit defective, by downregulating the checkpoint. With regard to these possibilities, we note the interesting finding that nocodazole-treated cultured cells with many unattached kinetochores spontaneously escape the checkpoint sooner when they contain residual astral microtubules (Andreassen and Margolis, 1994).

The functional characteristics of the inhibitory influence produced by unattached kinetochores and the location of its target are not fully understood. Rieder et al. (1997) investigated these issues by following mitosis in PtK1 cells containing two spindles in a common cytoplasm. Although adjacent to each other, the spindles were independent of one another, and their microtubule arrays did not overlap. The rationale for this approach was that the variability in the time for the completion of kinetochore attachment for any one spindle should produce instances where one spindle had completed chromosome congression while the other still had one or more unattached kinetochores. These workers found that multiple unattached kinetochores in one spindle did not inhibit anaphase onset in the neighboring spindle that was "mature," i.e., in which all chromosomes had established bipolar attachments. As in normal cells with only one spindle, anaphase started in the "mature" spindle on average 24 minutes after the last monooriented chromosome established bipolar connections. In no case did the unattached kinetochores in the "immature" spindle delay anaphase onset in the leading or "mature" spindle. Thus, the inhibitory influence produced by unattached kinetochores acts locally only at the level of the individual spindle; therefore, the target of the inhibitor is in the spindle, not the cytoplasm. Importantly, at least seven unattached kinetochores on a monopolar spindle approximately 20 μm away from a bipolar spindle did not prevent anaphase onset in the latter. In comparison, however, a single unattached kinetochore near one of the poles on a large multipolar PtK1 spindle inhibits anaphase onset in all parts of the spindle, even those located greater than 20 μm from the monooriented chromosome (Sluder et al., 1997). Thus, the inhibitory influence can propagate greater distances within a spindle than between adjacent independent spindles. Together these observations suggest that the inhibitor of the metaphase–anaphase transition produced by unattached kinetochores becomes structurally associated with the spindle containing such kinetochores.

This study also revealed that the molecular changes of the metaphase–anaphase transition propagate throughout the cell and are dominant over the inhibitory activity of unattached kinetochores (also see Sluder et al., 1994). In all cases the lagging spindle initiated anaphase on average 9 minutes after anaphase onset in the first spindle to congress all of its chromosomes. The key finding was that in eight cases the lagging or "immature" spindles initiated anaphase on average 9 minutes after the leaders even though the former contained one or more unattached kinetochores. This observation, together with the finding that the inhibitor produced by unattached kinetochores acts just at the level of the individual spindle, suggests that the molecular events of the metaphase–anaphase transition are initiated within the spindle, not the cytoplasm.

Apparent Checkpoint for Spindle Assembly

A number of observations have led to the popular hypothesis that mitotic cells possess a "spindle assembly" checkpoint that is distinct from that for unattached kinetochores (reviewed in Murray, 1992; Murray and Mitchison, 1994; Earnshaw and MacKay, 1994; Gorbsky, 1995; Wells, 1996). Complete inhibition of spindle microtubule assembly significantly delays or indefinitely blocks the metaphase–anaphase transition in every cell type tested (reviewed in Rieder and Palazzo, 1992) with the arguable exception of the early cleavage stage *Xenopus* zygotes (cf. Hara et al., 1980; Shinagawa, 1983). Even subtle changes in spindle assembly caused by low doses of microtubule poisons (Colcemid, nocodazole, or vinblastine), or the microtubule-stabilizing agent taxol, lead to significant delays in the metaphase–anaphase transition even though all kinetochores appear to establish functional attachments to the bipolar spindle and move to the metaphase plate (Zieve et al., 1980; Jordan et al., 1991, 1992, 1993; Rieder et al., 1994; McEwen et al., 1997). Importantly, the duration of mitosis is also greatly increased by defects in the spatial arrangement, or architecture, of spindle microtubules. In both vertebrate somatic cells and sea urchin zygotes mitosis is prolonged approximately two- to threefold if the cell assembles a monopolar spindle (Bajer, 1982; Wang et al., 1983; Sluder and Begg, 1985; Jensen et al., 1987). Furthermore, when the spindle in a sea urchin zygote is microsurgically cut into two half spindles, the duration of mitosis is tripled even though the cell contains the normal number of spindle poles and the normal complement of astral microtubules (Sluder and Begg, 1983). These results support the idea that the controls for the metaphase–anaphase transition can detect defects in the bipolar organization of the spindle or the presence of unpaired spindle poles that are not interacting with each other. However, the existence of a distinct "spindle assembly" checkpoint is open to question, because disruption of spindle organization can also produce unattached kinetochores and perturbations of microtubule–kinetochore interactions that lead to less than normal tension at one or more kinetochores.

To test whether gross defects in spindle architecture per se activate a checkpoint, Sluder et al. (1997) characterized the duration of mitosis in cells containing multipolar spindles with all chromosomes bioriented between pairs of spindle poles. The results of this study revealed that sea urchin zygotes containing tripolar or tetrapolar spindles progressed from NEB breakdown to anaphase onset with normal timing. Also, the presence of supernumerary, unpaired spindle poles did not prolong mitosis. Similarly, observation of PtK1 cells that formed tripolar or tetrapolar spindles revealed that they progressed through mitosis on average at the normal rate. More importantly, the interval between the bipolar attachment of the last monooriented chromosome and anaphase onset was normal. Thus, these cells do not functionally have a separate checkpoint that delays the onset of anaphase when there are too many spindle poles or when the spindle lacks a bipolar symmetry. As long as a cell assembles spindle microtubules when it comes into mitosis, the completion of kinetochore attachment is the event that limits when it will intiate the metaphase–anaphase transition.

The fact that spindle multipolarity inevitably leads to aneuploidy raises the question of why there is no checkpoint to monitor this important parameter. A possible answer is that checkpoint mechanisms offer a direct selective advantage only for defects that the cell can ultimately resolve. For example, chromosome monoorientation is a normal feature of mitosis and can be eventually corrected. In contrast, cells typically are not able to correct for the presence of too many spindle poles and consequently multipolar divisions always ensue (see Keryer et al., 1984; Sellitto and Kuriyama, 1988; Rieder et al., 1986). Thus, a checkpoint for the metaphase–anaphase transition that monitors bipolar spindle symmetry would serve no functional purpose.

Molecules Involved in the Checkpoint Signal Transduction Pathway

Genetic analyses of yeast mutants has identified six genes that code for proteins involved with the checkpoint: MAD1, MAD2, MAD3, BUB1, BUB2, and BUB3. These gene products are not involved in spindle assembly or function because yeast will divide normally with mutations in these genes. However, when spindle assembly is compromised by benomyl (an inhibitor of microtubule assembly), the mutant cells fail to arrest in mitosis, as normal cells do, and rapidly die. Combined biochemical and genetic analyses have defined a signal transduction pathway that puts BUB1 and BUB3 upstream of MAD1 and MAD2, with BUB2 and MAD3 downstream of these genes (Hardwick and Murray, 1995; Elledge, 1996; Hardwick et al., 1996; Rudner and Murray, 1996; Wells, 1996). Another kinase important for the checkpoint in budding yeast, MPS1, appears to act immediately upstream of MAD1 (Weiss and Winey, 1996). Recent work with the vertebrate homologues of these proteins has revealed that MAD2 and BUB1 localize to kinetochores of mammalian cells and are essential for the function of the checkpoint that detects unattached kinetochores (Chen et al., 1996; Li and Benezra, 1996; Taylor and McKeon, 1997 (Waters et al., 1998; Gorbsky et al., 1998)). Noteworthy are the findings that MAD2 and BUB1 immunoreactivities are lost from the kinetochores of chromosomes that have established bipolar connections to the spindle, while unattached kinetochores retain strong immunoreactivity. These observations suggest that when these two key components of the signal transduction pathway come off of the last unattached kinetochore, the checkpoint is relieved and the cell divides to produce two genetically identical daughters, thereby fulfilling the purpose of the cell cycle.

ACKNOWLEDGMENTS

The authors thank Dr. Alexey Khodjakov for help in preparing the figures. Work cited from their laboratories was supported by NIH GM 30758 to G.S., NIH GM 40198 to C.L.R., and NIH NCRR-01219, awarded by the DHHS/PHS, which supports the Wadsworth Center Biological Microscopy and Image Reconstruction Facility as a National Biotechnological Resource. E.H.H. is supported by a Cell Biology of Development training grant (NIH HD07312).

REFERENCES

Andreassen PR, Margolis RL (1994): Microtubule dependency of p34cdc2 inactivation and mitotic exit in mammalian cells. J Cell Biol 127:789–802.

Ault JG, DeMarco AJ, Salmon ED, Rieder CL (1991): Studies on the ejection properties of asters: Astral microtubule turnover influences the oscillatory behavior and positioning of mono-oriented chromosomes. J Cell Sci 99:701–710.

Ault JG, Rieder CL (1994): Centrosome and kinetochore movement during mitosis. Curr Opin Cell Biol 6:41–49.

Bailly E, Pines J, Hunter T, Bornens M (1992): Cytoplasmic accumulation of cyclin B1 in human cells: Association with a detergent-resistant compartment of the centrosome. J Cell Sci 101:529–545.

Bajer A (1959): Change of length and volume of mitotic chromosomes in living cells. Hereditas 45:579–596.

Bajer AS (1965): Subchromatid structure of chromosomes in the living state. Chromosoma 17:291–302.

Bajer AS (1982): Functional autonomy of monopolar spindle and evidence for oscillatory movement in mitosis. J Cell Biol 93:33–48.

Bajer AS, Mole-Bajer J (1956): Cine-micrographic studies on mitosis in endosperm II. Chromosoma 7:558–607.

Bajer AS, Mole-Bajer J (1972): Spindle Dynamics and Chromosome Movements. New York: Academic Press.

Bernat RL, Borisy GG, Rothfield NF, Earnshaw WC (1990): Injection of anticentromere antibodies in interphase disrupts events required for chromosome movement at mitosis. J Cell Biol 111:1519–1533.

Bradbury EM (1992): Reversible histone modifications and the chromosome cell cycle. BioEssays 14:9–16.

Brinkley BR (1985): Microtubule organizing centers. Annu Rev Cell Biol 1:145–172.

Brown KD, Wood KW, Cleveland DW (1996): The kinesin-like protein CENP-E is kinetochore-associated throughout poleward chromosome segregation during anaphase-A. J Cell Sci 109:961–969.

Campbell MS, Gorbsky GJ (1995): Microinjection of mitotic cells with the 3F3/2 antiphosphoepitope antibody delays the onset of anaphase. J Cell Biol 129:1195–1204.

Cassimeris L, Rieder CL, Salmon ED (1994): Microtubule assembly and kinetochore directional instability in vertebrate monopolar spindles: Implications for the mechanism of chromosome congression. J Cell Sci 107:285–297.

Cassimeris LU, Walker RA, Pryer NK, Salmon ED (1987): Dynamic instability of microtubules. Bioessays 7:149–154.

Chen RH, Waters JC, Salmon ED, Murray AW (1996): Association of spindle assembly checkpoint component XMAD2 with unattached kinetochores. Science 274:242–246.

Dunphy WG, Newport JW (1989): Fission yeast p13 blocks mitotic activation and tyrosine dephosphorylation of the *Xenopus* cdc2 protein kinase. Cell 58:181–191.

Earnshaw WC, Bernat RL, Cooke CA, Rothfield NF (1991): Role of the centromere/kinetochore in cell cycle control. Cold Spring Harbor Symp Quant Biol 56:675–685.

Earnshaw WC, MacKay AM (1994): Role of nonhistone proteins in the chromosomal events of mitosis. FASEB J 8:947–956.

Eigsti OJ, Dustin P (1995): Colchicine in Agriculture, Medicine, Biology, and Chemistry. Ames: The Iowa State College Press.

Elledge SJ (1996): Cell cycle checkpoints: Preventing an identity crisis. Science 274:1664–1672.

Fields, AP, and LJ Thompson (1995). The regulation of mitotic nuclear envelope breakdown: a role for multiple lamin kinases. Prog Cell Cycle Res 1:271–86.

Fishkind DJ, Wang YL (1993): Orientation and three-dimensional organization of actin filaments in dividing cultured cells. J Cell Biol 123:837–848.

Gabrielli BG, DeSouza CPC, Tonks ID, Clark JM, Hayward NK, Ellem KAO (1996): Cytoplasmic accumulation of cdc25B phosphatase in mitosis tritters centrosomal microtubule nucleation in HeLa cells. J Cell Sci 109:1081–1093.

Gaglio T, Dionne MA, Compton DA (1997): Mitotic spindle poles are organized by structural and motor proteins in addition to centrosomes. J Cell Biol 138:1055–1066.

Gallant P, Nigg EA (1992): Cyclin B2 undergoes cell cycle–dependent nuclear translocation and, when expressed as a non-destructible mutant, causes mitotic arrest in HeLa cells. J Cell Biol 117:213–224.

Gerace AP, Foisner F (1994): Integral membrane proteins and dynamic organization of the nuclear envelope. Trends Cell Biol 4:127–131.

Glotzer M, Murray AW, Kirschner MW (1991): Cyclin in degraded by the ubiquitin pathway. Nature 349:132–138.

Gorbsky, GJ, Chen RH, and Murray AW (1998). Microinjection of Antibody to Mad2 Protein into Mammalian Cells in Mitosis Induces Premature Anaphase. J. Cell Biol. 141:1193–1205.

Gorbsky GJ (1995): Kinetochores, microtubules and the metaphase checkpoint. Trends Cell Biol 5:143–148.

Gorbsky GJ, Ricketts WA (1993): Differential expression of a phosphoepitope at the kinetochores of moving chromosomes. J Cell Biol 122:1311–1321.

Hara K, Tydeman P, Kirschner M (1980): A cytoplasmic clock with the same period as the division cycle in *Xenopus* eggs. Proc Natl Acad Sci USA 77:462–466.

Hardwick KG, Murray AW (1995): Mad1p, a phosphoprotein component of the spindle assembly checkpoint in budding yeast. J Cell Biol 131:709–720.

Hardwick KG, Weiss E, Luca FC, Winey M, Murray AW (1996): Activation of the budding yeast spindle assembly checkpoint without mitotic spindle disruption. Science 273:953–956.

Hartwell LH, Weinert TA (1989): Checkpoints: Controls that ensure the order of cell cycle events. Science 246:629–634.

Heald R, McLoughlin M, McKeon F (1993): Human Wee1 maintains mitotic timing by protecting the nucleus from cytoplasmically activated Cdc2 kinase. Cell 74:463–474.

Heald R, Tournebize R, Blank T, Sandaltzopoulos R, Becker P, Hyman A, Karsenti E (1996): Self-organization of microtubules into bipolar spindles around artificial chromosomes in *Xenopus* egg extracts. Nature 382:420–425.

Heald R, Tournebize R, Habermann A, Karsenti E, Hyman A (1997): Spindle assembly in *Xenopus* egg extracts: Respective roles of centrosomes and microtubule self-organization. J Cell Biol 138:615–628.

Hendzel MJ, Wei Y, Mancini MA, Van Hooser A, Ranalli T, Brinkley BR, Bazett-Jones DP, Allis CD (1997): Mitosis-specific phosphorylation of histone H3 initiates primarily within pericentromeric heterochromatin during G_2 and spreads in an ordered fashion coincident with mitotic chromosome condensation. Chromosoma 106:348–360.

Heneen WK (1975): Kinetochores and microtubules in multipolar mitosis and chromosome orientation. Exp Cell Res 91:57–62.

Hinchcliffe EH, Cassels GO, Rieder CL, Sluder G (1998): The coordination of centrosome reproduction with nuclear events during the cell cycle in the sea urchin zygote. J Cell Biol J Cell Biol. 140:1417–1426.

Hirano T, Kobayashi R, Hirano M (1997): Condensins, chromosome condensation protein complexes containing XCAP-C, XCAP-E and a *Xenopus* homologue of the *Drosophila* barren protein. Cell 89:511–521.

Hirano T, Mitchison TJ (1994): A heterodimeric coiled-coil protein required for mitotic chromosome condensation in vitro. Cell 79:449–458.

Hogan CJ, Cande WZ (1990). Antiparallel microtubule interactions: Spindle formation and anaphase B. Cell Motil Cytoskel 16:99–103.

Hogan CJ, Wein H, Wordeman L, Scholey JM, Sawin KE, Cande WZ (1993): Inhibition of anaphase spindle elongation in vitro by a peptide antibody that recognizes kinesin motor domain. Proc Natl Acad Sci USA 90:6611–6615.

Holloway SL (1995): Sister chromatid separation in vivo and in vitro. Curr Opin Genet Dev 5:243–248.

Holloway S, Glotzer M, King RW, Murray AW (1993): Anaphase is initiated by proteolysis rather than by the inactivation of maturation-promoting factor. Cell 73:1393–1402.

Hoyt MA, Totis L, Roberts BT (1991): *S. cerevisiae* genes required for cell cycle arrest in response to loss of microtubule function. Cell 66:507–517.

Hyman AA, Karsenti E (1996): Morphogenetic properties of microtubules and mitotic spindle assembly. Cell 84:401–410.

Inoue S, Salmon ED (1995): Force generation by microtubule assembly/disassembly in mitosis and related movements. Mol Biol Cell 6:1619–1640.

Jensen CG, Davison EA, Bowser SS, Rieder CL (1987): Primary cilia cycle in PtK1 cells: Effects of colcemid and taxol on cilia formation and resorption. Cell Motil Cytol 7:187–197.

Jordan MA, Thrower D, Wilson L (1991): Mechanism of inhibition of cell proliferation by Vinca alkaloids. Cancer Res 51:2212–2222.

Jordan MA, Thrower D, Wilson L (1992): Effects of vinblastine, podophyllotoxin and nocodazole on mitotic spindles. Implications for the role of microtubule dynamics in mitosis. J Cell Sci 102:401–416.

Jordan MA, Toso RJ, Thrower D, Wilson L (1993): Mechanism of mitotic block and inhibition of cell proliferation by taxol at low concentrations. Proc Natl Acad Sci USA 90:9552–9556.

Karsenti E, Kobayashi S, Mitchison T, Kirschner M (1984a): Role of the centrosome in organizing the interphase microtubule array: Properties of cytoplasts containing or lacking centrosomes. J Cell Biol 98:1763–1776.

Karsenti E, Newport J, Hubble R, Kirschner M (1984b): Interconversion of metaphase and interphase microtubule arrays, as studied by the injection of centrosomes and nuclei into *Xenopus* eggs. J Cell Biol 98:1730–1745.

Keating TJ, Peloquin JG, Rodionov VI, Momcilovic D, Borisy GG (1997): Microtubule release from the centrosome. Proc Natl Acad Sci USA 94:5078–5083.

Keryer G, Ris H, Borisy GG (1984): Centriole distribution during tripolar mitosis in Chinese hamster ovary cells. J Cell Biol 98:2222–2229.

Khodjakov A, Cole RW, Bajer AS, Rieder CL (1996): The force for poleward chromosome motion in *Haemanthus* cells acts along the length of the chromosome during metaphase but only at the kinetochore during anaphase. J Cell Biol 132:1093–1104.

Khodjakov A, Rieder CL (1996): Kinetochores moving away from their associated pole do not exert a significant pushing force on the chromosome. J Cell Biol 135:315–327.

Kimura K, Hirano T (1997): ATP-dependent positive supercoiling of DNA by 13S condensin: A biochemical implication for chromosome condensation. Cell 90:625–634.

King RW, Deshaies RJ, Peters J-M, Kirschner MW (1996): How proteolysis drives the cell cycle. Science 274:1652–1659.

Koshland DE, Mitchison TJ, Kirschner MW (1988): Polewards chromosome movement driven by microtubule depolymerization in vitro. Nature 331:499–504.

Li J, Meyer AN, Donoghue DJ (1997): Nuclear localization of cyclin B1 mediates its biological activity and is regulated by phosphorylation. Proc Natl Acad Sci USA 94:502–507.

Li R, Murray AW (1991): Feedback control of mitosis in budding yeast. Cell 66:519–531.

Li X, Nicklas RB (1995): Mitotic forces control a cell-cycle checkpoint. Nature 373:630–632.

Li X, Nicklas RB (1997): Tension-sensitive kinetochore phosphorylation and the chromosome distribution checkpoint in praying mantid spermatocytes. J Cell Sci 110:537–545.

Li Y, Benezra R (1996): Identification of a human mitotic checkpoint gene: hsMAD2. Science 274:246–248.

Mastronade DN, McDonald KL, Ding R, McIntosh JR (1993): Interpolar spindle microtubules in PtK cells. J Cell Biol 123:1475–1489.

Mazia D (1961): Mitosis and physiology of cell division. In Brachet J, Mirsky A. (eds): The Cell, vol III. New York: Academic Press, pp 77–412.

Mazia D (1974): The cell cycle. Sci Am 230:53–64.

Mazia D, Harris P, Bibring T (1960): The multiplicity of the mitotic centers and the time-course of their duplication and separation. Biophys Biochem Cytol 7:1–20.

McEwen BF, Heagle AB, Cassels GO, Buttle KF, Rieder CL (1997): Kinetochore fiber maturation in PtK1 cells and its implications for the mechanisms of chromosome congression and anaphase onset. J Cell Biol 137:1567–1580.

McIntosh JR (1991): Structural and mechanical control of mitotic progression. Cold Spring Harb Symp Quant Biol 56:613–619.

McIntosh JR, Pfarr CM (1991): Mitotic motors. J Cell Biol 115:577–585.

Mitchison TJ (1990): Mitosis. The kinetochore in captivity. Nature 348:14–15.

Mitchison TJ, Salmon ED (1992): Poleward kinetochore fiber movement occurs during both metaphase and anaphase-A in newt lung cell mitosis. J Cell Biol 119:569–582.

Murray AW (1992): Creative blocks: Cell-cycle checkpoints and feedback controls. Nature 359:599–604.

Murray AW (1995): Cyclin ubiquitination: The destructive end of mitosis. Cell 81:149–152.

Murray AW, Mitchison TJ (1994): Mitosis. Kinetochores pass the IQ test. Curr Biol 4:38–41.

Nicklas RB (1967): Chromosome micromanipulation. II. Induced reorientation and the experimental control of segregation in meiosis. Chromosoma 21:17–50.

Nicklas RB (1988): The forces that move chromosomes in mitosis. Annu Rev Biophys Chem 17:431–449.

Nicklas RB, Arana P (1992): Evolution and the meaning of metaphase. J Cell Sci 102:681–690.

Nicklas RB (1989): The motor for poleward chromosome movement in anaphase is in or near the kinetochore. J Cell Biol 109:2245–55.

Nicklas RB, Koch CA (1969): Chromosome micromanipulation. III. Spindle fiber tension and the reorientation of mal-oriented chromosomes. J Cell Biol 43:40–50.

Nicklas RB, Staehly CA (1967): Chromosome micromanipulation. I. The mechanics of chromosome attachment to the spindle. Chromosoma 21:1–16.

Nicklas RB, and Ward SC (1994): Elements of error correction in mitosis: Microtubule capture, release, and tension. J Cell Biol 126:1241–1253.

Nicklas RB, Ward SC, Gorbsky GJ (1995): Kinetochore chemistry is sensitive to tension and may link mitotic forces to a cell cycle checkpoint. J Cell Biol 130:929–939.

Oegema K, Mitchison TJ (1997): Rappaport rules: Cleavage furrow induction in animal cells. Proc Natl Acad Sci USA 94:4817–4820.

Ohsumi K, Katagiri C, Kishimoto T (1993): Chromosome condensation in *Xenopus* mitotic extracts without histone H1. Science 262:2033–2035.

Parker LL, Sylvestre PJ, Byrnes MJ, Liu F, Piwnica-Worms H (1995): Identification of a 95-kDa WEE1 like tyrosine kinase in HeLa cells. Proc Natl Acad Sci USA 92:9638–9642.

Pederson T (1972): Chromatin structure and the cell cycle. Proc Natl Acad Sci USA 69:2224–2228.

Pederson T, Robbins E (1972): Chromatin structure and the cell division cycle. Actinomycin binding in synchronized HeLa cells. J Cell Biol 55:322–327.

Peng C-Y, Graves PR, Thoma RS, Wu Z, Shaw AS, Piwnica-Worms H (1997): Mitotic and G_2 checkpoint control: Regulation of 124-3-3 protein binding by phosphorylation of Cdc25C on Serine-216. Science 277:1501–1505.

Pines J, Hunter T (1991): Human cyclins A and B1 are differentially located in the cell and undergo cell cycle–dependent nuclear transport. J Cell Biol 115:1–17.

Rappaport R (1969): Aster-equatorial surface relations and furrow establishment. J Exp Zool 171:59–68.

Rieder CL (1982): The formation, structure, and composition of the mammalian kinetochore and kinetochore fiber. Int Rev Cytol 79:1–58.

Rieder CL (1990): Formation of the astral mitotic spindle: Ultrastructural basis for the centrosome-kinetochore interaction. Electron Microsc Rev 3:269–300.

Rieder CL, Alexander SP (1990): Kinetochores are transported poleward along a single astral microtubule during chromosome attachment to the spindle in newt lung cells. J Cell Biol 110:81–95.

Rieder CL, Ault JG Eichenlaub-Ritter U, Sluder G (1993): Morphogenesis of the mitotic and meiotic spindle: Conclusions obtained from one system are not necessarily applicable to the other. In Vig BK (ed): Chromosome Segregation and Aneuploidy. NATO-ASI Series, vol 72. New York: Springer-Verlag, pp 183–197.

Rieder CL, Cole RW, Khodjakov A, Sluder G (1995): The checkpoint delaying anaphase in response to chromosome monoorientation is mediated by an inhibitory signal produced by unattached kinetochores. J Cell Biol 130:941–948.

Rieder CL, Davison EA, Jensen LC, Cassimeris L, Salmon ED (1986): Oscillatory movements of monooriented chromosomes and their position relative to the spindle pole result from the ejection properties of the aster and half-spindle. J Cell Biol 103:581–591.

Rieder CL, Khodjakov A, Paliulis LV, Fortier TM, Cole RW, Sluder G (1997): Mitosis in vertebrate somatic cells with two spindles: Implications for the metaphase/anaphase transition checkpoint and cleavage. Proc Natl Acad Sci USA 94:5107–5112.

Rieder CL, Palazzo RE (1992): Colcemid and the mitotic cycle. J Cell Sci 102:387–92.

Rieder CL, Salmon ED (1994): Motile kinetochores and polar ejection forces dictate chromosome position on the vertebrate mitotic spindle. J Cell Biol 124:223–233.

Rieder CL, Schultz A, Cole R, Sluder G (1994): Anaphase onset in vertebrate somatic cells is controlled by a checkpoint that monitors sister kinetochore attachment to the spindle. J Cell Biol 127:1301–1310.

Rudner AD, Murray AW (1996): The spindle assembly checkpoint. Curr Opin Cell Biol 8:773–778.

Salmon ED (1989): Cytokinesis in animal cells. Curr Opin Cell Biol 1:541–547.

Satterwhite LL, Lohka MJ, Wilson KL, Cisek TY Corden LJ, Pollard TD (1992): Phosphorylation of myosin-II regulatory light chain by cyclin-p34cdc2. A mechanism for the timing of cytokinesis. J Cell Biol 118:595–605.

Sawin KE, Endow SA (1993): Meiosis, mitosis and microtubule motors. BioEssays 15:399–407.

Sawin KE, Mitchison TJ (1994): Microtubule flux in mitosis is independent of chromosomes, centrosomes, and antiparallel microtubules. Mol Biol Cell 5:217–226.

Sellitto C, Kuriyama R (1988): Distribution of pericentriolar material in multipolar spindles induced by colcemid treatment in Chinese hamster ovary cells. J Cell Sci 89:57–65.

Shaw SL, Yeh E, Maddox P, Salmon ED, Bloom K (1997): Astral microtubule dynamics in yeast: A microtubule-based searching mechanism for spindle orientation and nuclear migration into the bud. J Cell Biol 139:985–994.

Shinagawa A (1983): The interval of the cytoplasmic cycle observed in non-nucleate egg fragments is longer than that of the cleavage cycle in normal eggs of *Xenopus laevis*. J Cell Sci 64:147–162.

Skibbens RV, Rieder CL, Salmon ED (1995): Kinetochore motility after severing between sister centromeres using laser microsurgery: Evidence that kinetochore directional instability and position is regulated by tension. J Cell Sci 108:2537–2548.

Skibbens RV, Skeen VP, Salmon ED (1993): Directional instability of kinetochore motility during chromosome congression and segregation in mitotic newt lung cells: A push-pull mechanism. J Cell Biol 122:859–875.

Sluder G (1979): Role of spindle microtubules in the control of cell cycle timing. J Cell Biol 80:674–691.

Sluder G, Begg DA (1983): Control mechanisms of the cell cycle: Role of the spatial arrangement of spindle components in the timing of mitotic events. J Cell Biol 97:877–886.

Sluder G, Begg DA (1985): Experimental analysis of the reproduction of spindle poles. J Cell Sci 76:35–51.

Sluder G, Miller FJ, Thompson EA Wolf DE (1994): Feedback control of the metaphase–anaphase transition in sea urchin zygotes: Role of maloriented chromosomes. J Cell Biol 126:189–198.

Sluder G, Rieder CL (1985): Experimental separation of pronuclei in fertilized sea urchin eggs: Chromosomes do not organize a spindle in the absence of centrosomes. J Cell Biol 100:897–903.

Sluder G, Thompson EA, Miller FJ, Hayes J, Rieder CL (1997): The checkpoint control for anaphase onset does not monitor excess numbers of spindle poles or bipolar spindle symmetry. J Cell Sci 110:421–429.

Solomon MJ, Glotzer M, Lee TH, Philippe M, Kirschner MW (1990): Cyclin activation of p34cdc2. Cell 63:1013–1024.

Spencer F, Hieter P (1992): Centromere DNA mutations induce a mitotic delay in *Saccharomyces cerevisiae.* Proc Natl Acad Sci USA 89:8908–8912.

Steffen W, Linck RW (1992): Evidence for a non-tubulin spindle matrix and for spindle components immunologically related to tektin filaments. J Cell Sci 101:809–822.

Straight AF, Belmont AS, Robinett CC, Murray AW (1996): GFP tagging of budding yeast chromosomes reveals that protein-protein interactions can mediate sister chromatid cohesion. Curr Biol 6:1599–1608.

Taylor SS, McKeon F (1997): Kinetochore localization of murine Bub1 is required for normal mitotic timing and checkpoint response to spindle damage. Cell 89:727–735.

Thrower DA, Jordan MA, Wilson L (1996): Modulation of CENP-E organization at kinetochores by spindle microtubule attachment. Cell Motil Cytol 35:121–133.

Tokai N, Fujimoto-Nishiyama A, Toyoshima Y, Yonemura S, Tsukita S, Inoue J, Yamamoto T (1996): KID, a novel kinesin-like DNA binding protein, is localized to chromosomes and the mitotic spindle. EMBO J 15:457–467.

Tugendreich S, Tomkiel J, Earnshaw W, Hieter P (1995): CDC27Hs colocalizes with CDC16Hs to the centrosome and mitotic spindle and is essential for the metaphase to anaphase transition. Cell 81:261–268.

Vaisberg EA, Koonce MP, McIntosh JR (1993): Cytoplasmic dynein plays a role in mammalian mitotic spindle formation. J Cell Biol 123:849–858.

Vandre DD, Davis FM, Rao PN, Borisy GG (1984): Phosphoproteins are components of mitotic microtubule organizing centers. Proc Natl Acad Sci USA 81:4439–4443.

Wang RJ, Wissinger W, King EJ, Wang G (1983): Studies on cell division in mammalian cells. VII. A temperature-sensitive cell line abnormal in centriole separation and chromosome movement. J Cell Biol 96:301–306.

Wang S-Z, Adler R (1995): Chromokinesin: A DNA-binding, kinesin-like nuclear protein. J Cell Biol 128:761–768.

Waters JC, Cole RW, Rieder CL (1993): The force-producing mechanism for centrosome separation during spindle formation in vertebrates is intrinsic to each aster. J Cell Biol 122:361–372.

Waters JC, Chen RH Murray AW, Salmon ED (1998): Localization of Mad2 to Kinetochores Depends on Microtubule Attachment, Not Tension. J Cell Biol 141:1181–1191.

Waters JC, Mitchison TJ, Rieder CL, Salmon ED (1996a): The kinetochore microtubule minus-end disassembly associated with poleward flux produces a force that can do work. Mol Biol Cell 7:1547–1558.

Waters JC, Skibbens RV, Salmon ED (1996b): Oscillating mitotic newt lung cell kinetochores are, on average, under tension and rarely push. J Cell Sci 109:2823–2831.

Weiss E, Winey M (1996): The *Saccharomyces cerevisiae* spindle pole body duplication gene MPS1 is part of a mitotic checkpoint. J Cell Biol 132:111–123.

Wells WA (1996): The spindle assembly checkpoint: Aiming for a perfect mitosis, every time. Trends Cell Biol 6:228–234.

Wheatley SP, Hinchcliffe EH, Glotzer M, Hyman AA, Sluder G, Wang Y-L

(1997): CDK 1 inactivation regulates anaphase spindle dynamics and cytokinesis in vivo. J Cell Biol 138:385–393.

White JG, Borisy GG (1983): On the mechanisms of cytokinesis in animal cells. J Theor Biol 101:289–316.

Whitehead CM, Winkfein RJ, Rattner JB (1996): The relationship of HsEg5 and the actin cytoskeleton to centrosome separation. Cell Motil Cytoskel 35:298–308.

Yen TJ, Compton DA, Wise D, Zinkowski RP, Brinkley BR, Earnshaw WC Cleveland DW (1991): CENP-E, a novel human centromere-associated protein required for progression from metaphase to anaphase. EMBO J 10:1245–1254.

Yen TJ, Schaar BT (1996): Kinetochore function: molecular motors, switches and gates. Curr Opin Cell Biol 8:381–8.

Yen TJ, Li G, Schaar BT, Szilak I, Cleveland DW (1992): CENP-E is a putative kinetochore motor that accumulates just before mitosis. Nature 359:536–539.

Zieve GW, Turnbull D, Mullins M, McIntosh JR (1980): Production of large numbers of mitotic mammalian cells by use of the reversible microtubule inhibitor nocodazole. Exp Cell Res 126:397–405.

Zirkle RE (1970a): Involvement of the prometaphase kinetochore in prevention of precocious anaphase. J Cell Biol 47:235a.

Zirkle RE (1970b): UV-microbean irradiation of newt-cell cytoplasm: Spindle destruction, false anaphase, and delay of true anaphase. Radiat Res 41:516–537.

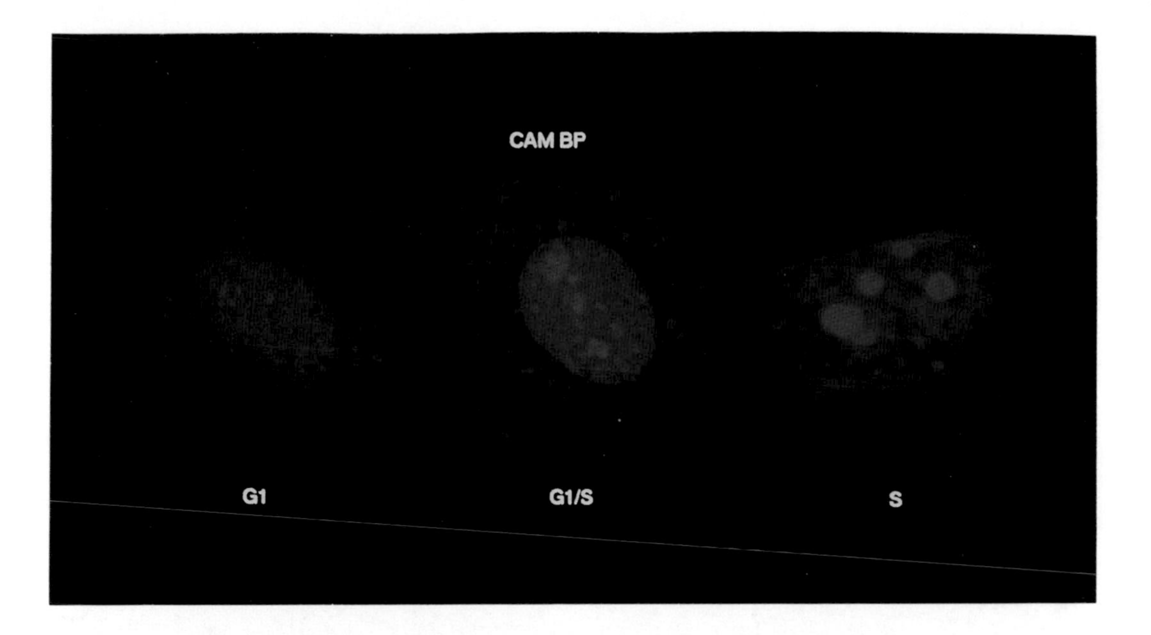

Figure 3.5. Immunocytochemical localization of CaM-BP68 in synchronized Chinese hamster embryo fibroblast (CHEF/18) cells. CHEF/18 cells grown on coverslips were synchronized by isoleucine deprivation as described by Reddy (1989). Zero hour (G_1), 6 hours (G_1/S), and 12 hours (S) following the release from isoleucine blocks, cells on the coverslips were extracted with buffer A (300 mM sucrose, 100 mM NaCl, 10 mM PIPES, 3 mM $MgCl_2$, 1 mM EGTA, pH 6.8, and 0.5% Triton X-100). This extraction procedure removes soluble cellular material, leaving cytoskeleton and nuclear architecture intact. Following extraction, cells were fixed on ice with 3.7% formaldehyde and subjected to immunostaining with rabbit polyclonal antibodies raised against purified CaM-BP68 and rhodamine-conjugated goat antirabbit IgG using a standard procedure. Cells were counterstained with DAPI, and photomicrography was performed with a Zeiss Axiofot microscope equipped for epifluorescence with a ×100 objective and single and multiple bandpass filter sets. Although a single stained cell is shown for each of the phases (G_1, G_1 phase; G_1/S, G_1/S boundary; and S, mid-to-late S phase), more than 85% of the cells on the coverslips representing each of the phases exhibited similar staining patterns under the synchronization conditions employed in these studies (E.G. Fey and G.P.V. Reddy, unpublished data).

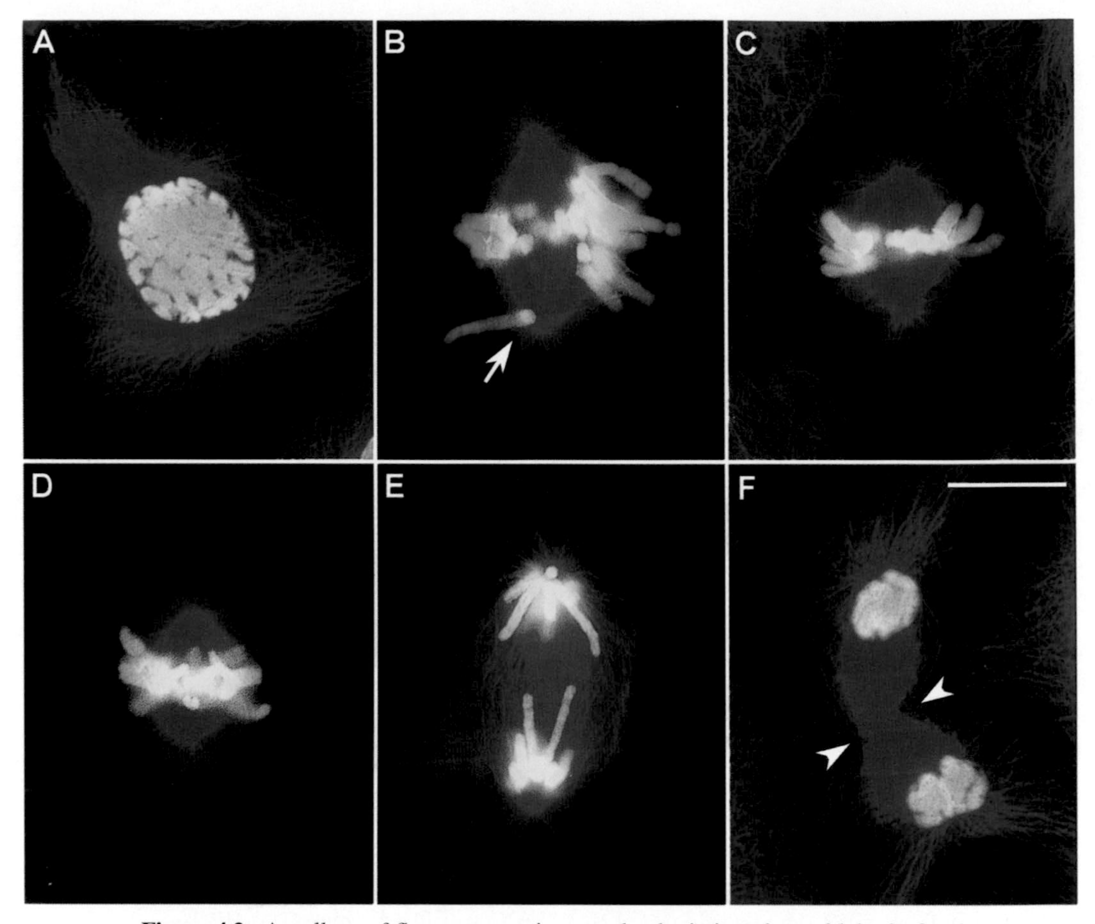

Figure 4.2. A gallery of fluorescent micrographs depicting glutaraldehyde-fixed and lysed PtK1 cells in various stages of mitosis. In these examples the microtubules (red) were stained using indirect immunofluorescent methods, while the chromosomes (green) were stained with the DNA probe Hoechst 33342. **(A)** A *late prophase* cell in which the two centrosomes and their associated radial arrays of astral microtubules have separated to opposite sides of the nucleus. Note that there are still many cytoplasmic microtubules in this cell that are not associated with the asters. **(B)** A *mid-prometaphase* cell that contains one monooriented chromosome (white arrow) and several congressing chromosomes. By this time, all of the cytoplasmic microtubules have disassembled, and most of the astral microtubules have been incorporated into the spindle. **(C)** A *metaphase* cell in which all of the chromosomes are aligned on the spindle equator. **(D)** A cell just entering *anaphase* in which the chromatids are disjoining. Note the compact nature of the spindle (cf. C and D). **(E)** A *late anaphase* cell in which the two groups of chromosomes are already at the spindle poles, which themselves are moving farther apart (anaphase B). **(F)** A *telophase* cell in which the two groups of well-separated chromosomes are reforming nuclei and in which cytokinesis (between the white arrowheads) is almost complete. The prominent bundle of microtubules between the two nuclei participates in cytokinesis and is known as the *midbody*. Bar in F = 15 μm. (Courtesy of Dr. Alexey Khodjakov.)

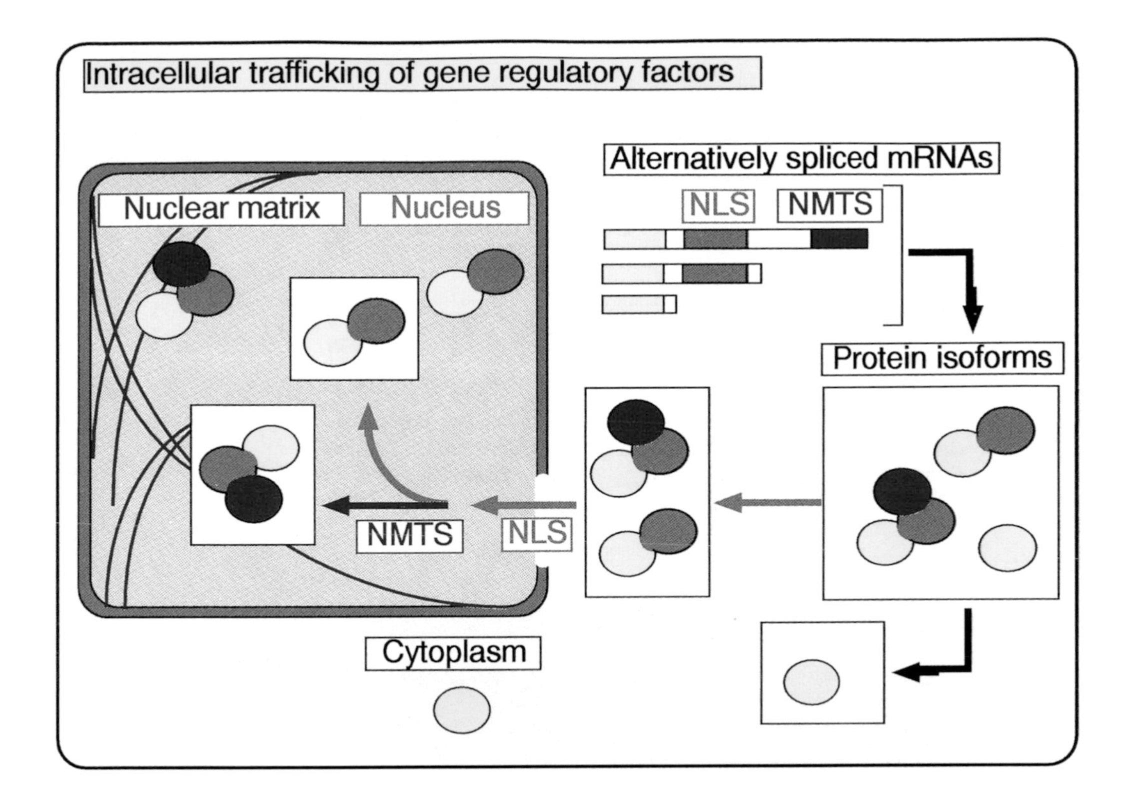

Figure 5.9. Intracellular trafficking of gene regulatory factors. The routing of transcription factors to distinct subcellular compartments is important for controlling the gene regulatory activities of these proteins. Presence (or functionality) of a nuclear localization signal (NLS, blue oval) dictates whether a protein (yellow oval) can be imported into the nucleus. A nuclear matrix targeting signal (NMTS, red oval) represents another key trafficking sequence that mediates association of regulatory proteins with the nuclear matrix. The paradigm for the diagram shown here is the AML/PEBP/CBFα class of transcription factors. These factors are encoded by several different genes that generate a series of alternatively spliced mRNAs. These mRNAs are translated into many protein isoforms that differ in the presence or absence of intracellular trafficking signals, determining subnuclear compartmentalization.

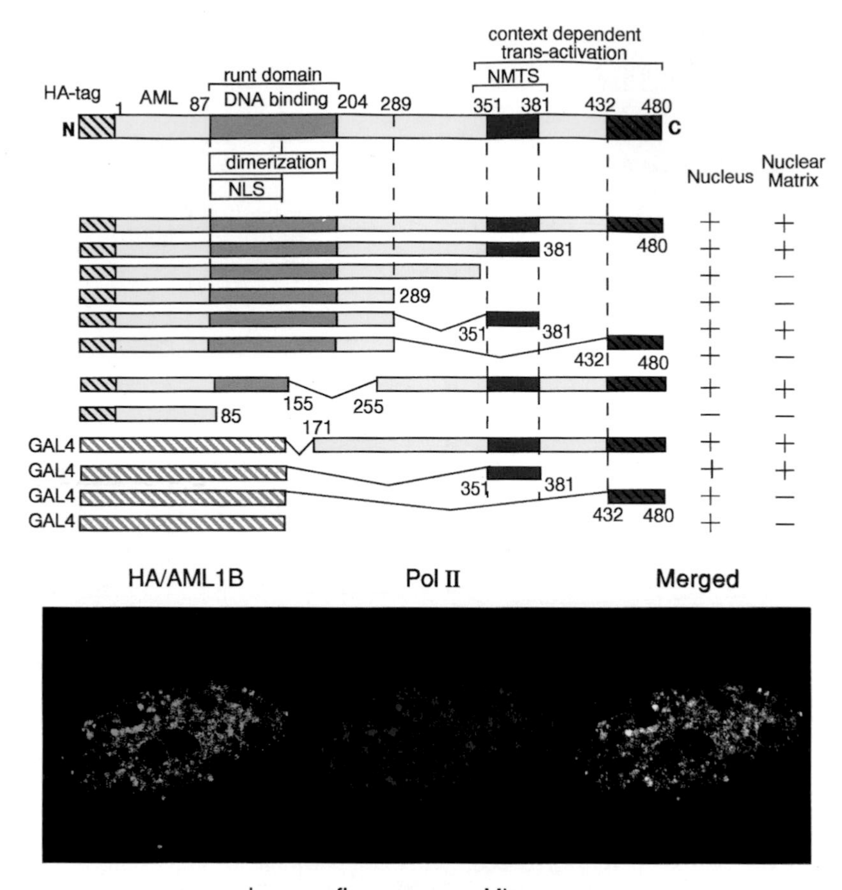

Figure 5.10. (Top): Delineation of a nuclear matrix targeting signal in AML-1B. The panel shows the relative location of functional domains regions (e.g., runt homology DNA binding domain) in the AML-1B transcription factor. The 3′ ultimate exons of the AML-1B and AML-3 genes encode for a C-terminal extension (between aa 250 and 480) which contains the NMTS and trans-activation domain. Deletion mutants of AML-1B, as well as chimeric GAL4/AML fusion proteins were analyzed for nuclear localization and intranuclear trafficking to the nuclear matrix (columns with plus and minus signs). (Bottom): AML-1B is targeted to transcriptionally active nuclear matrix domains. A subset of AML-1B and RNA polymerase II is colocalized in the nuclear matrix. Human osteosarcoma SAOS-2 cells were transiently transfected with a construct expressing HA-tagged CBFα2/AML-1B. In situ nuclear matrix samples were prepared 24 hrs after transfection. HA/AML-1B was detected with an antibody against HA (green) (Left) and RNA polymerase II_0 with the B3 anti-RNA polymerase II antibody (red) (Center). The merged image is shown in the Right panel; yellow indicates colocalization of AML-1B with RNA polymerase II_0. The colocalization of AML-1B with RNA polymerase II_0 was evaluated by confocal microscopy with a series of optical sections (0.5 μm intervals) through a single cell.

CHAPTER 5

GENE EXPRESSION: THE REGULATORY AND REGULATED MECHANISMS

GARY S. STEIN, ANDRÉ J. VAN WIJNEN,
BARUCH FRENKEL, DENNET HUSHKA, JANET L. STEIN,
and JANE B. LIAN
Department of Cell Biology and Cancer Center, University of Massachusetts Medical Center, 55 Lake Ave. North, Worcester, MA 01655

INTRODUCTION

Competency for proliferation and cell cycle progression is functionally linked to expression of genes associated with growth control. As the complexity and interdependency of parameters mediating growth control become increasingly apparent, the demarcations between regulated and regulatory components of mechanisms that promote or inhibit proliferation are eroding. There is recognition of both cause and effect relationships between the activities of factors that modulate the cell division cycle, reflecting multidirectional signaling between segments of regulatory cascades that are selectively operative in specific cells and tissues. The necessity of accommodating the integration of positive and negative growth regulatory signals is now appreciated in a broad spectrum of biological contexts. These include but are not restricted to (1) repeated traverse of the cell cycle for cleavage divisions during the initial stages of embryogenesis and continued renewal of stem cell populations; (2) stimulation of quiescent cells to proliferate for tissue remodeling and wound healing; and (3) exit from the cell cycle with the option to subsequently proliferate or terminally differentiate. Equally important is an appreciation for the cell cycle regulatory mechanisms that have been compromised in transformed and tumor cells and in nonmalignant disorders where abnormalities in cell cycle and/or growth control are operative.

In this chapter, we focus on transcriptional control of gene expression

The Molecular Basis of Cell Cycle and Growth Control, Edited by G.S. Stein, R. Baserga, A. Giordano, and D.T. Denhardt
ISBN 0-471-15706-6, pages 183–224.

during the cell cycle from the perspectives of rendering cells capable to replicate DNA and subsequently divide mitotically. Determinants for activation and/or suppression of genes at a series of checkpoints during the cell cycle are considered. Transcriptional control is discussed in relation to involvement of checkpoints as surveillance mechanisms that monitor fidelity of the genome and cell cycle regulatory components. Here, editing can be invoked if necessary followed by resumption of progression through the cell cycle or default to apoptotic pathways dependent on rectification of DNA damage as well as the structural and functional properties of the transcriptional machinery.

Promoters of genes and cognate factors are addressed as blueprints for physiological responsiveness to growth regulatory signals. We do not restrict consideration of proliferation-dependent transcriptional control to a catalogue of promoter elements and sequence-specific protein–DNA and protein–protein interactions. These components of transcriptional control are necessary but insufficient to understand regulation of proliferation within the context of expanding knowledge of nuclear architecture. The emerging evidence for nuclear structure–gene expression interrelationships, which supports gene localization and the concentration as well as targeting of regulatory factors to nuclear domains that are optimal for transcription, is discussed.

PROLIFERATION-DEPENDENT TRANSCRIPTIONAL CONTROL: THE EVOLUTION OF FUNDAMENTAL CONCEPTS AND EXPERIMENTAL CONTRIBUTIONS

Subdivision of the Cell Cycle Into Functional, Biochemically Defined Stages

A historical perspective of cell cycle and growth control provides a conceptual and experimental basis for our current understanding of gene expression, which supports the regulation of proliferation at the cellular, biochemical, and molecular levels.

The cornerstone for investigations into mammalian cell cycle control is the documentation by Howard and Pelc (1951) nearly four decades ago that proliferation of eukaryotic cells, analogous to that of bacteria, requires discrete periods of DNA replication (S phase) and mitotic division (M) with a postsynthetic, premitotic period designated G_2 and a postmitotic, presynthetic period designated G_1 (Fig. 5.1). The foundation for pursuit of biochemical regulatory mechanisms associated with growth control and cell cycle progression was provided by an elegant series of cell fusion and nuclear transplant experiments (reviewed by Prescott, 1976; Heichman and Roberts, 1994). Consequential influences of cytoplasm from various stages of the cell cycle on nuclei from other periods demonstrated the following basic principles of cell cycle control: (1) the onset of DNA synthesis is determined by cytoplasmic factors present throughout S phase but absent in pre-S phase; (2) a nuclear mechanism

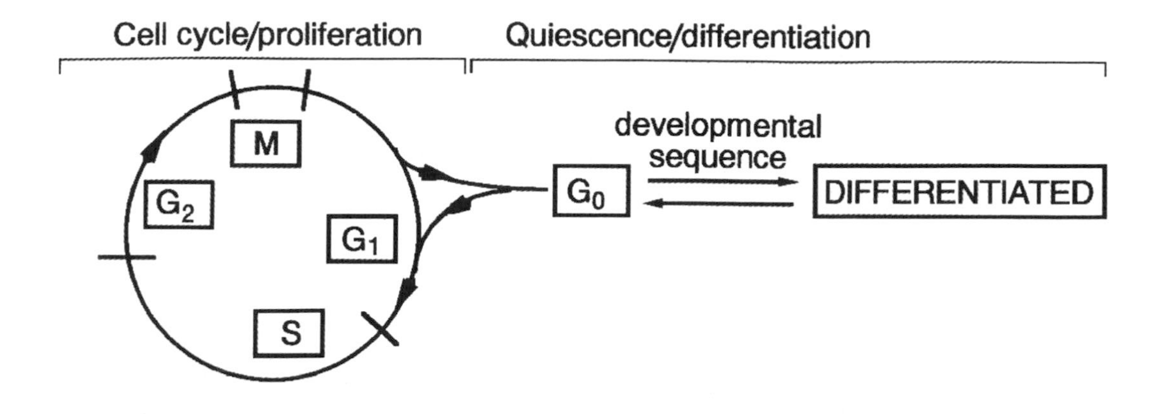

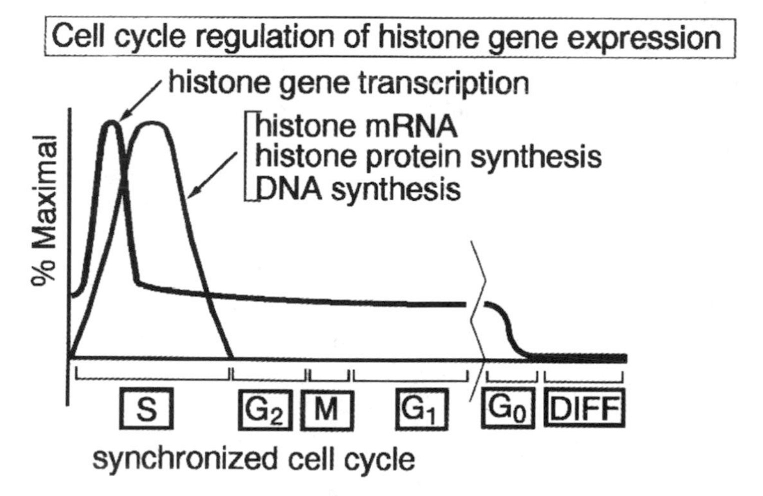

Figure 5.1. Cell cycle regulation of histone gene expression. **(Top)** The four stages of the somatic cell cycle (G_1, S, G_2, and M) support duplication of the genome and subsequent segregation of a diploid set of chromosomes into two progeny cells. Cells can exit the cell cycle into a quiescent nondividing state (G_0) with the option to re-enter the cell cycle or to differentiate into a committed cell expressing phenotypic markers characteristic of distinct tissue-specific lineages. **(Bottom)** The functional coupling between histone protein biosynthesis and DNA synthesis ensures the ordered packaging of the newly replicated genome into chromatin. Cells commencing S phase display a parallel increase in the accumulation of histone mRNAs, the level of histone protein synthesis, and the rate of DNA synthesis. The accumulation of histone mRNAs during S phase is mediated by a three- to fivefold enhancement in histone gene transcription during early S phase. Selective degradation of histone mRNAs in conjunction with a decline in DNA synthesis rates during late S phase is the result of histone mRNA destabilization. Histone mRNA destabilization may occur by an autonomous negative feedback loop involving regulatory interactions between histone mRNAs and the encoded histone proteins.

prevents re-replication of DNA without passage through mitosis; and (3) a dominant cytoplasmic factor in mitotic cells promotes mitosis in interphase cells irrespective of whether DNA replication has occurred. The broad biological relevance of cell cycle regulatory parameters is reflected by the phylogenetic conservation of control mechanisms in yeast, protozoans, echinoderms, amphibian oocytes, and mammalian cells.

It should be emphasized that while transcriptional control is a dominant component of growth-related gene regulation, post-transcriptional mechanisms are additionally operative. Biosynthesis of primary gene transcripts reflects responsiveness to a broad spectrum of regulatory signals. However, each stage in transcript processing, complexing with ribonucleoproteins, export from the nucleus, and association with polyribosomes and the cytoskeleton, is a potential target for modulation of expression. Messenger RNA stability is relevant to cell cycle–dependent control of gene expression. Transcripts from constitutively transcribed genes have finite half-lives. Histone gene expression is a striking example of the combined utilization of control at the levels of transcription, messenger RNA processing, and transcript stability (Fig. 5.1). Transcriptional upregulation occurs at the onset of S phase. Cell cycle stage-specific changes in transcript processing and turnover reflect the tight coupling of histone gene expression with DNA replication. Histone messenger RNAs are stable throughout S phase and are rapidly as well as selectively degraded at the S/G_2 transition or immediately following inhibition of DNA synthesis.

Requirements for Cell Cycle Stage-Specific Modifications in Gene Expression

The initial indication that modifications in gene expression are required to support entry into S phase and mitosis was obtained from inhibitor studies. First observed was the necessity of transcription and protein synthesis for DNA replication and mitotic division (Terasima and Yasukawa, 1966; Baserga et al., 1965). Restriction points late in G_1 and G_2 for competency to initiate S phase and mitosis were mapped (Sherr, 1993; Dowdy et al., 1993; Pardee, 1989). Subsequently, by the combined application of gene expression inhibitors and modulation of growth factor levels in cultured cells, a mitogen-dependent (growth factor/cytokine responsive) period was defined early in G_1 in which competency for proliferation is established, and a late G_1 restriction point was identified in which competency for cell cycle progression is attained (Pardee, 1989).

Identification of Cell Cycle Checkpoints

Checkpoints have been identified that govern passage through G_1 and G_2 (Fig. 5.2), where competency for cell cycle traverse is monitored (Nurse, 1994; Sherr, 1993, Dowdy et al., 1993). The first evidence for these checkpoints was provided by the observation of delayed entry into

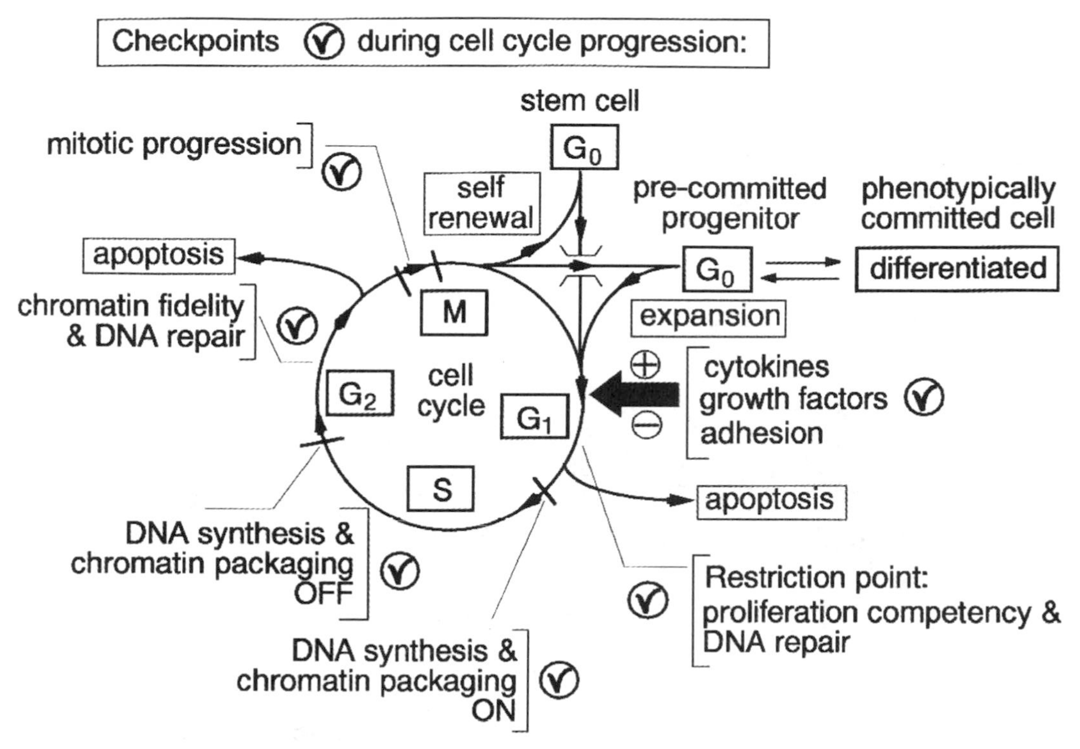

Figure 5.2. Multiple checkpoints control cell cycle progression. The cell cycle is regulated by several critical cell cycle checkpoints (indicated by checkmarks v) at which competency for cell cycle progression is monitored. Entry into and exit from the cell cycle (black lines and lettering) is controlled by growth regulatory factors (e.g., cytokines, growth factors, cell adhesion, and/or cell–cell contact) that determine self-renewal of stem cells and expansion of precommitted progenitor cells. The biochemical parameters associated with each cell cycle checkpoint are indicated by red lettering. Options for defaulting to apoptosis (blue lettering) during G_1 and G_2 are evaluated by surveillance mechanisms that assess fidelity of structural and regulatory parameters of cell cycle control. (See color plate.)

S phase or mitosis following exposure to radiation or carcinogens. Editing functions are operative, and decisions for continued proliferation, growth arrest, or apoptotic cell death are executed at these regulatory junctures (White, 1994) (Fig. 5.3). Here, long-standing fundamental questions deal with the requirement of proliferation for the onset of differentiation, as well as the extent to which proliferation and postproliferative expression of cell and tissue phenotypic genes are functionally interrelated or mutually exclusive. Knowledge of control that is operative at cell cycle checkpoints is rapidly occurring. The complexity of the surveillance mechanisms that govern decisions for cell cycle progression is becoming increasingly apparent. We are now aware of multiple checkpoints during S phase that monitor regulatory events associated with DNA replication, histone biosynthesis, and fidelity of chromatin assembly. Mitosis is similarly controlled by an intricate series of checkpoints that are responsive

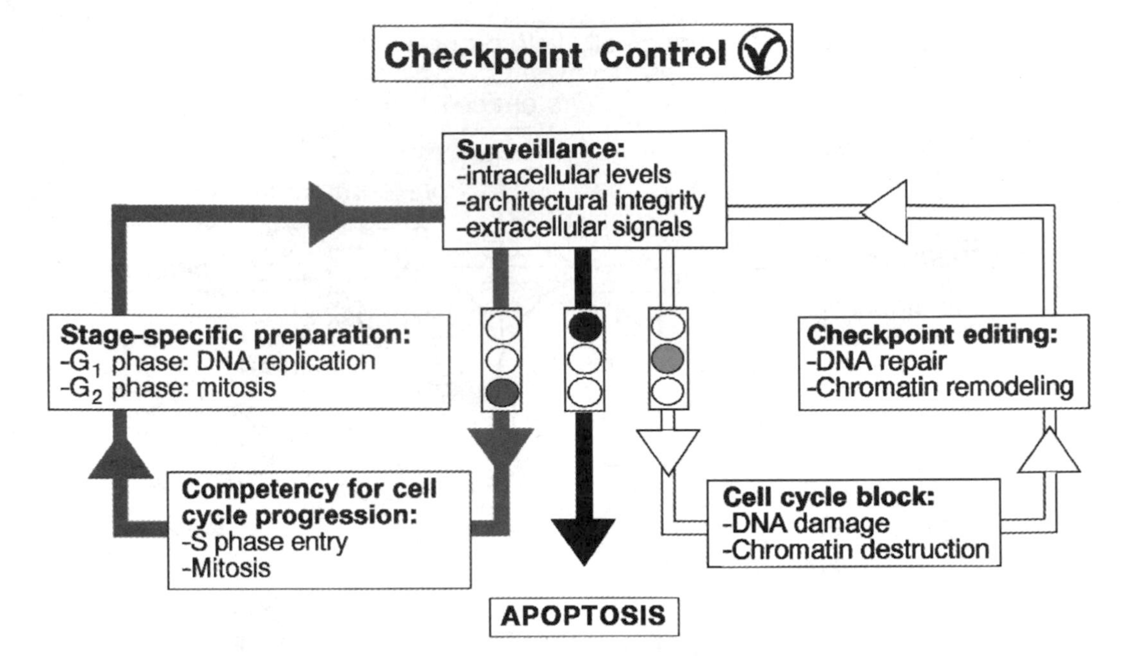

Figure 5.3. Surveillance and editing mechanisms mediating checkpoint control. **(Top)** Surveillance mechanisms monitor multiple biochemical and architectural parameters that control cell cycle progression. These parameters include the intracellular levels of regulatory proteins, structural and informational integrity of the genome, and extracellular signals governing cell cycle progression. The integration of this regulatory input can result in (1) competency for cell cycle progression (green traffic light and arrows), (2) cell cycle inhibition and activation of editing mechanisms (yellow traffic light and arrows), or (3) the active and regulated destruction of the cell in response to apoptotic signals (red traffic light and red lines). **(Middle)** Traverse of the cell cycle is regulated by a series of checkpoints at strategic positions within the cell cycle. Several major checkpoints (yellow arrows with checkmarks and blue lettering) warrant that a cell can only commit to a subsequent cell cycle stage upon satisfying essential biochemical and architectural criteria governing competency for cell cycle progression (green traffic lights). For example, at the "Restriction point" surveillance mechanisms (yellow traffic lights) integrate cell growth stimulatory and inhibitory signals, including growth factors, cell adhesion, and nutrient status (blue lettering). Checkpoints in G_1 and G_2 are necessary to ensure the integrity of the genome and, if necessary, to activate chromatin-editing mechanisms (blue lettering). **(Bottom)** Checkpoint control mechanisms monitor intracellular levels of cell cycle regulatory factors, as well as parameters of chromatin architecture. For example, the activation of cyclin-dependent kinases reflects the sensing of intracellular concentrations of the cognate cyclins. CDK activation is attenuated by CDK inhibitor proteins (CDIs), which inactivate CDK/cyclin complexes. Competency for cell cycle progression requires that cyclin levels reach a threshold (e.g., by exceeding the levels of available CDIs or by phosphorylation events altering the affinities of cyclins and CDIs for CDKs). As a consequence, activated CDK/cyclin complexes phosphorylate transcription factors that regulate expression of cell cycle stage-specific genes. Furthermore, key checkpoints in G_1 and G_2 monitor chromatin integrity and perform essential editing functions. DNA damage activates DNA repair mechanisms that fix informational errors in the genome and restore nucleosomal organization by chromatin remodeling. (See color plate.)

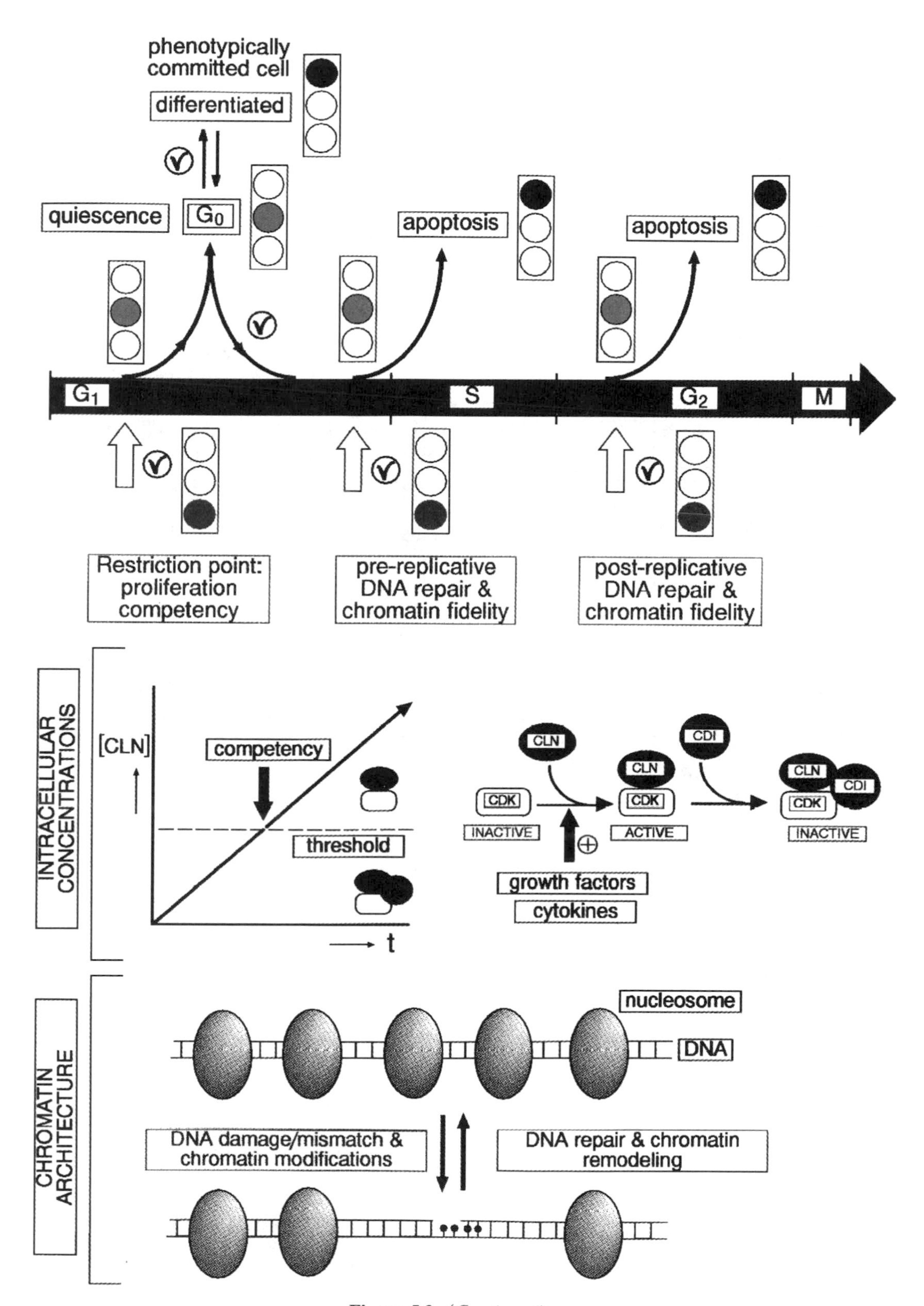

Figure 5.3. *(Continued)*

to biochemical and structural parameters of chromosome condensation, mitotic apparatus assembly, chromosome alignment, chromosome movement, and cytokinesis.

Multiple, Interdependent Cycles Operative During Proliferation

Several interdependent cycles are functionally linked to control of proliferation (Fig. 5.4). The first is a stringently regulated sequential series of biochemical and molecular parameters that support genome replication and mitotic division. The second is a cascade of cyclin-related regulatory factors that transduce growth factor–mediated signals into discrete phosphorylation events, controlling expression of genes responsible for both initiation of proliferation and competency for cell cycle progression. Other cell cycle–related regulatory loops involve chromosome condensation, spindle assembly, metabolism, and assembly of cdc2 and assembly/disassembly of DNA replication factor complexes (replicators and potential initiator proteins [Stillman, 1996; Hamlin et al., 1994; Muzi-Falconi and Kelly, 1995; Clyne and Kelly, 1995; Dubey et al., 1996; Gavin et al., 1995; Gossen et al., 1995; Leatherwood et al., 1996; Carpenter et al., 1996; Wohlgemuth et al., 1994]). It is becoming increasingly evident that each step in the regulatory cycles governing proliferation is responsive to multiple signaling pathways and has multiple regulatory options. The diversity in cyclin/cyclin-dependent kinase (CDK) complexes accommodates control of proliferation under multiple biological circumstances and provides functional redundancy as a compensatory mechanism. The regulatory events associated with the proliferation-related cycles support control within the contexts of (1) responsiveness to a broad spectrum of positive and negative mitogenic factors; (2) cell–cell and cell–extracellular matrix interactions; (3) monitoring sequence integrity of

---

Figure 5.4. Regulation of the cell cycle by cyclin-dependent kinases and tumor suppressor proteins. Competency for cell cycle progression is determined by cyclin-dependent kinases (CDKs; yellow rounded boxes) that monitor intracellular levels of cyclins (red ovals) and CDK inhibitory proteins (CDIs; blue circles). CDKs mediate phosphorylation of the pRb class of tumor suppressor proteins (i.e., pRb/p 105, p107 and p130), which results in activation of E2F and CDP/cut-homeodomain transcription factors (red boxes). These E2F-dependent and independent mechanisms induce expression of genes required for the G_1/S phase transition. The activities of CDKs are also influenced by phosphorylation (e.g., weel or CDK-activating kinase [CAK]), dephosphorylation (cdc25), ubiquitin-dependent proteolysis, and induction of CDIs by the tumor suppressor protein p53. Options for apoptosis are indicated within the context of cell cycle regulatory factors. Growth factors and cytokines induce the activities of CDKs, which mediate the G_0/G_1 transition (red arrow). Vitamin D– and transforming growth factor-β (TGF-β)–dependent cell signaling pathways upregulate CDIs (e.g., p21 and p27), which blocks cell cycle progression and supports differentiation in the presence of tissue-specific regulatory factors. (See color plate.)

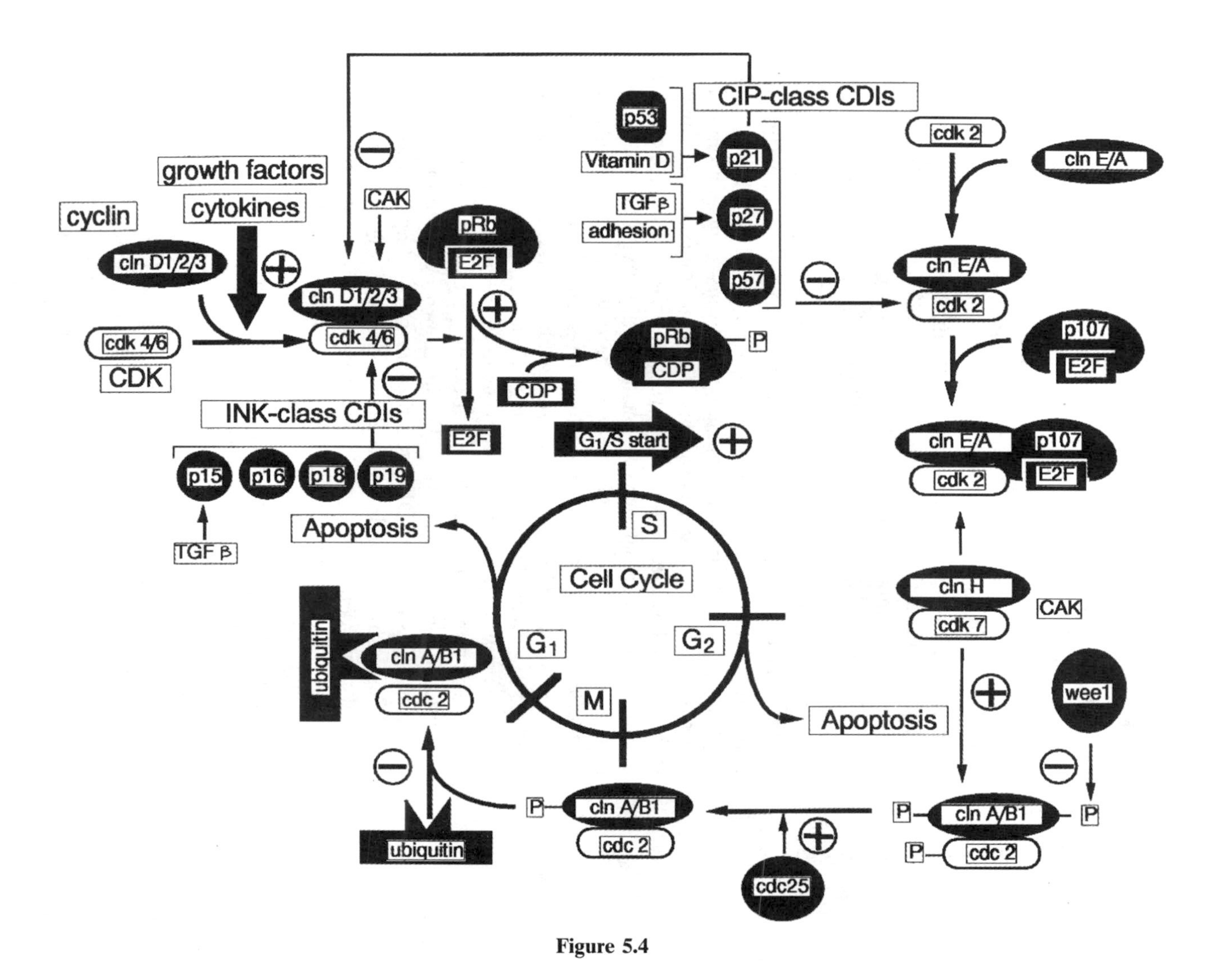

CIP-class CDIs
p53
Vitamin D
p21
TGFβ
adhesion
p27
p57
cdk 2
cln E/A
growth factors
cyclin
cytokines
CAK
pRb
E2F
cln D1/2/3
cdk 4/6
CDK
CDP
INK-class CDIs
E2F
G1/S start
p107
p15
p16
p18
p19
TGF β
Apoptosis
S
Cell Cycle
cln H
cdk 7
ubiquitin
cln A/B1
cdc 2
G1
G2
M
wee1
P
cdc25

Figure 5.4

the genome and invoking editing and/or apoptotic mechanisms if required; and (4) competency for differentiation.

Biochemical and Molecular Parameters of Cell Cycle Control

Characterization of the biochemical and molecular components of cell cycle and growth control emerged from systematic analyses of conditional cell cycle mutants in yeast (Hartwell et al., 1974; Hartwell and Weinert, 1989). These studies were the foundation for the concept that cell cycle competency and progression are controlled by an integrated cascade of phosphorylation-dependent regulatory signals. Cyclins are synthesized and activated in a cell cycle–dependent manner and function as regulatory subunits of CDKs. The CDKs phosphorylate a broad spectrum of structural proteins and transcription factors to mediate sequential parameters of cell cycle control. By complementation analysis, the genes for mammalian homologues of the yeast cell cycle regulatory proteins have been identified. In vivo overexpression, antisense, and antibody analyses have verified conservation of cell cycle–dependent regulatory activities and have validated functional contributions to control of cell cycle stage-specific events. The emerging concept is that the cyclins and CDKs are responsive to regulation by the phosphorylation-dependent signaling pathways associated with activities of the early response genes, which are upregulated following mitogen stimulation of proliferation (see Chapter 2, this volume) (reviewed in Hunter and Pines, 1994; Hartwell and Weinert, 1989; MacLachlan et al., 1995) (summarized in Fig. 5.4). Cyclin-dependent phosphorylation is functionally linked to activation and suppression of both p53 and Rb-related tumor suppressor genes, which mediate transcriptional events involved with passage into S phase. The activities of the CDKs are downregulated by a series of inhibitors (designated CDIs) and mediators of ubiquitination, which signal destabilization and/or destruction of these regulatory complexes in a cell cycle–dependent manner. Particularly significant is the accumulating evidence for functional interrelationships between activities of cyclin/CDK complexes and growth arrest at G_1 and G_2 checkpoints, when editing and repair are monitored following DNA damage (Fig. 5.3). It is at these times, and in relation to these processes, that apoptotic cell death is invoked as a compensatory mechanism.

During G_1, expression of genes associated with deoxynucleotide biosynthesis are upregulated (e.g., thymidine kinase, thymidylate synthase, dihydrofolate reductase) in preparation for DNA synthesis (King et al., 1994; Pardee and Keyomarsi, 1992; Johnson, 1992). As cells progress through G_1, regulatory factors required for initiation of DNA replication are sequentially expressed and/or activated. Following stimulation of quiescent cells to proliferate, expression of the fos/jun-related early response genes is induced early in G_1, playing a pivotal role in activation of subsequent cell cycle regulatory events. In S phase, DNA replication is paralleled by and functionally coupled with histone gene expression, providing the necessary basic chromosomal proteins (H1, H4, H3, H2A, and H2B) for packaging newly replicated DNA into chromatin (Azizkhan et al., 1993; Plumb et al., 1983). During G_2, regulatory factors for mitosis are synthesized, and modifications of chromatin structure to support mitotic chromosome condensation occur (reviewed in MacLachlan et al., 1995). Mitosis involves a sequential remodeling of genome architecture from uncondensed chromatin to highly condensed chromosomes and back to chromatin; assembly and subsequent disassembly of the mitotic apparatus; breakdown and re-formation of the nuclear membrane; and modifications in activities of factors required for reinitiation of cell cycle progression, quiescence, or differentiation.

As the sophistication of experimental approaches for dissection of promoter elements and characterization of cognate regulatory factors increases, there is an emerging recognition of cyclic modifications in occupancy of promoter domains and protein–protein interactions that control cell cycle progression. The changes observed in the site II cell cycle regulatory promoter element of the histone gene illustrates such changes in factor occupancy and activities that are functionally linked to activation, suppression, and subtle modifications in levels of expression that are proliferation dependent (Fig. 5.5) (van Wijnen et al., 1994, 1996; Vaughan et al., 1995; Stein et al., 1989; Holthuis et al., 1990; Pauli et al., 1987).

Recently, considerable attention has been directed to experimentally addressing transcriptional regulatory mechanisms that are associated with cell cycle and growth control. However, the importance of control at post-transcriptional levels should not be underestimated. For example, cell cycle–dependent modifications in histone mRNA processing and stability contribute to linkage of histone gene expression with DNA replication and the S phase of the cell cycle (Harris et al., 1991; Marzluff and Pandey, 1988; Morris et al., 1991; Pelz et al., 1991; Pardee, 1989). Compartmentalization of cell cycle regulatory factors and/or cognate gene transcripts, as well as phosphorylation of regulatory proteins, are components of post-transcriptional growth control (Zambetti et al., 1987; Birnbaum et al., 1997).

Accommodation of Unique Cell Cycle Regulatory Requirements in Specialized Cells

Consistent with the stringent requirement for fidelity of DNA replication and DNA repair to execute proliferation, stage-specific modifications in control of cell cycle regulatory factors have been observed to parallel physiological changes and perturbations in growth control. Some striking examples of physiological changes are regulatory mechanisms that support developmental transitions during early embryogenesis, when DNA replication and mitotic division occur in rapid succession in the absence of significant G_1 or G_2 periods. In contrast, proliferation in somatic cells of the adult requires passage through a cell cycle with G_1, S, G_2, and mitotic periods that are operative and necessary. Often, a prolonged G_1 period provides support for long-term quiescence

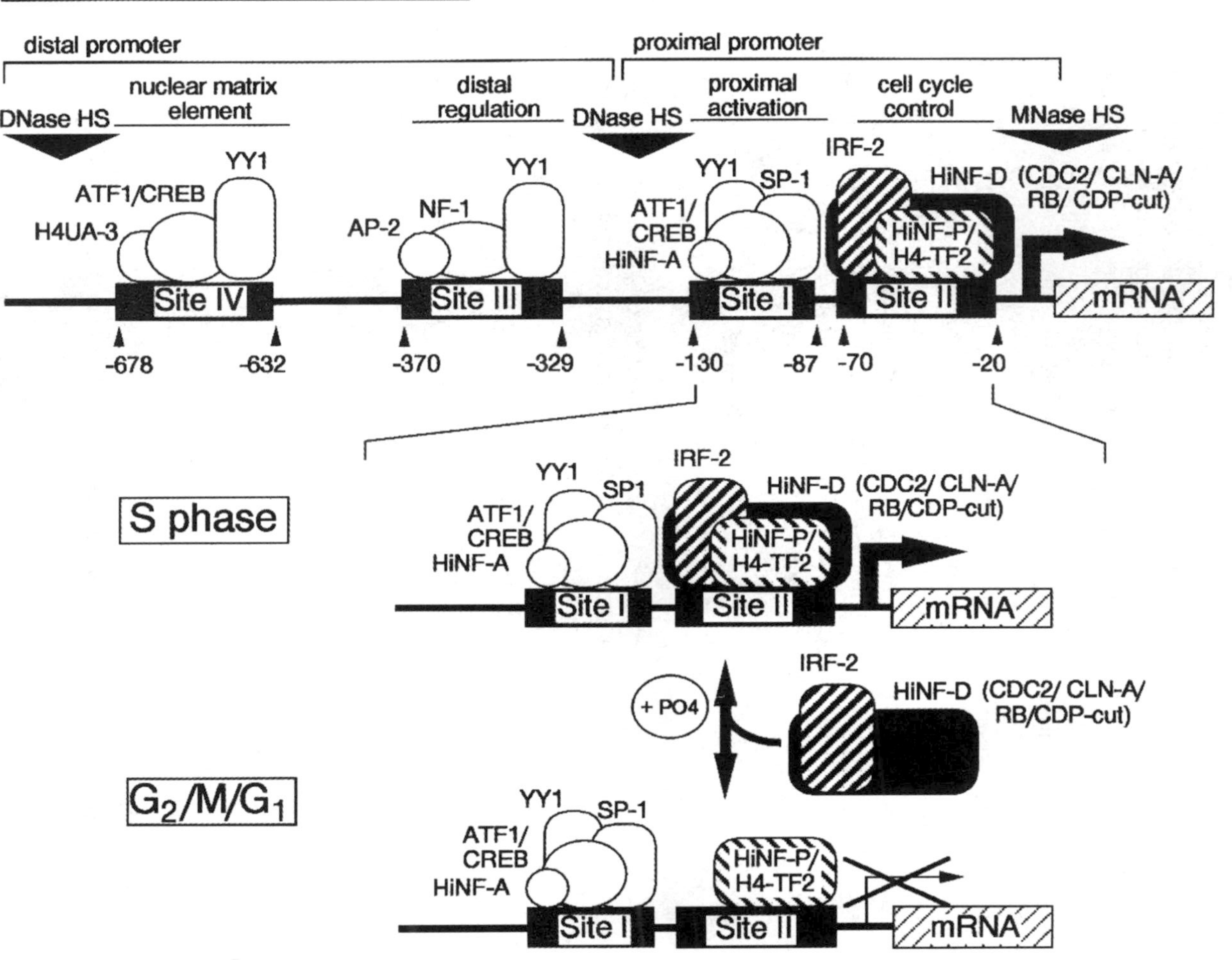
HISTONE GENE REGULATION
distal promoter
proximal promoter
DNase HS
nuclear matrix element
distal regulation
DNase HS
proximal activation
cell cycle control
MNase HS
H4UA-3
ATF1/CREB
YY1
AP-2
NF-1
YY1
ATF1/ CREB
HiNF-A
YY1
SP-1
IRF-2
HiNF-D (CDC2/ CLN-A/ RB/ CDP-cut)
HiNF-P/ H4-TF2
Site IV
Site III
Site I
Site II
mRNA
-678
-632
-370
-329
-130
-87
-70
-20
S phase
ATF1/ CREB
HiNF-A
YY1
SP1
IRF-2
HiNF-D (CDC2/ CLN-A/ RB/CDP-cut)
HiNF-P/ H4-TF2
Site I
Site II
mRNA
+ PO4
IRF-2
HiNF-D (CDC2/ CLN-A/ RB/CDP-cut)
G2/M/G1
ATF1/ CREB
HiNF-A
YY1
SP-1
HiNF-P/ H4-TF2
Site I
Site II
mRNA

Figure 5.5

of cells and tissues while retaining the competency to reinitiate proliferation for tissue remodeling and renewal. Stem cells require complex control of proliferation competency to modulate commitments for cell cycle progression, quiescence, or differentiation. Here, responsiveness to cell growth and tissue-specific regulatory factors must be stringently monitored.

The abrogated components of growth control in transformed and tumor cells are associated with and functionally linked to both the regulation and regulatory activities of cyclin CDK complexes. Characteristic alterations have been associated with progressive stages of neoplasia and specific tumors (reviewed in Hunter and Pines, 1994; Hartwell and Weinert, 1989). Frequently, the hallmark of tumor cells is co-expression of cell growth and tissue-specific genes rather than mutually exclusive expression in normal diploid cells. Consequently, tumor cells are providing us with valuable insight into rate-limiting regulatory steps in cell cycle and cell growth control. In addition, we are increasing our opportunity to therapeutically rectify proliferative disorders in a targeted manner. Particularly challenging is the possibility for restoring fidelity of regulatory mechanisms operative at cell cycle checkpoints, when responses to apoptotic signals prevent accumulation and phenotypic expression of mutations associated with growth control perturbations.

CONTRIBUTION OF CELL CYCLE REGULATORY FACTORS TO CONTROL OF DIFFERENTIATION

Because acquisition of tissue-specific phenotypes is normally associated with growth arrest, most of the attention to cell cycle regulatory factors within the context of differentiation has been focused on mechanisms that ensure exit from the cell cycle as a prerequisite for differentiation (Fig. 5.4). These mechanisms include hypophosphorylation of the Rb transcription factor and decreased representation or activity of cyclin/ CDK complexes, with recent emphasis on the upregulation of CDIs during differentiation. While this concept and its significance to carcinogenesis have been addressed in several recent reviews (Weinberg, 1995; MacLachlan et al., 1995; Kranenburg et al., 1995; Marks et al., 1996), here we present representative evidence demonstrating that many factors

---

Figure 5.5. Transcriptional control of histone gene expression. Histone H4 gene transcription is controlled by proximal (sites I and II) and distal (sites III and IV) gene regulatory domains. The site II domain (red square) is responsible for cell cycle regulation of transcription during S phase. Sites I, III, and IV are required for maximal activation of histone gene transcription. All four domains represent multipartite promoter regions that interact with cell cycle regulatory (red and striped rounded boxes) and constitutive transcription factors (yellow ovals). Cell cycle regulation of H4 gene transcription is reflected by modifications in phosphorylation-dependent protein–protein and protein–DNA interactions at the site II regulatory element during S phase and outside of S phase (G_2/M/ G_1). (See color plate.)

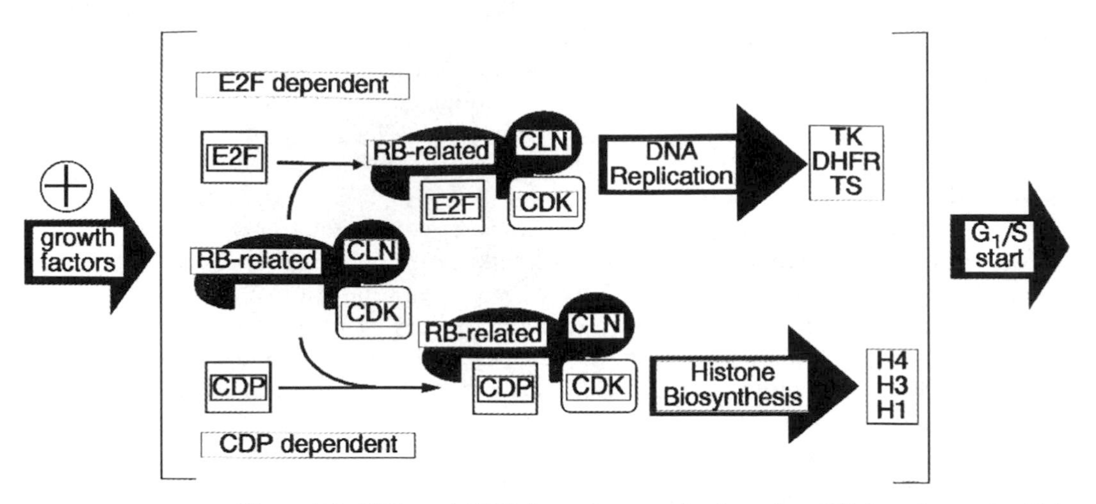

Figure 5.6. E2F- and CDP-dependent mechanisms for pRB function at the G_1/S phase transition point. The model depicts two regulatory routes (indicated by black arrows) by which a multitude of cyclins (CLN), CDKs, and pRb-related higher order transcription factor complexes may be recruited to the promoters of cell cycle–controlled genes following growth stimulation (indicated by left red arrow). This recruitment of pRb-related cell cycle mediators by E2F or CDP at defined stages of the cell cycle may activate or repress gene transcription. There are several genes encoding proteins involved in DNA synthesis or nucleotide metabolism, including thymidine kinase (TK), dihydrofolate reductase (DHFR), thymidylate synthase (TS), ribonucleotide reductase, DNA polymerase α, cdc2, cyclin A, proliferating cell nuclear antigen, c-myc, and B-myb. These DNA replication–related genes are regulated at least in part by an E2F-dependent mechanism, whereas histone genes are controlled by a CDP-dependent mechanism. (See color plate.)

controlling cell cycle progression acquire additional roles postproliferatively and are in fact active in the differentiation process of various cell types.

pRB and Related Proteins

The best characterized function of the Rb gene is the association of hypophosphorylated pRb with E2F transcription factors. Consequently, phosphorylation of pRb by CDKs during middle and late G_1 results in the timely transcriptional activation of E2F-regulated genes (reviewed in Weinberg, 1995; Nevins, 1992; La Thangue, 1994). However, it is now clear that other cell cycle regulatory transcription factors can interact with pRb (Fig. 5.6). Equally important, Rb function is not confined to the control of cell cycle progression and maintenance of quiescence. Numerous investigators have reported high abundance of pRb in a variety of postmitotic cell types, where it is mostly found hypophosphorylated (Yen et al., 1993; Gu et al., 1993; Kiyokawa et al., 1993; Szekely et al., 1993; Cordon-Cardo and Richon, 1994; reviewed in Weinberg, 1995).

Perhaps the best evidence for postproliferative roles for pRb is provided by pRb gene ablation studies. Cell proliferation in early mouse embryos lacking functional pRb initially appears normal, possibly attributable to functional redundancy with the pRb-related proteins p107 and/or p130. However, at midgestation, abrogation of neural development and erythropoiesis occur, resulting in prenatal lethality (Jacks et al., 1992; Clarke et al., 1992; Lee et al., 1992; reviewed in Weinberg, 1995; Slack and Miller, 1996; Lee et al., 1995). Biochemical and cellular analyses of the pRb-deficient cells confirmed maturation defects and p53-dependent premature apoptosis of specific neuron populations (Lee et al., 1994; Morgenbesser et al., 1994). In addition, pRb plays a role in myoblast differentiation, as demonstrated by the abrogation of this process in pRb-deficient cells (Schneider et al., 1994; Novitch et al., 1996; reviewed in Wiman, 1993). Similarly, in vitro adipocyte differentiation in Rb-deficient fibroblasts is blocked and is restorable upon re-expression of pRb (Chen et al., 1996). Finally, pRb has been implicated in sustaining the differentiation state of HL60 human leukemia cells (Yen and Varvayanis, 1994). pRb-induced differentiation has been attributed to its ability to interact with and modulate activity of transcription factors, such as myoD (Gu et al., 1993), adipogenic inducers of the C/EBP family (Chen et al., 1996), NF-IL6, the glucocorticoid receptor (Singh et al., 1995), ATF-2 (reviewed by Sellers and Kaelin, 1996; Kouzarides, 1995; Chen et al., 1995), and possibly some newly identified (Buyse et al., 1995) and yet to be characterized transcription factors. From the standpoint of new functions that cell cycle regulatory molecules acquire during differentiation, it is particularly interesting that Rb family members may function postproliferatively as stable complexes with E2F transcription factors, which in some cases are observed predominantly in terminally differentiated cells (e.g., Corbeil et al., 1995; Jiang et al., 1995; Shin et al., 1995).

Like pRb, the related proteins p107 and p130 interact with E2F transcription factors via their "pocket domain." Unlike pRb, they also interact with cyclin/CDK complexes via a sequence resembling a domain found in CDIs. However, as with pRb, mice lacking both p107 and p130 do not exhibit a generalized cell cycle defect, but rather specific defects in chondrocytic growth and limb development, with neonatal lethality (Cobrinik et al., 1996). Thus, while pRb plays a role in neural and hematopoietic development, p107 and p130 become critical later, during skeletal development. In vitro, upregulation of p130 in differentiating L6 cells has been implicated in myotube formation (Kiess et al., 1995a), and p107 has been recently found in a transcription factor complex bound to the bone-specific osteocalcin promoter (van Gurp et al., 1997). Thus, all three members of the Rb gene family seem to play tissue-specific differentiation-related roles in addition to their more traditional regulatory roles in cell cycle progression. Inhibition of apoptosis by pRb family members as a mechanism that maintains a differentiated phenotype is discussed elsewhere in this chapter.

Cyclin-Dependent Kinases

Whereas most CDKs normally function during active cell proliferation, two members of this family of enzymes, CDK5 and CDK7, are clearly different (Table 5.1). CDK5 is expressed in the central nervous system in a specific spatial and temporal fashion. In situ analyses indicate that it is excluded from mitotic neurons, and functional assays demonstrate postmitotic kinase activity associated with CDK5 in differentiating neurons (Tsai et al., 1993). Finally, manipulations of CDK5 and its p35 partner protein in vitro resulted in direct effects on neurite outgrowth in cortical cultures (Nikolic et al., 1996), and targeted deletion of CDK5 in mice resulted in defective brain development and perinatal lethality (Ohshima et al., 1996). Thus, CDK5 plays a critical role in central nervous system development. Unlike CDK5, the CDK-activating kinase CDK7 is abundant in diverse cell types, both proliferating and differentiated (Bartkova et al., 1996). Other CDKs (e.g., cdc2, CDK2, CDK4) have been occasionally observed in some postproliferative cells (Table 5.1), usually without, but in some cases with (Bartkova et al., 1996; Dobashi et al., 1996; Smith et al., 1997; Gao et al., 1995; Kranenburg et al., 1995; Jahn et al., 1994) associated kinase activity. The requirement of these kinases for cell cycle progression makes it difficult to directly address their role in cell differentiation with gene ablation approaches.

Cyclins

There is increasing evidence for cell type–specific postproliferative retention and even upregulation of cyclins in differentiating cells (Table 5.1). However, in most cases no kinase activity is associated with these postproliferative cyclins, and therefore their function may be related to association with other, nonkinase, proteins, such as pRB (e.g., Chen et al., 1995) or, as recently suggested, nuclear hormone receptors (Zwijsen et al., 1997). In one case, the postproliferative upregulation of cyclin E has been shown to support an osteoblast differentiation-related kinase activity, suppressible by inhibitory activity residing in proliferating osteoblasts (Smith et al., 1997). Additional evidence for involvement of cyclins in cell differentiation comes from experiments manipulating their levels. Transgenic mice overexpressing cyclin D1 under the control of an immunoglobulin enhancer contained fewer mature B and T cells (Bodrug et al., 1994). The myeloid cell line 32D fails to differentiate in the presence of overexpressed cyclin D2 or D3, probably due to their interaction with pRb and/or p107 (Kato and Sherr, 1993). Cyclin D1 gene ablation in mice resulted in breast- and nervous tissue–specific defects (Fantl et al., 1995; Sicinski et al., 1995). Consistent with this result, overexpression of cyclin D1 in mouse mammary epithelial cells resulted in a more differentiated phenotype (Han et al., 1996). These findings may be attributed to the CDK-independent activation of the estrogen receptor by cyclin D1 (Zwijsen et al., 1997). In other cases, however, cyclins may support the induction or maintenance of a differentiated phenotype by

TABLE 5.1. Persistence or Upregulation of Cyclins and CDKs in Differentiating Cells

Cyclins/Kinases	Cells
A	NGF-induced PC12 pheochromocytoma cells (Buchkovich and Ziff, 1994) DMSO-induced HL60 cells (Burger et al., 1994)
B	Lens fiber cells (Gao et al., 1995) Primary rat calvarial osteoblasts (Smith et al., 1995) DMSO-induced HL60 cells (Burger et al., 1994)
D1	Hepatocytes (Loyer et al., 1994) Kidney cells (Godbout and Andison, 1996) Senescent WI-38 human diploid fibroblasts (Lucibello et al., 1993) NGF-induced PC12 pheochromocytoma cells (Dobashi et al., 1995; Yan and Ziff, 1995; Tamaru et al., 1994; van Grunsven et al., 1996) TPA-induced HL60 cells (Horiguchi-Yamada et al., 1994; Burger et al., 1994) Megakaryocytes (Dami, HEL, and K562 cell lines) (Wilhide et al., 1995) p19 embryonal carcinoma (Kranenburg et al., 1995)
D2	Specific neural populations (Ross et al., 1996; Ross and Risken, 1994) p19 embryonal carcinoma (Kranenburg et al., 1995)
D3	L6, L8, G8, and C2C12 myoblasts (Kiess et al., 1995b; Rao and Kohtz, 1995; Jahn et al., 1994) HMBA-induced MEL cells (Kiyokawa et al., 1994)
E	Intestinal epithelial cells (Chandrasekaran et al., 1996) Hepatocytes (Loyer et al., 1994) Primary rat calvarial osteoblasts (Smith et al., 1995, 1997) Senescent WI-38 human diploid fibroblasts (Lucibello et al., 1993) L6 myoblasts (Kiess et al., 1995a,b) NGF-induced PC12 pheochromocytoma cells (Dobashi et al., 1995)
cdc2	Lens fiber cells (Gao et al., 1995) Sertoli cells (Rhee and Wolgemuth, 1995) C2C12 (Jahn et al., 1994)
CDK2	Sertoli cells (Rhee and Wolgemuth, 1995) L6 myoblasts (Kiess et al., 1995a,b) NGF-induced PC12 pheochromocytoma cells (Yan and Ziff, 1995)
CDK4	Sertoli cells (Rhee and Wolgemuth, 1995) Intestinal epithelial cells (Chandrasekaran et al., 1996) L6 myoblasts (Kiess et al., 1995a,b) NGF-induced PC12 pheochromocytoma cells (Dobashi et al., 1996; Yan and Ziff, 1995) p19 embryonal carcinoma (Kranenburg et al., 1995)
CDK5	Embryonic mouse neurons (Tsai et al., 1993) Embryonic *Xenopus* neurons (Gervasi and Szaro, 1995)
CDK7	Quiescent cells (various types) (Bartkova et al., 1996)
PCTAIRE 1	Sertoli cells (Rhee and Wolgemuth, 1995)
PCTAIRE 3	Sertoli cells (Rhee and Wolgemuth, 1995)

NGF, nerve growth factor; DMSO, dimethylsulfoxide; TPA, 12-o-tetradecanoyl 13-acetate; HMBA, hexamethylene bisacetamide activation of CDKs (Bartkova et al., 1996; Dobashi et al., 1996; Smith et al., 1997; Gao et al., 1995; Kranenburg et al., 1995; Jahn et al., 1994) or association with differentiation-related pRB/E2F complexes (Kiyokawa et al., 1994).

In summary, the similarities of cell cycle regulatory molecules among various tissues, as well as across species, suggest common mechanisms underlying cell cycle control in diverse cell types. However, it is becoming clear that specific cells have unique cell cycle requirements. This is evidenced by cell type–specific representation of cell cycle regulatory factors and by cell type–specific effects observed following over- and underexpression of these molecules. Moreover, as cells acquire specific phenotypic properties, some cell cycle regulatory factors persist or even become more abundant postproliferatively. Not only Rb family members but also cyclins and CDKs seem to play roles in cell type–specific differentiation processes. We are currently witnessing the initial steps in a growing field, investigating how these cell cycle regulatory molecules acquire new functions during differentiation, including interactions among themselves, with E2F and other transcription factor families, and with transcriptional elements of growth- and differentiation-related genes.

TRANSCRIPTIONAL CONTROL DURING THE CELL CYCLE

Insight into transcriptional control at strategic points during the cell cycle has been provided by characterization of promoters and cognate factors that regulate expression of genes associated with competency for proliferation, cell cycle progression, and mitotic division. The modular organization of these gene promoter elements offers a blueprint of responsiveness to a broad spectrum of physiological regulatory signals that determine levels of transcription. The overlapping sequence organization of each promoter domain and the multipartite complexes that involve protein–DNA and protein–protein interactions facilitate the convergence of growth regulatory signals to accommodate cell cycle and growth control under diverse biological circumstances. Transcriptional modulation of gene expression is required throughout the cell cycle and is linked to a temporal sequence of events that is necessary for proliferation. However, for clarity of presentation, we confine our considerations to examples of transcriptional control that are operative during G_1 and at the onset of S phase (Fig. 5.7).

Transcriptional Activation and Suppression of Genes Involved in Nucleotide Metabolism at the Restriction Point Preceding the G_1/S Transition

During the G_1/S phase transition, three critical events associated with activities of cell cycle checkpoints occur that prepare the cell for the duplication of chromatin. First, genes encoding enzymes involved in

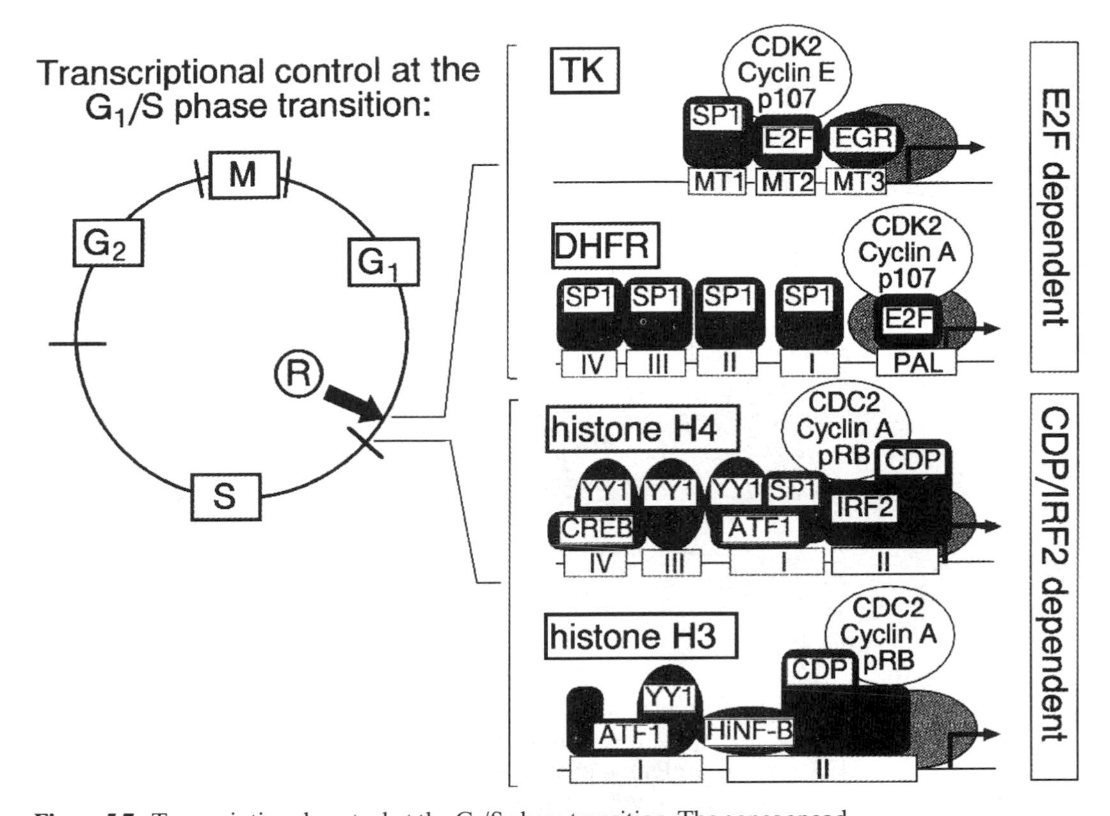

Figure 5.7. Transcriptional control at the G_1/S phase transition. The genes encoding enzymes involved in nucleotide metabolism (e.g., thymidine kinase [TK] and dihydrofolate reductase [DHFR]) and histone biosynthesis (e.g., H4 and H3) each are controlled by intricate arrays of promoter regulatory elements (open boxes) that influence transcriptional initiation by RNA polymerase II (grey oval). E2F elements in the promoters of the TK and DHFR genes interact with heterodimeric E2F factors that associate with CDKs, cyclins, and pRb-related proteins. In contrast, histone genes are controlled by the site II cell cycle regulatory element, which interacts with CDP–cut and IRF2 proteins. Analogous to E2F-dependent mechanisms, CDP–cut interacts with cdc2, cyclin A, and pRb, whereas IRF2 performs an activating function similar to "free" E2F. The presence of SP1 in the promoters of G_1/S phase–related genes provides a shared mechanism for further enhancement of transcription at the onset of S phase. (See color plate.)

nucleotide metabolism are activated to ensure that cellular deoxynucleotide triphosphate pools are adequate for the onset of DNA synthesis. Second, multiprotein complexes at DNA replication origins are assembled that regulate the initiation of DNA synthesis and prevent re-initiation at the same origin. Third, histone proteins are synthesized de novo to accommodate the packaging of newly replicated DNA into nucleosomes. Transcriptional activation of gene expression at the G_1/S phase transition represents the initial rate-limiting step for cell cycle progression into S phase.

The restriction point prior to the G_1/S phase transition integrates a multiplicity of cell signaling pathways that monitor growth factor levels, nutrient status, and cell–cell contact. This integration of positive and negative cell cycle regulatory cues culminates in the transcriptional upregulation of genes encoding enzymes and accessory factors that directly and indirectly control nucleotide metabolism and DNA synthesis (Fig. 5.7). Analysis of the thymidine kinase (TK) promoter and cognate promoter factors has revealed that maximal TK gene transcription involves at least three distinct cis-acting elements (MT1, MT2, and MT3) (Fridovich-Keil et al., 1993; Dou et al., 1994a,b; Dou and Pardee, 1996; Li et al., 1993; Good et al., 1995). These elements interact with cell cycle–dependent (e.g., Yi1 and Yi2) and constitutive (e.g., SP1) DNA-binding proteins. The Yi complexes interacting with the MT2 motif are associated with p107, as well as with cyclin- and CDK-related proteins (Dou et al., 1994a,b; Dou and Pardee, 1996; Li et al., 1993; Good et al., 1995). The Yi complexes are analogous to or identical with E2F-related higher order complexes containing cyclins, CDKs, and pRb-related proteins. Interestingly, cyclins A and E may represent the labile and rate-limiting restriction point proteins that were originally postulated based on results from early studies on cell growth control (Dou et al., 1993).

Each of the G_1/S phase genes is controlled by different arrays of cis-acting promoter elements and cognate factors. One unifying theme among many promoters of the R-point genes is the presence of E2F and SP1 consensus elements. Thus, one mechanism by which the cell achieves coordinate and temporal regulation of these genes at the G_1/S phase boundary is directly linked to the release of transcriptionally active E2F from inactive E2F/pRb complexes. The disruption of E2F/pRb is mediated by CDK4/CDK6-dependent phosphorylation of pRb in response to growth factor stimulation and cell cycle entry. Hence, the E2F-dependent activation of the R-point genes provides linkage between the onset of S phase and control of cell growth.

The E2F transcription factor represents a heterogenous class of heterodimers formed between one of five different E2F proteins (i.e., E2F-1 to E2F-5) and one of three distinct DP factors (DP1 to DP3). The various E2F factors may display preferences in promoter specificity, differ in the regulation of their DNA-binding activities during the cell cycle, and bind selectively to distinct pRb proteins (Meyers and Hiebert, 1995; Chen et al., 1995). The mechanism by which this multiplicity of E2F factors orchestrates transcriptional regulation of diverse sets of genes at the G_1/S phase transition is only beginning to be understood. Apart from the role of "free" E2F in activating genes at the G_1/S phase transition, promoter-bound complexes of E2F factors associated with pRb-related proteins, cyclin A, and CDK2 have active roles in repression of gene expression during early S phase (Krek et al., 1994).

E2F-responsive transcriptional modulation of R-point genes requires participation of the SP1 family of transcription factors (e.g., SP1 and SP3). For example, the TK promoter contains one E2F site and one SP1 site, and both are required for maximal transcriptional responsiveness at the G_1/S phase boundary (Karlseder et al., 1996). This synergistic enhancement involves direct protein–protein interactions between E2F and SP1. Consistent with the critical role of SP1 in cell cycle control of gene expression, protein–protein interactions between SP1 and pRb can also occur, suggesting that pRb can modulate the activities of E2F and SP1 in concert. Analogous to the TK promoter, the dihydrofolate reductase (DHFR) promoter is regulated by four SP1 elements that, together with E2F, mediate transcriptional upregulation at the G_1/S phase transition (Wade et al., 1995; Good et al., 1996; Schulze et al., 1994; Wells et al., 1996, 1997; Azizkhan et al., 1993; Schilling and Farnham, 1994, 1995). Interestingly, SP3 selectively represses SP1 activation of the DHFR promoter, but not the TK or histone H4 promoter (Birnbaum et al., 1995a). It appears that the cellular ratio of SP1 and SP3 levels may influence specific classes of cell cycle–regulated genes, but the physiological function of this regulatory mechanism remains to be elucidated.

Transcriptional Control at the G_1/S Phase Transition That Is Functionally Coupled with Initiation of DNA Synthesis

Conditions that establish competency for the initiation of DNA synthesis in vertebrates are monitored in part by the origin recognition complex (ORC) (Hickey and Malkas, 1997; Stillman, 1996). This complex appears to contain sequence-specific proteins that mark the location of DNA replication origins. Prior to S phase, the labile Cdc6p protein associates with ORC, which stages the subsequent binding of Mcm proteins ("licensing factors") to form large origin-bound pre-replication complexes. The mechanism by which these complexes facilitate the onset of template-directed synthesis of DNA remains to be established. However, activation of S phase–dependent CDKs is required for the initiation of DNA replication, but this event is also thought to prevent assembly of new pre-replication complexes (Stillman, 1996). This hypothesis provides a potential mechanism for stringent control of chromosomal duplication, which should occur only once during each somatic cell cycle. Thus, checkpoint controls at the onset of DNA synthesis serve to signal cellular competency for S phase entry and maintenance of the normal diploid genotype upon mitosis.

Once DNA synthesis has been initiated, replicative activity is confined to specific locations within the nucleus, referred to as *DNA replication foci.* DNA replication foci represent subnuclear domains that are thought to be highly enriched in multisubunit complexes ("DNA replication factories") containing enzymes involved in DNA synthesis, including DNA polymerases α and δ, PCNA, and DNA ligase (Hickey and Malkas, 1997; Leonhardt and Cardoso, 1995). The concentration of these factors at DNA replication foci that are associated with the nuclear matrix provides a solid-phase framework for understanding catalytic and regulatory components of DNA replication.

Coordinate Activation of Multiple DNA Replication-Dependent Histone Genes at the Onset of S Phase

The initiation of histone protein synthesis at the G_1/S phase transition is tightly coupled to the start and progression of DNA synthesis (Fig. 5.1). To prevent disorganization of nuclear architecture and chromosomal catastrophe during chromosome segregation at mitosis, it is critical that newly replicated DNA is packaged immediately into nucleosomes. Histones permit the precise packaging of 2 m of DNA into chromatin within each cell nucleus (diameter approximately 10 μm). This functional and temporal coupling poses stringent constraints on multiple parameters of histone gene expression, because somatic cells do not have storage pools for histone protein or histone mRNAs. The vast number of histone polypeptides that must be synthesized and the limited time of S phase allotted for this process necessitate a high histone protein synthesis rate. Mass production of each histone subtype occurs at an average rate of several thousand proteins per second throughout S phase. Moreover, because each 0.2 kb of DNA is packaged by nucleosomal octamers composed of histones H2A, H2B, H3, and H4, the stoichiometric synthesis of each of the histone subtypes is essential for efficient DNA packaging. Consequently, histone gene regulatory factors integrate a series of cell signaling pathways that monitor the onset of S phase and coordinate the expression of 50–100 distinct histone gene subtypes.

The first rate-limiting factor of histone gene expression is the enhancement of a low transcription rate that persists throughout the cell cycle (Fig. 5.1) (Plumb et al., 1983). Histone H4 gene transcription has been extensively studied, and a series of cis-acting elements and cognate factors have been identified in our laboratory (Fig. 5.5) (Vaughan et al., 1995; van Wijnen et al., 1994, 1996; Birnbaum et al., 1995b; Guo et al., 1995; Aziz et al., 1997; Stein et al., 1994). We first showed that genomic occupancy of histone gene promoter elements occurs throughout the cell cycle (Pauli et al., 1987), which was subsequently also shown for the R-point gene DHFR (Wells et al., 1996) by others. The constitutive occupancy of promoter regulatory elements is consistent with the concept that protein–protein interactions, post-translational modifications, and alterations in chromatin structure are important factors in modulating transcription of histone genes and other genes expressed during S phase (Chrysogelos et al., 1985, 1989; Moreno et al., 1986; Pemov et al., 1995; Ljungman, 1996). Similar to the R-point genes, the presence of SP1-binding sites is critical for maximum activation of histone genes (Fig. 5.7). However, unlike the R-point genes, the majority of histone genes do not contain E2F elements. Rather, a sophisticated and E2F-independent transcriptional mechanism has evolved for coordinate activation of histone genes.

As with E2F responsive genes, E2F-independent transcriptional control mechanisms must account for G_1/S phase–dependent enhancement of transcription, as well as attenuation of gene transcription at later stages of S phase (Fig. 5.5). The key cell cycle element for histone H4 genes is a highly conserved promoter domain designated *site II,* which encompasses binding sites for IRF2, the homeodomain-related "CCAAT Displacement Protein" CDP/cut, and the TATA-binding complex TFIID (Vaughan et al., 1995; van Wijnen et al., 1994, 1996; Aziz et al., 1997; Stein et al., 1994). IRF2 is required for maximal activation of histone gene transcription and appears to function at the G_1/S phase boundary in a manner analogous to "free" E2F by enhancing cell cycle–dependent transcription rates by about threefold (Vaughan et al., 1995). Phosphorylation of IRF2 in vivo occurs primarily on serine residues, which may be mediated by several ubiquitous kinases, including casein kinase II, protein kinase A, and protein kinase C (Birnbaum et al., 1997). Interestingly, IRF2 activity does not appear to be directly linked to phosphorylation by mitogen-activated protein (MAP) kinases or CDKs.

Involvement of the CDP/cut homeodomain protein in cell cycle control was initially established by the finding that this factor is a component of the HiNF-D complex (van Wijnen et al., 1996). In the multisubunit HiNF-D complex, the CDP/cut protein is associated with pRb, cyclin A, and CDK1/cdc2 (van Wijnen et al., 1994, 1997; Shakoori et al., 1995). HiNF-D may have a bifunctional role in H4 gene transcription. For example, binding of the HiNF-D complex to the H4 promoter is essential for maximal H4 gene promoter activity in cells nullizygous for IRF2. However, overexpression of the CDP/cut DNA-binding subunit of HiNF-D results in repression of H4 promoter activity (van Wijnen et al., 1996). CDP/cut in association with pRB, CDK1/cdc2, and cyclin A may perform a function very similar to that of the multiplicity of higher order E2F complexes. These CDP complexes bound to cyclins, CDKs, and pRb-related proteins attenuate the enhanced levels of histone gene transcription during mid-S phase, when physiological demand for histone mRNAs begins to diminish.

Similar to the R-point genes, histone gene promoters have auxiliary elements (e.g., site I) that support transcriptional activation during the cell cycle. For example, histone H4 genes contain binding sites for YY1 and SP1. The interaction of SP1 with site I modulates the efficiency of H4 gene transcription by an order of magnitude (Birnbaum et al., 1995b). The binding of YY1 to multiple sites in the histone H4 promoter may facilitate gene–nuclear matrix interactions (Guo et al., 1995; unpublished data). In addition, it has recently been shown that YY1 associates with the histone deacetylase rpd3. The possibility arises that post-translational modifications of histone proteins, when bound as nucleosomes to the H4 promoter, may parallel the modifications in chromatin structure that accompany modulations of histone H4 gene expression (Chrysogelos et al., 1985, 1989).

Stoichiometric synthesis of histone mRNAs and proteins requires coordinate control of histone gene expression at several gene regulatory levels. At the transcriptional level, coordinate activation of the five histone gene classes at the G_1/S phase transition may be mediated by CDP/cut, which has been shown to interact with the promoters of all major histone gene subtypes. The association of CDP/cut with pRb, cyclin A, and CDK1/cdc2 as components of the HiNF-D complexes interacting with DNA replication-dependent histone genes provides direct functional linkage between transcriptional coordination of histone gene expression and cyclin/CDK signaling mechanisms that mediate cell cycle progression.

The secondary levels at which histone gene expression is coordinated occur by post-transcriptional mechanisms, including transcript elongation and 3′ end processing, which produce mature histone mRNAs. Histone mRNAs do not have polyA tails, but instead contain a unique histone-specific hairpin-loop structure. Histone 3′ end processing is mediated by U7 snRNP complexes (Marzluff and Pandey, 1988; Schumperli, 1988), and accessory proteins that recognize histone mRNA 3′ ends are being characterized. It has been postulated that histone mRNA 3′ end processing is a key step in histone gene expression (Marzluff and Pandey, 1988; Schumperli, 1988; Harris et al., 1991), but the regulatory events that connect activation of this process to cell cycle progression remain to be established. However, because all histone mRNAs have highly similar structural elements at the 3′ end, the recognition of these structures by regulatory factors (Wang et al., 1996; Martin et al., 1997) may represent an important mechanism by which mature histone mRNAs are produced, transported to specific cytoplasmic locations, translated, and degraded.

Selective Downregulation of Histone Gene Expression Upon Cessation of DNA Replication at the S/G_2 Transition

When cells approach the S/G_2 transition, most of the genome has been replicated, and the demand for histone proteins to package newly replicated DNA diminishes (Fig. 5.1). Cells must ensure that histones do not accumulate in excess as these highly basic proteins would likely interfere with cellular and particularly nucleic acid metabolism. Therefore, histone mRNAs are selectively degraded during late S phase in concert with the completion of DNA synthesis. Molecular mechanisms have been elucidated that account for post-transcriptional control of histone gene expression during S phase by modulating mRNA stability. The histone mRNA-specific stem–loop structure plays a key role in regulating histone mRNA turnover. This stem–loop motif is present in all mRNAs encoding the live cell cycle–regulated histone subtypes. Therefore, this structure is considered pivotal in maintaining the stoichiometric balance of the five histone classes and coupling with DNA replication. Selective degradation of histone mRNA when DNA synthesis is halted is mediated by a 3′ exonuclease and requires active translation of histone mRNA bound to polyribosomes (Marzluff and Pandey 1988; Schumperli, 1988; Stein and Stein, 1984; Zambetti et al., 1987). Interestingly, mRNA destabilization does not occur when histone mRNA is targeted to membrane-bound ribosomes rather than to ribosomes associated with the nonmembranous cytoskeleton. Thus, histone mRNA degradation is a dynamic process that requires macromolecular complexes at specific subcellular locations.

It has been shown that histone proteins mediate histone mRNA degradation (Pelz and Ross, 1987) and that histone mRNA half-lives are modulated during the cell cycle (Morris et al., 1991). Because histone mRNAs are most stable when free histone protein concentrations are minimal and highly unstable when histone proteins accumulate in excess, it appears that selective downregulation of histone gene expression is achieved by an autoregulatory mechanism.

CONTROL MECHANISMS THAT MODULATE THE ACTIVITIES OF CELL CYCLE REGULATORY FACTORS

Cell Cycle Stage-Specific and Ubiquitin-Dependent Turnover of Gene Regulatory Factors

The activation and inactivation of cell cycle regulatory factors at specific stages of the cell cycle occur at multiple levels and are often achieved by a combination of control mechanisms. As discussed above, CDK-mediated phosphorylation pathways represent an important level of control. For example, the DNA-binding activity of the histone gene–regulatory HiNF-D complex is dramatically increased at the G_1/S phase boundary in stem cells (Shakoori et al., 1995; van Wijnen et al., 1997). This increase in HiNF-D activity occurs, although the levels of the DNA-binding subunit CDP/cut remain constant. Concomitantly, levels of CDK1/cdc2 and cyclin A are increased, and pRb becomes hyperphosphorylated (Shakoori et al., 1995; van Wijnen et al., 1997), which is known to occur in a CDK-dependent manner.

The inactivation of regulatory factors represents an equally important cell cycle control mechanism. Recently, many studies have focused on the role of ubiquitin-dependent proteolysis of factors by the 22 S proteasome (Tanaka and Tsurumi, 1997). The ubiquitin–proteasome system involves a large number of enzymes mediating ubiquitin activation (E1), ubiquitin conjugation (E2), or ubiquitin ligation (E3), which modulate turnover of cell cycle regulatory proteins (reviewed in King et al., 1997). For example, degradation of G_1 cyclins involves the CDC34 protein. CDC34 is a ubiquitin-conjugating enzyme and is conserved between yeast and vertebrates. The CDC34 gene is essential for the G_1/S phase transition. Ubiquitin-dependent degradation is also involved in constitutive turnover of cyclins throughout G_1, reflecting the labile nature of G_1 competency factors such as cyclin E.

Ubiquitin-dependent degradation also performs a key regulatory function during the G_2/M transition (King et al., 1997). The E2 enzymes encoded by UBC4 and UBC9 are involved in degradation of specific cyclins prior to the onset of mitosis. Similarly, completion of mitosis is regulated by the anaphase-promoting complex (APC). This high molecular weight ubiquitination complex is essential for chromosome segregation, but specific molecular targets have not been identified. Apart from degradation of cyclins at specific stages during the cell cycle, ubiquitin-dependent proteolysis may also be important for regulating the activities of oncoproteins, including c-fos, c-jun, and IRF2.

Parameters of Nuclear Architecture Influencing Cell Cycle Control

Nuclear Architecture Contributes to Transcriptional Regulation. There is a growing awareness of functional interrelationships mediating nuclear structure and function. Historically, there was a perceived dichotomy between regulatory mechanisms supporting gene expression and components of nuclear architecture. However, this parochial view is rapidly changing. The emerging concept is that both transcription and DNA synthesis occur in association with structural parameters of the nucleus. Consequently, it has become increasingly evident that the cellular and molecular mechanisms that contribute to both the regulated and regulatory relationships of nuclear morphology to the expression and replication of genes must be defined.

During the past several years, there has been an accrual of insight into the complexities of transcriptional control in eukaryotic cells. Our concept of a promoter has evolved from the initial expectation of a single regulatory sequence that determines transcriptional competency and level of expression. We now appreciate that transcriptional control is mediated by an interdependent series of regulatory sequences that reside 5′,3′ and within transcribed regions of genes. Rather than focusing on the minimal sequences required for transcriptional control to support biological activity, effects are being directed toward defining functional limits. Contributions of distal flanking sequences to regulation of transcription are being experimentally addressed. This is a necessity for understanding mechanisms by which activities of multiple promoter elements are responsive to a broad spectrum of regulatory signals, and the activities of these regulatory sequences are functionally integrated. Crosstalk between a series of regulatory domains must be understood under diverse biological circumstances where expression of genes supports cell and tissue functions. The overlapping binding sites for transcription factors within promoter regulatory elements and protein–protein interactions that influence transcription factor activity provide further components of the requisite diversity to accommodate regulatory options for physiologically responsive gene expression.

As the intricacies of gene organization and regulation are elucidated, the implications of a fundamental biological paradox become strikingly evident. How, with a limited representation of gene-specific regulatory elements and low abundance of cognate transactivation factors, can sequence-specific interactions occur to support a threshold for initiation of transcription within nuclei of intact cells. There is a growing appreciation that nuclear architecture provides a basis for support of stringently regulated modulation of cell growth and tissue-specific transcription that is necessary for the onset and progression of differentiation. Here, multiple lines of evidence point to contributions by three levels of nuclear organization to in vivo transcriptional control where structural parameters are functionally coupled to regulatory events (Fig. 5.8). The primary level of gene organization establishes a linear ordering of promoter regulatory elements. This representation of regulatory sequences reflects competency for responsiveness to physiological regulatory signals. However, interspersion of sequences between promoter elements that exhibit coordinate and synergistic activities indicates the requirement of a structural basis for integration of activities at independent regulatory domains. Parameters of chromatin structure and nucleosome organization are a second level of genome architecture that reduces the distance between promoter elements, thereby supporting interactions between the modular components of transcriptional control. Each nucleosome (approximately 140 nucleotide base pairs wound around a core complex of two each of H3, H4, H2, and H2B histone proteins) contracts linear spacing by sevenfold. Higher order chromatin structure further reduces nucleotide distances between regulatory sequences. Folding of nucleosome arrays into solenoid-type structures provides a potential for interactions that support synergism between promoter elements and responsiveness to multiple signaling pathways. A third level of nuclear architecture that contributes to transcriptional control is provided by the nuclear matrix.

The anastomosing network of fibers and filaments that constitutes the nuclear matrix supports the structural properties of the nucleus as a cellular organelle and accommodates structural modifications associated with proliferation, differentiation, and changes necessary to sustain phenotypic requirements of specialized cells. Regulatory functions of the nuclear matrix include but are by no means restricted to gene localization, imposition of physical constraints on chromatin structures that support formation of loop domains, RNA processing and transport of gene transcripts, and imprinting and modifications of chromatin structure. Taken together, these components of nuclear architecture facilitate biological requirements for physiologically responsive modifications in gene expression within the contexts of (1) homeostatic control involving rapid, short-term, and transient responsiveness; (2) developmental control that is progressive and stage specific, and (3) differentiation-related control that is associated with long-term phenotypic commitments to gene expression for support of structural and functional properties of cells and tissues.

We are just beginning to comprehend the significance of nuclear domains in the control of gene expression. However, it is already apparent that local nuclear environments that are generated by the multiple aspects of nuclear structure are intimately tied to developmental expression of cell growth and tissue-specific genes. From a broader perspective, reflecting the diversity of regulatory requirements as well as the phenotype-specific and physiologically responsive representation of nuclear structural proteins, there is a reciprocally functional relationship between nuclear structure and gene expression. Nuclear structure is a primary determinant of transcriptional control, and the expressed genes modulate the regulatory components of nuclear architecture. Thus, the power of addressing gene expression within the three-dimensional context of

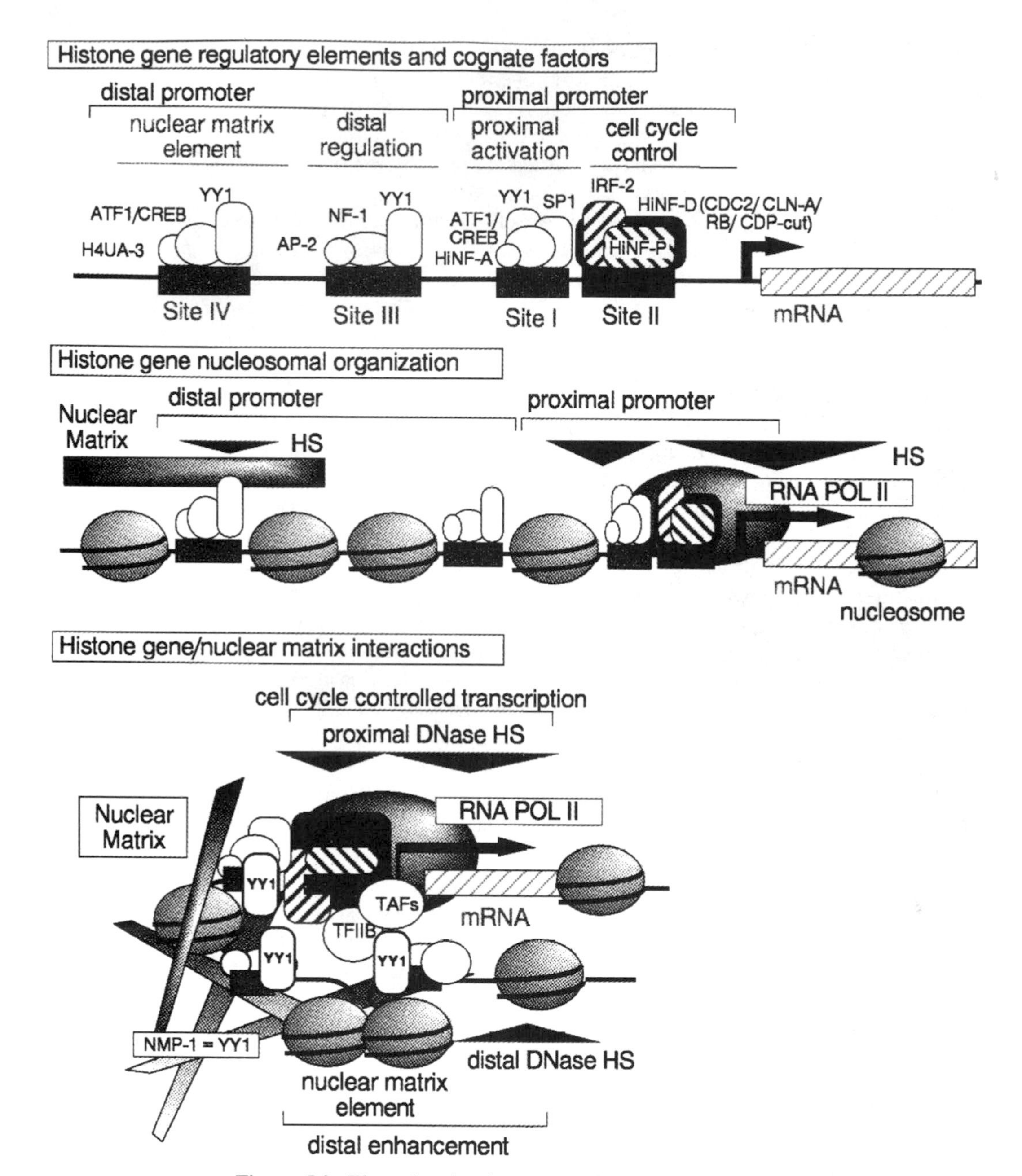

Figure 5.8. Three levels of gene regulatory organization determine transcriptional control of gene expression. The first level of organization is determined by the presence, number, distribution, orientation, and overlap of distinct promoter elements (blue and red boxes) that are recognized by cognate transcription factors (yellow and red ovals) **(top)**. The nucleosomal organization of gene promoters **(middle)** and association of genes and transcription factors with the nuclear matrix **(bottom)** represent key levels of organization. (See color plate.)

nuclear structure would be difficult to overestimate. Membrane-mediated initiation of signaling pathways that ultimately influence transcription have been recognized for some time. Here, the mechanisms that sense, amplify, dampen, and/or integrate regulatory signals involve structural as well as functional components of cellular membranes. Extending the structure–regulation paradigm to nuclear architecture expands the cellular context in which cell structure–gene expression interrelationships are operative.

It is becoming increasingly evident that proliferation and cell cycle–dependent transcriptional control to accommodate homeostatic regulation of cell and tissue function is modulated by the integration of a complex series of physiological regulatory signals. Fidelity of responsiveness necessitates the convergence of activities mediated by multiple regulatory elements of gene promoters. Our current knowledge of promoter organization and the repertoire of transcription factors that mediate activities provides a single dimension map of options for control of competency for proliferation and cell cycle progression. We are beginning to appreciate the additional structural and functional dimensions provided by chromatin structure, nucleosome organization, and subnuclear localization and targeting of both genes and transcription factors. Particularly exciting is increasing evidence for dynamic modifications in nuclear structure that parallel expression of genes during the cell cycle. The extent to which nuclear structure regulates and/or is regulated by modifications in gene expression remains to be experimentally established.

There is a necessity to mechanistically define how nuclear matrix–mediated subnuclear distribution of actively transcribed genes is responsive to nuclear matrix association of transcriptional and post-transcriptional regulatory factors. Furthermore, we cannot dismiss the possibility that association of regulatory factors with the nuclear matrix is consequential to sequence-specific interactions of transcriptionally active genes with the nuclear matrix. As these issues are resolved, we will gain additional insight into determinants of cause and/or effect relationships that interrelate specific components of nuclear architecture with proliferation and cell cycle–dependent gene expression at the transcriptional and post-transcriptional levels. However, it is justifiable to anticipate that while nuclear structure–gene expression interrelationships are operative under all biological conditions, situation-specific variations are the rule rather than the exception. As subtleties in the functional component of nuclear architecture are further defined, the significance of nuclear domains to transcription, RNA splicing, and processing of gene transcripts will be additionally understood.

A challenge that we face is to experimentally establish the rate-limiting nucleotide sequences and factors that integrate nuclear structure and gene expression. What are the sequence determinants and regulatory proteins that facilitate remodeling of chromatin structure and nucleosome organization to facilitate developmental and homeostatic requirements for proliferation-related transcription? How are structurally and

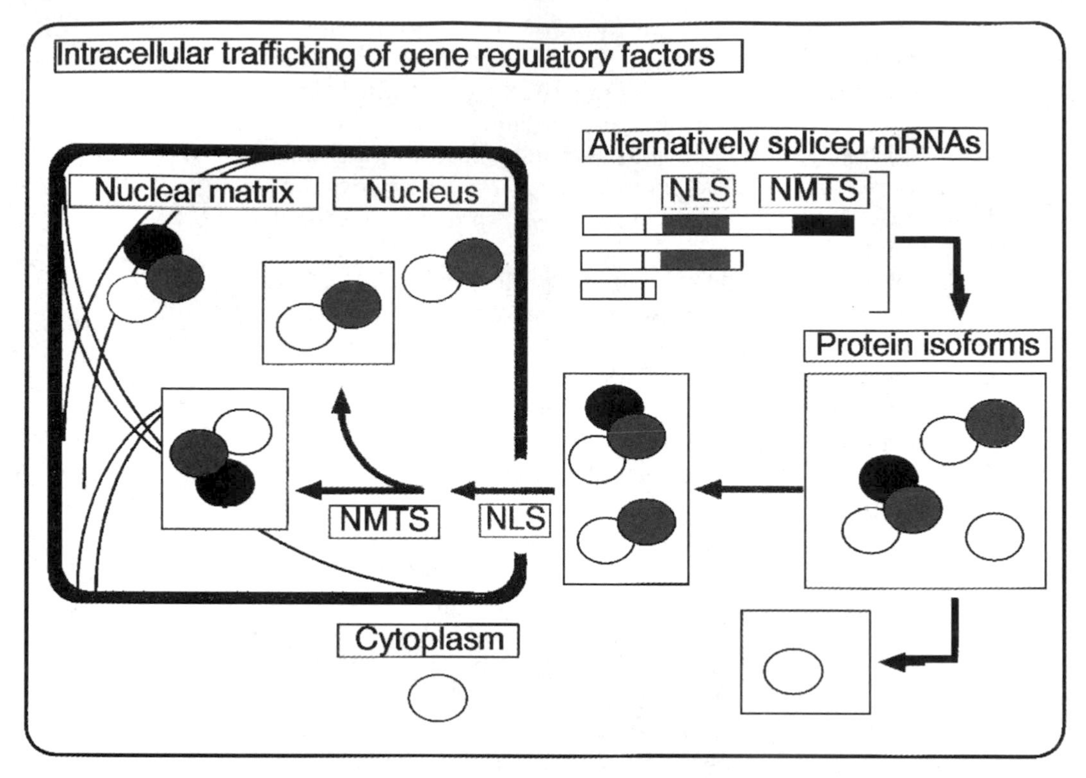

Figure 5.9. Intracellular trafficking of gene regulatory factors. The routing of transcription factors to distinct subcellular compartments is important for controlling the gene regulatory activities of these proteins. Presence (or functionality) of a nuclear localization signal (NLS, blue oval) dictates whether a protein (yellow oval) can be imported into the nucleus. A nuclear matrix targeting signal (NMTS, red oval) represents another key trafficking sequence that mediates association of regulatory proteins with the nuclear matrix. The paradigm for the diagram shown here is the AML/PEBP/CBFα class of transcription factors. These factors are encoded by several different genes that generate a series of alternatively spliced mRNAs. These mRNAs are translated into many protein isoforms that differ in the presence or absence of intracellular trafficking signals, determining subnuclear compartmentalization. (See color plate.)

functionally dynamic modifications in chromatin organization linked to proliferation and cell cycle–dependent interactions of genes with the nuclear matrix?

Nuclear Import and Intranuclear Trafficking. Intracellular targeting of cell cycle regulatory proteins may contribute to cell cycle control by concentrating the activities of these factors to specific intracellular domains (Fig. 5.9). For example, Cdc2/CDK1 displays a dispersed subnuclear distribution, but is also highly concentrated at the centrosomes (Pockwinse et al., 1997). This subnuclear location may regulate the assembly of the mitotic spindle. Association of cyclin B with membranes may be functionally linked to the breakdown of the nuclear membrane and/or subsequent formation of the nuclear envelope after mitosis (Jackman et al., 1995). The presence of the tumor suppressor protein p53 in the cytoplasm or nucleus has been correlated with cell growth control and the neoplastic phenotype of the cell (Moll et al., 1996). It has been well established that there are cell cycle–dependent shifts in the transfer of enzymes involved in DNA replication from the cytoplasm to the nucleus at the onset of S phase (see Chapter 3, this volume). The retinoblastoma protein pRb (Mancini et al., 1994), as well as transcription factors related to the c-fos and c-jun oncoproteins (van Wijnen et al., 1993), are associated with the nuclear matrix in a cell cycle– and/or cell growth–dependent manner. The association of DNA ligase I, DNA methyltransferase (Leonhardt and Cardoso, 1995), and DNA helicase (Molnar et al., 1997) with nuclear matrix–associated DNA replication domains also provides a mechanism for integrating enzymatic and regulatory functions at specific locations. Furthermore, the PML protein (Weis et al., 1994; Grande et al., 1996) and AML proteins (unpublished data) are redirected to distinct subnuclear domains as a consequence of structural modifications in these proteins due to chromosomal translocations in promyelocytic or myelogenous leukemias.

Because modifications in nuclear architecture are hallmarks of cancer cells, it is of considerable importance to define molecular principles governing the normal targeting of gene regulatory molecules to specific subnuclear domains (Fig. 5.9). Recently, we showed that transcriptionally active AML proteins, gene regulatory factors with a causative role in hemopoiesis, osteogenesis, and the progression of acute myelogenous leukemia, contains a 31 amino acid nuclear matrix targeting signal (NMTS) (Zeng et al., 1997b). The NMTS functions as a transactivator and directs AML factors to nuclear matrix–associated sites that support transcription (Zeng et al., 1997a) (Fig. 5.10). This NMTS is an autonomous protein domain that is absent in transcriptionally inactive AML isoforms. The NMTS is the frequent target of rearrangements due to chromosomal translocations that result in the separation of the NMTS, as well as a closely associated trans-activation domain, from the DNA-binding domain of AML factors. Hence, stringent control of intranuclear targeting may represent an important mechanism for maintaining the normal cell growth phenotype of somatic cells.

CONCLUSION

The necessity for stringent control of gene expression to support cell cycle and growth regulatory mechanisms is becoming increasingly appreciated. There is acute awareness of the required fidelity of growth control to regulate proliferation during development and tissue remodeling throughout the life of an organism. Significant inroads are being made into elucidation of transcriptional and post-transcriptional modifications in gene expression for both the stimulation of proliferation and the exit from the cell cycle to establish and maintain cells and tissues with

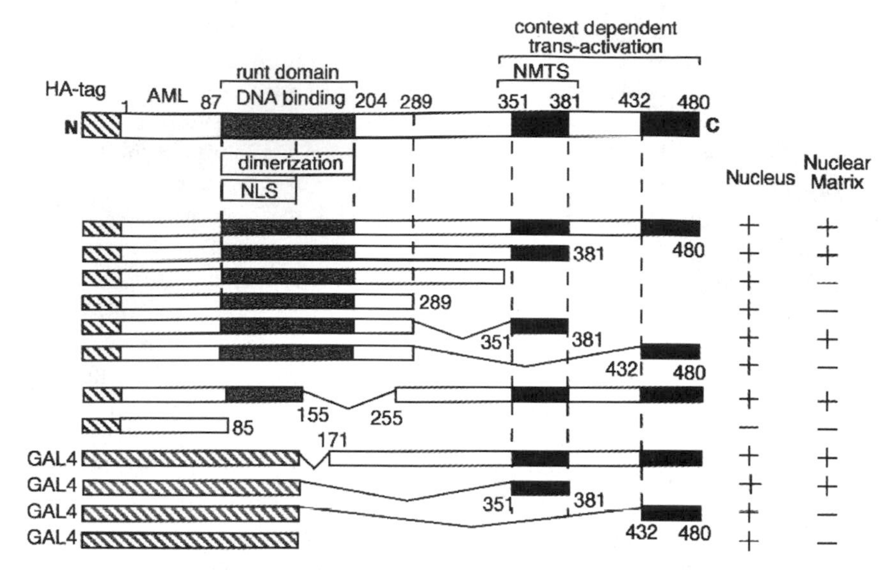

Figure 5.10. (Top): Delineation of a nuclear matrix targeting signal in AML-1B. The panel shows the relative location of functional domains regions (e.g., runt homology DNA binding domain) in the AML-1B transcription factor. The 3′ ultimate exons of the AML-1B and AML-3 genes encode for a C-terminal extension (between aa 250 and 480) which contains the NMTS and trans-activation domain. Deletion mutants of AML-1B, as well as chimeric GAL4/AML fusion proteins were analyzed for nuclear localization and intranuclear trafficking to the nuclear matrix (columns with plus and minus signs). (See color plate.)

specialized functions. Equally important, we are gaining insight into gene regulatory mechanisms that are operative to sustain pools of stem cells with competency to expand in response to physiological demands and commit to specific phenotypes.

Studies carried out during the past several years have provided indications of key regulatory points during the cell cycle that support competency for proliferation and cell cycle progression. A series of checkpoints have been functionally defined that monitor fidelity of the complex and interdependent regulatory events that control each cell cycle transition and initiation as well as cessation of proliferation. The extensive repertoire of growth regulatory factors that have been identified and characterized offer valuable clues to control of growth and differentiation. We are beginning to understand the interrelationships between growth-related gene regulatory mechanisms and components of cellular architecture. Additionally, we are learning effective ways to control proliferation under a broad spectrum of conditions. Consequently, new approaches are emerging for treatment of proliferation disorders, which include but are not restricted to cancer. Our capabilities for controlling the growth properties of stem cells without compromising options for phenotype development are expanding. One can be confident that additional insight into parameters of gene expression that mediate transcriptional and post-

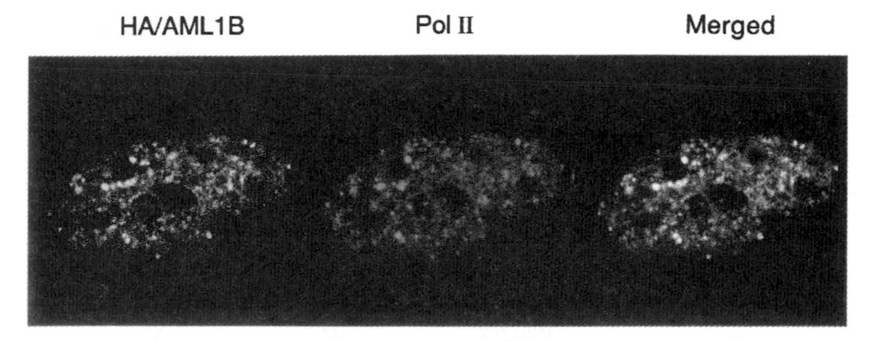

Figure 5.10. (Bottom): AML-1B is targeted to transcriptionally active nuclear matrix domains. A subset of AML-1B and RNA polymerase II is colocalized in the nuclear matrix. Human osteosarcoma SAOS-2 cells were transiently transfected with a construct expressing HA-tagged CBFα2/AML-1B. In situ nuclear matrix samples were prepared 24 hrs after transfection. HA/AML-1B was detected with an antibody against HA (green) (Left) and RNA polymerase II_0 with the B3 anti-RNA polymerase II antibody (red) (Center). The merged image is shown in the Right panel; yellow indicates colocalization of AML-1B with RNA polymerase II_0. The colocalization of AML-1B with RNA polymerase II_0 was evaluated by confocal microscopy with a series of optical sections (0.5 μm intervals) through a single cell. (See color plate.)

transcriptional control of cell proliferation and differentiation is forthcoming.

REFERENCES

Aziz F, van Wijnen AJ, Vaughan PS, Wu S, Shakoori AR, Lian JB, Soprano KJ, Stein JL, Stein GS (1998): The integrated activities of IRF-2 (HiNF-M), CDP/cut (HiNF-D), and H4TF-2 (HiNF-P) regulate transcription of a cell cycle controlled human histone H4 gene: Mechanistic differences between distinct H4 genes. Mol Biol Rep 25:1–12.

Azizkhan JC, Jensen DE, Pierce AJ, Wade M (1993): Transcription from TATA-less promoters: Dihydrofolate reductase as a model. Crit Rev Eukaryot Gene Expression 3:229–254.

Bartkova J, Zemanova M, Bartek J (1996): Expression of CDK7/CAK in normal and tumor cells of diverse histogenesis, cell-cycle position and differentiation. Int J Cancer 66:732–737.

Baserga R, Estensen RD, Petersen RO (1965): Inhibition of DNA synthesis in Ehrlich ascites cells by actinomycin D. II. The presynthetic block in the cell cycle. Proc Natl Acad Sci USA 54:1141.

Birnbaum MJ, van Wijnen AJ, Odgren PR, Last TJ, Suske G, Stein GS, Stein JL (1995a): Sp1 trans-activation of cell cycle regulated promoters is selectively repressed by Sp3. Biochemistry 34:16503–16508.

Birnbaum MJ, van Zundert B, Vaughan PS, Whitmarsh AJ, van Wijnen AJ, Davis

RJ, Stein GS, Stein JL (1997): Phosphorylation of the oncogenic transcription factor interferon regulatory factor 2 (IRF2) in vitro and in vivo. J Cell Biochem 66:175–183.

Birnbaum MJ, Wright KL, van Wijnen AJ, Ramsey-Ewing AL, Bourke MT, Last TJ, Aziz F, Frenkel B, Rao BR, Aronin N, Stein GS, Stein JL (1995b): Functional role for Sp1 in the transcriptional amplification of a cell cycle regulated histone H4 gene. Biochemistry 34:7648–7658.

Bodrug SE, Warner BJ, Bath ML, Lindeman GJ, Harris AW, Adams JM (1994): Cyclin D1 transgene impedes lymphocyte maturation and collaborates in lymphomagenesis with the myc gene. EMBO J 13:2124–2130.

Buchkovich KJ, Ziff EB (1994): Nerve growth factor regulates the expression and activity of p33cdk2 and p34cdc2 kinases in PC12 pheochromocytoma cells. Mol Biol Cell 5:1225–1241.

Burger C, Wick M, Muller R (1994): Lineage-specific regulation of cell cycle gene expression in differentiating myeloid cells. J Cell Sci 107:2047–2054.

Buyse IM, Shao G, Huang S (1995): The retinoblastoma protein binds to RIZ, a zinc-finger protein that shares an epitope with the adenovirus E1A protein. Proc Natl Acad Sci USA 92:4467–4471.

Carpenter PB, Mueller PR, Dunphy WG (1996): Role for a *Xenopus* orc2-related protein in controlling DNA replication. Nature 379:357–360.

Chandrasekaran C, Coopersmith CM, Gordon JI (1996): Use of normal and transgenic mice to examine the relationship between terminal differentiation of intestinal epithelial cells and accumulation of their cell cycle regulators. J Biol Chem 271:28414–28421.

Chen, PL, Riley DJ, Lee W-H (1995): The retinoblastoma protein as a fundamental mediator of growth and differentiation signals. Crit Rev Eukaryot Gene Expression 5:79–95.

Chen PL, Riley DJ, Chen Y, Lee WH (1996): Retinoblastoma protein positively regulates terminal adipocyte differentiation through direct interaction with C/EBPs. Genes Dev, 10:2794–2804.

Chrysogelos S, Riley DE, Stein GS, Stein JL (1985): A human histone H4 gene exhibits cell cycle dependent changes in chromatin structure that correlate with its expression. Proc Natl Acad Sci USA 82:7535–7539.

Chrysogelos S, Riley DE, Stein GS, Stein JL (1989): Fine mapping of the chromatin structure of a cell cycle–regulated human H4 histone gene. J Biol Chem 264:1232–1237.

Clarke AR, Maandag ER, van Roon M, van der Lugt NMT, van der Valk M, Hooper ML, Berns A, te Riele H (1992): Requirement for a functional Rb-1 gene in murine development. Nature 359:328–330.

Clyne RK, Kelly TJ (1995): Genetic analysis of an ARS element from the fission yeast *Schizosaccharomyces pombe.* EMBO J 14:6348–6357.

Cobrinik D, Lee M-H, Hannon G, Mulligan G, Bronson RT, Dyson N, Harlow E, Beach D, Weinberg RA, Jacks T (1996): Shared role of the pRB-related p130 and p107 proteins in limb development. Genes Dev 10:1633–1644.

Corbeil HB, Whyte P, Branton PE (1995): Characterization of transcription factor E2F complexes during muscle and neuronal differentiation. Oncogene 11:909–920.

Cordon-Cardo C, Richon VM (1994): Expression of the retinoblastoma protein is regulated in normal human tissues. Am J Pathol 144:500–510.

Dobashi Y, Kudoh T, Matsumine A, Toyoshima K, Akiyama T (1995): Constitutive overexpression of CDK2 inhibits neuronal differentiation of rat pheochromocytoma PC12 cells. J Biol Chem 270:23031–23037.

Dobashi Y, Kudoh T, Toyoshima K, Akiyama T (1996): Persistent activation of CDK4 during neuronal differentiation of rat pheochromocytoma PC12 cells. Biochem Biophys Res Commun 221:351–355.

Dou QP, Levin AH, Zhao S, Pardee AB (1993): Cyclin E and cyclin A as candidates for the restriction point protein. Cancer Res 53:1493–1497.

Dou QP, Molnar G, Pardee AB (1994a): Cyclin D1/cdk2 kinase is present in a G1 phase–specific protein complex Yi1 that binds to the mouse thymidine kinase gene promoter. Biochem Biophys Res Commun 205:1859–1868.

Dou QP, Pardee AB (1996): Transcriptional activation of thymidine kinase, a marker for cell cycle control. Prog Nucleic Acid Res Mol Biol 53:197–217.

Dou QP, Zhao S, Levin AH, Wang J, Helin K, Pardee AB (1994b): G1/S-regulated E2F-containing protein complexes bind to the mouse thymidine kinase gene promoter. J Biol Chem 269:1306–1313.

Dowdy SF, Hinds PW, Louie K, Reed SI, Arnold A, Weinberg RA (1993): Physical interaction of the retinoblastoma protein with human D cyclins. Cell 73:499–511.

Dubey DD, Kim SM, Todorov IT, Huberman JA (1996): Large, complex modular structure of a fission yeast DNA replication origin. Curr Biol 6:467–473.

Fantl V, Stamp G, Andrews A, Rosewell I, Dickson C (1995): Mice lacking cyclin D1 are small and show defects in eye and mammary gland development. Genes Dev 9:2364–2372.

Fridovich-Keil JL, Markell PJ, Gudas JM, Pardee AB (1993): DNA sequences required for serum-responsive regulation of expression from the mouse thymidine kinase promoter. Cell Growth Diff 4:679–687.

Gao CY, Bassnett S, Zelenka PS (1995): Cyclin B, p34cdc2, and H1-kinase activity in terminally differentiating lens fiber cells. Dev Biol 169:185–194.

Gavin KA, Hidaka M, Stillman B (1995): Conserved initiator proteins in eukaryotes. Science 270:1667–1671.

Gervasi C, Szaro BG (1995): The *Xenopus laevis* homologue to the neuronal cyclin-dependent kinase (cdk5) is expressed in embryos by gastrulation. Mol Brain Res 33:192–200.

Godbout R, Andison R (1996): Elevated levels of cyclin D1 mRNA in the undifferentiated chick retina. Gene 182:111–115.

Good L, Chen J, Chen KY (1995): Analysis of sequence-specific binding activity of cis-elements in human thymidine kinase gene promoter during G1/S phase transition. J Cell Physiol 163:636–644.

Good L, Dimri GP, Campisi J, Chen KY (1996): Regulation of dihydrofolate reductase gene expression and E2F components in human diploid fibroblasts during growth and senescence. J Cell Physiol 168:580–588.

Gossen M, Pak DT, Hansen SK, Acharya JK, Botchan MR (1995): A *Drosophila* homolog of the yeast origin recognition complex. Science 270:1674–1677.

Grande MA, van der Kraan I, van Steensel B, Schul W, de The H, van der Voort HT, de Jong L, van Driel R (1996) PML-containing nuclear bodies: Their spatial distribution in relation to other nuclear components. J Cell Biochem 63:280–291.

Gu W, Schneider JW, Condorelli G, Kaushal S, Mahdavi V, Nadal-Ginard B

(1993): Interaction of myogenic factors and the retinoblastoma protein mediates muscle cell commitment and differentiation. Cell 72:309–324.

Guo B, Odgren PR, van Wijnen AJ, Last TJ, Nickerson J, Penman S, Lian JB, Stein JL, Stein GS (1995): The nuclear matrix protein NMP-1 is the transcription factor YY1. Proc Natl Acad Sci USA 92:10526–10530.

Hamlin JL, Mosca PJ, Levenson VV (1994): Defining origins of replication in mammalian cells. Biochim Biophys Acta 1198:85–111.

Han EK, Begemann M, Sgambato A, Soh JW, Doki Y, Xing WQ, Liu W, Weinstein IB (1996): Increased expression of cyclin D1 in a murine mammary epithelial cell line induces p27kip1, inhibits growth, and enhances apoptosis. Cell Growth Differ 7:699–710.

Harris ME, Bohni R, Schneiderman MH, Ramamurthy L, Schumperli D, Marzluff WF (1991): Regulation of histone mRNA in the unperturbed cell cycle: Evidence suggesting control at two post-transcriptional steps. Mol Cell Biol 11:2416–2424.

Hartwell LH, Culotti J, Pringle JR, Reid BJ (1974): Genetic control of the cell division cycle in yeast. Science 183:46–51.

Hartwell LH, Weinert TA (1989): Checkpoints: Controls that ensure the order of cell cycle events. Science 246:629–634.

Heichman KA, Roberts JM (1994): Roles to replicate by. Cell 79:557–562.

Hickey RJ, Malkas LH (1997): Mammalian cell DNA replication. Crit Rev Eukaryot Gene Expression 7:125–157.

Holthuis J, Owen TA, van Wijnen AJ, Wright KL, Ransey-Ewing A, Kennedy MB, Carter R, Cosenza SC, Soprano KJ, Lian JB, Stein JL, Stein GS (1990): Tumor cells exhibit deregulation of the cell cycle histone gene promoter factor HiNF-D. Science 247:1454–1457.

Horiguchi-Yamada J, Yamada H, Nakada S, Ochi K, Nemoto T (1994): Changes of G1 cyclins, cdk2, and cyclin A during the differentiation of HL60 cells induced by TPA. Mol Cell Biochem 132:31–37.

Howard A, Pelc SR (1951): Nuclear incorporation of ^{32}P as demonstrated by autoradiography. Exp Cell Res 2:178–187.

Hunter T, Pines J (1994): Cyclins and cancer II: Cyclin D and CDK inhibitors come of age. Cell 79:573–582.

Jackman M, Firth M, Pines J (1995): Human cyclins B1 and B2 are localized to strikingly different structures: B1 to microtubules, B2 primarily to the Golgi apparatus. EMBO J 14:1646–1654.

Jacks T, Fazeli A, Schmitt EM, Bronson RT, Goodell MA, Weinberg RA (1992): Effects of an Rb mutation in the mouse. Nature 359:295–300.

Jahn L, Sadoshima J, Izumo S (1994): Cyclins and cyclin-dependent kinases are differentially regulated during terminal differentiation of C2C12 muscle cells. Exp Cell Res 212:297–307.

Jiang H, Lin J, Young SM, Goldstein NI, Waxman S, Davila V, Chellappan SP, Fisher PB (1995): Cell cycle gene expression and E2F transcription factor complexes in human melanoma cells induced to terminally differentiate. Oncogene 11:1179–1189.

Johnson LF (1992): G1 events and the regulation of genes for S-phase enzymes. Curr Opin Cell Biol 4:149–154.

Karlseder J, Rotheneder H, Wintersberger E (1996): Interaction of Sp1 with the growth- and cell cycle–regulated transcription factor E2F. Mol Cell Biol 16:1659–1667.

Kato JY, Sherr CJ (1993): Inhibition of granulocyte differentiation by G1 cyclins D2 and D3 but not D1. Proc Natl Acad Sci USA 90:11513–11517.

Kiess M, Gill RM, Hamel PA (1995a): Expression and activity of the retinoblastoma protein (pRB)-family proteins, p107 and p130, during L6 myoblast differentiation. Cell Growth Differ 6:1287–1298.

Kiess M, Gill RM, Hamel PA (1995b): Expression of the positive regulator of cell cycle progression, cyclin D3, is induced during differentiation of myoblasts into quiescent myotubes. Oncogene 10:159–166.

King RW, Deshaies RJ, Peters JM, Kirschner MW (1997): How proteolysis drives the cell cycle. Science 274:1652–1659.

King RW, Jackson PK, Kirschner MW (1994): Mitosis in transition. Cell 79:563–571.

Kiyokawa H, Richon VM, Rifkind RA, Marks PA (1994): Suppression of cyclin-dependent kinase 4 during induced differentiation of erythroleukemia cells. Mol Cell Biol 14:7195–7203.

Kiyokawa H, Richon VM, Venta-Perez G, Rifkind RA, Marks PA (1993): Hexamethylenebisacetamide-induced erythroleukemia cell differentiation involves modulation of events required for cell cycle progression through G1. Proc Natl Acad Sci USA 90:6746–6750.

Kouzarides T (1995): Transcriptional control by the retinoblastoma protein. Semin Cancer Biol 6:91–98.

Kranenburg O, de Groot RP, Van der Eb AJ, Zantema A (1995): Differentiation of P19 EC cells leads to differential modulation of cyclin-dependent kinase activities and to changes in the cell cycle profile. Oncogene 10:87–95.

Krek W, Ewen ME, Shirodkar S, Arany Z, Kaelin WG, Livingston DM (1994): Negative regulation of the growth-promoting transcription factor E2F-1 by a stably bound cyclin A–dependent protein kinase. Cell 78:161–172.

La Thangue NB (1994) DRTF1/E2F: An expanding family of heterodimeric transcription factors implicated in cell-cycle control. TIBS 19:108–114.

Leatherwood J, Lopez-Girona A, Russell P (1996): Interaction of Cdc2 and Cdc18 with a fission yeast ORC2-like protein. Nature 379:360–363.

Lee EY-HP, Chang C-Y, Hu N, Wang Y-CJ, Lai C-C, Herrup K, Lee W-H, Bradley A (1992): Mice deficient for Rb are nonviable and show defects in neurogenesis and haematopoiesis. Nature 359:288–294.

Lee EY, Hu N, Yuan SS, Cox LA, Bradley A, Lee WH, Herrup K (1994): Dual roles of the retinoblastoma protein in cell cycle regulation and neuron differentiation. Genes Dev 8:2008–2021.

Lee WH, Chen PL, Riley DJ (1995): Regulatory networks of the retinoblastoma protein. Ann NY Acad Sci 752:432–445.

Leonhardt H, Cardoso MC (1995): Targeting and association of proteins with functional domains in the nucleus: The insoluble solution. Int Rev Cytol 162B:303–335.

Li LJ, Naeve GS, Lee AS (1993): Temporal regulation of cyclin A-p107 and p33cdk2 complexes binding to a human thymidine kinase promoter element important for G1-S phase transcriptional regulation. Proc Natl Acad Sci USA 90:3554–3558.

Ljungman M (1996): Effect of differential gene expression on the chromatin structure of the DHFR gene domain in vivo. Biochim Biophys Acta 1307:171–177.

Loyer P, Glaise D, Cariou S, Baffet G, Meijer L, Guguen-Guillouzo C (1994): Expression and activation of cdks (1 and 2) and cyclins in the cell cycle progression during liver regeneration. J Biol Chem 269:2491–2500.

Lucibello FC, Sewing A, Brusselbach S, Burger C, Muller R (1993): Deregulation of cyclins D1 and E and suppression of cdk2 and cdk4 in senescent human fibroblasts. J Cell Sci 105:123–133.

MacLachlan TK, Sang N, Giordano A (1995): Cyclins, cyclin-dependent kinases and Cdk inhibitors: Implications in cell cycle control and cancer. Crit Rev Eukaryot Gene Expression 5:127–156.

Mancini MA, Shan B, Nickerson JA, Penman S, Lee WH (1994): The retinoblastoma gene product is a cell cycle–dependent, nuclear matrix–associated protein. Proc Natl Acad Sci USA 91:418–422.

Marks PA, Richon VM, Rifkind RA (1996): Cell cycle regulatory proteins are targets for induced differentiation of transformed cells: Molecular and clinical studies employing hybrid polar compounds. Int J Hem 63:1–17.

Martin F, Schaller A, Eglite S, Schumperli D, Muller B (1997): The gene for histone RNA hairpin binding protein is located on human chromosome 4 and encodes a novel type of RNA binding protein. EMBO J 16:769–778.

Marzluff WF, Pandey NB (1988): Multiple regulatory steps control histone mRNA concentrations. Trends Biochem Sci 13:49–52.

Meyers S, Hiebert SW (1995): Indirect and direct disruption of transcriptional regulation in cancer: E2F and AML-1. Crit Rev Eukaryot Gene Expression 5:365–383.

Moll UM, Ostermeyer AG, Haladay R, Winkfield B, Frazier M, Zambetti G (1996): Cytoplasmic sequestration of wild-type p53 protein impairs the G1 checkpoint after DNA damage. Mol Cell Biol 16:1126–1137.

Molnar GM, Crozat A, Kraeft SK, Dou QP, Chen LB, Pardee AB (1997): Association of the mammalian helicase MAH with the pre-mRNA splicing complex. Proc Natl Acad Sci USA 94:7831–7836.

Moreno ML, Chrysogelos SA, Stein GS, Stein JL (1986): Reversible changes in the nucleosomal organization of a human H4 histone gene during the cell cycle. Biochemistry 25:5364–5370.

Morgenbesser SD, Williams BO, Jacks T, DePinho RA (1994): p53-Dependent apoptosis produced by Rb-deficiency in the developing mouse lens. Nature 371:72–74.

Morris TD, Weber LA, Hickey E, Stein GS, Stein JL (1991): Changes in the stability of a human H3 histone mRNA during the HeLa cell cycle. Mol Cell Biol 11:544–553.

Muzi-Falconi M, Kelly TJ (1995): Orp1, a member of the Cdc18/Cdc6 family of S-phase regulators, is homologous to a component of the origin recognition complex. Proc Natl Acad Sci USA 92:12475–12479.

Nevins JR (1992): E2F: A link between the Rb tumor suppressor protein and viral oncoproteins. Science 258:424–429.

Nikolic M, Dudek H, Kwon YT, Ramos YF, Tsai LH (1996): The cdk5/p35 kinase is essential for neurite outgrowth during neuronal differentiation. Genes Dev 10:816–825.

Novitch BG, Mulligan GJ, Jacks T, Lassar AB (1996): Skeletal muscle cells lacking the retinoblastoma protein display defects in muscle gene expression and accumulate in S and G2 phases of the cell cycle. J Cell Biol 135:441–456.

Nurse P (1994): Ordering S phase and M phase in the cell cycle. Cell 79:547–550.

Ohshima T, Ward JM, Huh CG, Longenecker G, Veeranna, Pant HC, Brady RO, Martin LJ, Kulkarni AB (1996): Targeted disruption of the cyclin-dependent kinase 5 gene results in abnormal corticogenesis, neuronal pathology and perinatal death. Proc Natl Acad Sci USA 93:11173–11178.

Pardee AB (1989): G1 events and regulation of cell proliferation. Science 246:603–608.

Pardee AB, Keyomarsi K (1992): Cell multiplication. Curr Opin Cell Biol 4:141–143.

Pauli U, Chrysogelos S, Stein J, Stein G, Nick H (1987): Protein–DNA interactions in vivo upstream of a cell cycle regulated human H4 histone gene. Science 236:1308–1311.

Pelz SW, Brewer G, Bernstein P, Hart PA, Ross J (1991): Regulation of mRNA turnover in eukaryotic cells. Crit Rev Eukaryot Gene Expression 1:99–126.

Pelz SW, Ross J (1987): Autogenous regulation of histone mRNA decay by histone proteins in a cell-free system. Mol Cell Biol 7:4345–4356.

Pemov A, Bavykin S, Hamlin JL (1995): Proximal and long-range alterations in chromatin structure surrounding the Chinese hamster dihydrofolate reductase promoter. Biochemistry 34:2381–2392.

Plumb MA, Stein JL, Stein GS (1983): Coordinate regulation of multiple histone mRNAs during the cell cycle in HeLa cells. Nucleic Acids Res 11:2391–2410.

Pockwinse SM, Krockmalnic G, Doxsey SJ, Nickerson J, Lian JB, van Wijnen AJ, Stein JL, Stein GS, Penman S (1997): Cell cycle independent interaction of CDC2 with the centrosome, which is associated with the nuclear matrix–intermediate filament scaffold. Proc Natl Acad Sci USA 94:3022–3027.

Prescott DM (1976): Reproduction of Eukaryotic Cells. New York: Academic Press.

Rao SS, Kohtz DS (1995): Positive and negative regulation of D-type cyclin expression in skeletal myoblasts by basic fibroblast growth factor and transforming growth factor beta. A role for cyclin D1 in control of myoblast differentiation. J Biol Chem 270:4093–4100.

Rhee K, Wolgemuth DJ (1995): Cdk family genes are expressed not only in dividing but also in terminally differentiated mouse germ cells, suggesting their possible function during both cell division and differentiation. Dev Dynamics 204:406–420.

Ross ME, Carter ML, Lee JH (1996): MN20, a D2 cyclin, is transiently expressed in selected neural populations during embryogenesis. J Neurosci 16:210–219.

Ross ME, Risken M (1994): MN20, a D2 cyclin found in brain, is implicated in neural differentiation. J Neurosci 14:6384–6391.

Schilling LJ, Farnham PJ (1994): Transcriptional regulation of the dihydrofolate reductase/rep-3 locus. Crit Rev Eukaryot Gene Expression 4:19–53.

Schilling LJ, Farnham PJ (1995): The bidirectionally transcribed dihydrofolate reductase and rep-3a promoters are growth regulated by distinct mechanisms. Cell Growth Differ 6:541–548.

Schneider JW, Gu W, Zhu L, Mahdavi V, Nadal-Ginard B (1994): Reversal of terminal differentiation mediated by p107 in Rb−/− muscle cells. Science 264:1467–1471.

Schulze A, Zerfass K, Spitkovsky D, Henglein B, Jansen-Durr P (1994): Activation of the E2F transcription factor by cyclin D1 is blocked by p161INK4, the product of the putative tumor suppressor gene MTS1. Oncogene 9:3475–3482.

Schumperli D (1988): Multilevel regulation of replication-dependent histone genes. Trends Genet 4:187–191.

Sellers WR, Kaelin WG (1996): xRB as a modulator of transcription. Biochim Biophys Acta 1288:M1–M5.

Shakoori AR, van Wijnen AJ, Cooper C, Aziz F, Birnbaum M, Reddy GPV, Grana X, De Luca A, Giordano A, Lian JB, Stein JL, Quesenberry P, Stein GS (1995): Cytokine induction of proliferation and expression of CDC2 and cyclin A in FDC-P1 myeloid hematopoietic progenitor cells: Regulation of ubiquitous and cell cycle–dependent histone gene transcription factors. J Cell Biochem 59:291–302.

Sherr CJ (1993): Mammalian G1 cyclins. Cell 73:1059–1065.

Shin EK, Shin A, Paulding C, Schaffhausen B, Yee AS (1995): Multiple change in E2F function and regulation occur upon muscle differentiation. Mol Cell Biol 15:2252–2262.

Sicinski P, Donaher JL, Parker SB, Li T, Fazeli A, Gardner H, Haslam SZ, Bronson RT, Elledge SJ, Weinberg RA (1995): Cyclin D1 provides a link between development and oncogenesis in the retina and breast. Cell 82:621–630.

Singh P, Coe J, Hong W (1995): A role for retinoblastoma protein in potentiating transcriptional activation by the glucocorticoid receptor. Nature 374:562–565.

Slack RS, Miller FD (1996): Retinoblastoma gene in mouse neural development. Dev Genet 18:81–91.

Smith E, Frenkel B, MacLachlan T, Giordano A, Stein JL, Lian JB, Stein GS (1997): Postproliferative, cyclin E–associated kinase activity in differentiated osteoblasts: Inhibition by proliferating osteoblasts and osteosarcoma cells. J Cell Biochem 66:141–152.

Smith E, Frenkel B, Schlegel R, Giordano A, Lian JB, Stein JL, Stein GS (1995): Expression of cell cycle regulatory factors in differentiating osteoblasts: Postproliferative upregulation of cyclins B and E. Cancer Res 55:5019–5024.

Stein GS, Lian JB, Stein JL, Briggs R, Shalhoub V, Wright K, Pauli U, van Wijnen A (1989): Altered binding of human histone gene transcription factors during the shutdown of proliferation and onset of differentiation in HL-60 cells. Proc Natl Acad Sci USA 86:1865–1869.

Stein GS, Stein JL (1984): Is human histone gene expression autogenously regulated? Mol Cell Biochem 64:105–110.

Stein GS, Stein JL, van Wijnen AJ, Lian JB (1994): Histone gene transcription: A model for responsiveness to an integrated series of regulatory signals mediating cell cycle control and proliferation/differentiation interrelationships. J Cell Biochem 54:393–404.

Stillman B (1996): Cell cycle control of DNA replication. Science 274:1659–1664.

Szekely L, Jin P, Jiang WQ, Rosen A, Wiman KG, Klein G, Ringertz N (1993): Position-dependent nuclear accumulation of the retinoblastoma (RB) protein during in vitro myogenesis. J Cell Physiol 155:313–322.

Tamaru T, Okada M, Nakagawa H (1994): Differential expression of D type cyclins during neuronal maturation. Neurosci Lett 168:229–232.

Tanaka K, Tsurumi C (1997): the 26S proteosome: Subunits and functions. Mol Biol Rep 24:3–11.

Terasima T, Yasukawa M (1966): Synthesis of G1 protein preceding DNA synthesis in cultured mammalian cells. Exp Cell Res 44:669.

Tsai LH, Takahashi T, Caviness VS Jr, Harlow E (1993): Activity and expression pattern of cyclin-dependent kinase 5 in the embryonic mouse nervous system. Development 119:1029–1040.

van Grunsven LA, Billon N, Savatier P, Thomas A, Urdiales JL, Rudkin BB (1996): Effect of nerve growth factor on the expression of cell cycle regulatory proteins in PC12 cells: Dissection of the neurotrophic response from the antimitogenic response. Oncogene 12:1347–1356.

van Gurp MF, Hoffman H, Tufarelli C, Stein JL, Lian BJ, Neufeld EJ, Stein GS, van Wijnen AJ (1997): The CDP/cut homeodomain protein represses osteocalcin gene transcription via a tissue-specific promoter element: Formation of proliferation-specific protein/DNA complexes with the pRB-related protein p107 and cyclin A. Manuscript submitted.

van Wijnen AJ, Aziz F, Grana X, De Luca A, Desai RK, Jaarsveld K, Last TJ, Soprano K, Giordano A, Lian JB, Stein JL, Stein GS (1994): Transcription of histone H4, H3 and H1 cell cycle genes: Promoter factor HiNF-D contains CDC2, cyclin A and an RB-related protein. Proc Natl Acad Sci USA 91:12882–12886.

van Wijnen AJ, Bidwell JP, Fey EG, Penman S, Lian JB, Stein JL, Stein GS (1993): Nuclear matrix association of multiple sequence specific DNA binding activities related to SP-1, ATF, CCAAT, C/EBP, OCT-1 and AP-1. Biochemistry 32:8397–8402.

van Wijnen AJ, Cooper C, Odgren P, Aziz F, De Luca A, Shakoori RA, Giordano A, Quesenberry PJ, Lian JB, Stein GS, Stein JL (1997): Cell cycle dependent modifications in activities of pRb-related tumor suppressors and proliferation-specific CDP/cut homeodomain factors in murine hematopoietic progenitor cells. J Cell Biochem 66:512–523.

van Wijnen AJ, van Gurp MF, de Ridder M, Tufarelli C, Last TJ, Birnbaum M, Vaughan PS, Giordano A, Krek W, Neufeld EJ, Stein JL, Stein GS (1996): CDP/cut is the DNA binding subunit of histone gene transcription factor HiNF-D: A mechanism for gene regulation at the G_1/S phase cell cycle transition point independent of transcription factor E2F. Proc Natl Acad Sci USA 93:11516–11521.

Vaughan PS, Aziz F, van Wijnen AJ, Wu S, Harada H, Taniguchi T, Soprano K, Stein GS, Stein JL (1995): Activation of a cell cycle regulated histone gene by the oncogenic transcription factor IRF2. Nature 377:362–365.

Wade M, Blake MC, Jambou RC, Helin K, Harlow E, Azizkhan JC (1995): An inverted repeat motif stabilizes binding of E2F and enhances transcription of the dihydrofolate reductase gene. J Biol Chem 270:9783–9791.

Wang ZF, Whitfield ML, Ingledue TC 3rd, Dominski Z, Marzluff WF (1996): The protein that binds the 3′ end of histone mRNA: A novel RNA-binding protein required for histone pre-mRNA processing. Genes Dev 10:3028–3040.

Weinberg RA (1995): The retinoblastoma protein and cell cycle control. Cell 81:323–330.

Weis K, Rambaud S, Lavau C, Jansen J, Carvalho T, Carmo-Fonseca M, Lamond A, Dejean A (1994): Retinoic acid regulates aberrant nuclear localization of PML-RAR alpha in acute promyelocytic leukemia cells. Cell 76:345–356.

Wells J, Held P, Illenye S, Heintz NH (1996): Protein–DNA interactions at the major and minor promoters of the divergently transcribed DHFR and rep3 genes during the Chinese hamster ovary cell cycle. Mol Cell Biol 16:634–647.

Wells JM, Illenye S, Magae J, Wu CL, Heintz NH (1997): Accumulation of E2F-4/DP-1 DNA binding complexes correlates with induction of DHFR gene expression during the G1 to S phase transition. J Biol Chem 272:4483–4492.

White E (1994): p53, guardian of Rb. Nature 371:21–22.

Wilhide CC, Van Dang C, Dipersio J, Kenedy AA, Bray PF (1995): Overexpression of cyclin D1 in the Dami megakaryocytic cell line causes growth arrest. Blood 86:294–304.

Wiman KG (1993): The retinoblastoma gene: Role in cell cycle control and cell differentiation. FASEB J 7:841–845.

Wohlgemuth JG, Bulboaca GH, Moghadam M, Caddle MS, Calos MP (1994): Physical mapping of origins of replication in the fission yeast *Schizosaccharomyces pombe.* Mol Biol Cell 5:839–849.

Yan GZ, Ziff EB (1995): NGF regulates the PC12 cell cycle machinery through specific inhibition of the Cdk kinases and induction of cyclin D1. J Neurosci 15:6200–6212.

Yen A, Varvayanis S (1994): Late dephosphorylation of the RB protein in G2 during the process of induced cell differentiation. Exp Cell Res 214:250–257.

Yen A, Varvayanis S, Platko JD (1993): 12-O-tetradecanoylphorbol-13-acetate and staurosporine induce increased retinoblastoma tumor suppressor gene expression with megakaryocytic differentiation of leukemic cells. Cancer Res 53:3085–3091.

Zambetti G, Stein JL, Stein GS (1987): Targeting of a chimeric human histone fusion mRNA to membrane-bound polysomes in HeLa cells. Proc Natl Acad Sci USA 84:2683–2687.

Zeng C, McNeil S, Nickerson J, Shopland L, Lawrence JB, Penman S, Hiebert S, Lian JB, van Wijnen AJ, Stein JL, Stein GS (1998): Intranuclear targeting of AML/CBFα regulatory factors to nuclear matrix–associated transcriptional domains. Proc Natl Acad Sci USA 95:1585–1589.

Zeng C, van Wijnen AJ, Stein JL, Meyers S, Sun W, Shopland L, Lawrence JB, Penman S, Lian JB, Stein GS, Hiebert SW (1997b): Identification of a nuclear matrix targeting signal in the leukemia and bone-related AML/CBF-α transcription factors. Proc Natl Acad Sci USA 94:6746–6751.

Zwijsen RML, Wientjens E, Klompmaker R, van der Sman J, Bernards R, Michalides RJAM (1997) CDK-independent activation of estrogen receptor by cyclin D1. Cell 88:405–415.

CHAPTER 6

SIGNAL TRANSDUCTION PATHWAYS AND REGULATION OF THE MAMMALIAN CELL CYCLE: CELL TYPE–DEPENDENT INTEGRATION OF EXTERNAL SIGNALS

DAVID T. DENHARDT
Department of Cell Biology and Neuroscience, Nelson Biological Laboratories, Rutgers University, Piscataway, NJ 08854

INTRODUCTION

"This paper provides evidence that normal animal cells possess a unique regulatory mechanism to shift them between proliferative and quiescent states. . . . The name restriction point is proposed for the specific time in the cell cycle at which this critical release event occurs. . . . Malignant cells are proposed to have lost their restriction point control" (Pardee, 1974). Research in the one-score-and-four years that have elapsed since that seminal publication have amply elaborated on and confirmed this view of the cell cycle. In this chapter I review our present understanding of the "unique regulatory mechanism" deduced by Pardee.

Given the central importance of the cell cycle in the life of the cell, it comes as no surprise that the cell's decision to proliferate, or to continue proliferation, is controlled by multiple regulatory pathways acting in different ways at various points of the cell cycle. Cell cycles in unicellular organisms are primarily influenced by light/dark periodicities and the presence or absence of food, mating partners, or a suitable habitat. Additional complexity is introduced in cells programmed to be part of a multicellular organism. These cells must be capable of rapid replication during the development of the organism, but then in most cases must enter a state of replicative quiescence as the cells differentiate. In certain

The Molecular Basis of Cell Cycle and Growth Control, Edited by G.S. Stein, R. Baserga, A. Giordano, and D.T. Denhardt
ISBN 0-471-15706-6, pages 225–304.

tissues, cells must continue to proliferate at a controlled rate, whereas in other tissues cell cycling may occur only in certain physiological states. Also some quiescent cell types must be able to resume replication if the tissue is injured. In this chapter, which deals primarily with events in mammalian cells, I have attempted to paint a broad picture of how this impressive feat is accomplished.

Before reviewing our current understanding of the signaling pathways regulating the mammalian cell cycle, I summarize here what is known about the molecular mechanisms effecting cell proliferation. This machinery is composed of proteins that bring about DNA replication and successful passage of the cell through a replicative cycle. This subject is covered in more depth in other chapters in this book, and the informed reader may want to skip it. Control over the proliferative process is exerted by an evolutionarily ancient set of proteins that includes the cyclins and cyclin-dependent kinases, their inhibitors, certain proteinases and phosphatases, and the transcription regulators Rb, p53, and E2F. These are the targets of the signaling pathways that oversee cell replication. The signaling pathways referred to in the first section are described in detail in later sections. As used here, *signal* is a message to perform a task.

SIGNAL IMPLEMENTATION: TARGETS OF THE SIGNALING PATHWAYS

Overview of the Cell Cycle

As detailed elsewhere in this volume (see Chapters 2 and 3), the conventional mammalian cell cycle, illustrated in Figure 6.1, typically consists of two obvious stages, S and M, during which DNA synthesis and mitosis occur, respectively, and two intervening phases, G_1 and G_2, in which the cell is generally preparing to initiate DNA replication or cell division, respectively. Quiescent cells not in the cell cycle are usually defined as being in the G_0 state. Figure 6.1 shows a representation of these phases along with some of the events that orchestrate progression through the cell cycle. The actual portion of the cycle devoted to each phase depends on the cell and conditions.

The cyclins that help regulate the cell cycle are expressed at specific periods in the cycle as the result of mitogen- or cell cycle–regulated transcription and/or activation. Cells normally adherent to the extracellular matrix, fibroblasts, for example, require signals generated both by soluble mitogens and by integrin-mediated interactions with components of the extracellular matrix in order to transit G_1 and enter S phase (Assoian, 1997). The cyclins activate specific cyclin-dependent kinases (CDKs), which typically are present constitutively in the nucleus; the resulting phosphorylation of a variety of target proteins on serine and threonine residues propels the cell through the cycle. Control over the activity of the cyclin CDK complexes is exerted by specific phosphoryla-

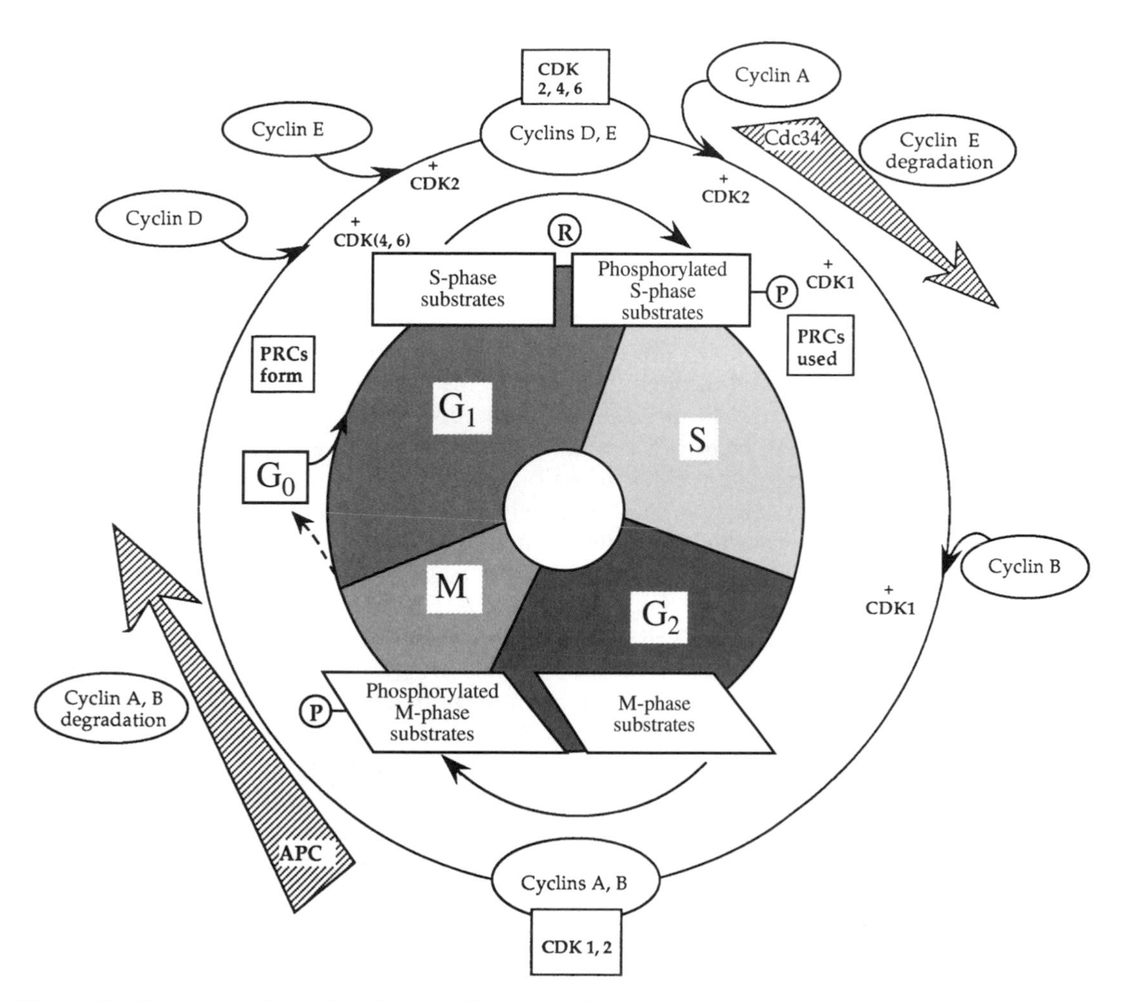

Figure 6.1. The mammalian cell cycle. As cells progress from G_0 into and through G_1, in response to mitogens, for example, the first cyclins to come into play are the D cyclins, which form complexes with CDK4 and CDK6, thereby enabling the kinases to phosphorylate Rb and E2F and to induce cyclin E expression. Cyclin E/CDK2, together with cyclin D/CDK(4,6), propels the cell through the restriction point (R) and into S phase. Cyclin E expression is maximal around the G_1/S border, and both cyclin E/CDK2 and cyclin A/CDK2 are necessary to initiate DNA replication. Cyclins D and E (the S phase cyclins) have short half-lives (~20 minutes life) possibly because of the presence of PEST sequences in cyclin D and sequences in cyclin E that facilitate degradation by the proteasome. Proteolysis of cyclin E, which is controlled by the ubiquitin-conjugating enzyme Cdc34, is necessary for passage through S into G_2. Cyclin A/CDK2 accumulates during S phase and promotes passage through S and G_2. Cyclin B/CDK1 complexes accumulate in G_2 and induce mitosis as the consequence of phosphorylation of target proteins such as the nuclear membrane lamins and various cytoskeletal proteins. Passage through the later stages of mitosis requires the degradation of several proteins including cyclins A and B, by proteolysis mediated by the anaphase-promoting complex (APC, cyclosome). Pre-replication complexes (PRCs) form on origins of DNA replication during early G_1, during S phase they are utilized for DNA replication and then inactivated.

tion (and dephosphorylation) events and by members of two families of CDK inhibitors.

Cyclins A, B, and E are permanently inactivated during each cycle by specific ubiquitination-controlled proteolysis that is tied to progression through the cell cycle. The cyclins, CDKs, CDK inhibitors, and the ubiquitination machinery are the major targets of the signaling pathways that regulate entrance into, and passage through, the cell cycle. Elevated expression of CDK inhibitors, notably p16, p21, and p27, can block cell proliferation and drive the cells into a quiescent, or G_0, phase.

The irrevocable decision to replicate is made at a point during G_1 referred to as the *restriction point* (Pardee, 1974). Progress of the cell up to and through this point in the cycle is usually driven by mitogenic "growth" factors that push the cell into a proliferative mode. Which growth factors or cytokines are actually mitogenic is determined by the lineage of the cell in question and its physiological state. Progress beyond the restriction point is largely controlled by a predetermined sequence of intracellular events, though a significant change in the extracellular environment may disrupt the process.

Cyclins and Cyclin-Dependent Kinases. Four cyclins (D, E, A, and B) are the major drivers of cell cycle progression (Pines, 1995; Sherr, 1996). They interact with one or more of a subset of the CDKs that includes CDK1 (also known as Cdc2), CDK2, CDK4, CDK5, and CDK6. These protein–protein interactions involve specific sequences in the cyclins and CDKs referred to respectively as the *cyclin box* and the *PSTAIR* region. Genes encoding the D cyclins are transcribed as long as mitogenic stimuli are maintained. The activity of the D cyclins is regulated by phosphorylation and proteolysis. In the absence of a proliferative signal, cyclin D1 is degraded by the Ca^{2+}-activated protease calpain (Choi et al., 1997). Cyclins E, A, and B are synthesized only during certain windows in the cycle and are then degraded. Induction of cyclin B1 transcription in S phase results from CDK2-dependent phosphorylation of the transcription factor NF-Y, which acts at two CCAAT elements in the proximal part of the promoter (Katula et al., 1997).

Cyclin D, actually a family of three related cyclins, D1, D2, and D3, is the first cyclin to be expressed as the cell passes through mid G_1 and becomes independent of the mitogenic stimuli that promote transcription of the cyclin D genes. Because the cyclin D proteins are metabolically unstable, withdrawal of mitogenic stimuli prior to passage through the restriction point terminates progression through the cell cycle. The D cyclins, which are induced by those specific mitogens to which the cell in question responds, stimulate CDK4 and CDK6 to phosphorylate the critical target proteins Rb and E2F, provided that inhibitory levels of the CDK inhibitors p16 and p21 are not present.

The importance of cyclin D1 is emphasized by the fact that it was also identified as the *bcl*1 protooncogene, which has been implicated as contributing significantly to a number of human malignancies. It can be oncogenic when stabilized or overexpressed (Hall and Peters, 1996), and it can enable fibroblasts to proliferate in an anchorage-independent manner (on poly[2-hydroxyethyl methacrylate] [polyHEMA]–coated plates) when ectopically expressed (Resnitzky, 1997). Though not identified as a protooncogene, cyclin E can synergize with Ha-Ras to confer a malignant phenotype on primary rat embryo fibroblasts in a Myc- and CDK4-dependent manner (Haas et al., 1997).

Some of the events that occur as the cell passes through the restriction point are illustrated in a simplified form in Figure 6.2. Cyclin D/CDK(4,6)–catalyzed phosphorylation of Rb and E2F, subsequently augmented by additional phosphorylation by cyclin E/CDK2, abrogates the ability of Rb/E2F-DP complexes to inhibit transcription of genes with E2F-binding sites in their promoter. Release of E2F/DP proteins from their association with Rb caused by phosphorylation of Rb (and E2F) may actively promote transcription, or it may alleviate transcriptional repression, depending on what other proteins are bound to the promoter (Jensen et al., 1997; Zwicker and Müller, 1997). In either case, the result is that the transcription of genes required for DNA synthesis and for continued progression into S phase (notably cyclins E and A) is turned on. Phosphorylation of histone H1, a known target of cyclin E/CDK2, is likely to open up chromatin structure and facilitate both transcription and replication. Inhibition of CDK activity by the p53-controlled inhibitors p21 (also p27 and p57 if present) is overcome, in part as the result of their phosphorylation by cyclin E/CDK2 and subsequent ubiquitin-mediated degradation. p27 is also able to induce the formation of E2F complexes that inhibit transcription of genes with E2F-binding sites (Shiyanov et al., 1997).

As diagrammed in Figure 6.1, the activity of cyclin D/CDK(4,6), followed closely by that of cyclin E/CDK2, is necessary for entrance into S phase. CDK2, complexed either with cyclin E or cyclin A, activates essential components of the DNA replication machinery by phosphorylation. CDK1, activated by cyclin A and subsequently by cyclin B, is responsible for driving the cell through the G_2 and M phases. CDK1 and CDK2 complexed with the mitotic cyclins A and B phosphorylate ser/thr residues at specific S/T-P sites in target proteins. Among these proteins are the nuclear lamins, caldesmon, and the microtubule-associated protein 4 (MAP4), which upon phosphorylation by CDK1 loses its ability to stabilize cytoplasmic microtubules (Ookata et al., 1997). Phosphorylation of serine or threonine residues immediately preceding a proline, at least in certain contexts, increases the affinity of the proline for the peptidyl-prolyl isomerase Pin1 (Yaffe et al., 1997). Pin1 activity is required for progression out of G_2 and through mitosis, possibly serving to activate or deactivate phosphorylated proteins, Cdc25, for example, as the result of a conformational change, a "switch," induced by proline isomerization.

The CDKs are regulated by phosphorylation, both inhibitory and activating, as shown in Figure 6.3. The activating phosphorylation, necessary for an effective interaction with the cyclin, is on a characteristic threonine residue near the C terminus; phosphorylation is catalyzed by

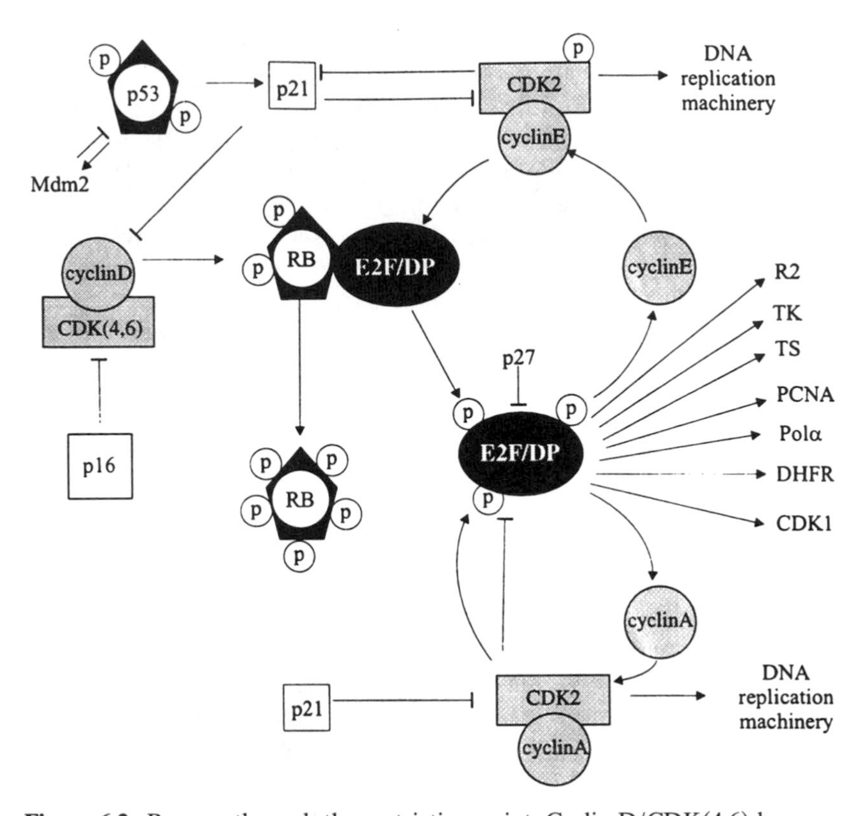

Figure 6.2. Passage through the restriction point. Cyclin D/CDK(4,6) becomes increasingly active in a mitogen-dependent manner in mid G_1, in part the result of the alleviation of inhibition by the specific cyclin D/CDK(4,6) inhibitor p16. Phosphorylation of target proteins, notably pRb and E2F, initially by cyclin D/CDK(4,6) and then by cyclin E/CDK2, frees the E2F/DP proteins to induce (or allow) transcription of various genes necessary for entrance into S phase. These include the R2 subunit of ribonucleotide reductase, thymidine kinase (TK), thymidylate synthetase (TS), DNA polymerase α (Polα), dihydrofolate reductase (DHFR), proliferating cell nuclear antigen (PCNA), cyclin A, cyclin E, and CDK1. Transcription of the cyclin E gene followed by the cyclin A gene begins in late G_1 and increases to maximal levels in S phase. Cyclins E and A activate CDK2, which phosphorylates various proteins, including some of those (E2F) phosphorylated by cyclin D/CDK4 and cyclin D/CDK6, providing positive feedback to push the cell into S phase. Subsequent phosphorylation of the DP proteins by cyclin A/CDK2 shuts down their transactivating potential. The CDKs are inhibited by inhibitors of the p21/p27 family, which if present in sufficient abundance can suppress activity of the kinase and block cell cycle progression. p21 synthesis is stimulated by p53, and p21 is itself inhibited by phosphorylation by cyclin E/CDK2. Transcription of the Mdm2 gene, whose product is an inhibitor of p53 activity, is stimulated by p53, thus serving as a negative feedback regulator.

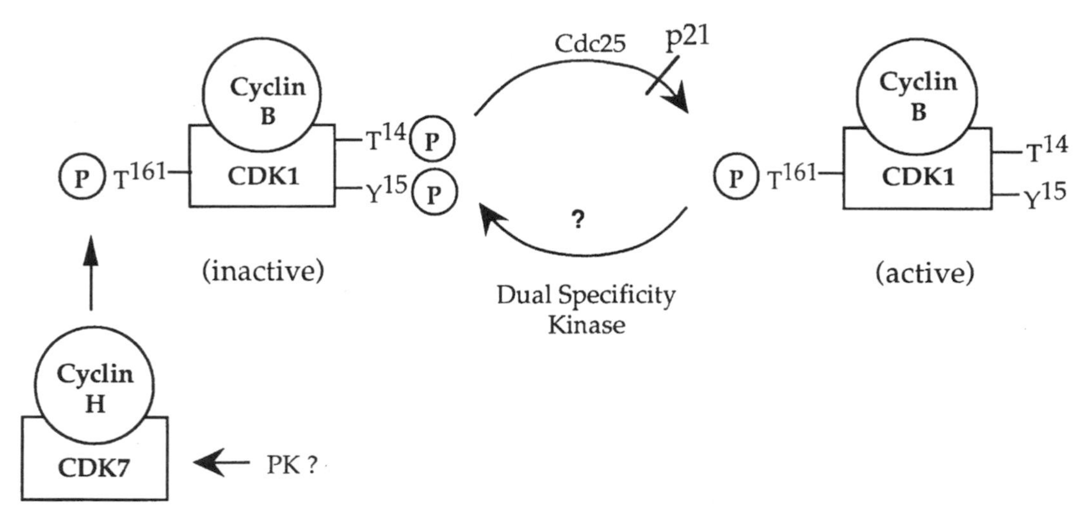

Figure 6.3. Phosphorylation and cyclin association control the activity of the CDKs. Phosphorylation of Thr-161 near the C terminus of CDK1 in this example, by the cyclin-activating kinase CAK (cyclin H/CDK7) is essential for CDK activity. Phosphorylation of Thr-14 and Tyr-15 by a dual specificity kinase inhibits the activity of the CDK, and this inhibition is released by dephosphorylation catalyzed by the cdc25 dual specificity phosphatase, which competes with p21 for access to the cyclin/CDK complex.

a CDK-activating kinase (CAK). CAK is itself a cyclin-dependent kinase (CDK7) dependent on the constitutively expressed cyclin H. It is found in association with the basal transcription factor TFIIH, which controls the phosphorylation by CAK of the C-terminal domain of RNA polymerase II (pol II) during initiation of synthesis and elongation of the pre-mRNA. The inhibitory phosphorylations of the CDKs occur on adjacent threonine and tyrosine residues near the N terminus, and removal of the phosphates is a function of the dual specificity phosphatase Cdc25, which possesses a cyclin-binding motif that competes with p21 for the cyclin (Saha et al., 1997).

There are three isoforms (A, B, C) of Cdc25 in metazoans, and they are differentially regulated during the cell cycle. Phosphorylation of the Cdc25 isoforms in part determines the extent to which they recognize and dephosphorylate various cyclin/CDK targets (Gabrielli et al., 1997). Cdc25A, which is a protooncogene in its own right, is a downstream effector of Myc and is essential for progression out of G_1. Evidence that Cdc25A activity is important in the overall process of cell cycle regulation is that it is not infrequently upregulated in human cancers, suggesting that excessive expression of this phosphatase can contribute to the malignant phenotype (Galaktionov et al., 1995, 1996). Cdc25B shows maximal expression in G_2, after which it is degraded by the proteasome by a process dependent on phosphorylation by cyclin A/CDK1 (Baldin et al., 1997). Cdc25C, which controls entry into mitosis by activating CDK1, appears to be kept in check at other times by the binding of a 14-3-3 protein to a phosphoserine residue; cell cycle–regulated dephosphorylation of the serine presumably activates Cdc25C (Peng et al., 1997). 14-3-3 proteins bind to phosphoserine residues in an RS×S*×P motif (Muslin et al., 1996).

Because the activities of most of the proteins involved in cell signaling, and the implementation of those signals, are regulated by phosphorylation, it follows that dephosphorylation must also be an important regulatory mechanism. However, except for Cdc25, we know little about which phosphatases are important in cell cycle regulation and how they are controlled. Many phosphatases appear to be constitutively present, generally functioning to subvert the signaling process. Some of the more widely studied include the phosphotyrosine phosphatases SHP1 and SHP2 and the ser/thr phosphoprotein phosphatases PP1, PP2A, PP2B (Hunter, 1995a; Cohen, 1997).

In addition to transcriptional regulation (Müller, 1995), timely degradation of the activating cyclins and the CDK inhibitors is also an elemental mode of control of the CDKs (King et al., 1996; Hoyt, 1977). The G_1/S cyclins D and E are relatively short-lived proteins, whereas the G_2/M cyclins A and B are more stable and are specifically targeted for destruction at mitosis. Cyclins E, A, and B are marked for degradation by ubiquitination, which is itself regulated by phosphorylation of proteins controlling the specificity of the ubiquitination process. Cdc34 (see Fig. 6.1) encodes a ubiquitin-conjugating enzyme that promotes the degradation of G_1 cyclins and is essential for the initiation of DNA replication (King et al., 1996). Yew and Kirschner (1997) found that in *Xenopus* embryos Cdc34 activity was required for initiation of DNA replication, possibly to facilitate the degradation of a cyclin E/CDK2 inhibitor (p27?) by the ubiquitin–proteasome pathway. Degradation of p27 by the proteasome appears to be the primary strategy for controlling this protein because p27 synthesis is relatively constant throughout the cycle (Pagano et al., 1995). Cyclin E is marked for destruction by the ubiquitin pathway when it is autophosphorylated by its associated CDK2 on Thr-380, thus providing a neat mechanism for regulating its own half-life (Won and Reed, 1996).

Degradation of cyclin B is mediated by a large multiprotein complex called the *anaphase-promoting complex* (APC), or cyclosome, which polyubiquitinates cyclin B and other proteins, thereby sensitizing them to proteasome degradation. Inhibitors of the metaphase–anaphase transition are also degraded by a similar process. The APC is activated by phosphorylation in a cell cycle–dependent manner. Its activity is regulated by a mitotic "checkpoint" protein, MAD2, which when activated, by spindle defects, for example, blocks the activity of the ubiquitin ligase (Li et al., 1997). Targets of APC-mediated degradation contain a conserved nine amino acid sequence (the destruction, or D, box). These examples indicate that proteolysis of specific proteins may be controlled by cell cycle–regulated phosphorylation of the target proteins (which sensitizes them to Cdc34-mediated ubiquitination, for example) or by cell cycle–regulated activation of the ubiquitination machinery (e.g., APC), which then degrades the target proteins (Pagano, 1997).

The p16/p19^{ARF} Family of CDK Inhibitors. p16 (which is also known as CDKN2, CDK4I, MTS-1, or INK4a) specifically inhibits cyclin D/CDK(4,6), opposing the activating effect of the cyclin on the kinases, in part by competing with the cyclin for a binding site near the N terminus of the CDK and in part by directly inhibiting the cyclin D/CDK4 complex (Sherr and Roberts, 1995; Coleman et al., 1997). A major function of p16, and other members of the family (p15, p18, and p19), is to inhibit CDK(4,6) and thereby to block phosphorylation of Rb and E2F (see Fig. 6.2). Consequently in $Rb^{+/+}$ cells p16 inhibits cell proliferation and can thus function as a tumor suppressor (Koh et al., 1995; Lukas et al., 1995). In the absence of Rb, however, p16 does not inhibit cell proliferation because Rb function is required to suppress progression through the restriction point (Fig. 6.2). p16 inhibits Ras-induced proliferation when overexpressed in Ras-transformed rat embryo fibroblasts. This inhibition can be counteracted by expression of a catalytically inactive CDK4 mutant, presumably because the inactive protein binds to the excess p16 and prevents it from inhibiting the endogenous active CDK4 (Serrano et al., 1995).

p16-deficient mice exhibit an exaggerated sensitivity to carcinogens and spontaneously develop soft tissue sarcomas at high frequency (Serrano et al., 1995). p16 is frequently mutated in human cancers, resulting in the loss of the ability to inhibit cyclin D/CDK(4,6). Deletion of the locus encoding p16 (9p21) is associated with hereditary melanoma. The fact that p16 is the only member of this family that appears able to act as a tumor suppressor suggests intriguing differences among the family members in the regulation of the normal cell cycle (Quelle et al., 1997). This story has recently become more complex with the report by Kamijo et al. (1997) that some or all of the properties of the p16-deficient cells may in fact be the result of their inability to produce p19^{ARF}, an alternative reading frame protein that despite having no amino acid homology with p16 is also a tumor suppressor that can, when overexpressed, arrest cells in G_1 or G_2. Of the three exons encoding both proteins, exon 1 is different, but exons 2 and 3 are the same except that they are read in different reading frames.

The levels of p16 mRNA and protein are elevated in human cells lacking Rb or in late passage senescent human cells than Rb^+ or early passage cells, consistent with the presence of a negative feedback loop by which the normally hypophosphorylated Rb inhibits p16 transcription (Hara et al., 1996). Hara et al. (1996) argue that the accumulation of p16 renders cells senescent by blocking phosphorylation of Rb by cyclin D/CDK(4,6) and that loss of p16 allows the development of a tumor phenotype. Consistent with this interpretation, primary fibroblast unable to make functional p16 proliferate rapidly, do not appear to senesce, form colonies in soft agar more efficiently, and can be readily transformed by Ha-*ras*^{val12}, but not by *myc* (Serrano et al., 1996). Uhrbom et al. (1997) used the tetracycline-repressible expression system to demonstrate that induced expression of p16 could render senescent a malignant human glioma cell line unable to make p16 because of a homozygous deletion of the gene.

In an interesting study, Serrano et al. (1997) presented evidence and arguments that expression of oncogenic Ras in primary human or rodent cells leads to an accumulation of both p53 and p16. This appeared to be the cause of an accompanying inhibition of cell proliferation reminiscent of cellular senescence because inactivation of either p53 or p16 aborted the senescence response and permitted continued proliferation of mouse (but not human) cells. This atypical antimitotic response to a Ras signal is not an immediate response to expression of oncogenic Ras, but rather a delayed response likely reflecting a homeostatic negative feedback process.

Mouse cells differ from human cells in that p16 mRNA and protein levels are not affected by the presence or absence of Rb (Palmero et al., 1997). However, p16 can still accumulate with repeated cell divisions (of primary fibroblasts in serum-containing medium) as the result of the stability of the mRNA and protein, reaching a level that results in suppression of proliferation of normal cells. Thus p16 mRNA and protein accumulate as mouse embryo fibroblasts are passed according to an immortalization protocol, a 3T3 procotol, for example. After 8–10 population doublings (5–10 weeks), the primary mouse embryo fibroblasts enter a crisis period where proliferation essentially ceases. Immortalized cells that escape crisis were found in most cases to contain even higher levels of p16 protein, suggesting that there must have been one or more additional mutations (or epigenetic changes in gene expression) that subverted the proliferation-suppressing ability of p16 (Zindy et al., 1997).

The p21 Family of CDK Inhibitors. p21 (which is also called WAF1, CIP1, or sdi1), and the related proteins p27 (Kip1) and p57 (Kip2), bind to free cyclin-dependent kinases CDK1, CDK2, CDK4, and CDK6 and their complexes with cyclins A, B, D, or E, inhibiting the kinase activity when present in sufficient excess (Sherr and Roberts, 1995; Zavitz and Zipursky, 1997). The various family members (the Cip/Kip family) exhibit tissue-specific expression and inhibit the different cyclin/CDKs with diverse efficiencies (Blain et al., 1997). They also vary in their response to growth factors; p21 expression is increased by mitogenic stimulation, whereas p27 tends to accumulate in quiescent cells. p21, and possibly p27, can compete with the Cdc25A dual phosphatase (see Fig. 6.3) for binding the tyrosine-phosphorylated CDK2 molecule, thereby preventing the phosphatase from activating the kinase (Saha et al., 1997). p21 has at least two domains: a CDK-binding domain that is particularly effective at suppressing entrance into S phase and a second domain that inhibits the activation of DNA polymerase δ by proliferating cell nuclear antigen (PCNA). PCNA, which confers processivity on DNA polymerases δ and ε by encircling the DNA and forming a "sliding clamp," can form complexes with a number of proteins involved in DNA synthesis and regulation of the cell cycle (Loor et al., 1997; Baker and Bell, 1998).

Control over the activity of these inhibitors is exerted at several levels. The relatively invariant levels of p27 mRNA and its rate of translation during the cell cycle suggest that it is the p27 protein itself that is the major target of regulation. Evidence for this includes the fact that phosphorylation of Thr-187 in p27 by cyclin E/CDK2 at late G_1 sensitizes the protein to proteolysis, likely involving ubiquitination and degradation via the proteasome (Morisaki et al., 1997; Pagano et al. 1995). Suppression of the ability of cyclin D/CDK(4,6) or cyclin E/CDK2 to phosphorylate various proteins required for the establishment of S phase, notably the E2Fs, prevents cells from exiting G_1. Progression through S phase is also blocked by elevated p21 expression, apparently as the result of the suppression of cyclin (A or E)/CDK–mediated phosphorylation of proteins required for DNA replication (Ogryzko et al., 1997). p21 (and p27) can also arrest cells in G_2, blocking entrance into mitosis, presumably by inhibiting cyclin A/CDK2 (Dulic et al., 1998; Niculescu et al., 1998). Decker (1995) observed that enhanced p21 expression in NIH3T3 cells resulting from activation of TrkA, the nerve growth factor (NGF) receptor, correlated with and perhaps was the cause of the NGF-induced growth arrest of these cells. Thus by various means an increase in p21 expression can suppress cell proliferation in a variety of contexts.

As discussed below, p21 is induced by growth factors controlling the Ras→Raf→MEK→MAPK(ERK) pathway. This suggests that p21 might provide a positive rather than a negative stimulus to passage through the cell cycle. Liu et al. (1996) proposed that at low concentrations p21 facilitates proliferation by promoting the assembly of p21/cyclin/CDK/PCNA complexes, whereas at higher concentrations, such as in response to severe stress, it becomes inhibitory as additional p21 molecules smother the complex. This is consistent with the earlier report of Zhang et al. (1994) that p21-containing cyclin/CDK/PCNA complexes exist in both catalytically active and inactive complexes in untransformed cells, with the inactive complexes containing a larger number of p21 subunits. In transformed cells in which many regulatory controls have been disrupted it appears that most of the CDKs are in simple binary complexes with a cyclin.

Shim et al. (1996) reported that p21 can inhibit the kinase activities of the SAPK/JNK and p38/RK families of mammalian MAP kinases (MAPKs), though how this inhibition is accomplished has not been elucidated. The ability of p21 to regulate these stress-activated signaling pathways (see Fig. 6.6, below) may perhaps indicate that it serves to suppress a stress response during cell proliferation. This would make sense if, for example, some of the intermediates generated during DNA replication (e.g., regions of single-stranded DNA) might otherwise induce a stress response. Arguing against this interpretation is the fact that p21 is an important DNA damage–inducible inhibitor of cell cycle progression. Its synthesis is elevated as the result of a p53-dependent enhancement of transcription in cells that have suffered certain types of DNA damage, for example that caused by ionizing radiation.

Other transcription factors also control p21 expression, since UV radiation (254 nm) can induce its expression via a p53-independent pathway, at least in diploid human fibroblasts (Loignon et al., 1997). Primary human diploid fibroblasts lacking p21 as the consequence of targeted disruption of the gene are unable to arrest their progression through the cell cycle after γ- ray–induced DNA damage and are capable of sustaining more population doublings than cells possessing an active p21 molecule. They bypass normal senescence, but enter a crisis phase at about 30 population doublings, possibly as the result of telomere shortening and subsequent loss of ability to replicate their DNA completely (Brown et al., 1997).

p21 appears to serve as a checkpoint by coordinating S phase with mitosis. Evidence for this is that cells unable to make p21 undergo repeated cycles of DNA replication without intervening mitoses when their DNA is damaged (Waldman et al., 1996). Although p21-deficient mice appear normal and do not exhibit increased frequencies of malignancies, keratinocytes derived from these mice, when transformed with Ha-ras, form more aggressive tumors than do keratinocytes expressing p21 (Missero et al., 1996). This behavior correlated with a reduction in the expression of the late keratinocyte marker keratin 1.

p27- and p57-deficient mice are also viable at birth but subsequently exhibit interesting defects and develop malignancies in specific tissues (Zavitz and Zipursky, 1997). The $p27^{-/-}$ mice grow to a larger size, and those organs that express the highest levels of p27 (thymus, spleen, testis) show the greatest increment in size. The $p57^{-/-}$ mice die shortly after birth, exhibiting developmental defects in tissues (lens of the eye, endochondral bone, kidney, muscle, intestine) that normally express p57. That tumors develop in these animals in a manner suggestive of the Beckwith-Wiedemann syndrome in humans indicates that p57 should be considered a tumor suppressor (Zhang et al., 1997). Overexpression of p27 in malignant cells arrests the cells in G_1/S and can induce apoptosis (Katayose et al., 1997).

Tumor Suppressor Rb. The retinoblastoma protein (Rb) is a 110 kDa nuclear phosphoprotein that binds to various cellular proteins as a function of its phosphorylation status. It and its two relatives (p130 and p107) are known as *pocket proteins* because of a structural motif involved in binding to cyclin D, cAbl, Mdm2 (murine double mutant 2), and the transcription factors E2F and SP1 (Taya, 1997; Herwig and Strauss, 1997). The hypophosphorylated forms found in G_1 act to inhibit cell cycle progression by suppressing the transactivating activity of E2F, which is necessary for the transcription of genes required for DNA replication (Fig. 6.2). The Rb family members exhibit different specificities for the various E2F family members, with Rb itself exhibiting specificity for E2F1, -2 and -3. The timing of their appearance during the cell cycle also differs. Hypophosphorylated Rb also binds to histone deacetylase 1. This results in localizing the enzyme to E2F-associated promoters and stimulating deacetylation of the adjacent chromatin (DePinho, 1998). Removal of the charged acetyl moieties from the histones is thought to generate a more compact chromatin structure, one less amenable to transcriptional activation.

Un- or hypophosphorylated Rb, found, for example, in quiescent G_0

cells and in differentiated cells, also inhibits RNA pol I and pol III, enzymes responsible for synthesis of ribosomal RNA and the small nuclear RNAs, respectively. The ability of Rb to inhibit synthesis of these RNAs accounts for the ability of Rb to suppress cell growth (White, 1997). Rb binds to the upstream binding factor (UBF), blocking its ability to bind to DNA and thereby abrogating transcription by RNA pol I (Cavanaugh et al., 1995; Voit et al., 1997). It represses transcription by RNA pol III, probably via an interaction with the essential general transcription factor TFIIIB (White et al., 1996). The action of Rb on transcription from pol II–dependent promoters is mediated by an interaction with the $TAF_{II}250$ component of TFIID; this may have positive or negative effects on transcription, depending on what other proteins are present (Shao et al., 1997). Via interactions with various transcription factors Rb can promote cell differentiation while suppressing cell proliferation. For example, unphosphorylated Rb proteins can interact with the helix-loop-helix domain of MyoD to induce transcription of muscle-specific genes (Gu et al., 1993), and it can inhibit the paired-like homeodomain protein Pax-3 (Wiggan et al., 1998).

Hypophosphorylated Rb, the product of cyclin D/CDK(4,6) action in early G, maintains its ability to bind to E2F and to mediate transcriptional repression (Ezhevsky et al., 1997). However, further phosphorylation of Rb (e.g., resulting from diminished p16 inhibition) by cyclin D/CDK(4,6) and by cyclin E/CDK2 in late G_1 causes dissociation of Rb from E2F1/DP1 (see Fig. 6.2). There are approximately a dozen ser/thr residues found in sequences in Rb that closely match consensus sequences for phosphorylation by the different CDKs. These sequences may specify interactions with specific target proteins. For example, phosphorylation of Thr-821 and Thr-826 prevents Rb binding to proteins with the LXCXE motif (viral oncoproteins), phosphorylation of Ser-807 and Ser-811 blocks binding of cAb1, and phosphorylation of these together with other sites in the protein disrupts binding to E2F (Knudsen and Wang, 1997).

This cumulative phosphorylation of Rb, initiated by cyclin D/CDK(4,6) and augmented by cyclin E/CDK2, eliminates the transcriptional repression imposed by the Rb/E2F complexes, allowing E2F/DP complexes to stimulate the transcription of various genes required for entrance into S phase (Lundberg and Weinberg, 1998). E2F/DP also enhances expression of cyclins A and E, which further stimulate CDK2 to phosphorylate E2F, thereby providing a positive feedback loop that amplifies the activation of E2F. Later in S phase, cyclin A/CDK2 forms a complex with E2F/DP1 and phosphorylates DP1, resulting in the release of E2F1/DP1 from the various promoters it is activating and the cessation of E2F1-driven transcription of various genes, including itself (Krek et al., 1994). Dephosphorylation of Rb by an anaphase-specific phosphatase at the end of mitosis reinstitutes the Rb/E2F-mediated inhibition of target genes as the cell enters G_1 (Ludlow et al., 1993).

In addition to their ability to associate with and inhibit E2F4 and -5, the Rb-related p130 and p107 proteins can, when suitably phosphorylated, also form complexes with cyclin A/CDK2 and cyclin E/CDK2, substantially inhibiting their kinase activity as assessed on histone H1. E2F is also found associated with some of the p130/cyclin A/CDK2 complexes, which exhibit a low level of kinase activity (Woo et al., 1997). Hauser et al. (1997) observed that association of cyclin A/CDK2 with p107 or p130 modified the substrate specificity of the cyclin-dependent kinase, inhibiting its ability to phosphorylate histone H1 efficiently but not its ability to phosphorylate Rb family members. This led these authors to propose that association of cyclin A/CDK2 with Rb proteins forms distinct pools of cyclin A/CDK2 complexes with physiologically significant alterations of substrate specificity. Additionally, such complexes would be insensitive to inhibition by the p21/p27/p53 proteins because their interaction with the CDK inhibitor is blocked by p107 or p130.

Tumor Suppressor p53. p53 is a transcription factor best known for its ability to suppress the development of cancer, in part by inducing cell death in metabolically deranged cells. Roughly 50% of human cancers possess a mutation in p53 that inhibits its efficacy as a tumor suppressor. Perturbations to cellular metabolism that stabilize the normally unstable wild-type protein induce either growth arrest or apoptosis. In healthy cells an important function for p53 is to cause progression through the cell cycle to cease when damage to the cellular genome is sensed. This activity has been referred to as *checkpoint control,* which may be defined as a surveillance mechanism that blocks cell cycle transitions (Nasmyth, 1996). Cessation of cell proliferation in response to DNA damage requires cAbl, a sometimes nuclear protein tyrosine kinase that binds to p53, enhancing its ability to stimulate transcription at promoters containing p53-binding sites (Goga et al., 1995).

Activated (e.g., by phosphorylation in response to DNA damage) p53 stimulates transcription of a number of genes, including GADD45, Mdm2, p21, IGF-BP3, Bax, and cyclin G (Cox and Lane, 1995; Ko and Prives, 1996; Levine, 1997). GADD45 binds to PCNA, which is found in normal cells in multiprotein complexes with various cyclin/CDKs and p21, and it promotes DNA repair while inhibiting DNA replication (Smith et al., 1994; Zhang et al., 1994). Consequent to DNA damage, there is also an increase in the stability of the p53 protein, likely as the result of phosphorylation. p53 may be phosphorylated by a variety of kinases, including PKC, casein kinase II, CDK1, DNA-dependent protein kinase, ERK, JNK, and PKA, although which of these are most relevant physiologically remains to be determined. Stimulation of transcription of the insulin-like growth factor-I–binding protein 3 (IGF-BP3) gene may lead to suppression of the growth of cells whose DNA has been damaged (Buckbinder et al., 1995).

Mdm2 is a cellular protooncogene often found to be amplified in human tumors. In the normal cell Mdm2 negatively regulates p53 by forming a complex with it and eliciting its degradation in an ubiquitin-dependent, proteasome-mediated manner (Haupt et al., 1997; Kubbutat et al., 1997). Thus p53 stimulates the expression of a gene whose product shortens its own (p53) lifetime. p53 modified by certain mutations, or by stress-induced modifications, is resistant to Mdm2-provoked degradation, thereby accumulating to increased levels. Mutant p53 proteins unable to induce Mdm2 transcription are more stable and accumulate to high levels as seen in some tumors (Lane and Hall, 1997). Mdm2 also interacts functionally with Rb, E2F1, and DP1 to promote cell cycle progression, at least in certain cell contexts (Momand and Zambetti, 1997), and possibly stimulates their degradation. Thus inappropriate increases in Mdm2 levels can stimulate inappropriate cell proliferation and exert an oncogenic effect.

The DNA-dependent ser/thr protein kinase (DNA-dependent PK) phosphorylates and activates cAbl, which in a negative feedback loop phosphorylates the DNA-dependent PK, thereby preventing its binding to DNA (Kharbanda et al., 1997). Shieh et al. (1997) discovered that phosphorylation of Ser-15 and Ser-37 in p53 by DNA-dependent PK impaired the ability of Mdm2 to inhibit p53-dependent transactivation. Additionally, phosphorylation of Mdm2 by the DNA-dependent PK impedes its binding to p53 (Mayo et al., 1997). Different conformational forms of p53 may be phosphorylated via different signaling pathways, giving rise to versions of the tumor suppressor protein with different activities (Hupp and Lane, 1995; Adler et al., 1997). p53-induced p21 expression results in an inhibition of CDK4 (complexed with cyclin D), CDK2 (complexed with either cyclin A or cyclin E), and CDK1 (complexed with cyclin B), thus effectively blocking progression into and through S phase (Fig. 6.2). The arrest of cells in the G_2 phase of the cycle that is caused by ionizing radiation is mediated by p53 induction of 14-3-3σ, which binds to and inhibits a protein, possibly Cdc25C, essential for entry into mitosis (Hermeking et al., 1997). Repair of the damaged DNA alleviates the p53-induced checkpoint cell cycle arrest.

The multifunctional transcriptional regulator and histone acetyltransferase p300/CBP, a CREB-binding protein (CBP), is an important effector and potential regulator of p53-regulated transcription. Complexes of p300/CBP with p53 can have either positive or negative effects on transcription, depending on the promoter and the status of the p53 protein, wild type, or mutant in the complex (Avantaggiati et al., 1997). This likely underlies the observation of Y. Lee et al. (1997) that the inhibition of the thymidylate synthase promoter by p53 was the result of the sequestering of a general transcription factor. Cells exposed to a variety of stimuli, noxious or otherwise, may choose to undergo apoptosis, or programmed cell death, rather than cell proliferation. Bcl-2, which is found on the cytoplasmic surfaces of the outer mitochondrial membrane, the endoplasmic reticulum, and the nuclear envelope, is one of the cell's regulatory molecules that determines whether the cell will survive or undergo apoptosis. Huang et al. (1997) found that mutation of a tyrosine residue in the N terminus of Bcl-2 had no effect on the ability of Bcl-2 to enhance cell survival but did seem to accelerate the entrance into the cell cycle. This suggests that Bcl-2 may regulate transit of cells out of G_0.

How the various actions of Bcl-2 are invoked is not fully understood. A possible controlling factor may be the ratio of Bcl-2 family members that promote cell survival (Bcl-2, Bcl-X_L) to those that promote cell death (Bax, Bad, Bak, and Bid). These proteins mediate various protein–protein interactions and are involved in the formation of pores in mitochondrial membranes (Reed, 1997). If the genome is severely damaged, or if abnormal oncogene-mediated signaling is detected, p53 induces an apoptotic response, possibly by stimulating Bax gene expression and thus decreasing the Bcl-2/Bax ratio (Miyashita and Reed, 1995). In a *tour de force* Polyak et al. (1997) identified transcripts of a number of genes whose expression was increased by p53. Many of these genes encoded functions, e.g., oxidoreductases, that affect the redox level in cells, leading to the suggestion that p53 might induce cell death by stimulating the oxidative degradation of key cellular constituents. Changes in redox potential could well be the cause of other consequences of p53 activation.

It is in these various capacities, particularly that of inducing cell death, that p53 exerts its control over the development of a malignancy. As long as it is active, the ability of many otherwise oncogenic mutations to give rise to a cancer is thwarted. Pierzchalski et al. (1997) have investigated how p53 might induce apoptosis in heart myocytes, for example, after an ischemic episode. Their data suggest that p53 not only causes a decrease in the Bcl-2/Bax ratio but also stimulates the promoters of both the angiotensinogen and angiotensin II AT1 receptor subtype genes. Upregulation of the renin–angiotensin system appeared to initiate programmed cell death as a consequence of calcium influx and the activation of calcium-dependent endonucleases (Cigola et al., 1997). Haupt et al. (1995) observed that overexpression of p53 in HeLa cells induced apoptosis by a process that could be partially opposed by Rb, suggesting that in some circumstances Rb can function as a survival factor.

E2F Proteins. Five members of the E2F family of transcription factors (E2F1–5) are known, each of which forms a functional heterodimer with DP1 or DP2. Their transcription-activating ability, but not their DNA-binding ability, is inhibited by unphosphorylated or minimally phosphorylated Rb. Hyperphosphorylation of Rb relieves the inhibition, leading to increased transcription of genes such as dihydrofolate reductase, DNA polymerase α, thymidine synthase, thymidine kinase, PCNA, ribonucleotide reductase R2, cyclin A, cyclin E, CDK1, Myc, Rb, E2F1, and a component of the origin recognition complex (see below)—most of which are necessary for DNA synthesis and S phase progression (Sladek, 1997; Bernards, 1997). E2F was first recognized as a cellular factor necessary for the transcription of the adenovirus E2 gene.

The E2F proteins possess domains responsible for dimerization with DP proteins, for binding to E2F response elements in promoters, for transactivation of transcription, and for interactions with members of the Rb pocket protein family. E2F1, -2, and -3 interact with Rb, whereas E2F4 and -5 interact with p130 and p107, usually functioning as transcriptional repressors in quiescent cells. Although the DP proteins are structurally similar to the E2F proteins, they do not interact directly with DNA, cyclins, or Rb protein; instead, they enhance the transactivating ability of the E2F proteins. Overexpression of dominant negative mutants of DP1 or DP2 in osteosarcoma cells caused the cells to arrest in G_1 (Wu et al., 1996).

E2F family members have distinct functions in regulating the cell cycle. When expressed from recombinant adenovirus vectors, E2F1, -2, and -3 are each able to induce quiescent rat embryo fibroblasts to enter S phase, and, in the case of E2F1, but not E2F2 or E2F3, the cells undergo an apoptotic response (DeGregori et al., 1997). Philips et al. (1997) showed that stimulation of DNA synthesis and apoptosis were separable functions of E2F1 in that DNA synthesis, but not the induction of apoptosis, required transcriptional transactivation. They suggested that free E2F1 can induce expression of "apoptotic genes" in the absence of normal cell cycle progression. E2F1, -2, and -3 also possess a motif for interaction with cyclin A/CDK2, possibly accounting for the fact that it is these three members of the E2F family that are the most efficient inducers of the cell cycle.

Each of the different members of the E2F family exhibits specificity with regard both to how their expression is regulated and to the set of genes whose expression they regulate. However, not all genes that contain E2F-binding sites exhibit cell cycle–dependent regulation of transcription. van Ginkel et al. (1997) showed that this was in part dependent on cooperation with other appropriately located transcription factors, for example, NF-Y, YY1, and Sp1. It seems that E2F sites located very close to the transcriptional start site in promoters lacking TATA boxes (e.g., in genes encoding Cdc25, CDK1, and cyclin A) generally act as repressor elements, whereas E2F sites located further upstream mediate transcriptional activation in genes, such as those encoding DNA polymerase α, Myc, or TK, that have TATA boxes (Tommasi and Pfeifer, 1997). DeGregori et al. (1997) have elegantly demonstrated the differences among the various E2F family members in the efficiency with which they can induce transcription of target genes by infecting fibroblasts with adenoviral vectors expressing different E2F family members. Thus, whereas DNA polymerase α mRNA levels were approximately equivalent in cells expressing each of E2F1, -2, -3 or -4, the mRNA levels for thymidine kinase and DHFR were preferentially induced by E2F2, CDK2 mRNA was elevated specifically by E2F3, and $p19^{ARF}$ mRNA was augmented specifically by E2F1.

Expression of E2F1 family members is normally regulated by the different cyclin/CDKs in complex ways that have been partially elucidated by Dynlacht et al. (1997). For example, the E2F1/DP1 heterodimer is phosphorylated by both cyclin A/CDK2 and cyclin B/CDK2; however, only the former results in downregulation of activity. This could reflect differences in the specific amino acids that are phosphorylated such that the formation of a stable, inactive complex could form with cyclin A/ CDK2 but not with cyclin B/CDK2. The conclusion from their studies was that the E2F1, -2, and -3, but not E2F4, 5, or E2F5, could be efficiently downregulated by cyclin A/CDK2, suggesting that these two subsets of E2F proteins have distinct functional properties. Control over nuclear localization appears to distinguish the E2F1–3 group from the E2F4–5 group. Nuclear localization of E2F4, a major E2F species, is controlled in a cell cycle–dependent manner, possibly by DP1. E2F/p130 and E2F/p107 complexes are predominantly cytoplasmic, whereas E2F/pRb is largely nuclear (H. Müller et al., 1997; Verona et al., 1997).

DNA Replication

It has been appreciated for some time that eukaryotic DNA is with few exceptions replicated only once during the S phase of the cell cycle and that replication can initiate at many locations in the DNA constituting each chromosome. How this feat is accomplished, however, has been a challenge to discover. (For a more detailed discussion than is presented here, see other chapters in this volume.)

In the last few years various lines of research have coalesced into a reasonably satisfying model that is consistent with the bulk of the experimental evidence (Wuarin and Nurse, 1996; Chong et al., 1996; Stillman, 1996; Jallepalli and Kelly, 1997; Dutta and Bell, 1997). The essence of the model is that after mitosis and through much of G_1, protein complexes form at specific sequences, known as origins, in the DNA. These origins, sequences more specifically known as *replicators,* are associated throughout much of the cell cycle with a set of proteins making up the origin recognition complex (ORC). After mitosis, additional proteins are recruited to form a pre-replication complex (PRC), which provides the foundation for the DNA replicative apparatus to assemble and initiate replication of the DNA, usually bidirectionally, shortly after the decision to enter S phase is made. The number of these PRCs that form on the chromosome appears to be a function of how rapidly the DNA is to be replicated. Chromatin in the cells of a developing embryo will have a higher density of PRCs forming along the chromosome than in cells from more slowly growing tissues. There is a hierachy of sites with a progressively diminishing capacity for binding origin recognition proteins and serving as sites for the formation of PRCs.

The PRCs are composed of a number of proteins, including Cdc6 and the MCM/P1 proteins. The MCM/P1 proteins are known in yeast as minichromosome maintenance proteins and among other functions appear to provide a helicase activity to the replication fork (Thommes et al., 1997; Kubota et al., 1997; Baker and Bell, 1988). The MCM proteins are much more abundant than the ORC proteins, and, unlike the ORC proteins, they bind to chromatin in a cell cycle–dependent manner. Cdc6 functions to load MCM proteins onto chromatin in early G_1; late in G_1 it is phosphorylated by CDK2 and likely degraded by a Cdc34-mediated process, thereby enabling the PRCs to initiate DNA replication (Hua and Newport, 1998). DNA replication commences when a fully functional replication complex has assembled on the PRC and has been appropriately activated by phosphorylation by the S phase cyclin/CDKs. After participating in the initiation of DNA replication, the phosphorylated MCM proteins become nonfunctional and in some systems migrate to the cytoplasm, possibly to be degraded, upon dissolution of the nuclear membrane at mitosis.

Re-replication of already replicated DNA is prevented by virtue of the fact that once an origin is utilized a "licensing factor" becomes nonfunctional, presumably because of cyclin-dependent phosphorylation (Mahbubani et al., 1997). The replication licensing factor-M, which is a complex of six MCM/P1 proteins, is essential for a PRC to be functional, and, in the absence of functional licensing factors, new PRCs cannot normally be formed. Inhibition of ser/thr kinase activity by 6-dimethylaminopurine after completion of DNA synthesis permits the MCM proteins in G_2 nuclei to reform functional PRCs, consistent with the hypothesis that it is phosphorylation of the MCM proteins that renders them nonfunctional (Coverley et al., 1998). Only after the cells have completed mitosis and degraded the mitotic cyclins can licensing factors initiate formation of new PRCs. This occurs during the $M \rightarrow G_1$ transition when there is no nuclear membrane. Reformation of the nuclear membrane prevents any additional licensing factors from accessing the DNA. The low cyclin-dependent kinase activity in early G_1 (cyclins A, B, and E have been degraded) permits Cdc6, replication licensing factor-B, and the MCM proteins, pre-existing or newly synthesized, to move back into the nucleus and reform competent PRCs at the ORCs.

Section Summary

The cell cycle is driven by the periodic synthesis of key proteins or their activation/deactivation by (de)phosphorylation and protein–protein interactions. Notable among these key proteins are the cyclins, whose activity—activation of specific cyclin-dependent kinases—drives the cell through various transitions. Irreversible transitions are ensured by positive feedback loops (e.g., E2F enhancement of cyclin E synthesis, cyclin E/CDK2–stimulated degradation of p21/p27) and by proteolysis of cyclins and other proteins whose activities are no longer required. Many of the proteolytic events are controlled by ubiquitination of target proteins, and the proteolytic machinery targeting ubiquitinated proteins is itself subject to cell cycle regulation.

Expression of specific functions is also controlled at the transcriptional level and by protein localization. Activity of the cyclin-dependent kinases is regulated not only by specific cyclins but also by phosphorylation and by specific inhibitors in the p16 and p21 families. Mutations in p16, p21, Rb, or p53 may contribute to the development of malignancies by removing brakes that regulate the proliferative status of the cell or by subverting mechanisms that would induce apoptosis in the (premalignant) cell. Hence the introduction of genes encoding the wild-type forms of these genes into cancer cells may offer an effective form of gene therapy. This approach will succeed, however, only in those cancer cells that have maintained the necessary endogenous cell cycle control pathways (Costanzi-Strauss et al., 1998).

SIGNAL PROCESSING: INTEGRATION OF SIGNALING PATHWAYS

This section deals with some of the intracellular events known or believed to intervene between the stimulus proximal receptors, discussed below, and the distal cell cycle effector proteins, discussed above, which execute the command to traverse the cell cycle. Two major types of signaling pathways that play critical roles in mitogenic signaling pathways have been characterized. One includes the Ras superfamily of $p21^{GTPase}$ proteins (the Ras, Rho, Rac, Ral, and Cdc42 minifamilies), which communicate with each other via GTPase-activating proteins (GAPs) and guanine nucleotide exchange factors (GEFs). Among other functions, they integrate cell cycle progression with cytoskeletal organization and cell differentiation. The second type of pathway consists of protein modules composed of members of the MAPK (ERKs, JNKs/SAPKs, p38/RKs), MAPKK (MEKs, SEKs, MKKs), and MAPKKK (Rafs, MEKKs, MTKs, MLKs) families of protein kinases. Acronyms and abbreviations are defined in Table 6-1. The terminology is confusing because different names have sometimes been used for the same protein, and there may be several similar proteins that bear an uncertain relation to each other. Activation of these pathways, the basic elements of which are conserved among yeasts, plants, invertebrates, and vertebrates, results in the phosphorylation of various enzymes, structural proteins, and transcription

TABLE 6.1. Abbreviations and Acronyms

Receptors and Ligands
- αAR, α-adrenergic receptor
- βAR, β-adrenergic receptor
- CSF, colony-stimulating factor
- cKit, stem cell factor receptor
- Epo(R), erythropoietin (receptor)
- ET, endothelin
- EGF(R), epidermal growth factor (receptor)
- FGF(R), fibroblast growth factor (receptor)
- G-CSF(R), granulocyte colony–stimulating factor (receptor)
- GPCR, G protein–coupled receptor
- IGF(R), insulin-like growth factor (receptor)
- LPA, lysophosphatidic acid
- mAChR, muscarinic acetylcholine receptor
- PDGF(R), platelet derived growth factor (receptor)
- RTK, receptor tyrosine kinase
- TrkA, "tropomyosin receptor kinase" is the receptor for NGF (nerve growth factor)

MAKKKs (mitogen-activated protein kinase kinase kinase: ser/thr-specific)
- MEKK, MEK kinase
- MTK, MAP three kinase
- MLK, multilineage kinase
- Raf-1, A-Raf, B-Raf, Ras-activated factors

MAKKs (mitogen-activated protein kinase kinase: dual specificity kinases that phosphorylate threonine and tyrosine)
- JNKK, JNK kinase
- MEK, (MAP/ERK kinase)
- MKK, MAP Kinase kinase
- SEK1, SAPK/ERK kinase 1

MAPKs (mitogen-activated protein kinases: proline-directed ser/thr kinases)
- ERK, extracellular regulated kinase
- JNK, Jun N-terminal kinase
- p38, RK (reactivating kinase)
- SAPK, stress-activated protein kinase

TABLE 6.1. (*Continued*)

Nonreceptor Protein Tyrosine Kinases
- Btk, Bruton's tyrosine kinase
- cAbl, protooncogene of Abelson leukemia virus
- Crk, chicken retroviral kinase
- Csk, C-terminal Src kinase
- FAK, focal adhesion kinase
- JAK, Janus kinase

Nonreceptor Protein ser/thr kinases
- Akt/PKB, product of v-*akt* oncogene/protein kinase B
- CAK, cyclin-activating kinase
- CDK, cyclin-dependent kinase
- DNA-PK, DNA-dependent protein kinase
- GSK, glycogen synthase kinase
- ILK, integrin-linked kinase
- MAPKAPK (MAPK-activated protein kinase)
- PAK, p65^{PAK}, p21-activated kinase
- PKA, protein kinase A
- PKB, protein kinase B; see Akt above
- PKC, protein kinase C
- PKN, protein kinase N
- p70^{S6K}, 70 kD ribosomal S6 kinase; also p70^{RSK}
- p90^{rsk}, 90 kD ribosomal S6 kinase
- ROK, Rho kinase
- p160^{ROCK}, Rho-associated coiled-coil kinase

Phosphatases
- Cdc25, dual specificity phosphatase (targets CDKs)
- MKP, a dual specificity phosphatase (targets ERKs)
- PTP, protein tyrosine phosphatase
- SHP-1, -2 (SH domain–containing protein tyrosine phosphatases)

Domains
- CRIB, Cdc42 and Rac interactive binding motif
- Dbl, dibble (prototype of a Rho family GEF)
- SH, Src homology (SH1, the tyrosine kinase domain; SH2, binds protein tyrosine phosphates; SH3, binds proline-rich sequences; SH4, sites of lipid modifications)
- PTB, phosphotyrosine binding
- PH, pleckstrin homology (affinity for phosphatidylinositol phosphates and $\beta\gamma$ subunits)

Transcription Factors
- AP1, activating protein 1 (Fos/Jun homo- and heterodimers)
- CREB, cAMP response element binding protein
- E2F/DP, Rb-regulated transcription factors
- Elk-1, Ets-like transcription factor
- Ets, avian erythroblastosis virus–E26 protooncogene
- p53, tumor suppressor
- Rb, retinoblastoma gene tumor suppressor
- SRF, serum response factor
- SAP1, SRF accessory protein 1, stress-activated protein 1
- STAT, signal transducer and activator of transcription
- TCFs (ternary complex factors)

Adaptor proteins
- p130^{CAS}, 130 kD Crk-associated substrate
- Shc, SH2-containing α_2 collagen-related
- Grb2, growth factor receptor bound protein 2

Cell Cycle/DNA Replication
- APC, anaphase-promoting complex
- DHFR, dihydrofolate reductase
- ORC, origin replication complex
- PCNA, proliferating cell nuclear antigen
- PRC, prereplication complex
- TK, thymidine kinase
- TS, thymidine synthase

Phospholipases
- PLC, phospholipase C
- PLD, phospholipase D
- PC-PLC, phosphatidylcholine phospholipase C

p21^{GTPASE} regulators
- GEF, guanine nucleotide exchange factor
- GAP, GTPase activating protein
- GDI, guanine nucleotide dissociation inhibitor
- GDS, guanine nucleotide dissociation stimulator
- SOS, son of sevenless, a Ras GEF

Miscellaneous
- DAG, diacylglycerol
- ECM, extracellular matrix
- IP_3, inositol-(1,4,5)trisphosphate
- PIP, phosphatidylinositol phosphate
- PIP_2, phosphatidylinositol-(4,5)bisphosphate
- PI3-K, phosphatidylinositol 3-kinase (not 3′)
- RGS, regulator of G protein signaling

factors. Although DNA replication is the usual focus of cell cycle studies, one cannot ignore the fact that to complete a cell cycle there also has to be an orderly duplication of all the other cellular components.

The Ras Cycle

Members of the p21^{GTPase} superfamily of proteins (e.g., Ras, Rho, Rac, Ral, Cdc42), which are typically associated with the inner face of the plasma membrane, with cellular membranes, or with cytoskeletal proteins, are essential upstream elements in many signaling pathways (Macara et al., 1996; Denhardt, 1996a). As illustrated for Ras in Figure 6.4, they are activated when an associated GDP is replaced by GTP, an exchange catalyzed by a GEF. A well-characterized GEF is the SOS (son of sevenless) protein that is often involved in Ras activation, usually in a complex with Grb2. Grb2 is an SH2-containing adaptor protein

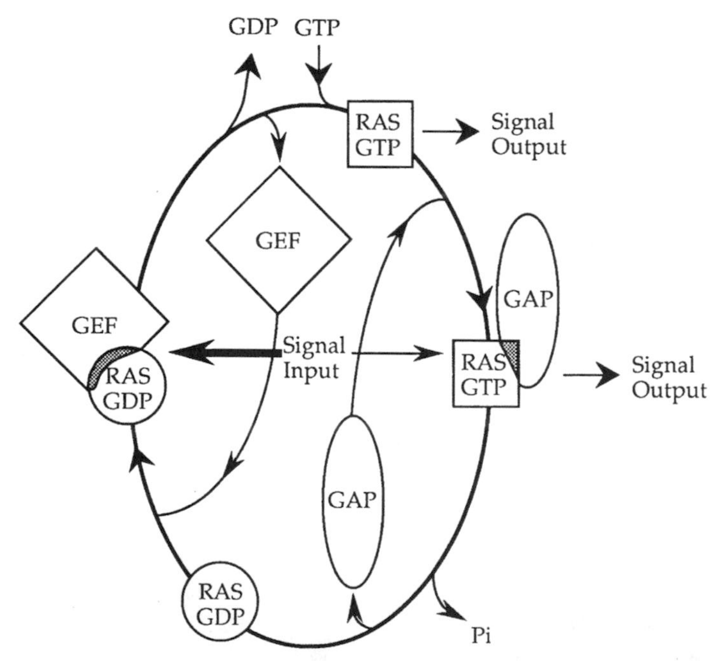

Figure 6.4. Cycling of Ras between an active GTP-bound form and an inactive GDP-bound form. Inactive, nucleotide-free Ras associates with GTP and is activated as a consequence. It continues to interact with and activate effector proteins until GAP forces hydrolysis of the GTP by the weak intrinsic GTPase activity of Ras. The inactive Ras/GDP complex associates with a GEF (guanine nucleotide exchange factor), which when activated evicts the GDP, permitting GTP to bind and again activating Ras. The signal from activated RAS is transmitted by protein–protein interactions, which may be different depending on whether GAP is bound or not. Input signals leading to Ras inactivation are thought to be transmitted more often via GEF and less often via GAP. (Reproduced from Denhardt, 1996a, with the permission of the publisher.)

that serves to mediate the interaction of SOS with various tyrosine-phosphorylated membrane receptors, causing Grb2/SOS to relocate from the cytosol to the plasma membrane. This allows SOS to directly access the membrane-associated Ras (see Figs. 6.7, 6.8, and 6.10, below). Relocation of Grb2/SOS to the plasma membrane may also be mediated by the adaptor protein Shc, which binds via its SH2 (Src homology 2) domain to a phosphotyrosine in the receptor, or in a protein associated with the receptor. Subsequent tyrosine phosphorylation of Shc provides a binding site for Grb2. Adaptor proteins like Shc and Grb2 do not possess enzymatic activity; rather, they generate physical connections between two other proteins via specific protein–protein interacting domains, for example SH2 and phosphotyrosine or SH3 and proline-rich PXXP sequences.

The change in the conformation of the $p21^{GTPase}$ protein induced by the substitution of GTP for GDP endows it with the capacity to interact with various effector proteins. Known mediators (see Figs. 6.5, 6.6, below) of $Rasp21^{GTPase}$ signaling include Raf, P13-K, Ral-BP, PKCζ, and Ras^{GAP} proteins and several MEKKs (Marshall, 1996). Most of these proteins are members of small families of closely related isoforms. GAP proteins, several of which are known, may also act as downstream signaling elements. Although their primary role is to hasten the inactivation of Ras, they may paradoxically enhance Ras signaling by suppressing unproductive interactions of Ras·GTP with an activated GEF (Giglione et al., 1997). GRF (guanine nucleotide-releasing factor) is a GEF that activates Ras in a Ca^{2+}/calmodulin-dependent manner. Two members of the GRF family have been identified: a 140 kDa species restricted to brain neurons (Farnsworth et al., 1995) and a 135 kDa species that is more widely distributed (Fam et al., 1997). They are clearly related to, but distinct from, the SOS class of GEFs. GEFs for the Rho family of $p21^{GTPase}$ proteins are also known as Dbl proteins because, as the cause of diffuse B-cell lymphoma, "dibble" was the first member of the Rho family of GEFs to be identified (Cerione and Zheng, 1996).

The $p21^{GTPase}$ Whirlpool

Some of the downstream effectors of the Ras, Ral, cdc42, Rac, and Rho proteins (each of which has one or more closely related isoforms) are indicated in Figure 6.5. The complexity of this figure is intended to convey in a simplified manner some of the potential interactions that may occur in the cell (Symons, 1996; Denhardt, 1996b). Which interactions actually occur will depend on which proteins the cell actually expresses. A key point is that the various $p21^{GTPase}$ proteins communicate via specific GAPs, GEFs, GDIs (guanine nucleotide dissociation inhibitors), GDSs (guanine nucleotide dissociation stimulators), and other binding proteins. This $p21^{GTPase}$ "whirlpool" somehow orchestrates a coordinated response to the multiple signaling inputs the cell receives. Many of these responses involve changes in the cytoskeleton and affect the ability of the cell to spread on a surface, to migrate, or to penetrate a barrier.

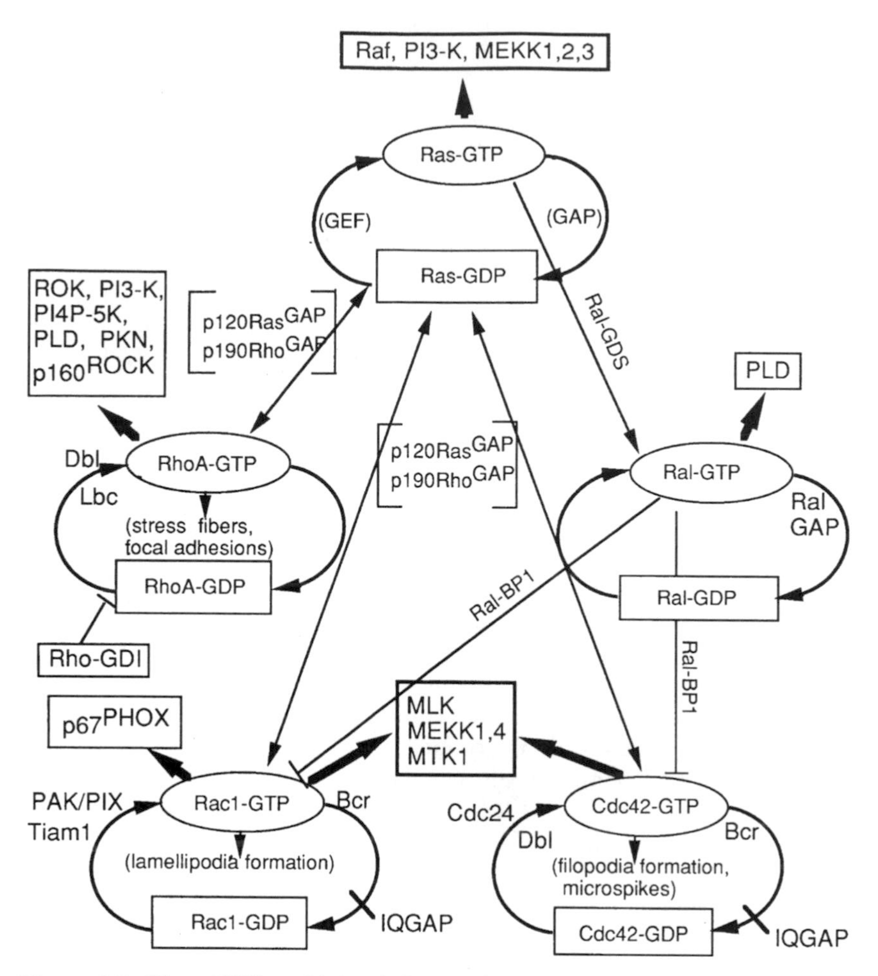

Figure 6.5. The $p21^{GTPase}$ whirlpool. Interactions among Ras/Rho/Rac/Cdc42/Ral family members involve GEFs and GAPs. Each of the five illustrated $p21^{GTPases}$ is closely related and regulated in a roughly similar manner, cycling between an active and an inactive form. The active GTP-bound signaling species is formed as the result of the action of a GEF on the inactive GDP-bound $p21^{GTPase}$, causing GDP to be replaced by GTP. For the Rho family members these proteins (e.g., Dbl, Lbc, Tiam1, PIX, Cdc24) have a "dibble" domain responsible for the interaction with the $p21^{GTPase}$ (Li and Zheng, 1997; Manser et al., 1998). The activated $p21^{GTPase}$ is deactivated by an intrinsic GTPase activity that can be stimulated by a GAP, whose activity is also regulated (Bcr, $p120Ras^{GAP}$, $p190Rho^{GAP}$). Some of the downstream effectors of the activated $p21^{GTPase}$ proteins are indicated by thick arrows pointing generally upward from the GTP-bound form. Interactions among the different $p21^{GTPase}$ proteins are shown by interconnecting arrows and appear to involve either GEFs or GAPs. Ral-binding protein 1 (Ral-BP1, which stimulates the GTPase activity of Cdc42 and Rac) and Ral–guanine nucleotide dissociation stimulator (Ral-GDS, a GEF for Ral) mediate the inhibitory and stimulatory reactions shown (Feig et al., 1996). Negative regulation is indicated by a —┤. ROK, Rho kinase; PI4P-5K, phosphatidylinositol 4-phosphate-5 kinase; PLD, phospholipase D; PKN, protein kinase $Np160^{ROCK}$, rho-associated coiled-coil kinase; IQGAP, a GAP protein with isoleucine/glutamine motifs. (Modified from Denhardt, 1996b, with permission of the publisher.)

Activation of any one of the $p21^{GTPase}$ proteins usually impacts on the activity of a number of target proteins. Rho activates several kinases, some of which phosphorylate inositol derivatives and regulate the interaction of various cytoskeletal proteins (Machesky and Hall, 1996). These include Rho kinase (ROK), phosphatidylinositol 4-phosphate 5-kinase (PI4P-5K), protein kinase N (PKN), and a 160 kDa Rho-associated coiled coil kinase ($p160^{ROCK}$) (references and additional details can be found in Denhardt, 1996b). Ral and Rho activate phospholipase D, thereby liberating arachidonic acid, the precursor of the prostaglandin family of potent signaling molecules. Rac and Cdc42 preferentially activate MEKK1, MEKK4, the multilineage kinases (MLKs), and the MAP three kinase (MTK1). These kinases stimulate pathways that are often involved in stress responses (see Fig. 6.6, below). In phagocytic cells Rac activates a membrane-associated NADPH oxidase via the 67 kDa PHOX subunit. Constitutively active forms of Rho, Rac, and Cdc42 can each stimulate DNA synthesis in quiescent 3T3 cells, and inhibition of signaling by any one of them by a dominant negative mutant inhibits serum-induced DNA synthesis (Olson et al., 1995). This last result is a strong indication that the function of each one of them is required for entrance into S phase, at least in serum-stimulated adherent fibroblasts.

By microinjecting various $p21^{GTPase}$ proteins into 3T3 cells, Lamarche et al. (1996) showed that stimulation by either Rac or Cdc42 of the $p65^{PAK}$–JNK stress response on the one hand or cytoskeletal rearrangements and G_1 progression on the other hand was controlled by different downstream signals emanating from different sites, as defined by mutations in the $p21^{GTPase}$. In other words, changing one amino acid at a specific effector site in the $p21^{GTPase}$ affected either JNK activation or cytoskeletal rearrangements and DNA synthesis, but not both. Thus the effector site mutation Y40C (mutating tyrosine to cysteine at position 40 in either protein) abrogated the interaction with $p65^{PAK}$ and aborted activation of the JNK/SAPK pathway by either $p21^{GTPase}$. This mutation defines the site that interacts with the CRIB motif, the *C*dc42 and *R*ac *i*nteractive *b*inding motif found in proteins that interact with Rac and Cdc42.

The Y40C mutation (in Rac or Cdc42) did not prevent Rho-mediated stimulation of DNA synthesis or interfere with the formation of lamellipodia by Rac or filopodia by Cdc42. The effector site mutation F37A (phenylalanine→alanine) in Rac, but not in Cdc42, abrogated interaction with the ser/thr kinase $p160^{ROCK}$ and aborted G_1 progression and the formation of lamellipodia. Interaction with $p65^{PAK}$ was not affected. The F37A mutation also disrupts crosstalk between Cdc42, Rac, and Rho. There are three members of the mammalian PAK family, and they are all directly activated by Cdc42 or Rac. They interact with a protein, PIX (PAK-interacting exchange factor) that is likely a GEF that mediates crosstalk between Cdc42 and Rac1 (Manser et al., 1998). PAK1 appears to be an essential intermediate in Ras signaling, as judged by oncogenic transformation of Rat-1 fibroblasts, but intriguingly does not appear to be essential for Ras transformation of mouse NIH3T3 cells (Tang et al., 1997). Presumably this is because of differences in the specific signaling pathways that are activated in these two cell lines.

In another approach, Keely et al. (1997) engineered expression of activated Cdc42, Rac-1 or PI3-K in the T47D mammary epithelial cell line. Their results suggested that persistent signaling from Cdc42 or Rac-1 disrupted the polarized cytoskeletal organization that was induced in cells maintained in a three-dimensional collagenous matrix and promoted a motile, invasive phenotype mediated by PI3-K activation. Qiu et al. (1997) demonstrated that in Rat-1 fibroblasts Cdc42 activity was essential for Ras transformation and that it controlled anchorage-independent growth in a Rac-independent manner. Rac activity, on the other hand, was essential for the proliferative response. As cell migration is incompatible with mitosis, it is not surprising that activation of pathways controlling one of these activities would be accompanied by suppression of the activation of pathways signaling the other. It is important that these signaling pathways, which are now described, be coordinated.

The MAPKKK–MAPKK–MAPK Cascade

Entrance into the cell cycle often occurs when the Ras-regulated signaling pathway involving the Raf→MEK→ERK protein phosphorylation cascade module dominates, as it usually does in response to mitogenic signals. Ras·GTP activates Raf-1 by attracting it to the inner surface of the plasma membrane and inducing conformational changes as the result of protein–protein interactions (Denhardt, 1996a; Morrison and Cutler, 1997). Phosphorylation of ser/thr residues by PKA, PKC, or other kinases may be inhibitory (PKA) or activating (PKC) and may determine specific interactions with the adaptor protein 14-3-3 that can lead to the formation of higher order protein complexes. Raf-1 can also be activated by Src- or JAK2-mediated tyrosine phosphorylation, and this may enhance the activity of the membrane-associated Raf-1 (Stokoe and McCormick, 1997).

In general it appears that only a small fraction of the Raf-1 in a cell is translocated to the membrane and activated, and only transiently, making biochemical studies difficult. Also, Raf-1 may exhibit different degrees of activity depending on its precise phosphorylation state. Activated Raf-1 phosphorylates the dual specificity MEK (MAPK/ERK kinase) on ser/thr residues, thereby inducing it to phosphorylate its unique targets ERK1 and ERK2 on threonine and tyrosine residues. Unlike the Rafs and MEKs, which have an extremely limited set of substrates, the activated ERKs are capable of phosphorylating many cytoplasmic and nuclear proteins on ser/thr-proline motifs. The entire signaling complex extending from the receptor through Ras to the downstream ERK appears to be organized in a functional form in membrane structures called *caveolae* (Liu et al., 1997).

The Raf→MEK→ERK phosphorylation cascade is one of three cascades, illustrated in Figure 6.6, that have been thoroughly documented in the mammalian cell (Denhardt, 1996b; Robinson and Cobb, 1997). Figure 6.6 is intended to convey the message that although multiple inputs converge on a small number of signaling modules, which overlap in terms of the transcription factors they activate, each input is accompanied also by additional signals that uniquely identify the stimulus. The result (differentiation or proliferation, for example) of a specific input will depend on the specific state of the cell and its lineage.

The three pathways are uniquely defined by the MAPK component ERK1,2; the JNK/SAPKs; and p38/RK, each of which has, as indicated in Figure 6.6, a TXY sequence in the kinase domain that is subject to dual phosphorylation on the threonine and tyrosine residues by the dual specificity kinases, including various MEKs, MKKs, and SEKs. The relationships among some of these MAPKKs, defined generally in rodent and human cells, remain to be clarified. There is some overlap in terms of the target proteins that are phosphorylated, but, because the efficiency with which a particular target protein is phosphorylated in vivo varies with the MAPK, the apparent overlap among the cascades may not be as great as it seems. Also, other signals that accompany activation of a particular cascade may augment or inhibit the activation of any particular downstream effector. Robinson and Cobb (1997) have enumerated the various mitogens that can activate the ERK pathway (and presumably stimulate proliferation) in different cell types, and they have reviewed evidence suggesting additional signaling modules in mammalian cells.

Johnson and colleagues (see Fanger et al., 1997) have performed a confocal immunofluorescence study of the MEK kinases MEKK1, -2, -3, and -4, revealing that specific MEKKs are localized to different cellular compartments. However, it should be noted that lesser amounts of any of the MEKKs, or immunologically distinct forms, could be found in other locations also. MEKK1 was found in the nucleus and post-Golgi vesicles, whereas MEKK2 and -4 were found in distinct Golgi-associated vesicles. The various members of the MEKK family were also differentially activated in COS cells by different upstream signals, and they exhibited distinct specificities for activation of downstream ERKs and JNKs/SAPKs. They did not activate p38/RK. Activation of ERK by a signal emanating from the EGF receptor (EGFR) was mediated largely by MEKK1, while JNK activation proceded via both MEKK1 and MEKK2. MEKK1 and -4 preferentially activate JNK in response to signals from Rac and Cdc42. MTK1 (MAP three kinase 1) is a major activator of p38/RK, and to a lesser extent JNK, in response to certain environmental stresses (osmotic shock, UV, inhibition of protein synthesis); it appears to act through MKK3, MKK6, and SEK1 (Takekawa et al., 1997). MKK7 is a recently described specific activator of JNK/SAPK under the control of Rac1 (Holland et al., 1997; Tournier et al., 1997).

As discussed in detail below, activation of specific cell surface receptors stimulates multiple signaling pathways, and some of the upstream signaling molecules (e.g., Ras, PI3-K, PKC, PLC, Src) in turn regulate more than one pathway, leading to an exploding cascade of signals.

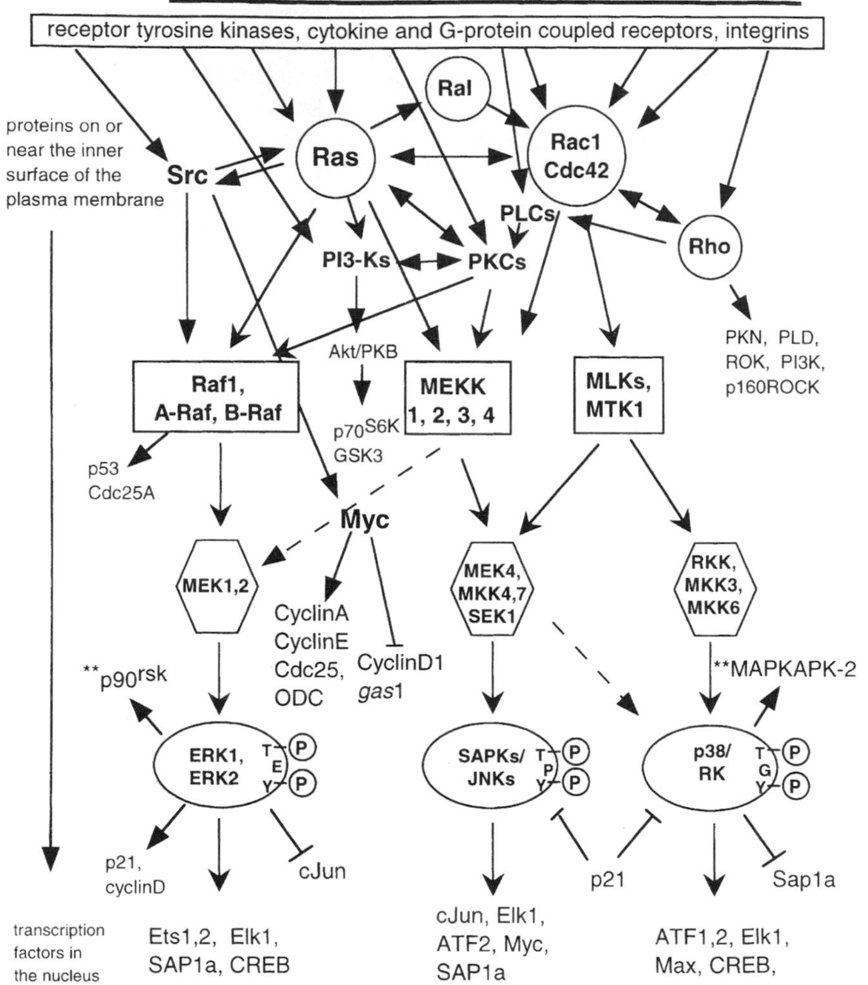

RESPONSE

Phosphorylation of nuclear and cytoplasmic proteins
Changes in gene expression and cytoskeletal structure

mitogenesis, metabolic regulation, cytoskeletal organization, apoptosis, differentiation, vesicular transport, growth inhibition, stress responses

Figure 6.6. Protein phosphorylation cascades initiated by various stimuli result in altered gene expression and cytoskeletal structure. Details of how individual $p21^{GTPase}$ (circles) are activated by specific inputs remains unclear in many cases. Proximal downstream mediators of $p21^{GTPase}$ signaling are indicated in rectangles; their immediate targets are usually the dual specificity kinases (in hexagons) belonging to the MAPKK family, which in turn target specific MAPKs (ovals). Some of the transcription factors that have been reported to be phosphorylated and activated by specific MAPKs in at least one cell type are indicated. Inhibitory phosphorylation is indicated by a ⊣. Considerable data suggest the existence of three reasonably discrete pathways, though with an unknown amount of

These must be controlled. One control mechanism is via negative feedback whereby phosphorylation of upstream elements by downstream kinases turns off the signal. Thus, phosphorylation of SOS, likely by an ERK, causes SOS to dissociate from Grb2 and terminates Ras signaling. Activation of the ERKs by an activated Raf does not by itself suffice to cause disassembly of SOS/Grb2 complexes, suggesting that an additional (growth factor or Ras-controlled?) signal is required (Klarlund et al., 1996). Signaling pathways are also shut down by dephosphorylation of those intermediates that are activated by phosphorylation.

Protein Kinases

As illustrated in Figures 6.5 and 6.6, ser/thr and tyr kinases, and dual specificity kinases, are major transducers of intracellular signals. Many are known, most of which are in small families of closely related proteins. Here we discuss several of them. Please note that some of the pathways and connections shown in Figures 6.5 and 6.6 are better established than others and that many interactions have been omitted for clarity.

Protein Kinase C. There are about 11 PKC isozymes that fall into three groups; conventional (α, β_1, β_2, and γ), novel (δ, ε, η, θ and μ), and atypical (λ and ξ) on the basis of their functional properties (Newton, 1995; Liu, 1996; Mufson, 1997). The different members are activated, translocated to a membrane in many cases, and downregulated in various ways. The conventional enzymes are activated by Ca^{2+} together with diacylglycerol (DAG) and phosphatidylserine; the novel enzymes are Ca^{2+} independent; and the atypical are not activated by Ca^{2+} or DAG. Many different proteins have been identified as substrates, and each is preferentially phosphorylated on ser/thr residues in particular sequence contexts by a subset of the isozymes (Nishikawa et al., 1997). For example, PKCδ phosphorylates and activates c-Raf, causing MEK and ERK activation in a Ras-independent manner (Ueda et al., 1996).

Studies by Schönwasser et al. (1998) have revealed that phorbol ester treatment of quiescent 3T3 cells activates ERK via MEK and stimulates DNA synthesis. These investigators showed that constitutively active

---

crosstalk (dashed arrows) and variable activation relative to each other depending on the stimulus. Each pathway is defined by the TXY motif in the MAPK component of the pathway. Arrows indicate some of the major routes of signal transmission. The double asterisks indicate two kinases that are not themselves transcription factors but that are activated by specific MAPKs and are responsible for the phosphorylation of certain transcription factors. As discussed in the text, localization of the components in the pathway and their recruitment into macromolecular complexes by factors that nucleate specific interactions (Xu et al., 1995) may endow the pathways with a specificity that is not apparent when efforts are made to dissect the pathways in vitro or by overexpression of a signaling molecule in vivo. (Modified from Denhardt, 1996b, with permission of the publisher.)

members of all three groups of PKCs, when transiently transfected into cells, could activate MEK and ERK and in addition that representatives of the conventional and novel groups acted through Raf while PKCζ activated MEK. In other situations PKC may act in an antiproliferative manner, as evidenced by its ability to attentuate mitogenic signaling generated by the EGFR (P. Chen et al., 1996). Many of the PKC isozymes can interact with pleckstrin homology (PH) domains found in other proteins, for example in Bruton's tyrosine kinase (Btk). This facilitates not only their interaction with each other but also their concerted action on target substrates (Yao et al., 1997). PH domains are ~100-residue domains conserved in a variety of signaling molecules that mediate protein–protein or protein–phosphoinositide interactions. They are structurally related to phosphotyrosine-binding domains (PTBs) and can bind certain inositol phosphates.

Phosphatidylinositol 3-Kinase. There are three classes of PI3-K enzymes, distinguished structurally and on the basis of their lipid substrate specificity (Toker and Cantley, 1997; Vanhaesebroeck et al., 1997). They all phosphorylate the 3-hydroxyl of a membrane-bound phosphatidylinositol in response to a suitable stimulus. Some of the isoforms are stimulated by receptor tyrosine kinases, Ras, or $G_{\beta\gamma}$ proteins, probably in part by being attracted to the inner surface of the plasma membrane where their substrates are located. Association of previously cytosolic proteins with a membrane-bound lipid causes conformational changes that can expose sequestered domains and render them accessible to additional phosphorylation by other kinases with the consequence that they become fully competent, for example to phosphorylate substrate proteins in the case of kinases.

Phosphorylation of phosphatidylinositol (4,5)bisphosphate produces phosphatidylinositol(3,4,5) trisphosphate [PI(3,4,5)P_3], which can attract additional proteins to the membrane via PH/PTB or SH2 domains (Rameh et al., 1995; Klippel et al., 1996; Ravichandran et al., 1997). These include Shc, the p85 subunit of PI3-K, the ser/thr kinase Akt/PKB, which may promote cell survival, and PKC-ε and -λ, which may activate the Raf→MEK→ERK pathway and stimulate cell proliferation (McIlroy et al., 1997). Akt/PKB is related to a retroviral oncogene (v-*akt*) and has homology with both PKC and PKA. A mediator of growth factor signaling, it can be activated by phosphorylation and binding to PI(3,4,5)P_3. Downstream targets include glycogen synthase kinase 3 (GSK3) and p70^{S6K} (Marte and Downward, 1997).

The PH domains found in various proteins have distinct preferences for different phosphoinositides and often prefer a phosphate in the D-5 position. The PH domains in Btk, SOS, and Tiam1 bind with higher affinity and specificity to PI(3,4,5)P_3 than to PI(4,5)P_2, whereas PLCδ and the β-adrenergic receptor kinase bind both equivalently (Rameh et al., 1997). Btk, a nonreceptor kinase essential for B-cell development, is activated by Src-mediated tyrosine phosphorylation and by binding via its PH domain to PI(3,4,5)P_3 (Li et al., 1997). PLCγ binds via its PH domain to $PI(3,4,5)P_3$ and is thereby activated (Falasca et al., 1998). Logan et al. (1997) have established that PI3-K is an important activator of JNK in EGF-stimulated cells, consistent with the thesis that JNK mediates an antiapoptotic response generated by PI3-K in these circumstances. The importance of this is that activation of PI3-K results in the activation of a variety of downstream effectors as a consequence of its ability to generate $PI(3,4,5)P_3$. Some PI3-K family members are also able to phosphorylate some proteins, though the significance of this activity is obscure (Hunter, 1995b).

Src. Src is the prototype of a family of about nine (Src, Fyn, Yes, Fgr, Lyn, Lck, Hck, Blk, and Yrk) nonreceptor protein tyrosine kinases, some of which are widely expressed (Src, Fyn, Yes), whereas others are generally restricted to hematopoietic and lymphoid lineages (Lck, Lyn, Hck, Blk) (Brown and Cooper, 1996). From the N terminus to the C terminus, five characteristic regions can be identified: (1) an N terminus that is usually myristoylated and, for some Src relatives, also palmitoylated on Cys-3, modifications that specify membrane localization; (2) an SH3 domain consisting of a β-barrel module responsible for binding to polyproline-II motifs; (3) a central SH2 domain that specifies binding both to certain tyrosine phosphates found in characteristic amino acid contexts and to certain inositol phospholipids; (4) an SH1 protein tyrosine kinase domain that can be stimulated by autophosphorylation; and (5) a tyrosine in a C-terminal domain that when phosphorylated (by the C-terminal Src kinase Csk) forms an intramolecular complex with the SH2 domain, thereby suppressing the activity of the kinase domain. Palmitoylation, which occurs only on myrisotylated molecules, anchors Src family members on the plasma membrane in caveolae close to various other receptors and potential substrates.

Src family members are generally found in an inactive state as the result of the SH2–phosphotyrosine and the SH3–polyproline motif intramolecular interactions that mask the SH1 domain (Xu et al., 1997). Activation occurs when the kinase is unmasked, for example when the Src SH2 domain or the SH3 domain binds, respectively, to an external tyrosine phosphate or a proline motif with higher affinity than the internal binding site or when the C-terminal regulatory tyrosine phosphate is dephosphorylated. Somani et al. (1997) showed, for instance, that the phosphatase SHP-1 can activate Src by dephosphorylating the C-terminal tyrosine phosphate. As discussed elsewhere in this chapter, these nonreceptor tyrosine kinases are major players in many signaling pathways.

Regulation of Cyclin D/CDK(4,6) Activity

The D cyclins and CDK inhibitors of the p16 and p21 families are primary regulators of entrance into the cell cycle. Their activity is controlled by endogenous signals and by mitogenic signal transduction pathways activated by external signals. As one example, Lukas et al. (1996) have shown that signals initiated by estradiol (via the estrogen nuclear receptor), EGF (via the one-pass tyrosine kinase EGFR), and thyrotropin (via its G protein–coupled receptor) all converge on and activate cyclin D/CDK(4,6), thereby stimulating proliferation of the target cell. These upstream "signal initiators" are described in more detail in the next section. Cyclin D1 expression, and consequently proliferative ability, is decreased in hamster lung fibroblasts when the stress-activated $p38^{MAPK}$ pathway is stimulated (Lavoie et al., 1996). This may serve to allow the cells time to recover from the stress. Molnar et al. (1997) found that microinjection of Cdc42 and other upstream activators (SEK1, MKK3, MKK6) of p38 into mouse 3T3 fibroblasts inhibited entrance into S phase via a mechanism dependent on p38.

One approach to assessing the requirement for Ras activity in cell cycle progression is to inactivate it, for example either with the neutralizing monoclonal antibody (Y13-259) or by expression of a dominant negative Ras allele (e.g., Ras^{N17}). By such means, Ras activity has been shown to be required by quiescent, serum-deprived cells to exit G_0 and enter G_1 in response to growth factor stimulation. This effect of Ras is Rb independent and is observed, therefore, in both $Rb^{+/+}$ and $Rb^{-/-}$ fibroblasts. However, when Ras is inactivated in asynchronously cycling cells, $Rb^{+/+}$ and $Rb^{-/-}$ cells yield different results. $Rb^{+/+}$ cells cease to progress into S phase and accumulate in G_1, while $Rb^{-/-}$ cells continue to cycle and replicate their DNA. Inhibition of Ras activity by the dominant negative Ras^{N17} mutant (thereby downregulating cyclin D1) abrogates activation of the G_1 CDKs and the resulting phosphorylation of Rb, thus preventing activation of E2F and arresting $Rb^{+/+}$ cells in G_1. The $Rb^{-/-}$ fibroblasts do not require Ras activity to pass the restriction point because, there being no Rb, E2F is not inhibited. Instead there is continuous expression of the Rb-regulated genes required for cell cycle progression (Peeper et al., 1997; Leone et al., 1997).

In immortal NIH3T3 fibroblasts expressing an inducible dominant negative Ras^{N17}, Aktas et al. (1997) found that Ras function was required for induction by serum of cyclin D1 mRNA accumulation in serum-starved 3T3 cells. They showed also that constitutive expression of cyclin D1 driven by a retroviral promoter was able to circumvent the requirement for Ras-mediated signaling. Ras activity was also required in late G_1 to downregulate p27 levels. This downregulation was accomplished both by suppressing p27 synthesis and by stimulating degradation of the protein (Takuwa and Takuwa, 1997). If p27 activity was not suppressed, then the activities of cyclin D/CDK(4/6) and cyclin E/CDK2 complexes were inhibited, resulting in G_1 arrest.

Transcriptional Control

ts13. The ts13 Syrian hamster ovary fibroblast cell line contains a temperature-sensitive (ts) mutation that prevents the cell from progressing into S phase at the restrictive temperature (Sekiguchi et al., 1991). This mutation defines the cell cycle control gene 1(CCG1), which encodes the TATA-binding factor $TAF_{II}250$, the largest subunit of the transcription factor TFIID (Ruppert et al., 1993; Hisatake et al., 1993). In the mutant cell, transcription driven by the cyclin A and cyclin D1 promoters is specifically reduced at the restrictive temperature (Wang et al., 1997). These researchers identified an enhancer element (TSRE) in the cyclin A promoter that binds to members of the ATF family of transcription factors and confers sensitivity to the ts13 mutation in $TAF_{II}250$. Rushton et al. (1997) found that in the ts13 mutant at the restrictive temperature cyclin D1 mRNA and protein levels were decreased, whereas p21 expression was increased. Expression of cyclin E and cyclin A (in contrast to what was found by Wang et al., 1997) were not affected by the temperature shift. The properties of the ts13 mutant provide compelling evidence, should such still be needed, for the existence of specific transcription factors that bind to elements in the promoters of genes that regulate the cell cycle. When these factors are activated, by phosphorylation for example, they may stimulate or inhibit transcription by interacting with the TFIID subunit $TAF_{II}250$.

Myc. The basic helix-loop-helix, leucine-zipper transcription factor Myc is a member of a family of nuclear phosphoproteins that regulate the expression of genes controlling cell proliferation, apoptosis, and differentiation (Henriksson and Lüscher, 1996). Often it acts to stimulate cell proliferation (or apoptosis) and to inhibit differentiation. Typically short-lived, it forms a heterodimer with another bHLHZip protein, Max, and recognizes a sequence, the E box (consensus = CANNTG), in various promoters.

Myc activates transcription of genes encoding cyclins A and E, while suppressing transcription of cyclin D1 (Jansen-Durr et al., 1993; Philipp et al., 1994) and the growth arrest–specific gene *gas*1, whose product can block cell proliferation (TC Lee et al., 1997). Transcription of the gene encoding cyclin D3 is increased by Myc, at least in HeLa cells, and this sensitizes the cells to TNF-α–induced apoptosis (Janicke et al., 1996). Myc also activates transcription of the gene encoding the dual specificity protein phosphatase Cdc25 (see Fig. 6.3), which is able to activate CDKs by dephosphorylating inhibitory threonine and tyrosine phosphates (Galaktionov et al., 1996). Other genes whose expression is increased by Myc include osteopontin (Castagnola et al., 1991), ornithine decarboxylase (Bello-Fernandez et al., 1993), and Rcl (Lewis et al., 1997). Rcl, which was identified along with 20 other proteins by a representational difference analysis of the mRNAs of nonadherent Rat-1a and Rat-1a Myc cells, had the ability to induce anchorage-independent growth of Rat-1a fibroblasts.

Several studies point to p27 inhibition of cyclin E/CDK2 activity as one of the important targets of Myc. Results obtained from the introduction of a dominant inhibitor of Myc in NIH3T3 cells led Berns et al. (1997) to conclude that Myc functioned to alleviate inhibition of cyclin E/CDK2 by p27. D. Müller et al. (1997) reported that phosphorylation of Thr-187 in p27 in Rat-1 cells was required for the release of p27 from cyclin E/CDK2 and its subsequent Myc-dependent sequestration and degradation. Vlach et al. (1996) obtained evidence that cyclin E/CDK2 activity was enhanced by Myc as a consequence of Myc-induced sequestration of an inhibitor of p27.

Other basicHLH proteins, MyoD, for example, control many many aspects of differentiation (e.g., Jan and Jan, 1993). ID proteins, which are induced by mitogenic signals, are HLH proteins that lack a basic DNA-binding domain. They can form dimers with HLH proteins, thereby antagonizing their transcriptional activation functions, which are particularly important in stimulating differentiation pathways and suppressing proliferation. Peters, Norton, and colleagues (Hara et al., 1997; Deed et al., 1997) have reported that CDK2, associated with either cyclin E or cyclin A, can phosphorylate both ID2 and ID3 during the G_1S transition. This phosphorylation alters the ability of the ID proteins to modulate bHLH–E box interactions, presumably facilitating the G_1S transition by serving as a switch between a proliferative response and cell differentiation.

Ras Control of the Cell Cycle

With important exceptions, most cells in the body are quiescent, existing either in the G_0 state or in the G_1 phase of the cell cycle, arrested somewhere before the restriction point. Different cell lineages require different stimuli to cause them to enter the cell cycle. For example, serum-starved fibroblasts in cell culture are considered to be in the G_0 state. Various growth factors, such as PDGF or lysophosphatidic acid (LPA), in serum drive the quiescent cells into G_1 and then through the restriction point and into S phase. As long as the culture conditions are suitable the (immortal) cells will continue to cycle. Certain cell types in the body, for instance stem cells in the bone marrow, or in epithelial tissues, continually proliferate, though it is not known how their continuously proliferating state is maintained. In other tissues, breast or uterus, for example, cell proliferation occurs only in response to specific physiological stimuli.

Many clues regarding how entrance into the cell cycle is regulated have come from studies on cancer cells, and our knowledge about both the positive signals that drive cells to proliferate and the negative signals that oppose proliferation has become fairly sophisticated. Although cells are often encouraged to proliferate by an active stimulus, there are occasions where removal of an inhibitory signal is required. For example, cells (hepatocytes) in the liver normally do not proliferate. However, if a portion of the liver is surgically removed, then the remaining liver cells, or at least some of them, will enter the cycle and proliferate until the mass of the liver is restored to what it was before. The nature of the substance that suppresses the proliferative state of the hepatocyte in the mature liver is not known.

Winston et al. (1996) used a glucocorticoid-inducible Ha-*ras* expression vector to investigate the relationship in 3T3 fibroblasts between Ras expression and mitogenic factors in plasma and serum. Serum-starved (G_0) cells could progress up to mid G_1 when Ras was induced, but progress beyond that point required unidentified plasma factors. Ras was capable of inducing synthesis of cyclin D, and the assembly of cyclin D/CDK complexes, but CDK activity was very low, apparently because of inhibition by p27. One or more factors in plasma were required in a second pathway to reduce p27 levels to the point where the cyclin D–dependent kinases could function. Ras also enhanced cyclin A and cyclin E expression, but to a lesser extent that cyclin D.

One of the downstream mediators of Ras signaling is Raf, and Lloyd et al. (1997) showed that activation of Raf (as part of an estrogen inducible Raf/estrogen receptor fusion protein) in rat Schwann cells caused increased expression of p21, which inhibited cyclin A/CDK2 and cyclin E/CDK2, and blocked passage through G_1. When SV40 large T antigen or a dominant negative mutant of p53 was expressed simultaneously with inducible Raf, p21 induction and growth arrest were not observed. As Raf behaves similarly to Ras in the Schwann cells, inducing distinct morphological changes and cell cycle arrest in a large-T reversible manner, these authors suggested that the Raf signaling pathway is responsible for the effects of Ras on the cell cycle.

Immortalizing oncogenes, (E1A, Myc) are commonly observed to suppress CDK inhibitor proteins and in this way cooperate with transforming oncogenes such as Ras or Raf. The state of the p53 tumor suppressor protein determines whether Raf provides a growth inhibitory or growth stimulatory signal. Thus, in NIH3T3 cells, which have a mutant p53 gene, ectopic expression of c-Raf-1 repressed p27, induced cyclin D1, and triggered a weak mitogenic response (Kerkhoff and Rapp, 1997). In contrast, in primary Schwann cells with a nonmutant p53, direct constitutive activation of the Raf pathway mimics a DNA-damage signal, which is interpreted by wild-type p53 as a signal to suppress proliferation via p21 (Lloyd et al., 1997). Activation of Raf via Ras is typically accompanied by an assortment of other signals (depending on how Ras was activated), which serve to modify the specific signal conveyed through Raf. For instance, because phosphorylation of Raf by protein kinase A can inhibit Raf activity (Hafner et al., 1994), stimuli that activate adenylate cyclase, via $G_{\alpha s}$, for example, will act to suppress the mitogenic response.

Diverse studies, some cited above, have shown that Ras signaling is required for cells to transverse the cell cycle. As one more example, Dobrowolski et al. (1994) demonstrated that input from Ras was needed at two or more points in the G_0 to G_1 S phase transition in BALB/c 3T3 cells. When Ras is activated, cyclin D mRNA levels are enhanced in various cell types (Filmus et al., 1994), the result of increased transcription of the cyclin D genes, possibly via ERK-induced activation of the c-Ets2 transcription factor (Albanese et al., 1995; Lavoie et al., 1996). Clearly, the major signaling pathway linking cell surface receptors to transit through the restriction point is the Ras→Raf→MEK→ERK→ cyclin D→Rb pathway together with accompanying Ras- and Raf-activated signals that among other consequences suppress inhibitors of cell cycle progression (see Fig. 6.5, 6.6).

Duplication, differentiation, or death are three quite distinct, and mutually exclusive, responses a cell can exhibit under any particular set of circumstances. CDKs, and many other signaling molecules, are intimately involved in each type of response (Gao and Zelenka, 1997). Induction of apoptosis by the withdrawal of trophic factors may be mediated by p38, whereas ERK-controlled pathways more often promote the proliferative response. Thus, in Rat-1 fibroblasts deprived of serum or in PC12 cells deprived of nerve growth factor, apoptosis is induced concomitantly with p38 activity, and inhibition of p38 activation by insulin or by the specific inhibitor PD169316 suppresses the apoptotic response (Kummer et al., 1997). In terminally differentiated cells (neurons), inhibition of CDK activity suppresses apoptosis (Park et al., 1997). Although stimulation of the Raf→MEK→ERK signaling cascade is usually associated with a proliferative response, several groups have shown in different model systems that prolonged and vigorous activation of this pathway can lead to cell cycle arrest, possibly caused by substantially increased p21 expression (Sewing et al., 1997; Woods et al., 1997; Pumiglia and Decker, 1997).

Section Summary

We are beginning to decipher the signaling pathways that control the cell's response to exogenous and endogenous signals. It is evident that the protein kinase cascades in the mammalian cell are derived from a primordial signaling cascade that has replicated itself and diverged to serve various purposes. As a consequence of this evolution, and because of some crosstalk among the pathways and repeated duplication of individual signaling elements, there is considerable redundancy built into the system. Studies of mice unable to make a particular protein, gene "knockouts", have revealed that not infrequently another protein can substitute for the absent protein or that the targeted protein's absence is felt in only very specific cell lineages. Whether a cell responds to a specific signal by duplicating, differentiating, or dying is determined by how it integrates the signals it is receiving with its intracellular state.

SIGNAL INITIATION: ACTIVATION OF SIGNALING PATHWAYS

After mitosis or meiosis, the stimulus to continue through another cell cycle can originate either endogenously or exogenously. Exogenous stimuli include fertilization of an egg, exposure to a growth factor, or removal of a repressing factor. An endogenous stimulus would be a genetic program that regulated an intrinsic signaling pathway to maintain the cell in a proliferative mode. Homeostatic mechanisms, involving, for example, tumor suppressor genes such as p53 or Rb, may generate inhibitory signals that oppose progression through the cell cycle. The decision to proliferate, which calls for growth and cell division, is determined by the balance between the positive and negative signals perceived by the cell. Many of the genes whose expression is regulated in a proliferation stimulus–responsive manner were reviewed by Hofbauer and Denhardt, (1991) (see also Chapter 5, this volume).

One-Pass Transmembrane Receptor Tyrosine Kinases

Platelet-derived growth factor, PDGF, is an extracellular peptide capable of stimulating quiescent fibroblasts to proliferate. It is released in massive amounts by platelets when a blood vessel in a vertebrate is injured. It and other cytokines and inflammatory mediators released at the wound site stimulate cells at the site to enter the cell cycle with the ultimate goal of repairing the injury. PDGF, a dimeric molecule, promotes dimerization of its receptor, a plasma membrane protein with a single transmembrane segment that is representative of a family of receptor tyrosine kinases (RTKs), including the receptors for EGF, NGF, FGF, CSF, insulin, and stem cell factor (van der Geer and Hunter, 1994). Dimerization, presumably accompanied by conformational changes in the receptor, rouses a latent protein kinase in the receptor to phosphorylate tyrosine residues in the cytoplasmic portion of the receptor.

As illustrated in Figure 6.7, cytoplasmic proteins with specific SH2 or PTB domains are attracted to specific tyrosine phosphates on the PDGF receptor (Pawson, 1995; Harrison, 1996). The amino acid context in which each tyrosine phosphate is embedded determines which SH2 or PTB domains can dock with it. Importantly, because not all of the proteins shown in Figure 7 can bind simultaneously to the same receptor, there will be competition among them for binding sites. As an example, Src (also Fyn and Yes) can bind to either Y579 or Y581, while Shc binds Y579 (the preferred site), Y740, Y751, or Y771 (Yokote et al., 1994). Phosphorylation of the alternatively spliced 66 kD isoform of Shc on ser/thr residues impairs its ability to associate with the tyrosine-phosphorylated receptor but not with Grb2, thereby setting up a situation where the larger isoform acts as a dominant negative inhibitor of signaling mediated by the 52 and 46 kD isoforms of Shc (Okada et al., 1997).

Receptor-mediated relocation of cytoplasmic proteins to the inner surface of the plasma membrane, possibly accompanied by subsequent tyrosine phosphorylation, results in the stimulation of multiple signaling pathways. Examples include activation of Ras (by SOS associated with Grb2 or Grb2/Shc), stimulation of P13-K activity, and enhancement of PLCγ activity. SHP-2 (also known as PTP1D, Syp) is a tyrosine phosphatase that appears to mediate mitogenic signals from both PDGF and thrombin receptors (Rivard et al., 1995). It could do this by dephosphorylating a negative regulator, for example, the inhibitory tyrosine phosphate in Src. Alternatively, receptor-stimulated tyrosine phosphorylation of SHP-2 may provide another binding site for Grb2/SOS (Rivard et al., 1995). Some of the activated PDGF-β and insulin receptors become associated with the $\alpha_v\beta_3$ integrin, perhaps evidence of a synergism in their proliferation-promoting ability (Schneller et al., 1997).

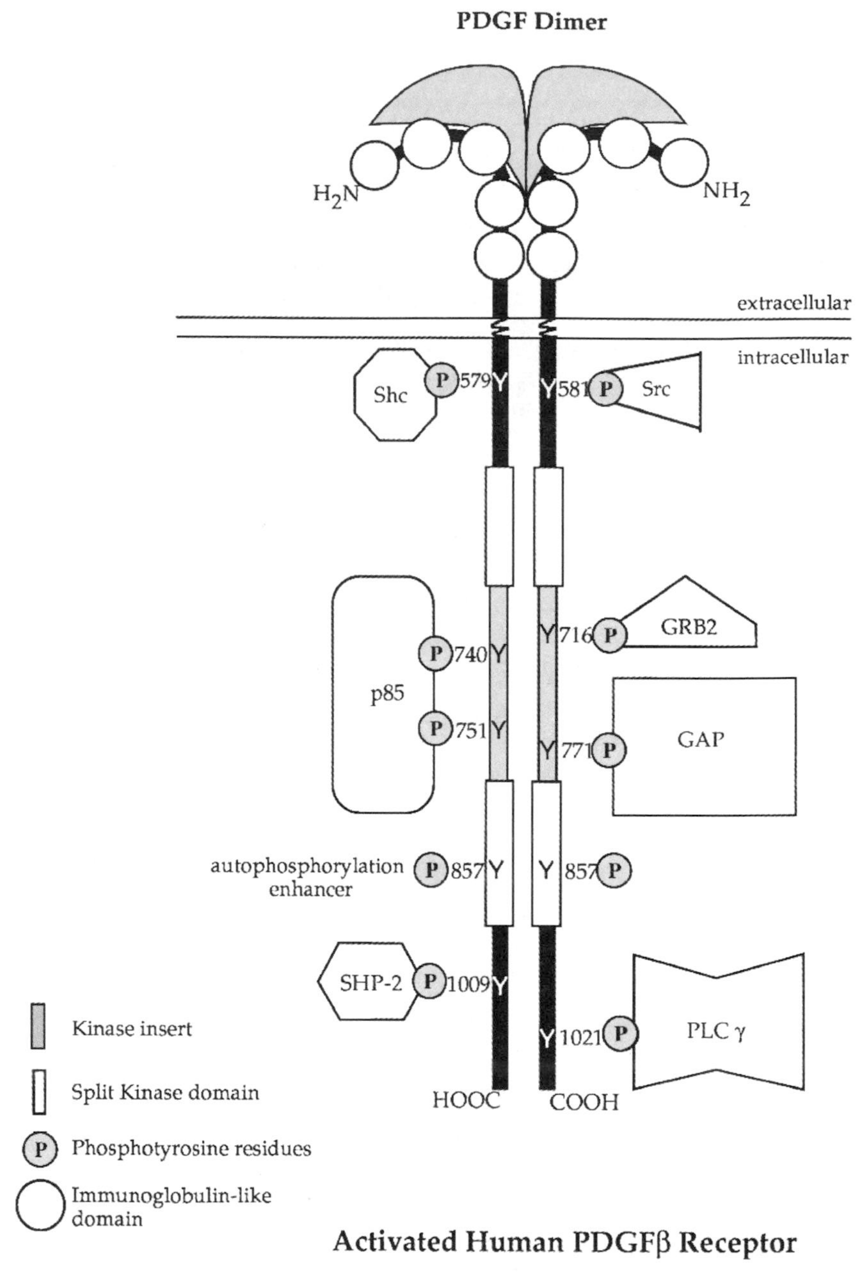

Figure 6.7. The activated PDGF receptor. A PDGF dimer brings two PDGF receptor monomers into close proximity, allowing multiple tyrosine residues to become phosphorylated. Phosphorylation of Tyr-857 enhances the kinase activity. Various proteins are able to bind to the phosphotyrosines with a specificity determined by amino acids neighboring the phosphotyrosine. p85 is the regulatory subunit that combines with the p110 catalytic subunit to form PI3-K. Proteins binding to the activated receptor may also be phosphorylated by the receptor. Because the stoichiometry of receptor autophosphorylation is not high, it is likely that on average any one receptor molecule will only be phosphorylated at a few sites and be able to bind only a subset of all possible proteins. (Modified from Denhardt, 1996a, with permission of the publisher.)

Weber et al. (1997b,c) showed that Chinese hamster embryo fibroblasts (IIC9) required sustained activation by mitogens (specifically PDGF) of ERK1 activity to maintain cyclin D1 expression and activity. Inhibition of mitogen-stimulated ERK1 with the MEK inhibitor PD98059 or a dominant negative ERK mutant at any time throughout early G_1 phase diminished cyclin D1 mRNA and protein levels and retarded entrance into S phase. This was possible because of the short half-life of cyclin D1 mRNA and protein. Concomitantly, proteolytic degradation of p27 was stimulated via a pathway dependent on Ras and RhoA, but not ERK. In this system regulation of the level of cyclin D1 expression and activity, and not the levels of CDK2 or CDK4, appeared primarily responsible for G_1 progression. There was no evidence in these cells of activation of JNK/SAPK, and inhibition of JNK/SAPK activity by a dominant negative mutant had no effect on cell cycle progression.

In mouse or human fibroblasts engineered to express the PDGF-β receptor, any one of several of the Janus kinases can associate with the activated receptor, become phosphorylated, and activate both STAT1 and STAT3 (signal transducer and activator of transcription, discussed further below) (Vignais et al., 1996). In vascular smooth muscle cells PDGF can stimulate tyrosine phosphorylation of STAT1α (p91), resulting in its translocation to the nucleus and binding to SIE (sis-inducible element, in the c-*fos* gene for example) and GAS (γ-interferon–activated site, e.g., in the NOSII gene) sites in promoters of target genes (Yamamoto et al., 1996). This recent finding simply emphasizes, should such emphasis still be necessary, the number and variety of signaling pathways that a given receptor can stimulate. However, the importance of any specific pathway still requires verification.

Bost et al. (1997) showed that in certain lung carcinoma cell lines the JNK/SAPK pathway rather than the ERK pathway was primarily responsible for the stimulation of proliferation by EGF. They did this by demonstrating (1) that antisense oligonucleotides targeting JNK or a dominant inhibitor of c-Jun could block the EGF-induced cell growth and (2) that the specific MEK inhibitor PD98059 suppressed ERK activation but not the EGF-stimulated proliferative response.

In one approach to assess the importance of the different signaling proteins in cell cycle control, Barone and Courtneidge (1995) used dominant negative forms of Ras, Src, and Myc to investigate their specific involvement in growth factor (PDGF, CSF-1, EGF) stimulation of DNA synthesis in NIH3T3 cells. Cells unable to proliferate when Src activity was blocked could be rescued by Myc expression, but not by Fos or Jun expression. (Fos and Jun are transcription factors that contribute to mitogen-induced gene expression.) Enhancement of Fos or Jun expression, but not Myc expression, could rescue cells blocked by a dominant negative Ras allele. This work suggests that the receptors for PDGF, EGF, and CSF-1 activate at least two pathways, a Ras-dependent pathway enhancing Fos/Jun activity and an Src-dependent pathway resulting in increased *myc* transcription. Activation of both pathways is required for 3T3 cells to respond mitogenically to receptor stimulation. A similar conclusion can be drawn from the studies of Gotoh et al. (1997) showing that in response to EGF stimulation Myc and Ras were activated as the consequence of signaling mediated by tyrosine phosphates found in two different sequence contexts in the receptor. Tyrosine phosphorylation of Y239 and Y240 led to activation of a pathway including Myc, whereas phosphorylation of Y317 activated the Ras/ERK pathway.

Seven-Pass Transmembrane G Protein–Coupled Receptors

This is a large superfamily of cell surface receptors, each of which is a single polypeptide chain containing seven transmembrane helices that is coupled to a heterotrimeric $G_{\alpha\beta\gamma}$ protein (Gutkind, 1998: Hamm, 1998). Many of them are sensory receptors responsible for vision, smell/taste, or touch/pain, while others respond to various neurotransmitters, hormones, and chemokines. The heterotrimeric G proteins consist of a monomeric G_α subunit that binds guanine nucleotides and a stable heterodimeric $G_{\beta\gamma}$ subunit. Binding of a ligand to a $G_{\alpha\beta\gamma}$ protein–coupled receptor causes conformational changes that result in the release of GDP from the inactive receptor-associated $G_\alpha{}^{GDP}{}_{\beta\gamma}$ protein and its replacement by GTP. The $G_\alpha{}^{GTP}$ and $G_{\beta\gamma}$ components then dissociate and become active, membrane-bound (via farnesyl and isoprenyl moieties on the α and γ subunits, respectively) signaling molecules (Bourne, 1977; Wess, 1997). Only rarely have oncogenic mutations in G_α proteins been identified (Seuwen and Pouysségur, 1992; Chen et al., 1997).

Figure 6.8, which serves as a basis for the following discussion, illustrates some of the signaling pathways these G protein–coupled receptors (GPCRs) can stimulate in cells containing the requisite factors. Although the great majority of signals transmitted by these receptors to intracellular signaling pathways do not stimulate the cell to proliferate, there are some that do (Post and Brown, 1996). Seuwen and Pouysségur (1991) have reviewed early research that implicated some $G_{\alpha\beta\gamma}$-mediated signaling pathways in regulating cell proliferation. Second messengers such as cAMP, DAG, and inositol phosphates may stimulate or inhibit DNA synthesis depending on the cell and the accompanying signals. When cell proliferation does occur, it is typically associated with activation of the ERKs, often mediated by Ras (Gutkind, 1998). However, activation of Ras or ERK does not by itself necessarily initiate a cell cycle; appropriate supporting signals may also be required. The question we address here is how a mitogenic signal is generated.

Roles of the G_α Subunits

The G_α subunits (about 20 have been identified) define four families of $G_{\alpha\beta\gamma}$ receptors on the basis of their prototypic α subunit: α_s and α_{olf}; α_i and α_o; α_q; and α_{12-13}. These differ in aspects of the signals they deliver and their sensitivity to cholera toxin (α_s) or pertussis toxin (α_i, α_o). The α_s (and α_{olf}) subunits activate adenylate cyclase and Ca^{2+} channels; the α_i and α_o subunits inhibit adenylate cyclase, regulate K^+ and Ca^{2+} chan-

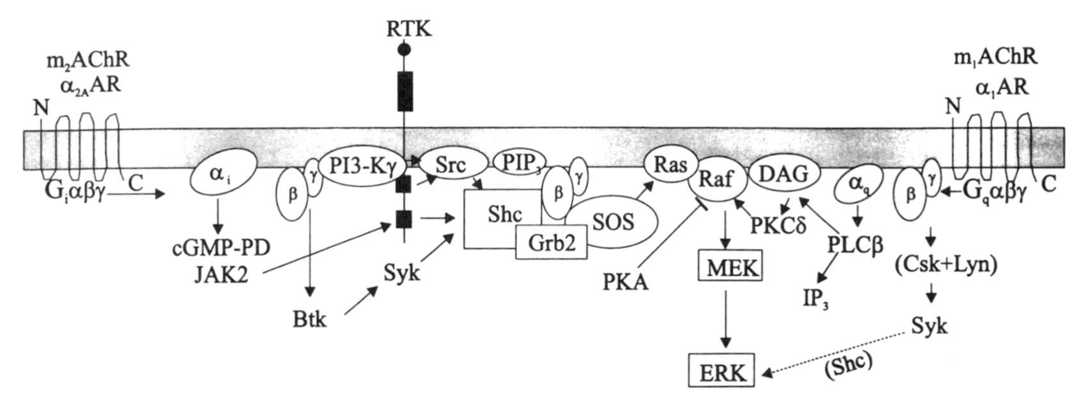

Figure 6.8. Proliferative signaling downstream of G_i and G_q protein-coupled receptors. This figure is an attempt to synthesize the conclusions of a number of investigators working with different cell types as discussed in the text. Only certain of the pathways illustrated are likely to be active in any given cell type. The M2 muscarinic acetylcholine receptor (m_2AChR; and also the α_{2A}-adrenergic receptor) couples to G_i, whereas the M1 muscarinic receptor (m_1AChR and also the α_1-adrenergic receptor) couples to G_q. G_i proteins can stimulate ERK via a pathway including Bruton's tyrosine kinase (Btk) and Syk, whereas G_q utilizes a pathway requiring both Csk and Lyn to activate Syk (Wan et al., 1997). The dotted line connecting Syk to ERK via Shc represents the Shc/Grb2/SOS/Ras/Raf/MEK pathway. The $\beta\gamma$ subunits may interact with proteins with PH domains, for example, Btk, PI3-Kγ, Shc, SOS, PLC, and the β-adrenergic receptor kinase (Pitcher et al., 1995; Coso et al., 1996). Activation of JAK2 by α_i, which also activates a cyclin GMP phosphodiesterase, may phosphorylate tyrosines on a receptor tyrosine kinase, thereby potentially activating a signaling pathway downstream of an RTK. ERK activation can be mediated via both a Ras-dependent pathway, likely involving $\beta\gamma$ proteins and Shc/Grb2/SOS complexes, and Ras-independent pathway, likely via PKCδ activation of Raf (van Biesen et al., 1996). PKCδ, a novel PKC, is activated by the diacylglycerol generated from phosphatidylinositol-(4,5)diphosphate as the consequence of PLCβ stimulation by α_q.

nels, and activate cGMP phosphodiesterase; the α_q subunit activates phospholipase Cβ; and the α_{12-13} subunits stimulate Na^+H^+ exchange (Neer, 1995; Dhanasekaran and Dermott, 1996). van Biesen et al. (1996) have shown that in CHO cells expressing the M1 muscarinic acetylcholine receptor (mAChR) that the pertussis toxin–sensitive $G_{o\alpha}$ (and also $G_{q/11}$) subunit could activate ERK via a pathway that required PKC but was independent of Ras and $\beta\gamma$. They caution that precisely how specific receptors couple via $G_{\alpha\beta\gamma}$ to downstream signaling elements is dependent on the cell, presumably a function of which signaling elements are actually expressed.

The α subunit has an intrinsic GTPase activity, related to that found in Ras (Fig. 6.4), that limits the lifetime of the activated protein. Furthermore, just as GAP proteins further constrain the lifetime of activated Ras, so do proteins known as RGS (regulators of G protein signaling) regulate GPCR signaling by accelerating hydrolysis of the bound GTP so that the inactive heterotrimeric $G_{\alpha\beta\gamma}$ can reconstitute itself (Dohlman and Thorner, 1997). Certain of the RGS proteins are very effective at suppressing G protein signaling; for example, RGS1 and RGS2 can attenuate ERK activation by G_q or G_i (Berman and Gilman, 1998).

***Roles of the $G_{\beta\gamma}$* Subunits.** The $G_{\beta\gamma}$ subunits (6 β and 12 γ identified so far) can interact, when freed of their association with G_α, with the C-terminal portion of PH domains found in a variety of proteins and stimulate their activity (Inglese et al., 1995; Hamm, 1998). Because the $\beta\gamma$ subunit is held to the plasma membrane by the prenylated γ subunit, the formation of complexes with target proteins results in the relocation of the target protein to the cell membrane and a conformational change that generates a signal. For example, Crespo et al. (1994), working with COS-7 cells engineered to express the M1 and M2 mAChRs, which are coupled, respectively, to G_q and G_i (see Fig. 6.8), found evidence that ERK activation was through a pathway stimulated by the $\beta\gamma$ subunits and requiring Ras, but not PKC, activity. The JNK/SAPK subgroup of MAP kinases, which typically is not associated with a mitogenic response, can also be activated by GPCRs. As one example, stimulation of mAChRs expressed in COS cells resulted in JNK activation as a consequence of $\beta\gamma$-mediated Ras/Rac-1 signaling (Coso et al., 1996).

Clues to some details of this Ras-dependent pathway were uncovered by Lopez-Ilasaca et al. (1997), who demonstrated that P13-Kγ could activate ERK by inducing tyrosine phosphorylation of the adaptor protein Shc. Their suggestion was that $\beta\gamma$ recruited PI3-Kγ to the membrane and there stimulated a Src kinase to phosphorylate Shc and subsequently activate Ras via an interaction with Grb2/SOS. Bokoch (1996) has reviewed the interplay between the $p21^{ras}$ and $G_{\alpha\beta\gamma}$ proteins and discussed the possible involvement of PH domains in mediating interactions between $\beta\gamma$ and Rho family GEFs.

Baldassare, Raben, and coworkers (Weber et al., 1997a; Cheng et al., 1997) have shed light on aspects of the involvement of $G_o\beta\gamma$ in cell cycle progression and tumor suppression. Their studies were performed in Chinese hamster embryo fibroblasts (IIC9) and derivatives (Goa cells) whose ability to produce $G_o\alpha$ had been abrogated by stable expression of an antisense RNA. In the Goa cell lines unable to express $G_o\alpha$, Cheng et al. (1997) observed that phosphatidylcholine-specific phospholipase C became constitutively expressed and conferred a transformed phenotype on the cells, possibly as the result of elevated levels of DAG and phosphorylcholine. Even in the absence of mitogens both the Ras/ERK pathway and cyclin D1/CDK complexes were constitutively activated in these cells, apparently as the result of elevated levels of free $\beta\gamma$ and consequent activation of Ras (Weber et al., 1997a). These investigators also observed that antisense-mediated transient ablation of G_q, G_{i2}, or G_{i3} (which is presumed to liberate the specific $\beta\gamma$ subunits associated with them) did not result in activation of the Ras/ERK pathway, suggesting that only the $\beta\gamma$ species associated with $G_o\alpha$ could activate Ras. Activation of Ras by $\beta\gamma$ could be the consequence of $\beta\gamma$-induced activation of a Src family kinase, perhaps by PI3-Kγ, followed by Shc phosphorylation (Bokoch, 1996; Lopez-Ilasaca et al., 1997). Precisely which β and γ subunits are associated with specific α subunits remains for the most part to be sorted out.

Role of Nonreceptor Src Family Kinases. Wan et al. (1996, 1997) investigated the coupling of the M1 and M2 mAChRs to the MEK→ERK pathway via members of the Src protein tyrosine kinase family in avian B lymphoma DT40 cells. These cells were selected for study because they support a high frequency of homologous recombination, making the derivation of gene "knockouts" relatively efficient. In these cells Syk activity was found to be essential for activation of the MEK→ERK pathway mediated by either the G_i-coupled M2 mAChR or the G_q-coupled M1 mAChR, whereas Lyn activity was required only by the latter receptor. Syk is an intracellular protein tyrosine kinase that along with members of the Src family has been implicated in signaling in lymphoid cells, for example, from the B-cell receptor. It is responsible for phosphorylation of a number of intracellular proteins (including itself and Shc) in response to engagement of the antigen receptor and may serve as a scaffold to assemble a macromolecular complex (Richards et al., 1996; Wan et al., 1997). Activation of the B cell receptor may lead to cell proliferation, cell activation, or apoptosis, depending on the developmental stage of the cell.

As diagrammed in Figure 6.8, ERK activation by the G_q-coupled pathway and Syk required both Csk and Lyn, whereas only Btk was required by the G_i-coupled pathway (Wan et al., 1996). Csk and Lyn were required for G_q but not G_i signaling, whereas Btk was required for G_i but not G_q signaling, in both cases as judged by activation of ERKs via mAChRs. Thus Csk, Lyn, and Syk are required for G_q-mediated activation of ERKs, whereas Btk and Syk are required for G_i-mediated activation of ERKs. There is thus a protein tyrosine kinase cascade linking the $G_{\alpha\beta\gamma}$ receptors G_i and G_q to ERK activation via Src family proteins and the common mediator Syk, which may stimulate Shc to activate Ras. Nagai et al. (1995) have shown that in B-cell receptor signaling Shc interacts with Syk to convey a signal from receptor-activated Lyn to Grb2/SOS and Ras via Lyn-activated Syk phosphorylation of Shc.

In a study of the angiotensin II receptor-1 (AT1) in rat aortic smooth muscle cells Schieffer et al. (1996) showed that angiotensin II activated Ras transiently by a mechanism dependent on Src. Evidence for this came from the observation that electroporation of Src antibodies, but not Yes or Fyn antibodies, into adherent cells blocked the ability of angiotensin II to stimulate the activation of Ras. In contrast to classic growth factors such as EGF, the magnitude of angiotensin II–induced Ras activation was less and Ras was more rapidly inactivated. Angiotensin II stimulation in these cells also leads to a Src-mediated tyrosine phosphorylation of $p120^{Ras\text{-}GAP}$ and $p190^{Rho\text{-}GAP}$. Although not directly demonstrated, it is possible that activation of $p120^{Ras\text{-}GAP}$ was responsible for the transient nature of Ras activation, possibly accounting for the fact that angiotensin II has only a weak growth promoting activity in these cells.

Role of Shc. The adaptor protein Shc possesses SH2, SH3, and PH/PTB domains as well as tyrosines that when phosphorylated provide docking sites for other SH2-containing proteins. Via its SH2 and PH/PTB domains it may also interact with specific tyrosine-phosphorylated proteins and with certain membrane-associated phosphatidylinositol phosphates (Ravichandran et al., 1997). It is a mediator of signaling via the thrombin receptor and a pertussis toxin–insensitive G protein, likely G_q. Thrombin activates its receptor by proteolytic cleavage of the external N-terminal domain of the GPCR. Following phosphorylation on tyrosine residues in thrombin-treated CCL39 hamster lung fibroblasts, possibly by an activated Src, Shc interacts with Grb2, thereby attracting Grb2/SOS complexes to the plasma membrane, where they can activate Ras (Y.-H. Chen et al., 1996). Proliferation of CCL39 cells in response to thrombin required, in addition to Ras activation of ERK, a signal from a pertussis toxin–sensitive G protein, either G_i or G_o, perhaps involving the protein tyrosine phosphatase SHP-2 (PTP1D). Rivard et al. (1995) showed that in the CCL39 cells SHP-2 was a positive mediator of mitogenic signals from both RTKs and GPCRs, likely because when tyrosine-phosphorylated it provides a binding site for Grb2/SOS.

Role of One-Pass RTKs. Several studies have implicated the one-pass RTKs as intermediates in the transmission of a signal from GPCRs to the ERK pathway. Sugden and Clerk (1997) have reviewed mechanisms by which the ERK subgroup of MAP kinases could be coupled to activation of the GPCRs. In addition to $\beta\gamma$-mediated activation, via P13-K and Src, of Shc/Grb2/Ras, these include the involvement of non-RTKs, various conventional and novel protein kinase C isotypes that can be activated by PLC-generated DAG, and possibly an RTK (see Fig. 6.8). This last suggestion is based on the fact that the EGFR is rapidly phosphorylated and activated when rat-1 fibroblasts are stimulated with endothelin (ET-1), thrombin, or LPA—all ligands for G protein–coupled receptors (Daub et al., 1996). Receptor phosphorylation leads to its association with Shc and Grb2 and results in ERK activation. Inhibition of EGFR signaling attenuated ERK activation by ET-1, thrombin, and LPA, suggesting that transactivation of EGFR by an unknown mechanism, possibly phosphorylation by Src, mediated GPCR activation of the ERK pathway.

Signaling via the insulin receptor is unique in several respects. As the receptor is already dimeric, ligand-induced conformational changes rather than receptor dimerization is the mechanism of signal transduction. Secondly, the major route of transmission of the signal is via insulin receptor substrate (IRS; two are known) molecules. These molecules become heavily phosphorylated upon activation of the insulin receptor and activate several signal transduction pathways in the cell, including those involving Grb2, PI3-K, and SHP-2 (Sun et al., 1995). Luttrell et al. (1995) reported that in Rat-1 cells the related heterotetrameric insulin and insulin-like growth factor I (IGF-I) receptors, which possess intrinsic ligand-stimulated protein tyrosine kinase activity, stimulated ERK phosphorylation via a mechanism that required $G_{\beta\gamma}$ and could be inhibited by pertussis toxin. One interpretation of this observation is that effective stimulation of ERK phosphorylation by insulin (or IGF-I) requires input from a GPCR, specifically a G_i-coupled receptor, for example, that for LPA. In support of this, ERK activation via the LPA receptor could be inhibited by inhibitors of tyrosine protein kinases. Apparently, mitogenic signals from these receptors synergize to provide effective Ras activation.

Roles of PKA, PKC, and PLC. Depending on the cell, there are a number of different pathways that the GPCRs can activate, and a given receptor may interact with more than one G protein subfamily, depending on the state of the cell and its lineage. PKC is generally thought to contribute to Raf activation, whereas PKA is inhibitory. Lefkowitz and his colleagues have explored in model systems the various pathways by which GPCRs may activate ERK in Ras-dependent and Ras-independent PKC-dependent pathways (Van Biesen et al., 1996; Luttrell et al., 1997; Daaka et al., 1997).

Figure 6.8 presents an attempt at a synthesis of these and other reports; the figure is flawed, however, in that it is impossible to illustrate the variety of different interactions that may occur depending on the specific receptor and the cell. As shown, ERK may be activated by stimulation of a G_q- (or G_o)-coupled receptor (e.g., mAChR) in a Ras-independent, PKC-dependent manner, possibly requiring α_q-induced PLCβ activation and subsequent phosphorylation of Raf by a PKC isoform (van Biesen et al., 1996). PLC is presumed to activate PKC isoforms via the DAG and inositol trisphosphate (IP_3) produced from phosphatidylinositol (4,5)bisphosphate (PIP_2). (IP_3 mobilizes Ca^{2+} from intracellular stores.) Various PLCβ isoforms are activated by different α and $\beta\gamma$ subunits (Dippel et al., 1996). Dippel et al. (1996) used antisense oligonucleotides directed against specific mRNAs to show that the M1 receptor activated PLCβ via α_q, α_{11}, β_1, β_4, and γ_4. ERK may also be activated in a Ras-dependent fashion by G_i-coupled receptors via $\beta\gamma$ and Shc (Crespo et al., 1994; Lopez-Ilasaca et al., 1997).

Phosphorylation of the β_2-adrenergic receptor (β_2AR) by PKA uncouples it from G_s and couples it to the pertussis toxin–sensitive G_i protein that activates Src and Ras in HEK293 cells (Daaka et al., 1997). Because stimulation of G_s typically activates adenylate cyclase, which via cyclicAMP causes PKA activation, PKA-induced phosphorylation of the β_2AR and its subsequent dissociation from G_s leads to a reduction in cAMP production in response to further stimulation. The enhanced coupling of the βAR to G_i gives rise to a feedback inhibition of adenylate cyclase, further suppressing cAMP production, while at the same time stimulating the Ras/ERK pathway via, the authors suggest, $\beta\gamma$ activation of Src and SOS. This may be a mechanism to drastically change the phenotype of the cell. Luttrell et al. (1997) report that in Rat-1 fibroblasts ERK activation correlates with Shc phosphorylation, and not with focal adhesion kinase (FAK) activation. These authors and Daaka et al. (1998) discuss the different mechanisms by which Ras may be activated by GPCRs in different cell types, and they note that inhibitors of receptor internalization block Raf/MEK-mediated activation of ERK1/2 by RTKs and GPCRs. Thus, it appears that, in contrast to activation of adenylate cyclase or PLC, activation of ERK mediated by the Shc/Ras pathway requires vesicle-mediated endocytosis into clathrin-coated pits.

Shapiro et al. (1996) observed that in rat tracheal smooth muscle cells activation of both the ERK and JNK/SAPK pathways accompanied the growth stimulatory action of either thrombin or endothelin, apparently via a Raf-1–independent pathway because Raf-1 was very poorly activated. Activation of either the endothelin or thrombin receptor caused a G protein–dependent stimulation of PLC isoforms, leading to release of DAG and activation of PKC. Perhaps an isoform of PKC can activate MEK. Forskolin, an activator of PKA, inhibited the ligand-induced DNA synthesis and phosphorylation of both ERK and JNK/SAPK, but whether this was the result of inhibition of Raf-l by PKA phosphorylation or some other mechanism, receptor desensitization, for example, is unclear. PKA activation, presumably accompanied by Raf inhibition, did not block PDGF-induced ERK phosphorylation. This may imply some subtle difference between the signaling pathways leading to ERK activation employed by the one-pass RTKs (PDGFR) and the seven-pass GPCRs.

Endothelin, which is a 21-amino acid peptide that is a potent vasoconstrictor, is suspected to play a role in proliferative glomerulonephritis because of its mitogenic action on kidney mesangial cells. Terada et al. (1998) have shown that in these cells the proliferative effect of endothelin was mediated by cyclin D, Rb, and p16. Stimulation of the endothelin receptor resulted in a rapid (2–5 min) activation of Ras and was accompanied by ERK phosphorylation; this was followed by a second activation (at 30–60 min) that stimulated the PI3-K→Akt→$p70^{S6K}$ pathway (Foschi et al., 1997). This biphasic response may be the consequence of two actions of ERK. One is the phosphorylation of SOS, causing it to dissociate from Ras, thereby terminating Ras signaling; the second is the induction of the dual specificity phosphatase MKP-1, which dephosphorylates and inactivates the ERKs. This latter event allows SOS to become dephosphorylated and once again interact with Grb2 and Shc to activate Ras. Details of how the endothelin receptor activates Ras are unclear but may involve a sustained phosphorylation of Shc by Src.

Cytokine Receptors

Cytokines are a diverse group of glycoproteins that includes growth factors, lymphokines, inflammatory mediators, and hematopoietic regulators distinct from the classic endocrine hormones. One strategy for classifying them could be on the basis of the type of receptor they use. Because of the variety of substances classified as cytokines (e.g., interferons, interleukins, colony-stimulating factors, erythropoietin, prolactin, and growth hormone), the cytokine receptors themselves are a heterogeneous but nevertheless functionally related group of membrane proteins. The cytokine receptors consist of from one to three polypeptide chains, each of which usually has a single transmembrane domain (a type I glycoprotein) that generally lacks any kinase activity; exceptions are cKit and CSF-1R, which have intrinsic protein tyrosine kinase domains, and IL-8R, which has a seven-transmembrane motif (Ihle, 1995; Taniguchi, 1995).

Specific cytokines promote the proliferation or survival of cells that express the appropriate repertoire of signaling elements. For example, interleukin (IL)-2, -4, -5, -6, -7, and -15 can induce lymphocyte proliferation (Taniguchi, 1995), and, at least in the case of IL-2 acting on the murine cytokine-dependent T-cell line CT-6, the mitogenic signaling pathway exhibits an uncommon requirement for p38 activation (Crawley et al, 1997). Receptors for some of the cytokines share common subunits, and in many cases there is significant overlap in the signaling elements activated by the different receptors; thus there is considerable redundancy in the signals generated and cell-dependent pleiotropism in the response to various cytokines. Signaling pathways common to these receptors employ various members of the STAT, Ras, and PI3-K families.

STAT Activation. One or more of the receptor proteins typically has a conserved membrane–proximal cytoplasmic sequence (box1/box2 motif) that associates with a JAK (Janus kinase; four are known), and they sort to the various receptors in different combinations. These kinases are activated, as illustrated in Figure 6.9, when the receptor is stimulated by its ligand to heterodimerize and/or to undergo a conformational alteration (Leaman et al., 1996a). Receptor activation results in the (cross)phosphorylation of tyrosine residues on the receptor, on JAK itself, and on associated proteins, notably the STATs. Phosphorylation of the STAT proteins (each interacts with a specific subset of the cytokine receptors) leads to their homo- or heterodimerization via SH2-phosphotyrosine interactions, translocation to the nucleus, and stimulation (usually) of transcription of target genes (Darnell, 1997). For example, STAT1 and STAT3 homo- and heterodimers bind to the SIE (c-sis inducible element) in the *cfos* promoter and thus contribute to *cfos* transcription that can be stimulated by several of the receptors (Leaman et al., 1996b). For some receptor–STAT combinations the STAT protein is associated with the inactive receptor while in other cases it requires the formation of a receptor tyrosine phosphate before the STAT will bind. Figure 6.9 shows a "generic" receptor–JAK–STAT signal transduction pathway; actual signaling pathways are variations on this fundamental theme and include additional signaling moieties. For instance, members of the Src family of nonreceptor protein kinases (Lck, Fyn, Lyn) may associate with the receptor, often at a membrane-proximal site, and contribute to cytokine signaling.

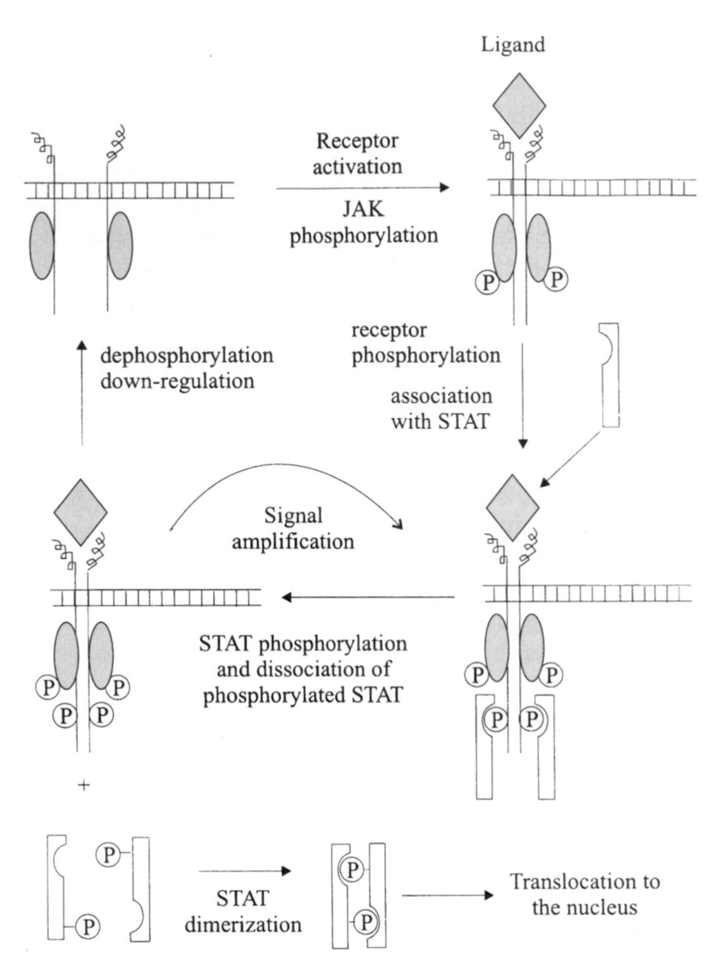

Figure 6.9. Cytokine signaling via Janus kinases (JAKs) and signal transducers and activators of transcription (STATs). Activation of a cytokine receptor by its ligand typically results in the aggregation of one or more types of receptor subunits and activation of specific JAK proteins associated with domains adjacent to the plasma membrane. Receptor aggregation and conformational alterations result in the phosphorylation (by JAK) of tyrosines in the JAK and receptor proteins. STATs and other proteins may bind via SH2 domains to receptor tyrosine phosphates and themselves become phosphorylated. (Some of these proteins may be associated with the receptor prior to activation because of other types of interactions.) Tyrosine phosphorylation of STAT proteins permits them to dissociate from the receptor, to dimerize, and to translocate to the nucleus where they can modify the expression of specific genes. The activated JAK proteins may phosphorylate other proteins besides STATs and the receptor, and individual phosphotyrosines on the receptors may stimulate additional signaling pathways.

Other signaling pathways may positively or negatively influence cytokine signaling. Phosphorylation of specific ser/thr residues in STAT proteins, by a MAPK or a PKC isozyme, for example, may promote STAT activation (David et al., 1995; Mufson, 1997). Chung et al. (1997) showed, on the other hand, that phosphorylation of Ser-727 in STAT3 induced by growth factor–stimulated ERK inhibited the tyrosine phosphorylation essential for dimerization and nuclear translocation of the protein. These authors make the point that these complex signaling pathways utilize differential phosphorylation of specific amino acids to regulate transcription factor activation and that the temporal relationships among the different phosphorylation events is critical to an understanding of the consequences of a specific phosphorylation event. Which STATs are activated by a particular ligand is determined by the receptor, not the associated JAK, and may be a function of the cell lineage also (Han et al., 1996). The specificity is ordained in part by the specific SH2-Y(P) interaction. STATs do not generally contribute strongly to a proliferative response; rather, they appear more concerned with cell survival and differentiation. However, in the mouse STAT5 is necessary for prolactin-dependent proliferation of mammary tissues and IL-2–dependent expansion of peripheral T cells (Ihle, personal communication).

Other receptors (e.g., those for EGF, PDGF, and angiotensin) have also been observed to activate certain JAKs and STATs, but the significance of this for signaling by those ligands is unclear. Leaman et al. (1996b) investigated the role of STAT1 and STAT3 in transducing a signal from the EGF and PDGF receptors. Their conclusions were that both STATs could be independently phosphorylated by the RTK, or a kinase controlled by the receptor, in the absence of any JAK activity. When JAK1 was present it could also be phosphorylated in response to EGF signaling. Although the PDGF receptor was able to phosphorylate all the JAK and STAT proteins tested (three and two, respectively), it did so with relatively low stoichiometry and with no obvious functional significance to the cell aside perhaps from stimulating c-fos expression via the SIE element (Vignais et al., 1996).

IL-2. IL-2, produced by activated T cells, stimulates the IL-2 receptors on those cells to promote cell proliferation. Signaling from the three-chain IL-2 receptor is particularly elaborate (Taniguchi, 1995). IL-2 activates STAT5 via JAK1, and JAK3 and has been reported to activate Shc, Raf-1, ERK, CDK2, CDK1, and PI3-K, all of which may contribute to a proliferative response in appropriately stimulated cells expressing the IL-2 receptor. Other downstream effectors of IL-2 include p27, Myc, c-Fos, and Bcl-2, proteins whose expression can promote cell survival. Beadling et al. (1996) have reported that two forms of STAT5 exist in IL-2–stimulated Kit225 cells that differ in the amount of serine phosphorylation. Serine phosphorylation, which appeared necessary for nuclear localization and transcriptional activity, did not seem to result from ERK or $p70^{S6K}$ activity.

IL-3. Evidence directly implicating STAT5 in the proliferative response induced by IL-3 in the IL-3–dependent pro-B-cell line BA/F3 was obtained by Mui et al. (1996). They showed that a dominant negative STAT5 protein inhibited the expression of several IL-3 early response genes (*cis*, *osm*, *Pim*-1, and to a lesser extent c-*fos*), but not c-*myc*, and partially inhibited IL-3–dependent growth. The α subunit of the receptor provided specificity for the ligand while the β subunit, shared with the IL-5 and GM-CSF receptors, enhanced the binding affinity. The membrane-distal portion of the β subunit was necessary for Ras→Raf→MEK→ERK activation and c-*fos* induction, whereas the membrane-proximal portion, to which JAK2 bound, was necessary for c-*myc*, *pim*-1 and *cis* induction. Activation of Ras and c-*fos* was necessary to promote the survival of the BA/F3 cells, possibly by upregulating *bcl*-2 expression, whereas c-*myc* and cyclin E expression correlated with induction of DNA synthesis (Kinoshita et al., 1995, 1997). Both of these signaling pathways (controlled by Ras and by Myc) appeared necessary for long-term proliferation of the cells. Suppression of apoptosis could be mediated either through a Raf/ERK pathway or through a rapamycin/wortmannin-sensitive pathway. In this research, mutants of Ras (Ras^{G12V}, $Ras^{G12V/T35S}$, and $Ras^{G12V/V45E}$) were utilized to distinguish individual Ras-controlled signaling pathways (Kinoshita et al., 1997). Sensitivity to rapamycin and wortmannin suggested the involvement of PI3-K and the Akt→$p70^{S6K}$ pathway in mediating the Raf-independent pathway.

In the BA/F3 cells, two different Ras-mediated signals, one transmitted through Raf→MEK→ERK and one transmitted via PI3-K and $p70^{S6K}$, were each independently capable of inhibiting the apoptotic response that occurred in the absence of IL-3 (Kinoshita et al., 1997). This is an example of two parallel pathways that when simultaneously activated serve to reinforce each other. Similarly, although stimulation of p38 has generally been associated in most cell types with a stress-induced inhibition of proliferation, Crawley et al. (1997) have shown in the murine T-cell line CT6 that the proliferative response induced by IL-2 or IL-7 can be inhibited by the pyrinidyl imidazole inhibitor SB203580, which is considered to be a specific inhibitor of the p38 MAP kinase. Thus in T cells p38 appears to be involved in transducing a mitogenic signal. Although ERK and JNK/SAPK are also activated, it appears that at least the former is not essential for T-cell proliferation.

Growth Hormone. Activation of the growth hormone receptor (GHR) in 3T3-F344A fibroblasts by growth hormone results in phosphorylation of both the receptor and JAK2, generating one or more tyrosine phosphates to which Shc can bind via its SH2 or PTB domains. This leads to phosphorylation of Shc and subsequent association with Grb2, presumably complexed with SOS (VanderKuur et al., 1995). Wang et al. (1995) showed that mutant GHR lacking most of its cytoplasmic domain (not box1 or box2, however) and all phosphorylatable tyrosines was still able to support JAK2 phosphorylation, STAT5 activation, and cell proliferation. These results suggest that JAK2 may alone be able to bind and phosphorylate Shc. Activation of ERK1,2 via the Ras→Raf→MEK→ERK cascade could occur when phosphorylated Shc interacts with Grb2/SOS, attracting it to the cell membrane where it can activate Ras. However, a binding site for Shc on JAK2 has not been demonstrated.

An alternative mechanism for ERK activation by growth hormone was suggested by Yamauchi et al. (1997). In their study, activation of the GHR stimulated JAK2 to phosphorylate the EGFR. Phosphorylation of the EGFR by JAK2 generates a tyrosine phosphate that serves as a docking site for Grb2 and leads to activation of the Grb2/SOS/Ras/Raf/MEK/ERK pathway. In this case, the kinase activity of the receptor was not required. These results indicate that receptors devoid of tyrosine kinase activity can stimulate a non-RTK (i.e., JAK2) to phosphorylate a RTK, with the result that signaling pathways characteristic of the RTK are stimulated independently of its own tyrosine kinase activity. The precise mix of pathways activated by the phosphorylated EGFR may differ, however, because JAK2 and the EGFR kinase may preferentially phosphorylate different tyrosines. IL-6 and γ-interferon, also activators of JAK2, were for reasons unknown unable to stimulate tyrosine phosphorylation of the EGFR.

Erythropoietin. Erythropoietin (Epo) regulates the proliferation and differentiation of cells of the erythroid lineage via its interaction with a single chain receptor, EpoR, which upon Epo-induced dimerization stimulates protein tyrosine phosphorylation by the protein tyrosine kinase receptor cKit and the cytosolic protein kinase JAK2 (Klingmüller, 1997). Both the receptors and associated proteins become phosphorylated, and multiple and redundant signaling pathways are stimulated. Recruitment of PI3-K to the Epo receptor and activation of ERKs appear to play a key role. Signals from cKit, induced by its ligand, stem cell factor, and from the EpoR are required for the proliferation of red blood cell precursors and their differentiation into mature erythrocytes. JAK2 has been implicated as an essential component of the mitogenic signaling pathway stimulated by Epo and also IL-3 (Zhuang et al., 1994). Although the proliferative response correlated well with the phosphorylation of Tyr-343 in the Epo receptor and the resultant activation of STAT5, receptors lacking any phosphorylatable tyrosines could still induce a mitogenic response, albeit less efficiently (Damen et al., 1995).

Quelle et al. (1996) found that both the membrane-distal portion of the Epo receptor (the C-terminal region) and STAT5 activation were not required for a mitogenic response in the DA-3 mouse myeloid line. What was necessary was the membrane-proximal portion, which was also necessary for JAK2 activation. This is evidence that a JAK2-dependent pathway, perhaps involving induction of c-*myc* transcription, was critical for the mitogenic response. In the cells in this study, those signaling elements (including PI3-K, PLCγ1, and Shc) activated by the membrane-distal portion of the receptor did not seem to be required for the mitogenic signal to be implemented. If Shc is not activated by this mutant receptor, then these results suggest that there must be another signaling pathway leading to mitogenesis, for example, via activation of a Src family kinase or a PKC isozyme. Mason-Garcia et al. (1995) reported that both erythropoietin and IL-3 could induce nuclear localization of PKCβ, where it presumably phosphorylates nuclear proteins and transcription factors. PKCβ, a conventional PKC that is activated by DAG phosphatidylserine, and Ca^{2+}, is also capable of activating Raf-1 and stimulating the proliferation of hematopoietic cells (Carroll and May, 1994).

Granulocyte Colony–Stimulating Factor. Upon activation, the granulocyte colony–stimulating (G-CSF) receptor, which resembles the IL-6 and Epo receptors, stimulates the tyrosine kinases Syk, the Src family kinase Lyn, and JAK. It is responsible for mediating G-CSF stimulation of the proliferation and differentiation of granulocyte precursors. Corey et al. (1998) have presented evidence that in avian B cells engineered to express the G-CSF receptor it is Lyn, and not Syk, JAK1, or JAK2, that is responsible for transmitting the mitogenic signal. PI3-K activity, tyrosine phosphorylation of Shc, and activation of the ERKs was defective in Lyn-deficient cells, suggesting that they might be the mediators of the G-CSF mitogenic signal.

Nuclear Hormone Receptors

This is a group of intracellular, sometimes intranuclear, receptors that bind small lipid-soluble ligands, which include glucocorticoids, retinoic acid, thyroid hormone and 1,25-dihydroxyvitamin D_3. These receptors are not generally looked upon as activators of a signaling pathway; rather, they are themselves DNA-binding proteins that recognize specific sequences in the promoters of target genes (Beato et al., 1995). Because of their hydrophobic character, most of the ligands are typically delivered to the cell in the custody of a specific binding protein. Once the ligand enters the cell it associates with its cognate receptor, which in the case of the glucocorticoid receptor subfamily is usually associated with one or more heat shock proteins. Conformational changes induced in the receptor abrogate interactions with the heat shock proteins, permitting the receptor to dimerize and to enter the nucleus. In the case of the thyroid hormone receptor and the retinoid receptor subfamilies, the receptors are already in the nucleus, often bound to the DNA even in the absence of ligand.

These transcription factors bind to specific sequences in the promoters of target genes, often acting as transactivating factors. However, they can also inhibit transcription either by interfering with the binding of other factors or by interacting negatively with other transcription factors (squelching) (Clark and Docherty, 1993). Clarke and Docherty (1993) have discussed mechanisms underlying the extensive amount of crosstalk between regulatory pathways controlling proliferation via AP1 (Fos, Jun) expression and differentiation via glucocorticoid receptors. For example, nuclear hormone receptors often interfere with transcriptional activation by AP1 complexes.

Many of these nuclear hormone receptors regulate differentiation, development, and many aspects of organ physiology. For example, retinoic acid and the retinoic acid receptors activate homeobox genes along the anterior-posterior axis during development (Gudas, 1994). Cell proliferation results only in those particular circumstances where the changes in gene transcription are sufficient to trigger a mitogenic response, for instance, by progesterone-induced upregulation of cyclin D1 expression in mammary gland tissue (Said et al., 1997). Some of the receptors, the glucocorticoid receptors, for example, possess amino acid sequences that are sufficiently similar to consensus sequences for the proline-directed CDKs that they are subject to cell cycle–dependent phosphorylation. The position of the cell in the cell cycle determines whether the receptors will be phosphorylated and subject to activation in response to hormone treatment. The response is generally maximal in cells in late G_1 and S because at this stage of the cell cycle the glucocorticoid receptor is minimally phosphorylated (Bodwell et al., 1996).

Much effort has gone toward understanding how estrogen promotes the proliferation of susceptible cells, and it is well established that increased activity of cyclin D1 is implicated. Increased cyclin D1/CDK(4,6) levels appear to activate cyclin E/CDK2 by alleviating p21-mediated inhibition, possibly because the p21 switches from binding cyclin E/CDK2 to binding cyclin D1/CDK(4,6), thereby allowing phosphorylation of Rb (by cyclin E/CDK2) sufficient to force MCF7 cells into S phase (Planas-Silva and Weinberg, 1997; Prall et al., 1997). The estrogen receptor can be phosphorylated and activated by cyclin A/CDK2, at least in cell lines expressing low levels of p27 (Trowbridge et al., 1997). In other cell lines cyclin D1 can activate the estrogen receptor in the absence of estrogen or a CDK by binding directly to the receptor and stimulating the transcription of estrogen-responsive genes (Zwijsen et al., 1997; Neuman et al., 1997). It appears that in some circumstances a positive feedback loop becomes established whereby estrogen via its receptor stimulates certain cyclin/CDKs, which in turn stimulate the estrogen receptor.

Migliaccio et al. (1996) studied in detail the ability of estradiol to activate signaling pathways in the estradiol-dependent human mammary carcinoma cells MCF-7. Their data indicated that estradiol, acting through the estradiol receptor, could stimulate phosphorylation of ERK, likely via the Ras signaling pathway. Shc and p190, a p21RAS^{GAP}-associated protein, became tyrosine phosphorylated as the result of estradiol-induced Src activation. The authors suggested that some of the estradiol receptors, which are largely nuclear, may be associated with the plasma membrane and that they accounted for the activation of Src at the cell membrane.

The peroxisome proliferator–activated receptor γ (PPARγ), which is an adipose-selective nuclear hormone receptor whose natural ligand appears to be a prostaglandin, induces withdrawal of cells from the cell cycle and promotes adipogenesis (Altiok et al., 1997). Increased phosphorylation and decreased DNA-binding activity of the E2F/DP transcription factors accompanied activation of PPARγ, apparently the consequence of decreased expression of the catalytic subunit of the serine/threonine phosphatase PP2A. This is consistent with the previously established ability of inhibitors of PP2A to inhibit cell cycle progression.

Integrins

Some red and white cell lineages do not require adhesion to the extracellular matrix to proliferate. Indeed, for these cells, matrix attachment may suppress replication and instead facilitate migration or differentiation. In contrast, untransformed endothelial, epithelial, and fibroblastic cells are dependent on anchorage to an extracellular matrix/basement membrane if they are to proliferate, or in many cases even survive. This is due to the fact that cell surface attachment factors convey signals to the cell that are essential for particular responses. These signals are transmitted via a family of receptors known as *integrins.* These molecules are unique by virtue of their ability to regulate, in response to internal signals, a cell's ability to attach to or detach from substrate, for example, during cell migration or mitosis. This is accomplished by changes in ligand *affinity,* resulting from conformational changes in the integrin, or ligand *avidity,* resulting from integrin clustering (Stewart and Hogg, 1996; Shattil and Ginsberg, 1997). The various integrins are involved in different kinds of interactions with extracellular macromolecules, and only some of these interactions can be touched on in this brief overview.

Role in Cytoskeletal Organization. As illustrated in the drawing in Figure 6.10, integrins are $\alpha\beta$-heterodimeric proteins, each subunit of which has a single transmembrane segment (Dedhar and Hannigan, 1996; Burridge and Chrzanowska-Wodnicka, 1996; Humphries, 1996; Giancotti, 1997). There are at least 16 α subunits and 8 β subunits, giving rise to about 22 characterized heterodimeric integrins. The extracellular domain of the α subunits contains several EF hand elements that bind Ca^{2+} with high affinity; the β subunit contains numerous cysteine disulfides. Together the two subunits recognize ligands with various motifs, the best known of which is the GRGDS sequence. The intracellular domains have no known enzymatic activity. Integrin-mediated signals are therefore conveyed by ligand-induced clustering of the integrins and associated molecules, accompanied by conformational changes of the interacting molecules. Attachment of cells to the extracellular matrix allows the integrins (particularly α_v and β_1 integrins) to become clustered on the cell surface into structures called *focal complexes* or *focal adhesions.* The clustering itself occurs in response to an increase in cell contractility and requires Rho/Rac $p21^{GTPase}$ activity (Hotchin and Hall, 1995; Chrzanowska-Wodnicka and Burridge, 1996).

Different integrins participate in various types of interactions at the cell surface. For example, $\alpha_6\beta_4$ is involved in hemidesmosome assembly,

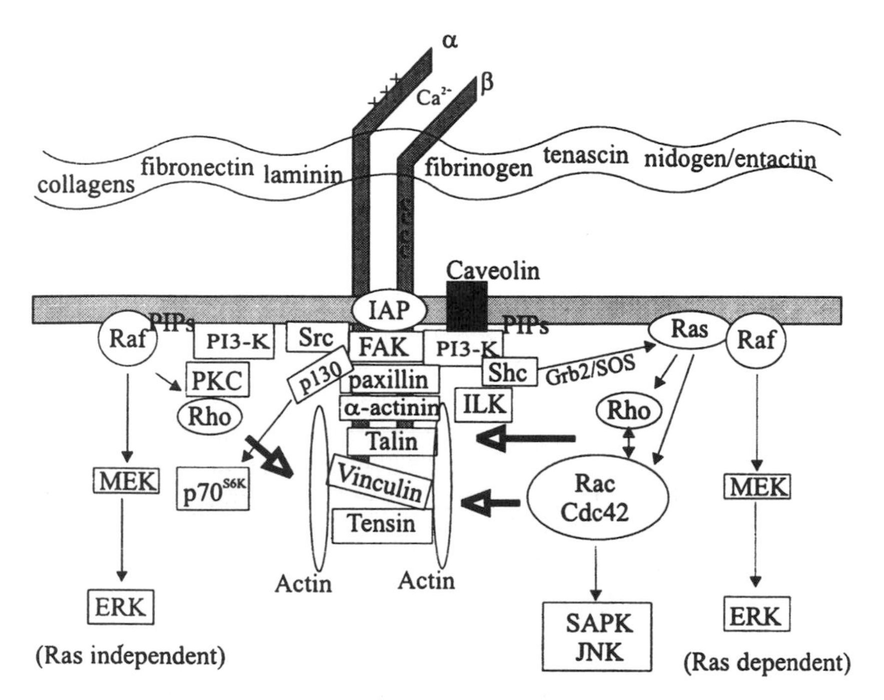

Figure 6.10. Integrin-mediated signaling. The large extracellular domains of the integrins interact with ligands such as fibronectin in the extracellular matrix. The smaller intracellular C-terminal domains interact with a variety of protein kinases and structural proteins. Essential points include the following: Certain proteins (FAK, paxillin, α-actinin, talin, and vinculin) are rather directly involved in interactions with the integrins, and α-actinin and talin along with vinculin and tensin mediate structural interactions with actin and the cytoskeleton. FAK (focal adhesion kinase) and Src are protein tyrosine kinases that phosphorylate various target proteins and thereby generate new binding sites for proteins containing SH2 or PTB domains. Activation of PI3-K can create docking sites at the plasma membrane for proteins able to interact with 3-phosphorylated phosphatidylinositol moieties. $p130^{CAS}$ (Crk-associated substrate) is an adaptor protein that interacts with FAK, becomes tyrosine phosphorylated, and localizes to focal adhesions. IAP is an integrin-associated transmembrane protein that may cooperate with certain integrins to bind various ligands and stimulate signaling pathways involving $G_{\alpha\beta\gamma}$ proteins. PKC, ILK (integrin linked kinase) and $p70^{S6K}$ are three ser/thr protein kinases also implicated in signaling at focal adhesions. Caveolin, which is held in the membrane by two transmembrane domains, is constitutively associated with β_1 integrins and mediates localization of Shc to the membrane. Activation of Ras-dependent and Ras-independent pathways utilizes mechanisms discussed earlier in this chapter.

and upon activation by laminins the β_4 subunit becomes tyrosine phosphorylated (by FAK or Src?) and recruits Shc and Grb2 to the complex (Mainiero et al., 1995). The $\alpha_3\beta_1$ integrins are found in focal complexes at the cell periphery rather than in focal adhesions, suggesting a role in cell motility, and they interact with the tetraspan membrane proteins CD63 and CD81, which are associated with phosphatidylinositol 4-kinase (Berditchevski et al., 1997).

In an impressive series of studies Hall and his associates have investigated how the Rho, Rac, and Cdc42 proteins control actin assembly and organization (Hall, 1994). Various types of focal adhesion complexes are induced when fibroblasts are microinjected with activated forms of these $p21^{GTPase}$ proteins. Cdc42 induces the formation of microspikes (filopodia); actin polymerization and lamellipodial protrusions are induced by Rac; and stress fiber assembly and integrin clustering at focal adhesions are induced by Rho (Nobes and Hall, 1995; Machesky and Hall, 1997). Some known or suspected interactions among these $p21^{GTPase}$ proteins are diagrammed in Figure 6.5.

On the cytoplasmic side of the focal adhesions, the association of cytoskeletal proteins such as vinculin, paxillin, α-actinin, and talin with the cytoplasmic domains of the integrins forms the site of attachment of actin stress fibers. Included among the proteins at the focal adhesion are the focal adhesion kinase (FAK), Src, and Csk. Autophosphorylation of FAK, associated with the β integrin subunit, is induced as a consequence of ligand-induced integrin clustering. A resulting tyrosine phosphate (Tyr-397 in FAK) binds and activates Src, stimulating it to phosphorylate target proteins. Effective interaction of FAK with Src-like kinases may also involve the Src SH3 domain (Thomas et al., 1998). The consequent tyrosine phosphorylation of proteins (notably FAK, paxillin, tensin, and $p130^{CAS}$) promotes their association with other proteins (e.g., Grb2/SOS, PI3-K) and the formation of a signaling complex (Miyamoto et al., 1995; Vuori et al., 1996; Parsons, 1996; Richardson et al., 1997). $p130^{CAS}$ in a complex with FAK and Src serves as a downstream mediator of FAK-promoted cell migration (Cary et al., 1998).

The recruiting of cytosolic proteins into focal adhesions as the result of ligand-induced integrin clustering is known as *outside→inside* signaling. The converse, *inside→outside* signaling, results from modifications to proteins associated with the intracellular domains of the integrins, or the integrins themselves. This may be induced by a signaling pathway stimulated by another receptor, the thrombin receptor, for example, and can influence the ligand-binding ability of the integrin. It is important, for example, in engineering the escape of monocytes and lymphocytes from the vascular system (Luscinskas and Lawler, 1994; Shattil and Ginsberg, 1997). Kolanus and Seed (1997) have reviewed the evidence that PKC and PI3-K have a major role in this process, for example, by subverting inhibitory mechanisms that maintain integrins in a dormant state. PKC can, for example, control FAK phosphorylation and the localization of cytoskeletal proteins induced by the engagement of the $\alpha_v\beta_5$ or $\alpha_{IIb}\beta_3$ integrin in hamster CS-1 melanoma cells or platelets, respectively (Lewis et al., 1996; Haimovich et al., 1996). Ras/Raf–mediated activation of ERK appears to suppress integrin activation, thus generating a negative feedback loop in situations where integrin activation leads to ERK activation (Hughes et al., 1997).

Burridge and Chrzanowska-Wodnicka (1996) have emphasized behavioral parallels between fibroblasts growing on a plastic surface in serum-containing culture medium and in a healing wound. Wound healing requires the proliferation of cells, induced by growth factors, followed by matrix attachment and contraction to close the wound. Fibroblasts in culture mimic this behavior by forming focal adhesions and contractile stress fibers in response to compounds such as PDGF, LPA, thrombin, and endothelin that activate the signaling pathways discussed above.

Role in the Cell Cycle. The inability of adhesion-dependent cells to progress through the cell cycle when they are not attached to an extracellular matrix is the consequence, at least in part, of an increase in the abundance of the p21 and/or p27 CDK inhibitors relative to the levels of cyclin/CDK complexes. Untransformed but immortal human fibroblasts (IMR90, KD) maintained in suspension in 1% methocel exhibit much reduced cyclin E/CDK2 activity relative to attached controls. This was shown to be due both to an increase (about fourfold) in the expression of p21 and p27 and to a decrease in phosphorylation (see Fig. 6.3) of the activating Thr-160 in CDK2 (Fang et al., 1996). Because cyclin D/CDK(4,6) and the CDK-activating kinase (CAK) activities were similar in both adherent and suspended cells, Fang et al. (1996) suggested that cyclin A expression was attenuated as a consequence of diminished cyclin E/CDK2 activity. In normal rat kidney (NRK) cells it was primarily cyclin A expression, and not cyclin D1– and cyclin E–dependent kinase activities, that was anchorage dependent (Guadagno et al., 1993; Fang et al., 1996).

Somewhat different results were obtained by Zhu et al. (1996), possibly because of differences in the methodology or cells (primary human fibroblasts and rodent cells rather than the immortal human cell lines used by Fang et al. [1996]). They observed that in the mortal human fibroblasts and immortal NIH3T3 mouse fibroblasts both cyclin D1 expression and cyclin E/CDK2 activity, and hence Rb hyperphosphorylation, were dependent on cell attachment. In the nonadherent cells, synthesis and translation of the cyclin D1 mRNA in the presence of mitogens was reduced in comparison with adherent cells. Also, cyclin E/CDK2 activity was inhibited by an increase in the ratio of p21 to cyclin E/CDK2, likely because of elevated p21 mRNA levels in the unattached cells. Regardless of cell-specific differences in detail, the important conclusion from these and other studies is that successful transit of adhesion-dependent cells through the cell cycle requires that multiple signaling pathways be activated both by essential soluble mitogens and by motifs in the insoluble extracellular matrix.

In Rat-6 embryo fibroblasts, the anchorage dependence of Rb hyperphosphorylation, cyclin E/CDK2 activity, and cyclin A expression can be circumvented by expression of v-H-Ras (Kang and Krauss, 1996). However, when oncogenic Ras was expressed in a variant (ER1–2) of these cells, only Rb hyperphosphorylation and cyclin E/CDK2 activity, but not the expression of cyclin A, were observed to be adhesion independent. This mutant was defective in *ras*-oncogene–mediated anchorage-independent growth, demonstrating that the adhesion dependence of cyclin A expression can be dissociated from the other two cell cycle–

regulated events, presumably because of a defect in one of the signaling pathways controlled by Ras.

Using NIH3T3 cells, Renshaw et al. (1997) investigated the importance of adhesion in growth factor activation of ERK2. They observed that ERK2 activation by (soluble) growth factors was dependent on adhesion; in the nonadherent cells suspended in agarose/methylcellulose activation of Ras and Raf, but not MEK or ERK2, occurred. These results suggest that for Raf to activate MEK an additional signal provided by the extracellular matrix is required, at least in these cells. In a different approach, Miyamoto et al. (1996) identified a variety of synergistic interactions between various growth factors and integrin ligation or aggregation. For example, EGF could transiently activate ERK in fibroblasts only when the cells' α_5 and β_1 integrins were engaged and clustered. Interestingly, ligand-induced integrin clustering caused various other growth factor receptors to aggregate transiently.

Radeva et al. (1997) have described an integrin-linked serine/threonine kinase (ILK) that when overexpressed could enhance the activities of cyclin D1/CDK4 and cyclin E/CDK2 and induce adhesion-independent proliferation of rat intestinal epithelial cells. Ras was not activated by ILK, however, and the cells remained dependent on serum stimulation of necessary signaling pathways to proliferate. Kawada et al. (1997) investigated the role of Ras in promoting anchorage-independent growth of rat embryo fibroblasts. In untransformed adherent fibroblasts p27 accumulated in the absence of serum and contributed to maintaining the cells in a quiescent state. p27 also accumulated in the presence of serum if the cells were unable to adhere (Fang et al., 1996; Zhu et al., 1996). Loss of substrate attachment resulted in an increase in the level of p27 mRNA in untransformed cells, whereas in transformed cells expressing oncogenic Ras there was an increase in the rate of degradation of the p27 protein. The evidence suggested that phosphorylation of p27 by the activated Raf/MEK/MAPK pathway inhibited its binding to CDK2 and accelerated its degradation (Kawada et al., 1997).

ERK, and in some cases JNK, activation in response to the activation of certain integrins by ECM proteins may be caused by the recruitment of the adaptor protein Shc to tyrosine-phosphorylated components in the focal adhesions followed by relocation of Grb2/SOS to the cell membrane and activation of Ras (Chen et al., 1994; Wary et al., 1996; Mainiero et al., 1997). However, Q. Chen et al. (1996) observed that dominant negative inhibitors of Ras failed to block ERK activation in response to integrin engagement. This suggests that there is an additional pathway leading to activation of MEK and ERK that does not require Ras activation. Such a pathway may involve phosphatidylinositol derivatives. Integrin-mediated activation of PI3-K, which increases the levels of $PI(3,4)P_2$ and $PI(3,4,5)P_3$, may activate Raf-1 directly (King et al., 1997). ERK activation is also independent of tyrosine phosphorylation of FAK (Lin et al., 1997).

Caveoli are small, pear-shaped invaginations of the plasma membrane that are found in most cell types. They contain various subtypes of caveolin, a two-pass transmembrane adaptor, as a defining feature. In an interesting study Wary et al. (1996) obtained evidence that the extracellular but membrane-proximal portion of the α_v, α_1, or α_5 subunit interacted with caveolin, apparently causing it to recruit Shc to the plasma membrane. Engagement of integrins linked to Shc facilitated growth factor–induced cell proliferation. In contrast, activation of $\alpha_2\beta_1$ by laminin did not activate Shc and instead caused the cell to exit the cell cycle and differentiate. Caveolin can also interact with a number of signaling molecules, including $G_{\alpha\beta\gamma}$-coupled receptors, Src, Ras, and the EGFR, whose ability to autophosphorylate is inhibited (Couet et al., 1997). Autoactivation of Src is inhibited by caveolin (S. Li et al., 1996).

In a study of the response of colon cancer cells to adherence to the ECM protein tenascin, which is recognized by a number of different integrins, Yokosaki et al. (1996) observed that the proliferative and spreading responses were always associated with the phosphorylation of FAK, paxillin, and ERK, whereas cell attachment alone was not associated with phosphorylation of any of these proteins. When serum-starved NIH3T3 cells were plated on fibronectin the cells spread, formed focal contacts, and activated the Ras/ERK signaling pathway. ERK activation, but not the cytoskeletal changes, could be inhibited by the dominant negative Ras^{N17}, indicating that under these conditions Ras activation was not required for the formation of focal contacts and stress fibers (Clark and Hynes, 1996). These results, and others reviewed by Lafrenie and Yamada (1996), led to the speculation that the complexity of the interactions occurring in the various types of focal clusterings induced by integrin aggregation (illustrated in Fig. 6.10) could be simplified if it were broken down into a cytoskeletal complex, involving interactions among cytoskeletal proteins, and a signaling complex, involving various kinases and GTPases. However, certain proteins (e.g., FAK, paxillin) are involved in both complexes, and they must still talk to each other. Schwartz (1997) has commented on the fact that integrins regulate many of the same pathways as growth factors in both synergistic and complementary ways and that signals from both the extracellular matrix and from soluble factors are often required for a proliferative response.

Section Summary

Different cell types present different combinations of receptors at their surface. When these receptors are activated by the appropriate ligands, the cell may initiate a proliferative response. (Other types of responses—differentiation, migration, or apoptosis—are alternatives.) In all cases the cell's response is determined by how it integrates the upstream signals with endogenous metabolic processes. Whether a particular ligand delivers a mitogenic signal depends on the particular cell, its lineage, and its physiological state (e.g., adherent or nonadherent). A given ligand may stimulate metabolic processes (e.g., insulin), promote differentiation (e.g., CSF-1), or inhibit cell death (e.g., neurotrophins). The same ligand may elicit different responses in different cell types or in the same cell under different conditions. How the cell responds to activation of a particular receptor, which typically stimulates several signaling pathways as a function of the specific motifs in the receptor and associated proteins that become phosphorylated, is determined by the intensity (the number of activated receptors) and the duration (a function of the specific receptor) of the signal and how it is interpreted by the cell.

CONCLUDING OVERVIEW

Regulation of the cell cycle is as complicated as life itself. From a primitive single-cell organism that replicated when conditions were favorable and did not replicate when conditions were not favorable, a complex multicellular organism has evolved. For this to happen multiple levels of controls were developed to allow cell proliferation in different tissues to respond to different signals. This was accomplished in part by gene duplication and divergence, creating families of controlling molecules that could be regulated by divergent stimuli. Safeguards (checkpoints) were also established to coordinate various cellular activities, and "fail-safe" backup systems (tumor suppressors) created to eliminate cells with abnormal or disrupted signaling. Prominent mechanisms of control include regulation of gene transcription and protein stability, both controlled by cycles of regulated phosphorylation and dephosphorylation of tyrosine, serine, and threonine residues in key proteins.

Central to the process of cell cycle control in higher eukaryotes is the phosphorylation of Rb and E2F by cyclin D/CDK (4,6) under the control of the $p21^{GTPase}$ family (notably Ras) and the cyclin-dependent kinase inhibitors of the p16 and p21 families. Although the primary site of regulation is at the restriction point, there are other sites of regulation also, notably at mitosis. In a very cell-specific manner, input from various ligand–receptor interactions is filtered through a signaling network that converges on the control elements that determine whether the cell will proliferate, differentiate, remain quiescent, or die. With modifications of detail but not of substance, the "unique regulatory mechanism" proposed by Pardee (1974) to control the proliferative state of the eukaryotic cell has in many respects been elucidated by the past generation of researchers.

ACKNOWLEDGMENTS

Research support from the National Institutes of Health, the Cancer Research Foundation, and the Charles and Joanna Busch endowment of Rutgers University is very much appreciated. Sincere thanks to Sid Auerbach, Herbert Geller, Lorraine Gudas, Beatrice Haimovich, Jim Ihle, Jerry Langer, John Lenard, Sid Pestka, and Leslie Petch for commenting on all or part of the manuscript and attempting to keep me honest. I acknowledge with pleasure the proficiency of Kathy Curtis and

Tracy Miller in producing the illustrations. Finally, my apologies to authors whose publications have not been cited due to space constraints and my own shortcomings.

REFERENCES

Adler V, Pincus MR, Minamoto T, Fuchs SY, Bluth MJ, Brandt-Rauf PW, Freidman FK, Robinson RC, Chen JM, Wang XW, Harris CC, Ronai Z (1997): Conformation-dependent phosphorylation of p53. Proc Natl Acad Sci USA 94:1686–1691.

Aktas H, Cai H, Cooper GM (1997): Ras links growth factor signaling to the cell cycle machinery via regulation of cyclin D1 and the Cdk inhibitor $p27^{KIP1}$. Mol Cell Biol 17:3850–3857.

Albanese C, Johnson J, Watanabe G, Eklund N, Vu D, Arnold A, Pestell RG (1995): Transforming $p21^{ras}$ mutants and c-Ets-2 activate the cyclin D1 promoter through distinguishable regions. J Biol Chem 270:23589–23597.

Altiok S, Xu M, Spiegelman BM (1997): PPARγ induces cell cycle withdrawal: Inhibition of E2F/DP DNA-binding activity via down-regulation of PP2A. Genes Dev 11:1987–1998.

Assoian RK (1997): Control of the G_1 phase cyclin-dependent kinases by mitogenic growth factors and the extracellular matrix. Cytokine Growth Factor Rev 8:165–170.

Avantaggiati ML, Ogryzko V, Gardner K, Giordano A, Levine AS, Kelly K (1997): Recruitment of p300/CBP in p53-dependent signal pathways. Cell 89:1175–1184.

Baker TA, Bell SP (1998): Polymerases and the replisome: Machines within machines. Cell 92:295–305.

Baldin V, Cans C, Knibiehler M, Ducommun B (1997): Phosphorylation of human Cdc25B phosphatase by CDK1–cyclin A triggers its proteasome-dependent degradation. J Biol Chem 272:32731–32734.

Barone MV, Courtneidge SA (1995): Myc but not Fos rescue of PDGF signaling block caused by kinase-inactive Src. Nature 378:509–511.

Beadling C, Ng J, Babbage JW, Cantrell DA (1996): Interleukin-2 activation of STAT5 requires the convergent action of tyrosine kinases and a serine/threonine kinase pathway distinct from the Raf1/ERK2 MAP kinase pathway. EMBO J 15:1902–1913.

Beato M, Herrlich P, Shutz G (1995): Steroid hormone receptors: Many actors in search of a plot. Cell 83:851–857.

Bello-Fernandez C, Packham G, Cleveland JL (1993): The ornithine decarboxylase gene is a transcriptional target of c-Myc. Proc Natl Acad Sci USA 90:7804–7808.

Berditchevski F, Tolias KF, Wong K, Carpenter CL, Hemler ME (1997): A novel link between integrins, transmembrane-4 superfamily proteins (CD63 and CD81), and phosphatidylinositol 4-kinase. J Biol Chem 272:2595–2598.

Berman DM, Gilman AG (1998): Mammalian RGS proteins: Barbarians at the gate. J Biol Chem 273:1269–1272.

Bernards R (1997) E2F: A nodal point in cell cycle regulation. Biochim Biophys Acta 1333:M33–M40.

Berns K, Hijmans EM, Bernards R (1997): Repression of c-Myc responsive genes in cycling cells causes G_1 arrest through reduction of cyclin E/CDK2 kinase activity. Oncogene 15:1347–1356.

Blain SW, Montalvo E, Massague J (1997): Differential interaction of the cyclin-dependent kinase (Cdk) inhibitor $p27^{Kip1}$ with cyclin A–Cdk2 and cyclin D2–Cdk4. J Biol Chem 272:25863–25872.

Bodwell JE, Hu J-M, Hu L-M, Munck A (1996): Glucocorticoid receptors: ATP and cell cycle dependence, phosphorylation, and hormone resistance. Am J Respir Crit Care Med 154:52–56.

Bokoch GM (1996): Interplay between Ras-related and heterotrimeric GTP binding proteins: Lifestyles of the big and little. FASEB J 10:1290–1295.

Bost F, McKay R, Dean N, Mercola D (1997): The JUN kinase/stress-activated protein kinase pathway is required for epidermal growth factor stimulation of growth of human A549 lung carcinoma cells. J Biol Chem 272:33422–33429.

Bourne HR (1997): How receptors talk to trimeric G proteins. Curr Opin Cell Biol 9:134–142.

Brown JP, Wei W, Sedivy JM (1997): Bypass of senescence after disruption of $p21^{CIP1/WAF1}$ gene in normal diploid human fibroblasts. Science 277:831–833.

Brown MT, Cooper JA (1996): Regulation, substrates and functions of src. Biochim Biophys Acta 1287:121–149.

Buckbinder L, Talbott R, Velasco-Miguel S, Takenaka I, Faha B, Seizinger BR, Kley N (1995): Induction of the growth inhibitor IGF-binding protein 3 by p53. Nature 377:646–649.

Burridge K, Chrzanowska-Wodnicka M (1996): Focal adhesions, contractility, and signaling. Annu Rev Cell Dev Biol 12:463–519.

Carroll MP, May WS (1994): Protein kinase C–mediated serine phosphorylation directly activates Raf-1 in murine hematopoietic cells. J Biol Chem 269:1249–1256.

Cary LA, Han DC, Polte TR, Hanks SK, Guan J-L (1998): Identification of $p130^{CAS}$ as a mediator of focal adhesion kinase-promoted cell migration. J Cell Biol 140:211–221.

Castagnola P, Bet P, Quarto R, Gennari M, Cancedda R (1991): cDNA cloning and gene expression of chicken osteopontin. J Biol Chem 266:9944–9949.

Cavanaugh AH, Hempel WM, Taylor LJ, Rogalsky V, Todorov G, Rothblum LI (1995): Activity of RNA polymerase I transcription factor UBF blocked by Rb gene product. Nature 374:177–180.

Cerione RA, Zheng Y (1996): The Dbl family of oncogenes. Curr Opin Cell Biol 8:216–222.

Chen P, Xie H, Wells A (1996): Mitogenic signaling from the EGF receptor is attenuated by a phospholipase C-γ/protein kinase C feedback mechanism. Mol Biol Cell 7:871–881.

Chen Q, Kinch MS, Lin TH, Burridge K, Juliano RL (1994): Integrin-mediated cell adhesion activates mitogen-activated protein kinases. J Biol Chem 269:26602–26605.

Chen Q, Lin TH, Der CJ, Juliano RL (1996): Integrin-mediated activation of mitogen-activated protein (MAP) or extracellular signal-related kinase (MEK) and kinase is independent of ras. J Biol Chem 271:18122–18127.

Chen Y-H, Grall D, Salcini AE, Pelicci PG, Pouyssegur J, Van Obberghen-Schilling E (1996): Shc adaptor proteins are key transducers of mitogenic signaling mediated by the G protein–coupled thrombin receptor. EMBO J 15:1037–1044.

Cheng J, Weber JD, Baldassare JJ, Raben DM (1997): Ablation of G_o α-subunit results in a transformed phenotype and constitutively active phosphatidylcholine-specific phospholipase C. J Biol Chem 272:17312–17319.

Choi YH, Lee SJ, Nguyen P, Jang JS, Lee J, Wu M-L, Takano E, Maki M, Henkart PA, Trepel JB (1997): Regulation of cyclin D1 by calpain protease. J Biol Chem 272:28479–28484.

Chong JPJ, Thommes P, Blow JJ (1996): The role of MCM/P1 proteins in the licensing of DNA replication. Trends Biochem Sci 21:102–106.

Chrzanowska-Wodnicka M, Burridge K (1996): Rho-stimulated contractility drives the formation of stress fibers and focal adhesions. J Cell Biol 133:1403–1415.

Chung J, Uchida E, Grammar TC, Blenis J (1997): STAT3 serine phosphorylation by ERK-dependent and -independent pathways negatively modulates its tyrosine phosphorylation. Mol Cell Biol 17:6508–6516.

Cigola E, Kajstura J, Li B, Meggs LG, Anversa P (1997): Angiotensin II activates programmed myocyte cell death in vitro. Exp Cell Res 231:363–371.

Clark AR, Docherty K (1993): Negative regulation of transcription in eukaryotes. Biochem J 296:521–541.

Clark EA, Hynes RO (1996): Ras activation is necessary for integrin-mediated activation of extracellular signal–regulated kinase 2 and cytosolic phospholipase A_2 but not for cytoskeletal organization. J Biol Chem 271:14814–14818.

Cohen PTW (1997): Novel protein serine/threonine phosphatases: Variety is the spice of life. Trends Biochem Sci 22:245–251.

Coleman KG, Wautlet BS, Morrissey D, Mulheron J, Sedman SA, Brinkley P, Price S, Webster KR (1997): Identification of CDK4 sequences involved in cyclin D1 and p16 binding. J Biol Chem 272:18869–18874.

Corey SJ, Dombrosky-Ferlan PM, Zuo S, Krohn E, Donnenberg AD, Zorich P, Romero G, Takata M, Kurosaki T (1998): Requirement of Src kinase Lyn for induction of DNA synthesis by granulocyte colony–stimulating factor. J Biol Chem 273:3230–3235.

Coso OA, Teramoto H, Simonds WF, Gutkind JS (1996): Signaling from G protein–coupled receptors to c-Jun kinase involves $\beta\gamma$ subunits of heterotrimeric G proteins acting on a ras and Rac1–dependent pathway. J Biol Chem 271:3963–3966.

Costanzi-Strauss E, Strauss BE, Naviaux RK, Haas M (1998): Restoration of growth arrest by $p16^{WAF1}$, $p21^{WAF1}$, pRB, and p53 is dependent on the integrity of the endogenous cell-cycle control pathways in human glioblastoma cell lines. Exp Cell Res 238:51–62.

Couet J, Sargiacomo M, Lisanti MP (1997): Interaction of a receptor tyrosine kinase, EGF-R, with caveolins. J Biol Chem 272:30429–30438.

Coverley D, Wilkinson HR, Madine MA, Mills AD, Laskey RA (1998): Protein kinase inhibition in G_2 causes mammalian Mcm proteins to reassociate with chromatin and restores ability to replicate. Exp Cell Res 238:63–69.

Cox LS, Lane DP (1995): Tumour suppressors, kinases and clamps: How p53 regulates the cell cycle in response to DNA damage. BioEssays 17:501–508.

Crawley JB, Rawlinson L, Lali FV, Page TH, Saklatvala J, Foxwell BMJ (1997):

T cell proliferation in response to interleukins 2 and 7 requires p38MAP kinase activation. J Biol Chem 272:15023–15027.

Crespo P, Xu N, Simonds WF, Gutkind JS (1994): Ras-dependent activation of MAP kinase pathway mediated by G-protein $\beta\gamma$ subunits. Nature 369:418–420.

Daaka Y, Luttrell LM, Ahn S, Della Roca GJ, Ferguson SSG, Caron MG, Lefkowitz RJ (1998): Essential role for G protein–coupled receptor endocytosis in the activation of mitogen-activated protein kinase. J Biol Chem 273:685–688.

Daaka Y, Luttrell LM, Lefkowitz RJ (1997): Switching of the coupling of the β_2-adrenergic receptor to different G proteins by protein kinase A. Nature 390:88–91.

Damen JE, Wakao H, Miyajima A, Krosl J, Humphries, RK, Cutler RL, Krystal G (1995): Tyrosine 343 in the erythropoietin receptor positively regulates erythropoietin-induced cell proliferation and Stat5 activation. EMBO J 14:5557–5568.

Darnell Jr JE (1997): STATs and gene regulation. Science 277:1630–1635.

Daub H, Weiss FU, Wallasch C, Ullrich A (1996): Role of transactivation of the EGF receptor in signaling by G-protein–coupled receptor. Nature 379:557–560.

David M, Petricoin III E, Benjamin C, Pine R, Weber MJ, Larner AC (1995): Requirement for MAP kinase (ERK2) activity in interferon α– and interferon β–stimulated gene expression through STAT proteins. Science 269:1721–1723.

Decker SJ (1995): Nerve growth factor–induced growth arrest and induction of $p21^{Cip1/WAF1}$ in NIH-3T3 cells expressing TrkA. J Biol Chem 270:30841–30844.

Dedhar S, Hannigan GE (1996): Integrin cytoplasmic interactions and bidirectional transmembrane signaling. Curr Opin Cell Biol 8:657–669.

Deed RW, Hara E, Atherton GT, Peters G, Norton JD (1997): Regulation of Id3 cell cycle function by Cdk-2–dependent phosphorylation. Mol Cell Biol 17:6815–6821.

DeGregori J, Leone G, Miron A, Jakoi L, Nevins JR (1997): Distinct roles for E2F proteins in cell growth control and apoptosis. Proc Natl Acad Sci USA 94:7245–7250.

Denhardt DT (1996a): Signal-transducing protein phosphorylation cascades mediated by Ras/Rho proteins in the mammalian cell: The potential for multiplex signaling. Biochem J 318:729–747.

Denhardt DT (1996b): Oncogene-initiated aberrant signaling engenders the metastatic phenotype: Synergistic transcription factor interactions are targets for cancer therapy. Oncogene 7:261–291.

DePinho RA (1998): The cancer-chromatin connection. Nature 391:533–536.

Dhanasekaran N, Dermott JM (1996): Signaling by the G_{12} class of G proteins. Cell Signal 8:235–245.

Dippel E, Kalkbrenner F, Wittig B, Schultz G (1996): A heterotrimeric G protein complex couples the muscarinic m1 receptor to phospholipase C-β. Proc Natl Acad Sci USA 93:1391–1396.

Dobrowolski S, Harter M, Stacey DW (1994): Cellular ras activity is required for passage through multiple points of the G_0/G_1 phase in BALB/c 3T3 cells. Mol Cell Biol 14:5441–5449.

Dohlman HG, Thorner J (1997): RGS proteins and signaling by heterotrimeric G proteins. J Biol Chem 272:3871–3874.

Dulic V, Stein GH, Far DF, Reed SJ (1998): Nuclear accumulation of $p21^{Cip1}$ at the onset of mitosis: A role at the G_2/M-phase transition. Mol Cell Biol 18:546–557.

Dutta A, Bell SP (1997): Initiation of DNA replication in eukaryotic cells. Annu Rev Cell Dev Biol 13:293–332.

Dynlacht BD, Moberg K, Lees JA, Harlow E, Zhu L (1997): Specific regulation of E2F family members by cyclin-dependent kinases. Mol Cell Biol 17:3867–3875.

Ezhevsky SA, Nagahara H, Vocero-Akbani AM, Gius DR, Wei MC, Dowdy SF (1997): Hypo-phosphorylation of the retinoblastoma protein (pRb) by cyclin D:Cdk4/6 complexes results in active pRb. Proc Natl Acad Sci USA 94:10699–10704.

Falasca M, Logan SK, Lehto VP, Baccante G, Lemmon MA, Schlessinger J (1998): Activation of phospholipase Cγ by PI 3-kinase–induced PH domain-mediated membrane targeting. EMBO J 17:414–422.

Fam NP, Fan W-T, Wang Z, Zhang L-J, Chen H, Moran MF (1997): Cloning and characterization of Ras-GRF2, a novel guanine nucleotide exchange factor for ras. Mol Cell Biol 17:1396–1406.

Fang F, Orend G, Watanabe N, Hunter T, Ruoslahti E (1996): Dependence of cyclin E–CDK2 kinase activity on cell anchorage. Science 271:499–502.

Fanger GR, Lassignal Johnson N, Johnson GL (1997): MEK kinases are regulated by EGF and selectively interact with Rac/Cdc42. EMBO J 16:4961–4972.

Farnsworth CL, Freshney NW, Rosen LB, Ghosh A, Greenberg ME, Feig LA (1995): Calcium activation of ras mediated by neuronal exchange factor ras-GRF. Nature 376:524–527.

Feig LA, Urano T, Cantor S (1996): Evidence for a Ras/Ral cascade. Trends Biochem Sci 21:438–441.

Filmus J, Robles AI, Shi W, Wong MJ, Colombo LL, Conti CJ (1994): Induction of cyclin D1 overexpression by activated ras. Oncogene 9:3627–3633.

Foschi M, Chari S, Dunn MJ, Sorokin A (1997): Biphasic activation of $p21^{ras}$ by endothelin-1 sequentially activates the ERK cascade and phosphatidylinositol 3-kinase. EMBO 16:6439–6451.

Gabrielli BG, Clark JM, McCormack AK, Ellem KAO (1997): Hyperphosphorylation of the N-terminal domain of Cdc25 regulates activity toward cyclin B1/Cdc2 but not cyclin A/Cdk2. J Biol Chem 272:28607–28614.

Galaktionov K, Chen X, Beach D (1996): Cdc25 cell-cycle phosphatase as a target of c-myc. Nature 382:511–517.

Galaktionov K, Lee AK, Eckstein J, Draetta G, Meckler J, Loda M, Beach D (1995): Cdc25 phosphatases as potential human oncogenes. Science 269:1575–1577.

Gao CY, Zelenka PS (1997): Cyclins, cyclin-dependent kinases and differentiation. BioEssays 19:307–315.

Giancotti FG (1997): Integrin signaling: Specificity and control of cell survival and cell cycle progression. Curr Opin Cell Biol 9:691–700.

Giglione C, Parrini MC, Baouz S, Bernardi A, Parmeggiani A (1997): A new function of p120-GTPase–activating protein. J Biol Chem 272:25128–25134.

Goga A, Liu X, Hambuch TM, Senechal K, Major E, Berk AJ, Witte OW,

Sawyers CL (1995): p53 dependent growth suppression by the c-Ab1 nuclear tyrosine kinase. Oncogene 11:791–799.

Gotoh N, Toyoda M, Shibuya M (1997): Tyrosine phosphorylation sites at amino acids 239 and 240 of Shc are involved in epidermal growth factor–induced mitogenic signaling that is distinct from ras/mitogen-activated protein kinase activation. Mol Cell Biol 17:1824–1831.

Gu W, Schneider JW, Condorelli G, Kaushal S, Mahdavi V, Nadal-Ginard B (1993): Interaction of myogenic factors and the retinoblastoma protein mediates muscle cell commitment and differentiation. Cell 72:309–324.

Guadagno TM, Ohtsubo M, Roberts JM, Assoian RK (1993): A link between cyclin A expression and adhesion-dependent cell cycle progression. Science 262:1572–1575.

Gudas LJ (1994): Retinoids and vertebrate development. J Biol Chem 269:15399–15402.

Gutkind JS (1998): The pathways connecting G protein–coupled receptors to the nucleus through divergent mitogen-activated protein kinase cascades. J Biol Chem 273:1839–1842.

Haas K, Johannes C, Geisen C, Schmidt T, Karsunky H, Blass-Kampmann S, Obe G, Möröy T (1997): Malignant transformation by cyclin E and Ha-Ras correlates with lower sensitivity towards induction of cell death but requires functional Myc and CDK4. Oncogene 15:2615–2623.

Hafner S, Adler HS, Mischak H, Janosch P, Heidecker G, Wolfman A, Pippig S, Lohse M, Ueffing M, Kolch W (1994): Mechanism of inhibition of Raf-1 by protein kinase A. Mol Cell Biol 14:6696–6703.

Haimovich B, Kaneshiki N, Ji P (1996): Protein kinase C regulates tyrosine phosphorylation of $pp125^{FAK}$ in platelets adherent to fibrinogen. Blood 87:152–161.

Hall A (1994): Small GTP-binding proteins and the regulation of the actin cytoskeleton. Annu Rev Cell Biol 10:31–54.

Hall M, Peters G (1996): Genetic alterations of cyclins, cyclin-dependent kinases, and Cdk inhibitors in human cancer. Adv Cancer Res 68:67–108.

Hamm HE (1998): The many faces of G protein signaling. J Biol Chem 273:669–672.

Han Y, Leaman DW, Watling D, Rogers NC, Groner B, Kerr IM, Wood WI, Stark GR (1996): Participation of JAK and STAT proteins in growth hormone–induced signaling. J Biol Chem 271:5947–5952.

Hara E, Hall M, Peters G (1997): Cdk2-dependent phosphorylation of Id2 modulates activity of E2A-related transcription factors. EMBO J 16:332–342.

Hara E, Smith R, Parry D, Tahara H, Stone S, Peters G (1996): Regulation of p^{CDKN2} expression and its implications for cell immortalization and senescence. Mol Cell Biol 16:859–867.

Harrison SC (1996): Peptide-surface association: The case of PDZ and PTB domains. Cell 86:341–343.

Haupt Y, Maya R, Kazaz A, Oren M (1997): Mdm2 promotes the rapid degradation of p53. Nature 387:296–299.

Haupt Y, Rowan S, Oren M (1995): p53-mediated apoptosis in HeLa cells can be overcome by excess pRB. Oncogene 10:1563–1571.

Hauser PJ, Agrawal D, Chu B, Pledger WJ (1997): p107 and p130 associated cyclin A has altered substrate specificity. J Biol Chem 272:22954–22959.

Henriksson M, Lüscher B (1996): Proteins of the myc network: Essential regulators of cell growth and differentiation. Cancer Res 68:108–182.

Hermeking H, Lengauer C, Polyak K, He T-C, Zhang L, Thiagalingam S, Kinzler KW, Vogelstein B (1997): 14-3-3α is a p53-regulated inhibitor of G_2/M progression. Mol Cell 1:3–11.

Herwig S, Strauss M (1997): The retinoblastoma protein: A master regulator of cell cycle, differentiation and apoptosis. Eur J Biochem 246:581–601.

Hisatake K, Hasegawa S, Takada R, Nakatani Y, Horikoshi M, Roeder RG (1993): The p250 subunit of native TATA box-binding factor TFIID is the cell-cycle regulatory protein CCG1. Nature 362:179–181.

Hofbauer R, Denhardt DT (1991): Cell cycle–regulated and proliferation stimulus-responsive genes. Crit Rev Eukaryot Gene Expression 1:247–300.

Holland PM, Magali S, Campbell JS, Noselli S, Cooper JA (1997): MKK7 is a stress-activated mitogen-activated protein kinase kinase functionally related to hemipterous. J Biol Chem 272:24994–24998.

Hotchin NA, Hall A (1995): The assembly of integrin adhesion complexes requires both extracellular matrix and intracellular rho/rac GTPases. J Cell Biol 131:1857–1865.

Hoyt MA (1997): Eliminating all obstacles: Regulated proteolysis in the eukaryotic cell cycle. Cell 91:149–151.

Hua XH, Newport J (1998): Identification of a preinitiation step in DNA replication that is independent of origin recognition complex and Cdc6, but dependent on cdk2. J Cell Biol 140:271–281.

Huang DCS, O'Reilly LA, Strasser A, Cory S (1997): The anti-apoptosis function of Bc1-2 can be genetically separated from its inhibitory effect on cell cycle entry. EMBO J 16:4628–4638.

Hughes PE, Renshaw MW, Pfaff M, Forsyth J, Keivens VM, Schwartz MA, Ginsberg MH (1997): Suppression of integrin activation: A novel function of a ras/raf-initiated MAP kinase pathway. Cell 88:521–530.

Humphries MJ (1996): Integrin activation: The link between ligand binding and signal transduction. Curr Opin Cell Biol 8:632–640.

Hunter T (1995a): Protein kinases and phosphatases: The Yin and Yang of protein phosphorylation and signaling. Cell 80:225–236.

Hunter T (1995b): When is a lipid kinase not a lipid kinase? When it is a protein kinase. Cell 83:1–4.

Hupp TR, Lane DP (1995): Two distinct signaling pathways activate the latent DNA binding function of p53 in a casein kinase II–independent manner. J Biol Chem 270:18165–18174.

Ihle JN (1995): Cytokine receptor signaling. Nature 377:591–594.

Inglese J, Koch WJ, Touhara K, Lefkowitz RJ (1995): $G_{\beta\gamma}$ interactions with PH domains and ras–MAPK signaling pathways. Trends Biochem Sci 20:151–156.

Jallepalli PV, Kelly TJ (1997): Cyclin-dependent kinase and initiation at eukaryotic origins: A replication switch? Curr Opin Cell Biol 9:358–363.

Jan YN, Jan LY (1993): HLH proteins, fly neurogenesis, and vertebrate myogenesis. Cell 75:827–830.

Janicke RU, Lin XY, Lee FHH, Porter AG (1996): Cyclin D3 sensitizes tumor cells to tumor necrosis factor-induced, c-Myc–dependent apoptosis. Mol Cell Biol 16:5245–5253.

Jansen-Durr P, Meichle A, Steiner P, Pagano M, Finke K, Botz J, Wessbecher

J, Draetta G, Eilers M (1993): Differential modulation of cyclin gene expression by MYC. Proc Natl Acad Sci USA 90:3685–3689.

Jensen DE, Black AR, Swick AG, Azizkhan JC (1997): Distinct roles for Sp1 and E2F sites in the growth/cell cycle regulation of the DHFR promoter. J Cell Biochem 67:24–31.

Kamijo T, Zindy F, Roussel MF, Quelle DE, Downing JR, Ashmun RA, Grosveld G, Sherr CJ (1997): Tumor suppression at the mouse INK4a locus mediated by the alternative reading frame product $p19^{ARF}$. Cell 91:649–659.

Kang J-S, Krauss RS (1996): Ras induces anchorage-independent growth by subverting multiple adhesion-regulated cell cycle events. Mol Cell Biol 16:3370–3380.

Katayose Y, Kim M, Rakkar ANS, Li Z, Cowan KH, Seth P (1997): Promoting apoptosis: A novel activity associated with the cyclin-dependent kinase inhibitor p27. Cancer Res 57:5441–5445.

Katula KS, Wright KL, Paul H, Surman DR, Nuckolls FJ, Smith JW, Ting JP-Y, Yates J, Cogswell JP (1997): Cyclin-dependent kinase activation and S-phase induction of the cyclin B1 gene are linked through the CCAAT elements. Cell Growth Differ 8:811–820.

Kawada M, Yamagoe S, Murakami Y, Suzuki K, Mizuno S, Uehara Y (1997): Induction of $p27^{Kip1}$ degradation and anchorage independence by Ras through the MAP kinase signaling pathway. Oncogene 15:629–637.

Keely PJ, Westwick JK, Whitehead IP, Der CJ, Parise LV (1997): Cdc42 and Rac1 induce integrin-mediated cell motility and invasiveness through PI(3)K. Nature 390:632–635.

Kerkhoff E, Rapp UR (1997): Induction of cell proliferation in quiescent NIH 3T3 cells by oncogenic c-raf-1. Mol Cell Biol 17:2576–2586.

Kharbanda S, Pandey P, Jin S, Inoue S, Bharti A, Yuan Z-M, Weichselbaum R, Weaver D, Kufe D (1997): Functional interaction between DNA–PK and c-Abl in response to DNA damage. Nature 386:732–735.

King RW, Deshaies RJ, Peters J-M, Kirschner MW (1996): How proteolysis drives the cell cycle. Science 274:1652–1658.

King WG, Mattaliano MD, Chan TO, Tsichlis PN, Brugge JS (1997): Phosphatidylinositol 3-kinase is required for integrin-stimulated AKT and raf-1/mitogen–activated protein kinase pathway activation. Mol Cell Biol 17:4406–4418.

Kinoshita T, Shirouzu M, Kamiya A, Hashimoto K, Yokoyama S, Miyajima A (1997): Raf/MAPK and rapamycin-sensitive pathways mediate the anti-apoptotic function of p21Ras in IL-3–dependent hematopoietic cells. Oncogene 15:619–627.

Kinoshita T, Yokota T, Arai K, Miyajima A (1995): Suppression of apoptotic death in hematopoietic cells by signaling through the IL-3/GM-CSF receptors. EMBO J 14:26–275.

Klarlund JK, Cherniack AD, McMahon M, Czech MP (1996): Role of the raf/mitogen-activated protein kinase pathway in $p21^{ras}$ desensitization. J Biol Chem 271:16674–16677.

Klingmüller U (1997): The role of tyrosine phosphorylation in proliferation and maturation of erythroid progenitor cells. Eur J Biochem 249:637–647.

Klippel A, Reinhard C, Kavanaugh WM, Apell G, Escobedo M-A, Williams LT (1996): Membrane localization of phosphatidylinositol 3-kinase is sufficient to activate multiple signal-transducing kinase pathways. Mol Cell Biol 16:4117–4127.

Knudsen ES, Wang JYJ (1997): Dual mechanism for the inhibition of E2F binding to RB by cyclin-dependent kinase-mediated RB phosphorylation. Mol Cell Biol 17:5771–5783.

Ko LJ, Prives C (1996): p53: Puzzle and paradigm. Genes Dev 10:1054–1072.

Koh J, Enders GH, Dynlacht BD, Harlow E (1995): Tumour-derived p16 alleles encoding proteins defective in cell-cycle inhibition. Nature 375:506–510.

Kolanus W, Seed B (1997): Integrins and inside-out signal transduction: Converging signals from PKC and PIP_3. Curr Opin Cell Biol 9:725–731.

Krek W, Ewen ME, Shirodkar S, Arany Z, Kaelin Jr WG, Livingston DM (1994): Negative regulation of the growth-promoting transcription factor E2F-1 by a stably bound cyclin A–dependent protein kinase. Cell 78:161–172.

Kubbutat MHG, Jones SN, Vousden KH (1997): Regulation of p53 stability by Mdm2. Nature 387:299–303.

Kubota Y, Mimura S, Nishimoto S, Masuda T, Nojima H, Takisawa H (1997): Licensing of DNA replication by a multiprotein complex of MCM/P1 proteins in *Xenopus* eggs. EMBO J 16:3320–3331.

Kummer JL, Rao PK, Heidenreich KA (1997): Apoptosis induced by withdrawal of trophic factors is mediated by p38 mitogen-activated protein kinase. J Biol Chem 272:20490–20494.

Lafrenie RM, Yamada KM (1996): Integrin-dependent signal transduction. J Cell Biochem 61:543–553.

Lamarche N, Tapon N, Stowers L, Burbelo PD, Aspenstrom P, Bridges T, Chant J, Hall A (1996): Rac and Cdc42 induce actin polymerization and G_1 cell cycle progression independently of $p65^{PAK}$ and the JNK/SAPK MAP kinase cascade. Cell 87:519–529.

Lane DP, Hall PA (1997): MDM2—arbiter of p53's destruction. Trends Biochem Sci 22:372–374.

Lavoie JN, L'Allemain G, Brunet A, Muller R, Pouyssegur J (1996): Cyclin D1 expression is regulated positively by the $p42/p44^{MAPK}$ and negatively by the $p38/HOG^{MAPK}$ pathway. J Biol Chem 271:20608–20616.

Leaman DW, Leung S, Li X, Stark GR (1996a): Regulation of STAT-dependent pathways by growth factors and cytokines. FASEB J 10:1578–1588.

Leaman DW, Pisharody S, Flickinger TW, Commane MA, Schlessinger J, Kerr IM, Levy DE, Stark GR (1996b): Roles of JAKs in activation of STATs and stimulation of c-fos gene expression by epidermal growth factor. Mol Cell Biol 16:369–375.

Lee TC, Li L, Philipson L, Ziff EB (1997): Myc represses transcription of the growth arrest gene gas1. Proc Natl Acad Sci USA 94:12886–12891.

Lee Y, Chen Y, Chang L-S, Johnson LF (1997): Inhibition of mouse thymidylate synthase promoter activity by the wild-type p53 tumor suppressor protein. Exp Cell Res 234:270–276.

Leone G, DeGregori J, Sears R, Jakoi L, Nevins JR (1997): Myc and Ras collaborate in inducing accumulation of active cyclin E/Cdk2 and E2F. Nature 387:422–426.

Levine AJ (1997): p53, the cellular gatekeeper for growth and division. Cell 88:323–331.

Lewis BC, Shim H, Li Q, Wu CS, Lee LA, Maity A, Dang CV (1997): Identification of putative c-myc–responsive genes: Characterization of rcl, a novel growth-related gene. Mol Cell Biol 17:4967–4978.

Lewis JM, Cheresh DA, Schwartz MA (1996): Protein kinase C regulates $\alpha v\beta 5$-dependent cytoskeletal associations and focal adhesion kinase phosphorylation. J Cell Biol 134:1323–1332.

Li R, Zheng Y (1997) Residues of the Rho family GTPases Rho and Cdc42 that specify sensitivity to Dbl-like guanine nucleotide exchange factors. J Biol Chem 272:4671–4679.

Li S, Couet J, Lisanti MP (1996): Src tyrosine kinases, G_α subunits, and H-Ras share a common membrane-anchored scaffolding protein, caveolin. J Biol Chem 271:29182–29190.

Li Y, Gorbea C, Mahaffey D, Rechsteiner M, Benezra R (1997): MAD2 associates with the cyclosome/anaphase-promoting complex and inhibits its activity. Proc Natl Acad Sci USA 94:12431–12436.

Li Z, Wahl MI, Eguinoa A, Stephens LR, Hawkins PT, Witte ON (1997): Phosphatidylinositol 3-kinase-γ activates Bruton's tyrosine kinase in concert with src family kinases. Proc Natl Acad Sci USA 94:13820–13825.

Lin TH, Aplin AE, Shen Y, Chen Q, Schaller M, Romer L, Aukhil I, Juliano RL (1997): Integrin-mediated activation of MAP kinase is independent of FAK: Evidence for dual integrin signaling pathways in fibroblasts. J Cell Biol 136:1385–1395.

Liu J-P (1996): Protein kinase C and its substrates. Mol Cell Endocrinol 116:1–29.

Liu Y, Martindale JL, Gorospe M, Holbrook NJ (1996): Regulation of $p21^{WAF1/CIP1}$ expression through mitogen-activated protein kinase signaling pathway. Cancer Res 56:31–35.

Liu P, Ying, Y-S, Anderson RGW (1997): Platelet-derived growth factor activates mitogen-activated protein kinase in isolated caveolae. Proc Natl Acad Sci USA 94:13666–13670.

Lloyd AC, Obermuller F, Staddon S, Barth CF, McMahon M, Land H (1997): Cooperating oncogenes converge to regulate cyclin/cdk complexes. Genes Dev 11:663–677.

Logan SK, Falasca M, Hu P, Schlessinger J (1997): Phosphatidylinositol 3-kinase mediates epidermal growth factor–induced activation of the c-Jun N-terminal kinase signaling pathway. Mol Cell Biol 17:5784–5790.

Loignon M, Fetni R, Gordon AJE, Drobetsky EA (1997): A p53-independent pathway for induction of $p21^{waf1cip1}$ and concomitant G_1 arrest in UV-irradiated human skin fibroblasts. Cancer Res 57:3390–3394.

Loor G, Zhang S-J, Zhang P, Toomey NL, Lee MYWT (1997): Identification of DNA replication and cell cycle proteins that interact with PCNA. Nucleic Acids Res 25:5041–5046.

Lopez-Ilsaca M, Crespo P, Pellici PG, Gutkind JS, Wetzker R (1997): Linkage of G protein–coupled receptors to the MAPK signaling pathway through PI 3-kinase γ. Science 275:394–397.

Ludlow JW, Glendening CL, Livingston DM, DeCaprio JA (1993): Specific enzymatic dephosphorylation of the retinoblastoma protein. Mol Cell Biol 13:367–372.

Lukas J, Bartkova J, Bartek J (1996): Convergence of mitogenic signaling cascades from diverse classes of receptors at the cyclin D–cyclin-dependent kinase-pRb–controlled G_1 checkpoint. Mol Cell Biol 16:6917–6925.

Lukas J, Parry D, Aagaard L, Mann DJ, Bartkova J, Strauss M, Peters G, Bartek J (1995): Retinoblastoma-protein–dependent cell-cycle inhibition by the tumour suppresser p16. Nature 375:503–506.

Lundberg AS, Weinberg RA (1998): Functional inactivation of the retinoblastoma protein requires sequential modification by at least two distinct cyclin–cdk complex. Mol Cell Biol 18:753–761.

Luscinskas FW, Lawler J (1994): Integrins as dynamic regulators of vascular function. FASEB J 8:929–938.

Luttrell LM, Daaka Y, Rocca GJD, Lefkowitz RJ (1997): G protein–coupled receptors mediate two functionally distinct pathways of tyrosine phosphorylation in rat 1a fibroblasts. J Biol Chem 272:31648–31656.

Luttrell LM, van Biesen T, Hawes BE, Koch WJ, Touhara K, Lefkowitz RJ (1995): $G_{\beta\gamma}$ subunits mediate mitogen-activated protein kinase activation by the tyrosine kinase insulin-like growth factor 1 receptor. J Biol Chem 270:16495–16498.

Macara IG, Lounsbury KM, Richards SA, McKiernan C, Bar-Sagi D (1996): The ras superfamily of GTPases. FASEB J 10:625–630.

Machesky LM, Hall A (1997): Role of actin polymerization and adhesion to extracellular matrix in Rac- and Rho-induced cytoskeletal reorganization. J Cell Biol 138:913–926.

Mahbubani HM, Chong JPJ, Chevalier S, Thömmes P, Blow JJ (1997): Cell cycle regulation of the replication licensing system: Involvement of a Cdk-dependent inhibitor. J Cell Biol 136:125–135.

Mainiero F, Murgia C, Wary KK, Curatola AM, Pepe A, Blumemberg M, Westwick JK, Der CJ, Giancotti FG (1997): The coupling of $\alpha_6\beta_4$ integrin to Ras-MAP kinase pathways mediated by Shc controls keratinocyte proliferation. EMBO J 16:2365–2375.

Mainiero F, Pepe A, Wary KK, Spinardi L, Mohammadi M, Schlessinger J, Giancotti FG (1995): Signal transduction by the $\alpha_6\beta_4$ integrin: Distinct β_4 subunit sites mediate recruitment of Shc/Grb2 and association with the cytoskeleton of hemidesmosomes. EMBO J 14:4470–4481.

Manser E, Loo T-H, Koh C-G, Zhao Z-S, Chen X-Q, Tan L, Tan I, Leung T, Lim L (1998): PAK kinases are directly coupled to the PIX family of nucleotide exchange factors. Mol Cell 1:183–192.

Marshall CJ (1996): Ras effectors. Curr Opin Cell Biol 8:197–204.

Marte BM, Downward J (1997): PKB/Akt: Connecting phosphoinositide 3-kinase to cell survival and beyond. Trends Biochem Sci 22:355–358.

Mason-Garcia M, Harlan E, Mallia C, Jeter JR (1995): Interleukin-3 or erythropoietin induced nuclear localization of protein kinase C β isoforms in hematopoietic target cells. Cell Prolif 28:145–155.

Mayo LD, Turchi JJ, Berberich SJ (1997): Mdm-2 phosphorylation by DNA-dependent protein kinase prevents interaction with p53. Cancer Res 57:5013–5016.

McIlroy J, Chen D, Wjasow C, Michaeli T, Backer JM (1997): Specific activation of p85-p110 phosphatidylinositol 3-kinase stimulates DNA synthesis by ras– and p70 S6 kinase–dependent pathways. Mol Cell Biol 17:248–255.

Migliaccio A, Di Domenico M, Castoria G, de Falco A, Bontempo P, Nola E Auricchio F (1996): Tyrosine kinase/$p21^{ras}$/MAP-kinase pathway activation by estradiol-receptor complex in MCF-7 cells. EMBO J 15:1292–1300.

Missero C, Di Cunto F, Kiyokawa H, Koff A, Paolo Dotto G (1996): The absence of $p21^{Cip1/WAF1}$ alters keratinocyte growth and differentiation and promotes ras-tumor progression. Genes Dev 10:3065–3075.

Miyamoto S, Teramoto H, Coso OA, Gutkind JS, Burbelo PD, Akiyama SK, Yamada KM (1995): Integrin function: Molecular hierarchies of cytoskeletal and signaling molecules. J Cell Biol 131:791–805.

Miyamoto S, Teramoto H, Gutkind JS, Yamada KM (1996): Integrins can collaborate with growth factors for phosphorylation of receptor tyrosine kinases and MAP kinase activation: Role of integrin aggregation and occupancy of receptors. J Cell Biol 135:1633–1642.

Miyashita T, Reed JC (1995): Tumor suppressor p53 is a direct transcriptional activator of the human bax gene. Cell 80:293–299.

Molnar A, Theodoras AM, Zon LI, Kyriakis JM (1997): Cdc42Hs, but not Rac1, inhibits serum-stimulated cell cycle progression at G_1/S through a mechanism requiring p38/RK. J Biol Chem 272:13229–13235.

Momand J, Zambetti GP (1997): Mdm-2: "Big brother" of p53. J Cell Biochem 64:343–352.

Morisaki H, Fujimoto A, Ando A, Nagata Y, Ikeda K, Nakanishi M (1997): Cell cycle–dependent phosphorylation of p27 cyclin-dependent kinase (Cdk) inhibitor by cyclin E/Cdk2. Biochem Biophys Res Commun 240:386–390.

Morrison DK, Cutler RE, Jr (1997): The complexity of Raf-1 regulation. Curr Opin Cell Biol 9:174–179.

Mufson RA (1997): The role of serine/threonine phosphorylation in hematopoietic cytokine receptor signal transduction. FASEB J 11:37–44.

Mui AL-F, Wakao H, Kinoshita T, Kitamura T, Miyajima A (1996): Suppression of interleukin-3–induced gene expression by a C-terminal truncated Stat5: Role of Stat5 in proliferation. EMBO J 15:2425–2433.

Müller D, Bouchard C, Rudolph B, Steiner P, Stuckmann I, Saffrich R, Ansorge W, Huttner W, Eilers M (1997): Cdk2-dependent phosphorylation of p27 facilitates its myc-induced release from cyclin E/cdk2 complexes. Oncogene 15:2561–2576.

Müller H, Moroni MC, Vigo E, Petersen BO, Bartek J, Helin K (1997): Induction of S-phase entry by E2F transcription factors depends on their nuclear localization. Mol Cell Biol 17:5508–5520.

Müller R (1995): Transcriptional regulation during the mammalian cell cycle. Trends Genet 11:173–178.

Muslin AJ, Tanner JW, Allen PM, Shaw AS (1996): Interaction of 14-3-3 with signaling proteins is mediated by the recognition of phosphoserine. Cell 84:889–897.

Nagai K, Takata M, Yamamura H, Kurosaki T (1995): Tyrosine phosphorylation of Shc is mediated through Lyn and Syk in B cell receptor signaling. J Biol Chem 270:6824–6829.

Nasmyth K (1996): Viewpoint: Putting the cell cycle in order. Science 274:1643–1645.

Neer EJ (1995): Heterotrimeric G proteins: Organizers of transmembrane signals. Cell 80:249–257.

Neuman E, Ladha MH, Lin N, Upton TM, Miller SJ, DiRenzo J, Pestell RG, Hinds PW, Dowdy SF, Brown M, Ewen ME (1997): Cyclin D1 stimulation of estrogen receptor transcriptional activity independent of cdk4. Mol Cell Biol 17:5338–5347.

Newton AC (1995): Protein kinase C: Structure, function, and regulation. J Biol Chem 270:28495–28498.

Niculescu III AB, Chen X, Smeets M, Hengst L, Prives C, Reed SI (1998): Effects of p21$^{Cip1/Waf1}$ at both the G_1/S and the G_2/M cell cycle transitions: pRb is a critical determinant in blocking DNA replication and in preventing endoreduplication. Mol Cell Biol 18:629–643.

Nishikawa K, Toker A, Johannes F-J, Songyang Z, Cantley LC (1997): Determination of the specific substrate sequence motifs of protein kinase C isozymes. J Biol Chem 272:952–960.

Nobes CD, Hall A (1995): Rho, Rac, and Cdc42 GTPases regulate the assembly of multimolecular focal complexes associated with actin stress fibers, lamellipodia, and filopodia. Cell 81:53–62.

Ogryzko VV, Wong P, Howard BH (1997): WAF1 retards S-phase progression primarily by inhibition of cyclin-dependent kinases. Mol Cell Biol 17:4877–4882.

Okada S, Kao AW, Ceresa BP, Blaikie P, Margolis B, Pessin JE (1997): The 66-kDa Shc isoform is a negative regulator of the epidermal growth factor–stimulated mitogen-activated protein kinase pathway. J Biol Chem 272:28042–28049.

Olson MF, Ashworth A, Hall A (1995): An essential role for Rho, Rac, and Cdc42 GTPases in cell cycle progression through G_1. Science 269:1270–1272.

Ookata K, Hisanaga S-I, Sugita M, Okuyama A, Murofushi H, Kitazawa H, Chari S, Bulinski JC, Kishimoto T (1997): MAP4 is the in vivo substrate for CDC2 kinase in HeLa cells: Identification of an M-phase specific and a cell cycle–independent phosphorylation site in MAP4. Biochemistry 36:15873–15883.

Pagano M (1997): Cell cycle regulation by the ubiquitin pathway. FASEB J 11:1067–1075.

Pagano M, Tam SW, Theodoras AM, Beer-Romero P, Del Sal G, Chau V, Yew PR, Draetta GF, Rolfe M (1995): Role of the ubiquitin-proteasome pathway in regulating abundance of the cyclin-dependent kinase inhibitor p27. Science 269:682–685.

Palmero I, McConnell B, Parry D, Brookes S, Hara E, Bates S, Jat P, Peters G (1997): Accumulation of p16^{INK4a} in mouse fibroblasts as a function of replicative senescence and not of retinoblastoma gene status. Oncogene 15:495–503.

Pardee AB (1974): A restriction point for control of normal animal cell proliferation. Proc Natl Acad Sci USA 71:1286–1290.

Park DS, Morris EJ, Greene LA, Geller HM (1997): G_1/S cell cycle blockers and inhibitors of cyclin-dependent kinases suppress camptothecin-induced neuronal apoptosis. J Neurosci 17:1256–1270.

Parsons JT (1996): Integrin-mediated signaling: Regulation by protein tyrosine kinases and small GTP-binding proteins. Curr Opin Cell Biol 8:146–152.

Pawson T (1995): Protein modules and signaling networks. Nature 373:573–580.

Peeper DS, Upton TM, Ladha MH, Neuman E, Zalvide J, Bernards R, DeCaprio JA, Ewen ME (1997): Ras signaling linked to the cell-cycle machinery by the retinoblastoma protein. Nature 386:177–181.

Peng C-Y, Graves PR, Thoma RS, Wu Z, Shaw AS, Piwnica-Worms H (1997): Mitotic and G_2 checkpoint control: Regulation of 14-3-3 protein binding by phosphorylation of Cdc25C on serine-216. Science 277:1501–1504.

Phillips AC, Bates S, Ryan KM, Helin K, Vousden KH (1997): Induction of DNA synthesis and apoptosis are separable functions of E2F-1. Genes Dev 11:1853–1863.

Philipp A, Schneider A, Vasrik I, Finke K, Xiong Y, Beach D, Alitalo K, Eilers M (1994): Repression of cyclin D1: A novel function of MYC. Mol Cell Biol 14:4032–4043.

Pierzchalski P, Reiss K, Cheng W, Cirielli C, Kajstura J, Nitahara JA, Rizk M, Capogrossi MC, Anversa P (1997): p53 induces myocyte apoptosis via the activation of the renin–angiotensin system. Exp Cell Res 234:57–65.

Pines J (1995): Cyclins and cyclin-dependent kinases: A biochemical view. Biochem J 308:697–711.

Pitcher JA, Touhara K, Payne ES, Lefkowitz RJ (1995): Pleckstrin homology domain-mediated membrane association and activation of the β-adrenergic receptor kinase requires coordinate interaction with $G_{\beta\gamma}$ subunits and lipid. J Biol Chem 270:11707–11710.

Planas-Silva MD, Weinberg RA (1997): Estrogen-dependent cyclin E–cdk2 activation through p21 redistribution. Mol Cell Biol 17:4059–4069.

Polyak K, Xia Y, Zweler JL, Kinzler KW, Vogelstein B (1997): A model for p53-induced apoptosis. Nature 389:300–305.

Post GR, Brown JH (1996): G protein–coupled receptors and signaling pathways regulating growth responses. FASEB J 10:741–749.

Prall OWJ, Sarcevic B, Musgrove EA, Watts CKW, Sutherland RL (1997): Estrogen-induced activation of Cdk4 and Cdk2 during G_1–S phase progression is accompanied by increased cyclin D1 expression and decreased cyclin-dependent kinase inhibitor association with cyclin E–Cdk2. J Biol Chem 272:10882–10894.

Pumiglia KM, Decker SJ (1997): Cell cycle arrest mediated by the MEK/mitogen-activated protein kinase pathway. Proc Natl Acad Sci USA 94:448–452.

Qiu R-G, Abo A, McCormick F, Symons M (1997): Cdc42 regulates anchorage-independent growth and is necessary for ras transformation. Mol Cell Biol 17:3449–3458.

Quelle DE, Cheng M, Ashmun RA, Sherr CJ (1997): Cancer-associated mutations at the INK4a locus cancel cell cycle arrest by $p16^{INK4a}$ but not by the alternative reading frame protein $p19^{ARF}$. Proc Natl Acad Sci USA 94:669–673.

Quelle FW, Wang D, Nosaka T, Thierfelder WE, Stravopodis D, Weinstein Y, Ihle JN (1996): Erythropoietin induces activation of Stat5 through association with specific tyrosines on the receptor that are not required for a mitogenic response. Mol Cell Biol 16:1622–1631.

Radeva G, Petrocelli T, Behrend E, Leung-Hagesteijn C, Filmus J, Slingerland J, Dedhar S (1997): Overexpression of the integrin-linked kinase promotes anchorage-independent cell cycle progression. J Biol Chem 272:13937–13944.

Rameh LE, Arvidsson A, Carraway III KL, Couvillon AD, Rathbun G, Crompton A, VanRenterghem B, Czech MP, Ravichandran KS, Burakoff SJ, Wang D-S, Chen C-S, Cantley LS (1997): A comparative analysis of the phosphoinositide binding specificity of pleckstrin homology domains. J Biol Chem 272:22059–22066.

Rameh LE, Chen C-S, Cantley LC (1995): Phosphatidylinositol $(3,4,5)P_3$ interacts with SH2 domains and modulates PI 3-kinase association with tyrosine-phosphorylated proteins. Cell 83:821–830.

Ravichandran KS, Zhou M-M, Pratt JC, Harlan JE, Walk SF, Fesik SW, Burakoff SJ (1997): Evidence for a requirement for both phospholipid and phosphotyrosine binding via the shc phosphotyrosine-binding domain in vivo. Mol Cell Biol 17:5540–5549.

Reed JC (1997): Double identity for proteins of the Bcl-2 family. Nature 387:773–776.

Renshaw MW, Ren X-D, Schwartz MA (1997): Growth factor activation of MAP kinase requires cell adhesion. EMBO J 16:5592–5599.

Resnitzky D (1997): Ectopic expression of cyclin D1 but not cyclin E induces anchorage-independent cell cycle progression. Mol Cell Biol 17:5640–5647.

Richards JD, Gold MR, Hourihane SL, DeFranco AL, Matsuuchi L (1996): Reconstitution of B cell antigen receptor–induced signaling events in a non-lymphoid cell line by expressing the Syk protein-tyrosine kinase. J Biol Chem 271:6458–6466.

Richardson A, Malik RK, Hildebrand JD, Parsons JT (1997): Inhibition of cell spreading by expression of the C-terminal domain of focal adhesion kinase (FAK) is rescued by coexpression of src or catalytically inactive FAK: A role for paxillin tyrosine phosphorylation. Mol Cell Biol 17:6906–6914.

Rivard N, McKenzie FR, Brondello J-M, Pouyssegur J (1995): The phosphotyrosine phosphatase PTP1D, but not PTP1C, is an essential mediator of fibroblast proliferation induced by tyrosine kinase and G protein–coupled receptors. J Biol Chem 270:11017–11024.

Robinson MJ, Cobb MH (1997): Mitogen-activated protein kinase pathways. Curr Opin Cell Biol 9:180–186.

Ruppert S, Wang EH, Tjian R (1993): Cloning and expression of human $TAF_{II}50$: A TBP-associated factor implicated in cell-cycle regulation. Nature 362:175–179.

Rushton JJ, Steinman RA, Robbins PD (1997): Differential regulation of transcription of p21 and cyclin D1 conferred by $TAF_{II}250$. Cell Growth Differ 8:1099–1104.

Saha P, Eichbaum Q, Silberman ED, Mayer BJ, Dutta A (1997): $p21^{CIP1}$ and Cdc25A: Competition between an inhibitor and an activator of cyclin-dependent kinases. Mol Cell Biol 17:4338–4345.

Said TK, Conneely OM, Medina D, O'Malley BW, Lydon JP (1997): Progesterone, in addition to estrogen, induces cyclin D1 expression in the murine mammary epithelial cell, in vivo. Endocrinology 138:3933–3939.

Schieffer B, Paxton WG, Chai Q, Marrero MB, Bernstein KE (1996): Angiotensin II controls $p21^{ras}$ activity via $pp60^{c\text{-}src}$. J Biol Chem 271:10329–10333.

Schneller M, Vuori K, Ruoslahti E (1997): $\alpha v\beta 3$ integrin associates with activated insulin and PDGFβ receptors and potentiates the biological activity of PDGF. EMBO J 16:5600–5607.

Schönwasser DC, Marais RM, Marshall CJ, Parker PJ (1998): Activation of the mitogen-activated protein kinase/extracellular signal–regulated kinase pathway by conventional, novel, and atypical protein kinase C isotypes. Mol Cell Biol 18:790–798.

Schwartz MA (1997): Integrins, oncogenes, and anchorage independence. J Cell Biol 139:575–578.

Sekiguchi T, Nohiro Y, Nakamura Y, Hisamoto N, Nishimoto T (1991): The human CCG1 gene, essential for progression of the G_1 phase, encodes a 210-kilodalton nuclear DNA-binding protein. Mol Cell Biol 11:3317–3325.

Serrano M, Gomez-Lahoz E, DePinho RA, Beach D, Bar-Sagi D (1995): Inhibition of Ras-induced proliferation and cellular transformation by $p16^{INK4}$. Science 267:249–252.

Serrano M, Lee H-W, Chin L, Cordon-Cardo C, Beach D, DePinho RA (1996): Role of the INK4a locus in tumor suppression and cell mortality. Cell 85:27–37.

Serrano M, Lin AW, McCurrach ME, Beach D, Lowe SW (1997): Oncogenic ras provokes premature cell senescence associated with accumulation of p53 and $p16^{INK4a}$. Cell 88:593–602.

Seuwen K, Pouysségur J (1992): G protein–controlled signal transduction pathways and the regulation of cell proliferation. Adv Cancer Res 58:75–94.

Sewing A, Wiseman B, Lloyd AC, Land H (1997): High-intensity raf signal causes cell cycle arrest mediated by $p21^{Cip1}$. Mol Cell Biol 17:5588–5597.

Shao Z, Siegert JL, Ruppert S, Robbins PD (1997): Rb interacts with $TAF_{II}250$/TFIID through multiple domains. Oncogene 15:385–392.

Shapiro PS, Evans JN, Davis RJ, Posada JA (1996): The seven-transmembrane–spanning receptors for endothelin and thrombin cause proliferation of airway smooth muscle cells and activation of the extracellular regulated kinase and c-Jun NH_{2}- terminal kinase groups of mitogen-activated protein kinases. J Biol Chem 271:5750–5754.

Shattil SJ, Ginsberg MH (1997): Perspectives series: Cell adhesion in vascular biology. J Clin Invest 100:1–5.

Sherr CJ (1996): Cancer cell cycles. Science 274:1672–1677.

Sherr CJ, Roberts JM (1995): Inhibitors of mammalian G_1 cyclin-dependent kinases. Genes Dev 9:1149–1163.

Shieh S-Y, Ikeda M, Taya Y, Prives C (1997): DNA damage–induced phosphorylation of p53 alleviates inhibition by MDM2. Cell 91:325–334.

Shim J, Lee H, Park J, Kim H, Choi E-J (1996): A non-enzymatic p21 protein inhibitor of stress-activated protein kinases. Nature 381:804–807.

Shiyanov P, Hayes S, Chen N, Pestov DG, Lau LF, Raychaudhuri P (1997): p26Kip1 induces an accumulation of the repressor complexes of E2F and inhibits expression of the E2F-regulated genes. Mol Biol Cell 8:1815–1827.

Sladek TL (1997): E2F transcription factor action, regulation and possible role in human cancer. Cell Prolif 30:97–105.

Smith ML, Chen I-T, Zhan Q, Bae I, Chen C-Y, Gilmer TM, Kastan MB, O'Connor PM, Fornace Jr AJ (1994): Interaction of the p53-regulated protein Gadd45 with proliferating cell nuclear antigen. Science 266:1376–1379.

Somani A-K, Bignon JS, Mills GB, Siminovitch KA, Branch DR (1997): Src kinase activity is regulated by the SHP-1 protein-tyrosine phosphatase. J Biol Chem 272:21113–21119.

Stewart M, Hogg N (1996): Regulation of leukocyte integrin function: Affinity vs avidity. J Cell Biochem 61:554–561.

Stillman B (1996): Cell cycle control of DNA replication. Science 274:1659–1663.

Stokoe D, McCormick F (1997): Activation of c-raf-1 by ras and src through different mechanisms: Activation in vivo and in vitro. EMBO J 16:2384–2396.

Sugden PH, Clerk A (1997): Regulation of the ERK subgroup of MAP kinase cascades through G protein–coupled receptors. Cell Signal 9:337–351.

Sun XJ, Wang L-M, Zhang Y, Yenush L, Myers Jr MG, Glasheen E, Lane WS, Pierce JH, White MF (1995): Role of IRS-2 in insulin and cytokine signaling. Nature 377:173–177.

Symons M (1996) Rho family GTPases: the cytoskeleton and beyond. Trends Biochem Sci 21:178–181.

Takekawa M, Posas F, Saito H (1997): A human homolog of the yeast Ssk2/Ssk22 MAP kinase kinase kinases, MTK1, mediates stress-induced activation of the p38 and JNK pathways. EMBO J 16:4973–4982.

Takuwa N, Takuwa Y (1997): Ras activity late in G_1 phase required for $p27^{kip1}$ downregulation, passage through the restriction point, and entry into S phase in growth factor–stimulated NIH 3T3 fibroblasts. Mol Cell Biol 17:5348–5358.

Tang Y, Chen Z, Ambrose D, Liu J, Gibbs JB, Chernoff J, Field J (1997): Kinase-deficient Pak1 mutants inhibit ras transformation of Rat-1 fibroblasts. Mol Cell Biol 17:4454–4464.

Taniguchi T (1995): Cytokine signaling through nonreceptor protein tyrosine kinases. Science 268:251–255.

Taya Y (1997): RB kinases and RB-binding proteins: New points of view. Trends Biochem Sci 22:14–17.

Terada Y, Inoshita S, Nakashima O, Yamada T, Tamamori T, Ito H, Sasaki S, Marumo F (1998): Cyclin D1, p16, and retinoblastoma gene regulate mitogenic signaling of endothelin in rat mesangial cells. Kidney Int 53:76–83.

Thomas JW, Ellis B, Boerner RJ, Knight WB, White II GC, Schaller MD (1998): SH2- and SH3-mediated interactions between focal adhesion kinase and src. J Biol Chem 272:577–583.

Thommes P, Kubota Y, Takisawa H, Blow JJ (1997): The RLF-M component of the replication licensing system forms complexes containing all six MCM/P1 polypeptides. EMBO J 16:3312–3319.

Toker A, Cantley LC (1997): Signaling through the lipid products of phosphoinositide-3-OH kinase. Nature 387:673–676.

Tommasi S, Pfeifer GP (1997): Constitutive protection of E2F recognition sequences in the human thymidine kinase promoter during cell cycle progression. J Biol Chem 272:30483–30490.

Tournier C, Whitmarsh AJ, Cavanaugh J, Barrett T, Davis RJ (1997): Mitogen-activated protein kinase kinase 7 is an activator of the c-Jun NH_2-terminal kinase. Proc Natl Acad Sci USA 94:7337–7342.

Trowbridge JM, Rogatsky I, Garabedian MJ (1997): Regulation of estrogen receptor transcriptional enhancement by the cyclin A/Cdk2 complex. Proc Natl Acad Sci USA 94:10132–10137.

Ueda Y, Hirai S-i, Osada S-i, Suzuki A, Mizuno K, Ohno S (1996): Protein kinase Cδ activates the MEK-ERK pathway in a manner independent of Ras and dependent on Raf. J Biol Chem 271:23512–23519.

Uhrbom L, Nister M, Westermark B (1997): Induction of senescence in human malignant glioma cells by $p16^{INK4A}$. Oncogene 15:505–514.

van Biesen T, Hawes BE, Raymonds JR, Luttrell LM, Koch WJ, Lefkowitz RJ (1996): G_o-protein α-subunits activate mitogen-activated protein kinase via a novel protein kinase C–dependent mechanism. J Biol Chem 271:1266–1269.

van der Geer P, Hunter T (1994): Receptor protein-tyrosine kinases and their signal transduction pathways. Annu Rev Cell Biol 10:251–337.

VanderKuur J, Allevato G, Billestrup N, Norstedt G, Carter-Su C (1995): Growth hormone–promoted tyrosyl phosphorylation of SHC proteins and SHC association with Grb2. J Biol Chem 270:7587–7593.

van Ginkel PR, Hsiao K-M, Schjerven H, Farnham PJ (1997): E2F-mediated growth regulation requires transcription factor cooperation. J Biol Chem 272:18367–18374.

Vanhaesebroeck B, Leevers SJ, Panayotou G, Waterfield MD (1997): Phosphoinositide 3-kinases: A conserved family of signal transducers. Trends Biochem Sci 22:267–272.

Verona R, Moberg K, Estes S, Starz M, Vernon JP, Lees JA (1997): E2F activity is regulated by cell cycle–dependent changes in subcellular localization. Mol Cell Biol 17:7268–7282.

Vignais M-L, Sadowski HB, Watling D, Rogers NC, Gilman M (1996): Platelet-derived growth factor induces phosphorylation of multiple JAK family kinases and STAT proteins. Mol Cell Biol 16:1759–1769.

Vlach J, Hennecke S, Alevizopoulos K, Conti D, Amati B (1996): Growth arrest by the cyclin-dependent kinase inhibitor $p27^{Kip1}$ is abrogated by c-Myc. EMBO J 15:6595–6604.

Voit R, Schafer K, Grumm I (1997): Mechanism of repression of RNA polymerase I transcription by the retinoblastoma protein. Mol Cell Biol 17:4230–4237.

Vuori K, Hirai H, Aizawa S, Ruoslahti E (1996): Induction of $p130^{cas}$ signaling complex formation upon integrin-mediated cell adhesion: A role for src family kinases. Mol Cell Biol 16:2606–2613.

Waldman T, Lengauer C, Kinzler KW, Vogelstein B (1996): Uncoupling of S phase and mitosis induced by anticancer agents in cells lacking p21. Nature 381:713–716.

Wan Y, Bence K, Hata A, Kurosaki T, Veillette A, Huang X-Y (1997): Genetic evidence for a tyrosine kinase cascade preceding the mitogen-activated protein kinase cascade in vertebrate G protein signaling. J Biol Chem 272:17209–17215.

Wan Y, Kurosaki T, Huang X-Y (1996): Tyrosine kinases in activation of the MAP kinase cascade by G-protein–coupled receptors. Nature 380:541–544.

Wang EH, Zou S, Tjian R (1997): $TAF_{II}250$-dependent transcription of cyclin A is directed by ATF activator proteins. Genes Dev 11:2658–2669.

Wang Y-D, Wong K, Wood WI (1995): Intracellular tyrosine residues of the human growth hormone receptor are not required for the signaling of proliferation or Jak-STAT activation. J Biol Chem 270:7021–7024.

Wary KK, Mainiero F, Isakoff SJ, Marcantonio EE, Giancotti FG (1996): The adaptor protein Shc couples a class of integrins to the control of cell cycle progression. Cell 87:733–743.

Weber JD, Cheng J, Raben DM, Gardner A, Baldassare JJ (1997a): Ablation of $G_0\alpha$ overrides G_1 restriction point control through ras/ERK/cyclin D1–CDK activities. J Biol Chem 272:17320–17326.

Weber JD, Hu W, Jefcoat SC, Raben DM, Baldasare JJ (1997b): Ras-stimulated extracellular signal-related kinase 1 and RhoA activates coordinate platelet-derived growth factor-induced G_1 progression through the independent regulation of cyclin D1 and $p27^{Kip1}$. J Biol Chem 272:32966–32971.

Weber JD, Raben DM, Phillips PJ, Baldassare JJ (1997c): Sustained activation of extracellular-signal–regulated kinase 1 (EK1) is required for the continued expression of cyclin D1 in G_1 phase. Biochem J 326:61–68.

Wess J (1997): G-protein–coupled receptors: Molecular mechanisms involved in receptor activation and selectivity of G-protein recognition. FASEB J 11:346–354.

White RJ (1997): Regulation of RNA polymerases I and III by the retinoblastoma protein: A mechanism for growth control? Trends Biochem Sci 22:77–80.

White RJ, Trouche D, Martin K, Jackson SP, Kouzarides T (1996): Repression of RNA polymerase III transcription by the retinoblastoma protein. Nature 382:88–90.

Wiggan O, Taniguchi-Sidle A, Hamel PA (1998): Interaction of the pRB-family proteins with factors containing paired-like homeodomains. Oncogene 16:227–236.

Winston JT, Coats SR, Wang Y-Z, Pledger WJ (1996): Regulation of the cell cycle machinery by oncogenic ras. Oncogene 12:127–134.

Won K-A, Reed SI (1996): Activation of cyclin E/CDK2 is coupled to site-specific autophosphorylation and ubiquitin-dependent degradation of cyclin E. EMBO J 15:4182–4193.

Woo MS-A, Sanchez I, Dynlacht BD (1997): p130 and p107 use a conserved domain to inhibit cellular cyclin-dependent kinase activity. Mol Cell Biol 17:3566–3579.

Woods D, Parry D, Cherwinski H, Bosch E, Lees E, McMahon M (1997): Raf-induced proliferation or cell cycle arrest is determined by the level of raf activity with arrest mediated by $p21^{Cip1}$. Mol Cell Biol 17:5598–5611.

Wu C-L, Classon M, Dyson N, Harlow E (1996): Expression of dominant-negative mutant DP-1 blocks cell cycle progression in G_1. Mol Cell Biol 16:3698–3706.

Wu J-R, Gilbert DM (1997): The replication origin decision point is a mitogen-independent, 2-aminopurine-sensitive, G_1-phase event that precedes restriction point control. Mol Cell Biol 17:4312–4321.

Wuarin J, Nurse P (1996): Regulating S phase: CDKs, licensing and proteolysis. Cell 85:785–787.

Xu S, Robbins D, Frost J, Dang A, Lange-Carter C, Cobb MH (1995): MEKK1 phosphorylates MEK1 and MEK2 but does not cause activation of mitogen-activated protein kinase. Proc Natl Acad Sci USA 92:6808–6812.

Xu W, Harrison SC, Eck MJ (1997): Three-dimensional structure of the tyrosine kinase c-Src. Nature 385:595–601.

Yaffe MB, Schutkowski M, Shen M, Zhou XZ, Stukenberg PT, Rahfeld J-U, Xu J, Kuang J, Kirschner MW, Fischer G, Cantley LC, Lu KP (1997): Sequence-specific and phosphorylation-dependent proline isomerization: A potential mitotic regulatory mechanism. Science 278:1957–1960.

Yamamoto H, Crow M, Cheng L, Lakatta E, Kinsella J (1996): PDGF receptor-to-nucleus signaling of p91 (STAT1α) transcription factor in rat smooth muscle cells. Exp Cell Res 222:125–130.

Yamauchi T, Ueki K, Tobe K, Tamemoto H, Sekine N, Wada M, Honjo M, Takahashi M, Takahashi T, Hirai H, Tushima T, Akanuma Y, Fujita T, Komuro I, Yazaki Y, Kadowaki T (1997): Tyrosine phosphorylation of the EGF receptor by the kinase Jak2 is induced by growth hormone. Nature 390:91–96.

Yao L, Suzuki H, Ozawa K, Deng J, Lehel C, Fukamachi H, Anderson WB, Kawakami Y, Kawakami T (1997): Interactions between protein kinase C and pleckstrin homology domains. J Biol Chem 272:13033–13039.

Yew PR, Kirschner MW (1997): Proteolysis and DNA replication: The Cdc34 requirement in the *Xenopus* egg cell cycle. Science 277:1672–1675.

Yokosaki Y, Monis H, Chen J, Sheppard D (1996): Differential effects of the integrins $\alpha 9\beta 1$, $\alpha v\beta 3$, and $\alpha v\beta 6$ on cell proliferative responses to tenascin. J Biol Chem 271:24144–24150.

Yokote K, Mori S, Hansen K, McGlade J, Pawson T, Heldin C-H, Claesson-Welsh L (1994): Direct interaction between Shc and the platelet-derived growth factor β-receptor. J Biol Chem 269:15337–15343.

Zavitz KH, Zipursky SL (1997): Controlling cell proliferation in differentiating tissues: Genetic analysis of negative regulators of $G_1 \rightarrow$S-phase progression. Cell Biol 9:773–781.

Zhang H, Hannon GJ, Beach D (1994): p21-containing cyclin kinases exist in both active and inactive states. Genes Dev 8:1750–1758.

Zhang P, Liegeois NJ, Wong C, Finegold M, Hou H, Thompson JC, Silverman A, Harper JW, DePinho RA, Elledge SJ (1997): Altered cell differentiation and proliferation in mice lacking $p57^{KIP2}$ indicates a role in Beckwith-Wiedemann syndrome. Nature 387:151–158.

Zhu X, Ohtsubo M, Bohmer RM, Roberts JM, Assoian RK (1996): Adhesion-dependent cell cycle progression linked to the expression of cyclin D1, activation of cyclin E-cdk2, and phosphorylation of the retinoblastoma protein. J Cell Biol 133:391–403.

Zhuang H, Patel SV, He T, Sonsteby SK, Niu Z, Wojchowski DM (1994): Inhibition of erythropoietin-induced mitogenesis by a kinase-deficient form of Jak2. J Biol Chem 269:21411–21414.

Zindy F, Quelle DE, Roussel MF, Sherr CJ (1997): Expression of the $p16^{INK4a}$ tumor suppressor versus other INK4 family members during mouse development and aging. Oncogene 15:203–211.

Zwicker J, Müller R (1997): Cell-cycle regulation of gene expression by transcriptional repression. Trends Genet 13:3–6.

Zwijsen RM, Wientjens E, Klompmaker R, van der Sman J, Bernards R, Michalides RJAM (1997): CDK-independent activation of estrogen receptor by cyclin D1. Cell 89:405–415.

CHAPTER 7

DIFFERENTIATION, DEVELOPMENT, AND PROGRAMMED CELL DEATH

M. CRISTINA CARDOSO and HEINRICH LEONHARDT
Franz Volhard Clinic/Max Delbrueck Center for Molecular Medicine, Humboldt University, 13122 Berlin, Germany

INTRODUCTION

A central issue in understanding the mechanisms of development is to elucidate how cellular division is controlled. In animal development, most structures are generated by a combination of cell division, cell migration, and cell death. Interestingly, in plants, cell migration plays no role, and cell death, though used for many purposes, does not contribute to organogenesis. During plant morphogenesis, controlled patterns and numbers of cell divisions constitute the main strategy (Jurgens, 1995; Meyerowitz, 1997). In this chapter, we focus on control of cell division and death in animal development, including epigenetic regulation.

CELL CYCLE REGULATION AND DIFFERENTIATION

In higher eukaryotes, one can distinguish three types of cells in terms of their proliferative properties: the ones that are always dividing throughout the lifetime of the organism (e.g., epithelium cells or hematopoietic stem cells); the ones that remain in a quiescent state (designated G_0) but can be stimulated to proliferate (e.g., T and B lymphocytes or fibroblasts); and the ones that are terminally differentiated and cannot start proliferating and do not divide throughout the life of the organism. The latter cell type, which includes skeletal myotubes, cardiac myocytes, adipocytes, and neurons, can grow (hypertrophy) in response to particular external stimuli but cannot divide (hyperplasia). They differ from quiescent cells in this additional block. Hypertrophy is actually the only mode of coping with increasing functional demands that is accessible to

The Molecular Basis of Cell Cycle and Growth Control, Edited by G.S. Stein, R. Baserga, A. Giordano, and D.T. Denhardt
ISBN 0-471-15706-6, pages 305–347.

several vital tissues such as myocardium, skeletal muscle, and nerves, which loose their proliferative capacity during ontogeny. This growth strategy ensues that such tissues have little to no regenerative capacity and cannot compensate for malfunction or loss of their structural units (reviewed in Goss, 1966). That raises the question of whether the terminally differentiated cells in these tissues irreversibly lost essential components of the replicative apparatus or whether this block to proliferation is an actively maintained state. Elucidating how terminally differentiated cells can grow but, at a certain stage of development or differentiation, loose their ability to divide, should indicate ways to reverse this state to obtain tissue regeneration and, in the opposite direction, to suppress malignancy in these tissues. In fact, uncontrolled cell proliferation leading to cancer results from an impediment in the execution of a particular differentiation program, and, consequently, cancers can be as distinct as the cell types from which they originate. This is not to say that incomplete differentiation will yield malignant tumors or that tumors cannot also contain differentiated cells. Cell fusion experiments between normal and malignant cells showed that the hybrids, in which malignancy was abrogated, executed the normal cell differentiation program (reviewed in Harris, 1990).

Cell cycle studies have been traditionally performed using quiescent cell systems such as fibroblast cells or primary lymphocyte cultures. The vast majority of the cells that form our most essential organs permanently lose their ability to proliferate during development and, therefore, belong not to the quiescent type but to the terminally differentiated cell type. Initially the work concentrated on how cell cycle progression is stimulated and how cellular transformation is achieved. Over a decade ago, researchers working in once separate fields, such as DNA tumor viruses, growth factors including their respective signaling pathways, oncogenes, and tumor suppressors, recognized that their target factors were the same. Also, work on very different systems spanning from yeast to human cells, including marine invertebrates, *Drosophila,* and *Xenopus,* established many basic principles that are remarkably conserved throughout evolution. Hence, cyclins and their respective cyclin-dependent kinases (see Chapter 2, this volume) were recognized as key positive regulators of cell cycle functioning even across species. The pursuit of factors that could bring about cell cycle arrest led then to a search for negative regulators, which included tumor suppressors such as the retinoblastoma-related proteins and the more recently isolated cyclin-dependent kinase inhibitors (see Chapter 2, this volume).

Recent observations from gene expression studies indicated that many cell cycle regulators exhibit differences in their tissue-specific abundance. The latter suggests, on the one hand, that members of the same family may function in a nonredundant manner in vivo and, on the other hand, that they might have differentiation-specific functions either in the establishment or in the maintenance of a particular differentiated tissue. This hypothesis has been recently supported by the tissue-specific effects of mice with null mutations in genes coding for cell cycle factors (e.g., cyclin

D1$^{-/-}$ mice show reduced body size, a hypoplastic retina, and failure to undergo steroid-induced proliferation of mammary epithelium during pregnancy [Sicinski et al., 1995], whereas cyclin D2$^{-/-}$ mice have hypoplastic ovaries and testes [Sicinski et al., 1996]). Coming from another research area, studies on the characterization of terminal differentiation had, more than three decades ago, clearly established that, upon appropriate environmental cues, determined progenitor cells stop proliferating and differentiate into the corresponding lineage (Konigsberg et al., 1960; Stockdale and Holtzer, 1961). In fact, cell proliferation and terminal differentiation are mutually exclusive phenomena, and both occur at specific times and places during development. In the past years, as several of the basic molecules that regulate cell cycle progression were uncovered as well as antibodies and probes for both activating and inhibiting regulators of cell cycle became available, much attention turned to how terminally differentiated cells establish and maintain their permanent cell cycle withdrawal at a specific time in development. Indeed, the answer to these questions is a pivotal issue in elucidating the mechanisms of development.

Among the different terminal differentiation systems, skeletal muscle stands as a paradigm, and, though much progress has recently been made with neuronal and cardiac cultures, the lack of appropriate cell lines has slowed down progress in these areas. Myogenesis became a model system in which to dissect the molecular basis of the terminal differentiation program and how it can be disrupted, mostly because of the availability of several in vitro culture systems that can reproduce all the basic features of myogenic differentiation. These systems include primary cultures of fetal myoblasts or newborn satellite cells, as well as established mammalian myogenic cell lines, and have been used to study terminal differentiation for over three decades. By now, an enormous wealth of information has accumulated about many different aspects of the differentiation process in this cell type. In particular, in the last few years a combination of molecular probes together with cell lines and animals with mutations in myogenic determination genes or in cell cycle regulators allowed direct testing of the key molecular determinants of cell cycle control during terminal differentiation in this cell type. Hypotheses mostly derived from work on skeletal muscle cells have inspired experiments to test whether similar mechanisms operate in other differentiated systems, including cardiogenesis and neurogenesis. Though among these cell types there are differences in the cell cycle regulators present, their roles are likely to be at least partially overlapping. We will, therefore, use the skeletal muscle system to present the current view of cell cycle regulation during differentiation.

About 30 years ago, different myogenic cell culture systems from avian and rodent origins were first generated and the basic phenomenology of the differentiation process established. Myoblast cells can terminally differentiate in vitro and form syncytial myotube cultures, in this manner recapitulating the developmental program of muscle formation. In vivo as in vitro, conversion of undifferentiated myoblasts into terminally

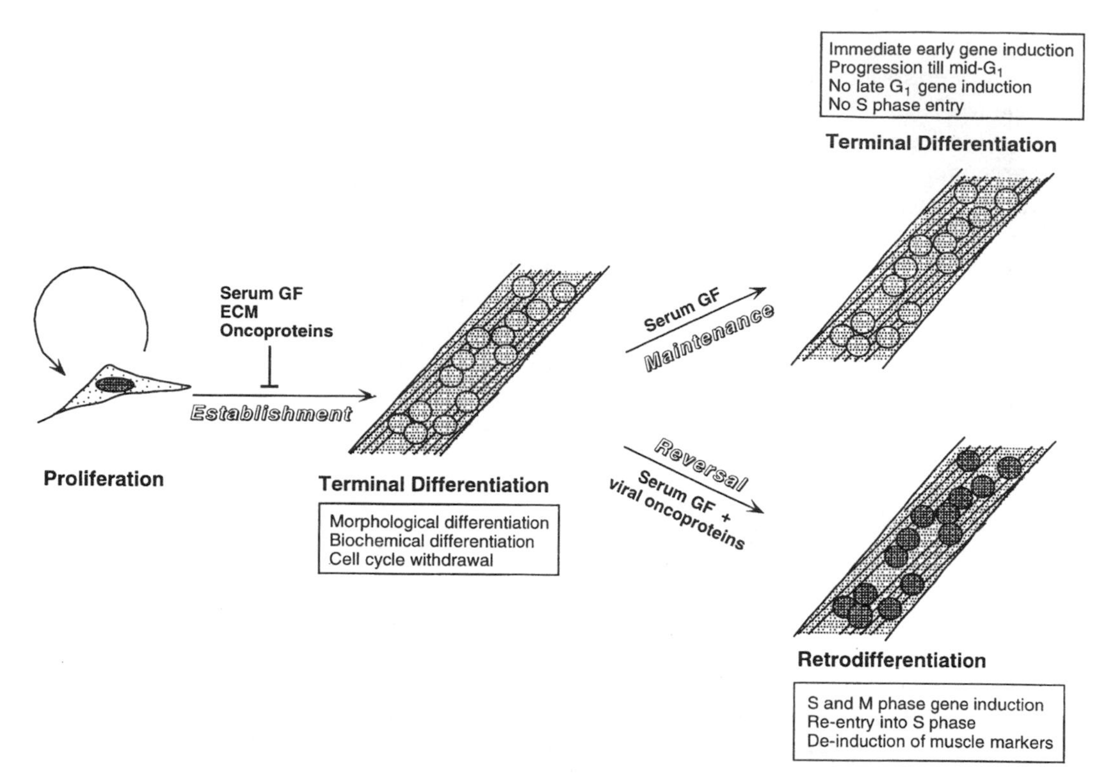

Figure 7.1. Establishment, maintenance, and reversal of terminal differentiation. Conversion of undifferentiated myoblasts into terminally differentiated myotubes is induced under low-mitogen conditions and entails a sequence of cellular events. These include induction of an array of muscle-specific genes (biochemical differentiation), fusion to form multinucleated myotubes (morphological differentiation) that in vivo will ultimately form functional myofibers, and permanent withdrawal from the cell cycle. Differentiated myotubes respond to mitogen stimulation by the expression of immediate-early genes and progression through mid G_1 but can neither express late G_1 markers nor re-enter S phase. This inability can be reversed by expression of viral oncoproteins in the myotubes, indicating that it is not an irreversible state but rather an actively maintained one. Upon viral oncoprotein expression, S phase is induced and muscle gene expression repressed.

differentiated myotubes entails a sequence of cellular events. These include induction of an array of muscle-specific genes (biochemical differentiation), fusion to form multinucleated myotubes (morphological differentiation) that in vivo will ultimately form functional myofibers, and permanent withdrawal from the cell cycle (Nadal-Ginard, 1978; and references therein) (Fig. 7.1).

By taking advantage of the availability of myogenic cell lines and of the molecular cloning and characterization of muscle-specific regulatory sequences, it was possible to identify two major classes of muscle transcription factors in vertebrates, the MyoD family and the myocyte-enhancer factor 2 (MEF2) family. The former includes four members:

MyoD (Davis et al., 1987), Myf5 (Braun et al., 1989), myogenin (Edmondson and Olson, 1989; Wright et al., 1989), and MRF4 (Rhodes and Konieczny, 1989; Braun et al., 1990; Miner and Wold, 1990). These myogenic basic helix-loop-helix (bHLH) proteins bind to a DNA sequence termed the *E box* as heterodimers with the ubiquitously expressed bHLH factors of the E-protein family (reviewed in Weintraub, 1993; Olson and Klein, 1994). The MEF2 family also encompasses four members, MEF2A–D (Pollock and Treisman, 1991; Yu et al., 1992; Breitbart et al., 1993; Martin et al., 1993, 1994; McDermott et al., 1993), which generate multiple isoforms through alternative splicing and bind an A/T rich DNA sequence termed the *MEF2 site* (reviewed in Olson et al., 1995). A combination of work in different vertebrate systems, including gene knockouts in mice, led to the following picture: Early in development, expression of MyoD and Myf5 in proliferating precursor cells (normally in different subsets of cells) is necessary for the determination of these cells to the myogenic lineage and to the subsequent initiation of a cascade of events leading to the expression of the MEF2 factors; these activate expression of myogenin, which, in addition to stimulating transcription of muscle structural proteins also stimulates MEF2 transcription, creating a positive feedback loop resulting in high levels of MEF2 and myogenin and, later in differentiation, of MRF4. Both of these transcription factor families synergistically activate muscle-specific promoters containing either or both DNA-binding sites. In addition to directing muscle-specific transcription and in a manner independent of its transcriptional activity, at least MyoD has been shown to be involved in cell cycle withdrawal during terminal differentiation (Crescenzi et al., 1990; Sorrentino et al., 1990; Thorburn et al., 1993), thereby creating a mechanism for crosstalk between differentiation and cell cycle machinery. Indeed, muscle differentiation is intimately linked to the cell cycle in that muscle gene transcription can only be initiated in myoblasts in G_0/G_1 phase. This suggests that factors expressed or functional only in this phase should be necessary for differentiation or that at other cell cycle phases inhibitors of myogenesis are present. Elucidating how the activity of these myogenic factors is regulated during the cell cycle should help to clarify this issue.

Cessation of proliferation is a prerequisite for myogenic cells (and other cell types) to switch on the terminal differentiation program, and indeed several viral and cellular oncogenes and growth factors were shown to block myogenic differentiation (Fig. 7.1) (reviewed in Alema and Tato, 1994; Olson, 1992). Peptide growth factors (see Chapter 6, this volume) have been shown to act as positive or as negative regulators of muscle differentiation. Fibroblast growth factor (FGF) (Clegg et al., 1987) and transforming growth factor type-β (TGF-β) (Massague et al., 1986; Olson et al., 1986) were reported to act as potent inhibitors of myogenic differentiation, whereas insulin-like growth factors (IGF) were shown to stimulate muscle differentiation (Ewton and Florini, 1981). Oncogenes from all functional groups, when expressed in myoblast cells, induce cellular transformation and block differentiation. Inhibition of differentiation was correlated with interference, at different levels, with myogenic transcription factor function. Proliferating myoblasts express at least one myogenic transcription factor (MyoD or Myf5) before differentiation, which under mitogen-deprivation conditions initiates a series of events including the induction of a repertoire of muscle-specific genes and the down regulation of proliferation-associated genes, ultimately leading to the formation of multinucleated myotubes. Several levels of regulation have been proposed to negatively control the activities of these factors in proliferating myoblasts. One level is the inhibition of the accumulation of myogenic factor transcripts by fos, jun, src, ras, and msx1 (Massague et al., 1986; Olson et al., 1986; Konieczny et al., 1989; Lassar et al., 1989; Falcone et al., 1991; Bengal et al., 1992; Song et al., 1992; Woloshin et al., 1995). Id, a dominant negative HLH protein, has been shown to competitively inhibit transcriptionally active heterodimer formation between myogenic factors and E2 proteins (Benezra et al., 1990; Jen et al., 1992). Another level of regulation is protein kinase C–mediated phosphorylation of the basic domain of myogenic transcription factors, which has been shown to interfere with their DNA binding in some cases, but it does not seem to do so in others (L. Li et al., 1992; Hardy et al., 1993). Recently, inhibition of MyoD and myogenin muscle gene activation by cyclin D1–dependent kinase was reported (Rao et al., 1994; Rao and Kohtz, 1995; Skapek et al., 1995, 1996; Guo and Walsh, 1997), providing another important piece in the molecular crosstalk between differentiation and cell cycle machinery (Fig. 7.2). Interestingly, the C-terminal acidic region of cyclin D1 was shown to be required for inhibition of myogenic factor transcriptional activity, but the N-terminal domain, which binds the retinoblastoma protein (pRb), was dispensable (Rao et al., 1994). Consistent with this finding, phosphorylation of pRb by cyclin D1–dependent kinase was not necessary for inhibition of MyoD transactivation of muscle transcription by this G_1 cyclin (Skapek et al., 1996).

The postmitotic state of myotubes differs significantly from the reversible G_0 cell cycle arrest of quiescent fibroblasts in that mitogen stimulation does not lead to proliferation (Konigsberg et al., 1960; Stockdale and Holtzer, 1961). Terminally differentiated myotubes can still respond to serum stimulation by expressing immediate-early genes and progressing until a mid G_1 state (Fig. 7.1) (Endo and Nadal-Ginard, 1986; Tiainen et al., 1996a). Therefore, myotubes have not permanently lost essential components of the signal transduction pathway that link cell membrane with the nucleus, though cell surface receptors are in general downregulated (Olwin and Hauschka, 1988; Hu and Olson, 1990).

After fusion, mitogen stimulation can lead to hypertrophy, but nuclear DNA synthesis and cell division are permanently inhibited (Fig. 7.1). This inability could be due to the loss or irreversible alteration of the basic cellular machinery necessary for DNA replication and mitosis during differentiation. This fundamental question was first addressed by infecting or microinjecting muscle cultures with transforming DNA and

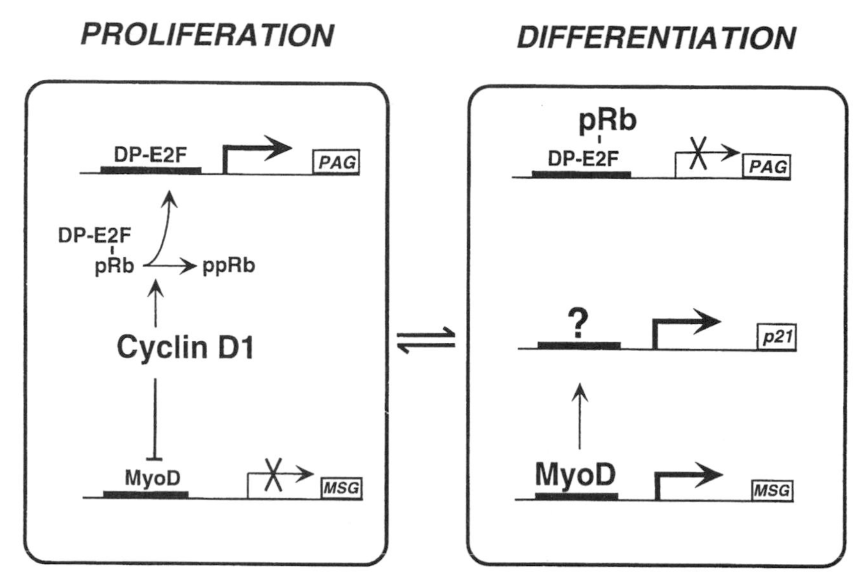

Figure 7.2. Proliferation and terminal differentiation are mutually exclusive. Upon growth factor withdrawal, cyclin D1 level decreases, and this leads to the accumulation of underphosphorylated retinoblastoma protein (pRb), which through its interaction with E2F-DP complexes shuts off transcription of proliferation-associated genes (PAG). Concomitantly, cyclin D1 downregulation releaves its inhibition of MyoD transcriptional activity and allows accumulation of MyoD and other myogenic factors, which induce transcription of muscle-specific genes (MSG). MyoD also activates expression of cyclin-dependent kinase inhibitors such as p21, leading to cell cycle withdrawal.

RNA viruses such as polyoma virus, simian virus 40 (SV40), and Rous sarcoma virus (RSV) (Fogel and Defendi, 1967; Yaffe and Gershon, 1967; Lee et al., 1968; Graessmann et al., 1973) because these viruses had been shown to induce host DNA synthesis and cell division in cells arrested in the cell cycle either by contact inhibition or by X-irradiation (Dulbecco et al., 1965). DNA synthesis as measured by tritiated thymidine incorporation could be detected at the time of fusion of the infected myoblasts into myotubes, but preformed myotubes were refractory to viral infection (Fogel and Defendi, 1967). Other reports, though, presented evidence of cellular DNA synthesis and mitotic figures in fully differentiated myotubes (Yaffe and Gershon, 1967; Lee et al., 1968; Graessmann et al., 1973). These results were questioned by other groups, which proposed that DNA synthesis and mitosis can be induced in the nuclei of cells that have just fused but not in the nuclei of mature myotubes (Holtzer et al., 1975; Falcone et al., 1984; Hu and Olson, 1990). It was only in 1989, with the generation and characterization of a myoblast cell line carrying a thermosensitive mutant form of SV40 T antigen under the control of an inducible promoter that this question was finally answered. In the course of studies on the molecular nature of the terminal differentiated state, the C2C12 mouse myogenic cell line

(Yaffe and Saxel, 1977; Blau et al., 1985) was stably transfected with a heavy metal–inducible, thermolabile SV40 T antigen (C2SVTts cells) (Endo and Nadal-Ginard, 1989; Cardoso et al., 1993; Gu et al., 1993). This cell line can differentiate into multinucleated muscle fibers under nonpermissive conditions for T antigen expression and activity, but, upon mitogen stimulation and shift to permissive conditions, it is able to reenter the cell cycle, as opposed to control "wild-type" muscle cultures (Fig. 7.1). The patterns of DNA replication foci observed in myotube nuclei induced to retrodifferentiate are indistinguishable from those of cycling myoblasts and fibroblasts (Cardoso et al., 1993). The visualization of typical early, middle, and late S phase replication patterns indicates that myotubes retain the potential to undergo a complete and well-coordinated S phase, which can be unlocked simply by expression of the viral SV40 T antigen (Cardoso et al., 1993) but not by a mutant T antigen form (K1 mutation), which is unable to bind pRb (Gu et al., 1993). Yet, continuous expression of large T is not required, suggesting a role in overriding a block and/or passing a restriction point, but not to actually drive DNA replication as in the viral infection (Cardoso et al., 1993), and supporting a view of terminal differentiation as a product of active and continuous regulation (Blau, 1992) rather than an irreversible default state.

In addition to SV40 large T antigen, adenovirus E1A 12S protein has been recently shown to induce cell cycle entry in terminally differentiated muscle cells (Crescenzi et al., 1995; Tiainen et al., 1996b). SV40 T antigen and adenovirus E1A are to date the only known proteins able to reactivate the cell cycle in terminally differentiated cells. Both of these DNA tumor virus oncogenes have no cellular counterparts. This differs from retroviral oncogenes whose expression can in many cases inhibit establishment of differentiation but whose cellular homologues remain inducible upon mitogenic stimulation in myotubes and cannot mediate retrodifferentiation.

A potential target of large T antigen and E1A during reversal of terminal differentiation is pRb. Elevated Rb mRNA levels, either due to transcription rate variation or to increased stability, as well as the hypophosphorylation of the pRb and tethering to the nucleus, have been correlated with cellular differentiation (Chen et al., 1989; Gu et al., 1993; Thorburn et al., 1993). Also, the hypophosphorylated pRb forms a complex with SV40 large T antigen (Ludlow et al., 1989), which correlates with the transforming properties of this viral oncogene (Cherington et al., 1988; DeCaprio et al., 1989). Rb phosphorylation as a consequence of large T antigen expression could lead to an increase in the free E2F-DP transcription factor complexes (see Chapter 5, this volume). The latter is presumably the transcriptionally active form (Hiebert et al., 1992; Weintraub et al., 1992) and could then stimulate transcription from promoters containing E2F sites, among which are several genes encoding proliferation-associated proteins (Fig. 7.2) (reviewed in Nevins, 1992; Muller, 1995). In agreement with this hypothesis, E2F and DP transcription factors have been found associated with pRb and p130 in extracts of differentiated muscle cultures, indicating that either of these tumor suppressor proteins or both may regulate E2F in muscle (Corbeil et al., 1995; Halevy et al., 1995; Kiess et al., 1995; Shin et al., 1995).

To directly test whether pRb inactivation by T antigen is necessary for cell cycle re-entry in terminally differentiated myotubes, a myogenic cell line derived from teratomas obtained from $Rb^{-/-}$ embryonic stem cells was generated (the CC42 cell line). The characterization of this cell line showed that the pRb-related protein p107 can substitute for pRb during establishment of muscle differentiation. Contrary to pRb, though, p107 is downregulated upon serum addition and cannot, thereafter, maintain the postmitotic state of myotubes (Schneider et al., 1994). Similar results have been independently obtained and extended to p130-deficient muscle cells. The latter was also shown to be unable to prevent cell cycle re-entry of myotubes (Novitch et al., 1996). Furthermore, studies of muscle differentiation in a mouse model expressing low levels of the pRb protein clearly demonstrated that during myogenesis in vivo pRb is essential for myoblast differentiation and survival (Zacksenhaus et al., 1996). Another possible target of E1A and SV40 T antigen is the nuclear protein p300, and, in fact, the ability of E1A to prevent skeletal muscle differentiation was shown to require sequences that bind p300 (Mymryk et al., 1992; Caruso et al., 1993). Recently, p300 was shown to potentiate MyoD and myogenin-dependent transcription activation and to be required for MyoD-dependent cell cycle arrest (Puri et al., 1997). Nevertheless, p300 is dispensable for the maintenance of the postmitotic state of myotubes and cannot substitute for pRb.

In addition to its role in the establishment and maintenance of permanent cell cycle withdrawal, pRb was shown to be required for activation of the muscle-specific transcription program in cooperation with MyoD (Gu et al., 1993). More specifically, pRb-deficient myogenic cells can induce the expression of early differentiation markers such as myogenin and the cyclin-dependent kinase inhibitor p21 but fail to elicit the expression of late differentiation markers such as myosin heavy chain (Novitch et al., 1996). At present, it is unclear how pRb cooperates with the myogenic factors to bring about the expression of the full differentiation program. Although ectopic expression of pRb stimulates MyoD-mediated transactivation of muscle promoters, pRb is not present in E-box DNA-binding complexes, which contain heterodimers of myogenic factors and E proteins (Gu et al., 1993). On the other hand, pRb and p130 are found in E2F-DP DNA-binding complexes in muscle extracts, but these complexes lack any associated myogenic factor. Thus, it is feasible that direct in vivo binding does not occur, and, alternatively, an indirect regulation takes place. Interestingly, in transient assays the Rb gene promoter has been shown to be transactivated by coexpressed MyoD by a mechanism independent of direct DNA binding (Martelli et al., 1994), suggesting an indirect mechanism for cross-regulation.

In addition to Rb, the transcription of the general cyclin-dependent kinase inhibitor p21 is also upregulated during myogenic differentiation in a manner that is independent of p53 but can be induced by MyoD (Fig. 7.2) (Halevy et al., 1995). This upregulation has also been observed in vivo during embryonic muscle differentiation (Parker et al., 1995). Even though p21 and p27 (as well as other cyclin-dependent kinase inhibitors such as p18) (Franklin and Xiong, 1996) are expressed at high levels in differentiated myotubes, they cannot maintain the postmitotic state of myotubes in the absence of functional pRb (Novitch et al., 1996). Furthermore, in spite of high p21 levels, pRb phosphorylation and re-entry into S phase occurred upon expression of E1A in pRb wild-type myotubes (Tiainen et al., 1996b). A primary function of cyclin-dependent kinase inhibitors in myotubes may be to maintain the hypophosphorylated, activated state of pRb, but they are not sufficient to maintain the block to cell cycle progression and prevent pRb phosphorylation as a consequence of SV40 T antigen or adenovirus E1A protein activity. An exception occurs in myotubes cultured from newts. In these amphibians, myotubes can dedifferentiate and yield proliferating mononucleated blastema cells in the regenerating limb (Lo et al., 1993). Although newt myotubes in culture have hypophosphorylated pRb, upon serum stimulation they undergo S phase. pRb is thought to be the physiological substrate of cyclin D/CDK4 complexes, which carry out the first inactivating phosphorylation steps of pRb. Cell cycle re-entry in newt myotubes also requires phosphorylation of pRb via a cyclin D/CDK4/6–dependent mechanism, but can be blocked by overexpression of cyclin-dependent kinase inhibitors (Tanaka et al., 1997). The difference in the regulation of pRb between newt and mammalian muscle might explain the amazing regenerative potential of these animals.

Our understanding of the molecular basis of cell cycle control during terminal differentiation is still limited, but several of the essential components and links have been identified, especially in the skeletal muscle system. As depicted in Figure 7.2, in mitogen-rich conditions, high levels of cyclin D1 mediate repression of (1) pRb growth suppressive activity and (2) MyoD transactivation of the general cyclin-dependent kinase inhibitor p21 and muscle-specific genes, thus keeping myoblasts in a proliferative state. Mitogen depletion (or other inductive cues) induces the downregulation of cyclin D1. This allows accumulation of underphosphorylated pRb, which represses via E2F-DP binding the transcription of proliferation-associated genes. In addition, release of cyclin D1 inhibition of MyoD transactivation leads to the induction of the muscle differentiation program and stimulation of p21 accumulation and cell cycle withdrawal. It is now possible to test multiple hypotheses derived from this model using skeletal muscle as well as other terminal differentiation systems.

In myogenic differentiation, as in other differentiation systems, the fine balance between opposing cellular signals determines whether the cell will divide or differentiate. When the correct control of the mutual exclusion between proliferative and differentiated states fails, the outcome is either neoplasia or apoptotic cell death.

APOPTOSIS DURING DEVELOPMENT

In the previous section, we focus on molecular mechanisms that are the basis for the decision to divide or to differentiate. In animal development, another important process regulating the appropriate numbers of each cell type is cell death by a controlled process termed *apoptosis.* Cell death during mammalian embryogenesis starts at the blastula stage and continues throughout the entire life as a means of keeping tissue homeostasis.

As a cell depends on extracellular signals to proliferate or to differentiate, it also does so to survive, and, in their absence, it activates an intrinsic program leading to death. This dependence on environmental signals for division and differentiation ensures that a particular cell type will divide or differentiate only when and where it is needed. The abrogation of this dependence on extracellular factors in the absence of cell death results in uncontrolled proliferation and ensuing neoplasia. To the decision whether to turn on the proliferation or the differentiation programs is added the decision to switch on the death program, which, from a certain point on at least, is an irreversible process. The balance among these three phenomena results in the development of specialized structures that execute particular functions in the organism.

A clear distinction between the processes underlying the death of cells during embryogenesis and tissue kinetics (and, also, some pathological states) and the one occurring in the center of acute conditions such as ischemia or hypoxia was established by Kerr et al. (1972). Morphological evidence indicated that in the latter cases nuclei become pycnotic (condensed, hyperchromatic nuclei), and cells and organelles swell and rupture, leaking their contents and provoking an inflammatory response. This cell death process is called *necrosis,* and it is characterized by general cell lysis due to physical or chemical insult to the cell. Necrosis is a pathological process, and it occurs as a consequence of impaired membrane transport with uncontrolled movement of ions and fluids. The cell death occurring during embryogenesis, metamorphosis, normal tissue turnover, and endocrine-dependent tissue atrophy called *apoptosis* entails, by contrast, cell shrinkage and membrane blebbing. Moreover, the plasma membrane and the membranes of organelles do not rupture; rather, they undergo changes in their composition, the chromatin condenses, the nuclear DNA becomes fragmented, and the nucleus collapses. One of the early events in the apoptotic program is the activation of an inside–outside phosphatidylserine (PS) translocase, which catalyzes the externalization of PS in the outer leaflet of the plasma membrane (Martin et al., 1995). This early change in the composition of the plasma membrane signals to the neighboring cells via their PS receptors that apoptosis is in progress, and, before there is leakage of cellular contents, the apoptotic cell is phagocytosed by its neighbors or by macrophages in the absence of inflammation.

Several methods, in addition to morphological observation, have been developed that allow detection of apoptotic cells. Two methods are based on the enzymatic or electrophoretic detection of DNA fragmentation, which is a hallmark of genetic death in apoptosis. In one, the cellular DNA is isolated and analyzed by agarose gel electrophoresis, giving the characteristic "ladder" of 200–300 bp fragments resulting from cleavage of the chromatin between the nucleosomes during apoptosis (oligonucleosomal ladder). The second method utilizes the enzyme terminal deoxynucleotidyltransferase (TdT) to catalyze the addition of labeled nucleotides to the 3′-OH ends of the fragmented cellular DNA, forming random heteropolymers of incorporated nucleotides, which can then be detected in situ or by flow cytometry. Two other very recent protocols rely on early changes in the execution of the apoptosis program, before morphological changes are visualized. These include the above-described externalization of PS in the plasma membrane and the activation of the ICE family of proteases (see below). The former takes advantage of the nonenergy-dependent specific binding of annexin V to PS and can be performed with living or fixed cells. The latter does not require intact cells and assays for the activity of an ICE-like protease (CPP32 or caspase 3) with a substrate conjugate that upon cleavage releases free fluorescent or chromogenic conjugate that can then be detected and quantified. All these methods exploit events that are common to the death program in most mammalian cells studied thus far.

Cell death was first described during amphibian metamorphosis more than 150 years ago and was later found to occur in most developing tissues of multicellular organisms (reviewed in Clarke and Clarke, 1996). It was only about a decade ago that the first genes required for apoptosis during nematode development were identified in genetic studies (Ellis and Horvitz, 1986). Subsequent studies showed that these genes had mammalian counterparts (e.g., the *Caenorhabditis elegans* Ced-9 and Ced-3 proteins are homologous to the mammalian Bcl-2 and ICE, respectively) (Yuan et al., 1993; Hengartner and Horvitz, 1994) and that the human genes could function in the nematode (Vaux et al., 1992; Hengartner and Horvitz, 1994), indicating that the death program is evolutionarily conserved, as it is the case for the cell cycle machinery.

In the past few years, studies from different research areas, including viral oncogenes, the genetics of *C. elegans* development, amphibian metamorphosis, thymocyte maturation, and cancer and the p53 protein, have culminated in a number of very important findings on the molecular basis of apoptosis. Genetic and biochemical experiments indicated that the apoptotic process is controlled by the Bcl-2 family of proteins, which either function as agonists (Bax or Bak) or as antagonists (Bcl-2 or Bcl-X_L) of apoptosis. At least 16 different genes of the Ced-9/Bcl-2 family have been identified to date (reviewed in Korsmeyer, 1995), and the various members can dimerize with each other. The ratio of apoptosis-promoting and -inhibiting factors in the cell determines whether the apoptotic program will be turned on (Fig. 7.3). A hint toward the mechanism of action of the Bcl-2 family members comes from structural studies suggesting that their three-dimensional structure is similar to bacterial colicins and that they may function, as the bacterial proteins, as pore-

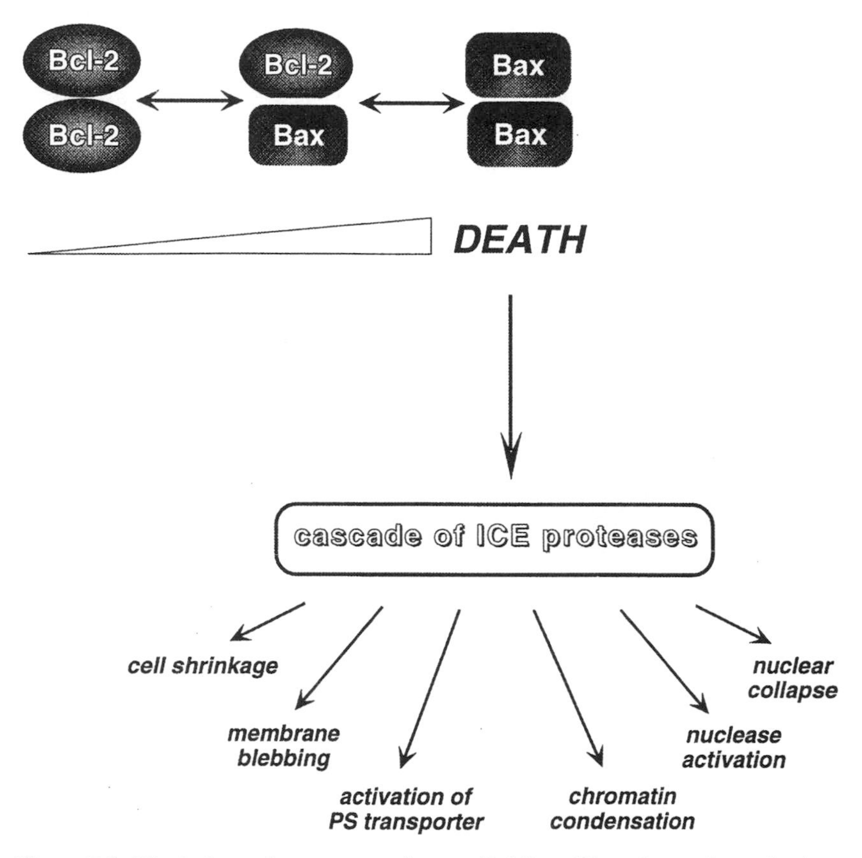

Figure 7.3. The balance between two factors, Bcl-2 and Bax, determines whether the cell dies or lives. Predominance of Bax homodimers activates a cascade of ICE family proteases, which cleave intracellular substrates after specific aspartate residues. Among the known substrates are lamin A, poly(ADP) ribose polymerase, rhoGDI, and actin. This causes a series of morphological and biochemical changes, including (1) cell shrinkage, (2) membrane blebbing and changes in its composition (externalization of phosphatidylserine), (3) chromatin condensation and internucleosomal cleavage, and (4) nuclear breakdown resulting from lamin cleavage.

forming transmembrane proteins (Muchmore et al., 1996). The various Bcl-2 family members have been found associated with the outer membrane of the mitochondria, the nuclear envelope, and the endoplasmic reticulum, and this association is mediated by a protein domain located at their C-terminal 22 amino acids. Though their exact mechanism of action is not known, genetic disruption of the Ced-9 in the nematode (Hengartner et al., 1992) and of Bcl-2 or Bcl-X in mice (Veis et al., 1993; Motoyama et al., 1995) clearly established their role in preventing cell death during development. Conversely, disruption of Bax in mice demonstrates its role in promoting neuronal cell death in development (Deckwerth et al., 1996).

Bcl-2 family members control the activation of a family of cysteine proteases, designated ICE proteases (see Fig. 7.3). These are related to the interleukin-1β–converting enzyme (ICE), which was identified as being homologous to the *C. elegans* protein Ced-3 that is required for all the cell deaths occurring during worm development (Yuan et al., 1993). Multiple members of this family have since been identified and in some cases implicated in apoptosis (reviewed in Nagata, 1997). They are all cysteine proteases that cleave their substrates after specific aspartate residues (therefore, they are designated *caspases,* for cysteine aspases) (Alnemri et al., 1996) and they are themselves activated by cleavage at specific aspartic acid residues. By now, there are at least 10 ICE members identified in humans, which can be subdivided into three groups (ICE-like, CPP32-like, and Ich1-like) based on their sequence homology (Alnemri et al., 1996). Specific recognition sequences have been identified for three caspases (1, 3, and 6) as well as some of their respective substrates. The latter include procaspases, poly(ADP) ribose polymerase (PARP), DNA-PK, rho-GDI, actin, and lamin A. The observation that, at least in vitro, caspases can activate themselves and also other caspases suggests that they function in a proteolytic cascade much as the cellular cascades of protein kinases that transduce proliferative or stress-activated signals. Because, in all animal cells studied, specific peptide or protein caspase inhibitors can block the cell death program, this family of proteins seems to play a central role in apoptosis by cleaving intracellular proteins from different subcellular compartments (nucleus, nuclear lamina, cytoskeleton, cytoplasm, and endoplasmic reticulum). A recent example of their fundamental role in a characteristic morphological change in cells undergoing apoptosis is the nuclear collapse. When mutant lamin A, carrying an amino acid substitution in a conserved position that prevents its cleavage by caspases, is overexpressed in mammalian cells and apoptosis induced, nuclear envelope breakdown and cell death is delayed by several hours, though other nuclear changes, such as nuclease activation, still take place at the normal pace (Rao et al., 1996).

Although the role of several of the cell deaths in development is not fully understood, the analysis of mutant organisms (from invertebrates to humans) begins to clarify the functions of the individual components either in tissue-specific or in generalized developmental defects. Cell death in development serves multiple purposes (reviewed in Sanders and Wride, 1995), among them sculpturing structures or deleting unneeded ones. Examples of the former are elimination of interdigital material during digit formation in mammals; creation of the preamniotic cavity in preimplantation mouse embryos; development of the presumptive neural tissue during gastrulation and formation of the neural plate during neurulation; and heart morphogenesis. In the latter case, one study reported 31 zones in the developing chick heart where apoptosis occurs (Pexieder, 1975), which clearly indicates an important role of cell death in the remodeling that occurs in this vital organ. During embryogenesis, some structures are created that are only required by one sex and are deleted at a certain time in the other sex by apoptosis. Other structures are remnants required by ancestral species or structures needed only at one stage of development, which are eliminated by apoptosis. As an example, during amphibian metamorphosis, there is massive muscle apoptotic cell death in the regressing tadpole tail and also in the remodeling of the dorsal body muscle during larval-to-adult maturation (Nishikawa and Hayashi, 1995).

Another function of apoptosis is the elimination of overproduced cells or of dangerous and nonfunctional cells. An example of the former happens in the differentiation of the nervous system, where half of the neurons and oligodendrocytes produced die (reviewed in Cowan et al., 1984). Overproduction followed by massive cell death may be a mechanism to match the number of innervating neurons to the target cells and to match the number of glial cells to the axons they myelinate. Neuronal survival is largely determined by the limited amount of neurotrophic factors produced by the target cells and by the competition for those factors with other neurons. Interestingly, inhibition of apoptosis by disrupting one of the ICE-like proteases (CPP32/caspase 3) yields mice that die around birth with a vast excess of cells in the central nervous system (Kuida et al., 1996). A very well-studied system that illustrates the role of apoptosis in the elimination of dangerous cells is the maturation of T and B lymphocytes, during which more than 95% of the cells produced die by apoptosis. The diversity of lymphocyte receptors generated via undirected genetic rearrangements during differentiation inevitably causes potentially harmful self-reactive receptors. These are eliminated by apoptosis in the thymus, and this maturation process requires the activity of death factor–receptor pairs. To date, three death factors (TNF, FasL, TRAIL) and four death factor receptors (TNFR1, Fas, DR3/Wsl-1, CAR1) have been isolated. Mice strains with mutations in FasL (gld) (Takahashi et al., 1994) and Fas (lpr) (Watanabe-Fukunaga et al., 1992; Adachi et al., 1995) clearly demonstrated the importance of this factor–receptor pair in the death of lymphocytes. Binding of the ligand FasL is thought to induce oligomerization of the receptor with transduction of the death signal by recruiting a protein called FADD/Mort 1, which in turn binds to caspase 8, initiating the protease cascade of the apoptotic program (reviewed in Nagata, 1997).

Abnormal or nonfunctional cells resulting from different cellular stresses (ultraviolet or ionizing radiation, heat shock, oxidative stress) are also removed by apoptosis. Though the pathway is not well understood, stress stimuli seem to involve initially a cascade of kinases belonging to the stress-activated protein kinases (SAPK) and the p38/RK family. It is not known how this signaling pathway interacts with the Bcl-2 family and/or with the ICE protease cascade, because most of the known targets of these kinases are transcription factors (reviewed in Cosulich and Clarke, 1996). In the case of DNA damage, the tumor suppressor p53 is often involved in the response process, and it can either cause cell cycle arrest while the damage is being repaired via the upregulation of cyclin-dependent kinase inhibitor p21 (El Deiry et al., 1993), or, possibly when the damage is too severe, it can induce apoptosis via upregulation of Bax (Miyashita and Reed, 1995). The upregulation of these two factors is mediated by p53 at the transcriptional level (reviewed in Levine, 1997).

Finally, cells that either do not complete or attempt to reverse terminal differentiation, most often activate the apoptosis program. The ones that do not die will likely give rise to tumors (see previous section). Two types of experimental approaches have consolidated this proposal. One has been the analysis of the phenotype of mice with homozygous deletion in cell cycle regulators and in transcription factors. The other has resulted from attempts to reinduce cell cycle progression in terminally differentiated cells such as cardiocytes and myotubes. The best example of the former case has been the analysis of pRb-deficient mice. These mice show extensive cell death in the tissues that normally express high levels of the pRb protein, which include liver, nervous system, lens, and skeletal muscle (Clarke et al., 1992; Jacks et al., 1992; Lee et al., 1992; Zacksenhaus et al., 1996).

These observations suggest that pRb is required to prevent apoptosis in the cell population committed to differentiate in addition to playing a role in the differentiation process itself. In accordance with this, ectopic expression of pRb has been reported to inhibit apoptosis induced by ionizing radiation (Haas-Kogan et al., 1995). One possibility is that apoptosis is activated upon inappropriate cell cycle entry, and the corollary of it would be that pRb prevents apoptosis by keeping cells out of the cell cycle. In support of this proposal, many genes involved in cell cycle progression are also associated with cell death. Examples of these are myc (Evan et al., 1992), cyclin D1 (Freeman et al., 1994), E1A (Debbas and White, 1993), and E7 (Pan and Griep, 1994). All of these proteins have been shown to bind and antagonize the function of pRb in cell cycle arrest. In addition, there is evidence suggesting that early steps in cell cycle progression also take place in the cell death pathway. Epithelial cells from the prostate were shown to synthetize DNA and induce proliferation markers such as proliferative cell nuclear antigen (PCNA) before undergoing apoptotic DNA fragmentation (Colombel et al., 1992). Another line of evidence comes from terminally differentiated cells induced to re-enter the cell cycle. Over 20 years ago, morphological observations of myotubes transformed with a temperature-sensitive mutant of the oncogenic Rous sarcoma virus and differentiated under nonpermissive conditions showed that, when these cultures were shifted back to the permissive temperature for viral function, myotubes vacuolated and degenerated completely within 72 hours (Holtzer et al., 1975). More recently, retrodifferentiation of myotubes by inducible expression of temperature-sensitive SV40 T antigen mutant not only achieved re-entering the cell cycle but also a certain percentage of the myotubes showed morphological features of apoptosis (Endo and Nadal-Ginard, 1989). A separate study also using the heat-labile SV40 T antigen mutant, reported that at nonpermissive conditions the cells underwent apoptosis. These cells expressed at permissive conditions high levels of wild-type p53 complexed with large T antigen, which after the temperature shift and T antigen degradation was not destabilized and was no longer in a complex with T antigen. Because overexpression of p53 was shown to induce apoptosis, the authors suggested that this accumulation of p53 could be responsible for the apoptosis in these cells (Yanai and Obinata, 1994). Whether this is also the cause of the cell death observed in part of the cells undergoing retrodifferentiation needs further investigation. SV40 T antigen expression in a different type of cells, the cerebellar Purkinje cells from transgenic mice, did not bring about tumorigenesis; rather, it correlated once again with cell ablation (Feddersen et al., 1992). In another terminally differentiated muscle cell type, cardiac myocytes, expression of E1A (Kirshenbaum and Schneider, 1995; Liu and Kitsis, 1996) or of E2F1 (Kirshenbaum et al., 1996) induced S phase re-entry and apoptosis. The kinetics of both processes suggested that the DNA synthesis preceded the apoptosis and that the cells did not complete the cell cycle (Liu and Kitsis, 1996). Expression of the adenoviral E1B protein together with E1A or with E2F1 avoids apoptosis and allows the cell cycle to progress (Kirshenbaum and Schneider, 1995; Kirshenbaum et al., 1996), most likely via inhibition of p53 activation of apoptosis (Debbas and White, 1993).

Altogether, these results reinforce the idea that the processes of cell cycle, cell death, and cell differentiation are coupled, and this is at least in part achieved by "sharing" some of the same regulatory molecules like pRb.

EPIGENETIC PATHWAYS IN DEVELOPMENT AND GROWTH CONTROL

One of the most fundamental discoveries in molecular biology was the deciphering of the genetic code. According to standard textbooks, the genetic alphabet is made out of four letters (nucleotides or bases): A (adenine), C (cytosine), G (guanine), and T (thymine). However, if we put this hypothesis to a simple test and analyze some of our own DNA, we find that there is in fact a fifth letter. About 1% of all bases in our DNA is 5-methylcytosine (^{m}C; Fig. 7.4). Now, where does this fifth nucleotide come from, how does it change our view of genetics, and what does it have to do with cell cycle and growth control? The purpose of this section is to address these questions. We first give a brief introduction to the basic knowledge and methods and then outline the role and regulation of DNA methylation.

This fifth, minor base, 5-methylcytosine, is generated by a postreplicative modification called *DNA methylation,* whereby a DNA methyltransferase (DNA MTase) transfers a methyl group to the target cytosine at the palindromic DNA sequence 5′-CG-3′ (CpG sites) (see Fig. 7.4). However, not all of these CpG sites are methylated. On average, only 60%–70% are methylated in mammalian cells, and the pattern of methylated and unmethylated CpG sites changes during development and disease.

The following brief overview of current knowledge in this rapidly

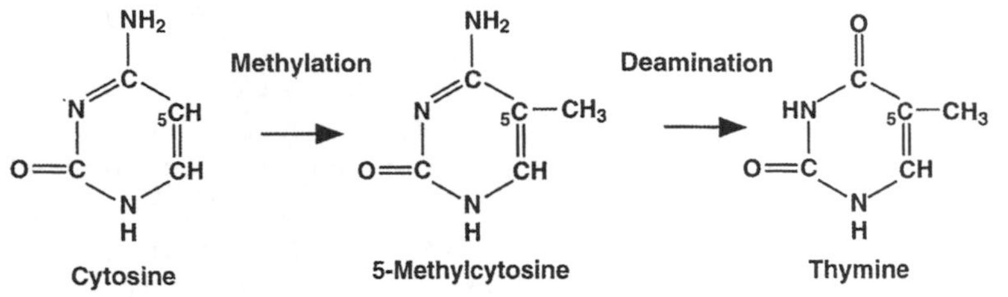

Figure 7.4. DNA methylation and deamination. The structure of the cytosine ring is shown. In the methylation reaction a proton (H^+) at the C5 position is replaced by a methyl group. The product, 5-methylcytosine, looks very similar to thymine and can be converted by a hydrolytic deamination. This structural similarity is the basis for C to T transitions, which occur at high frequency at CpG dinucleotides and represent mutational hotspots in the human genome.

progressing field shows that DNA methylation is anything but just a minor modification. By now, it is clear that DNA methylation is absolutely required for mammalian development and plays a critical role in the regulation of gene expression. This means, on the other hand, that errors in the methylation pattern can have far-reaching effects. DNA methylation can cause the erroneous activation or inactivation of genes and thus offset the balance of growth promoting and growth suppressing factors, just like changes in the DNA sequence itself. This potential of DNA methylation has long been overlooked or underestimated, but it finally caught broad attention and was recently found in many types of tumors. In addition to this direct effect on gene expression, DNA methylation can also indirectly affect the balance of growth control, because 5-methylcytosine residues are also hot spots for mutations. Altogether, these observations warrant a closer examination of the role of this fifth base in the regulation and deregulation of growth control in mammalian cells.

Detection and Regulation of DNA Methylation

A variety of techniques are available with which to analyze the methylation status of DNA (Grigg and Clark, 1994). The easiest and traditional method is based on digestion with methylation-sensitive restriction endonucleases. Most often used is the enzyme HpaII, which recognizes and cleaves the DNA sequence CCGG but not $C^{m}CGG$. For comparison the DNA is cut with its isoschizomer MspI, which does not discriminate between methylated and unmethylated sites and thus cleaves both. This assay can only be performed on the original genomic DNA because the methylation gets lost after subcloning and propagation in *Escherichia coli.* Therefore, the products of the HpaII and MspI digests are analyzed by Southern blotting techniques. Because one blot can be hybridized multiple times with different probes, the methylation levels of several regions of interest can be determined from one sample. The disadvantage of this approach is that only 1 out of 16 possible methylation sites (C^mCGG but not A^mCGG, and so forth) can be tested with this method.

Because methylation at a few single sites can often play a critical role in the regulation of gene expression, other methods are needed. The most comprehensive but also technically demanding one is the bisulfite method for selective base conversion. The basic principle is that cytosine residues in single-stranded DNA are efficiently deaminated to uracil in the presence of sodium bisulfite. Under these conditions, 5-methylcytosine residues do not react and remain unchanged (Frommer et al., 1992). The methylation status can then be determined with standard sequencing techniques because uracil is being read as a T (thymine) in place of the original cytosine. Moreover, the degree of methylation at any cytosine residue can be quantified by automated fluorescence-based genomic sequencing (Paul and Clark, 1996). This is also extremely sensitive because, after the selective base conversion in the first step, the template can be amplified by polymerase chain reaction and theoretically allows analysis of single cells.

Using any of the above methods to follow the DNA methylation pattern, one can find that the pattern of methylated and unmethylated sites is precisely maintained over many cell division cycles (Wigler et al., 1981). This process is often referred to as *maintenance methylation.* However, during development and differentiation dramatic changes, i.e., de novo methylation as well as loss of methylation, can be observed. (These processes are schematically outlined below in Fig. 7.6) Given the far-reaching effects of DNA methylation, it becomes clear that a precise maintenance as well as the controlled change of methylation pattern is crucial for the viability of mammals. The basic questions are (1) how, on the one hand, the methylation pattern is precisely maintained over many rounds of DNA replication and, on the other hand, changes are introduced at specific developmental stages; and (2) what goes wrong in some diseases.

We are still far from completely understanding the regulation of DNA methylation, but an important step in this direction was the cloning of the mouse and later the human DNA MTase (Bestor et al., 1988: Yen et al., 1992). This recombinant DNA MTase shows a strong preference for hemimethylated sites, which are generated by semiconservative DNA replication, and specifically transfers the methyl group to the newly synthesized strand at these sites. However, unmethylated sites are not recognized and remain unchanged. These enzymatic properties can in principle account for the maintenance of a given methylation pattern after DNA replication (Fig. 7.5). Still, an important question was how is this precision achieved in vivo; after all, there are on the order of 10^8 hemimethylated sites to be recognized and methylated during each cell cycle. A first clue came from studies on the subcellular localization of the DNA MTase. It has previously been shown that DNA replication occurs in a discrete pattern of nuclear foci that undergo characteristic changes throughout S phase (reviewed in Leonhardt and Cardoso, 1995).

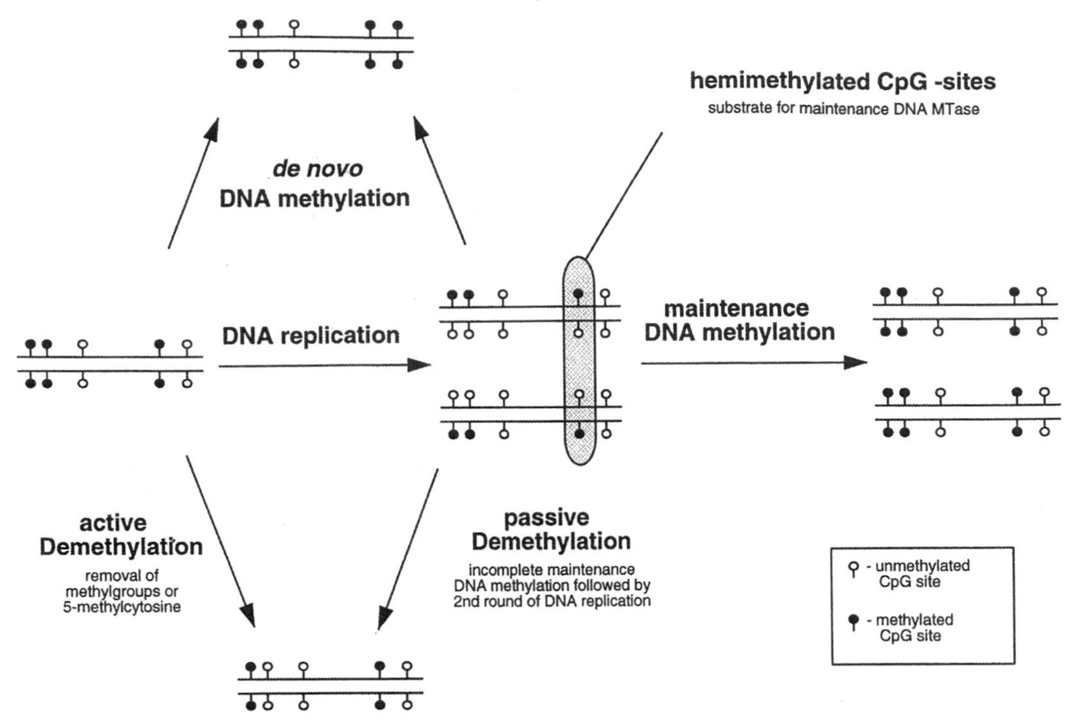

Figure 7.5. Regulation of DNA methylation patterns. The basic processes are outlined with the example of a specific DNA fragment containing five CpG sites of which three are methylated. In quantitative terms, the predominant process is the maintenance of a given methylation pattern after replication of the genome. The enzyme responsible for this reaction, DNA MTase (see also Fig. 7.6), has a strong preference for hemimethylated sites, i.e., sites that carry a methylgroup in the "old" strand and are unmethylated in the newly synthesized strand. Completely unmethylated sites are not modified, and thus the methylation pattern is precisely maintained. As discussed in the text, changes occur at certain stages of development and disease. At this moment, it is not yet known which enzymes are involved in the de novo methylation. On the other side, demethylation can occur by two different processes, either by active removal of 5-methylcytosine from the DNA or by a passive mechanism of failing to fill in a methyl group in the newly synthesized strand.

Several replication enzymes (PCNA, DNA polymerase α, and replication protein A) were found to be tightly associated with these foci (Bravo and Macdonald-Bravo, 1985; Cardoso et al., 1993; Hozak et al., 1993). It came as a surprise to find that DNA MTase was also localized at these foci, just like a replication enzyme. Further experiments identified a targeting sequence that is necessary and sufficient for localization at replication foci. This sequence is located in the N-terminal regulatory domain of DNA MTase and by itself can direct unrelated proteins like the β-galactosidase from *Escherichia coli* to these replication foci (Fig. 7.6). These results suggested an assembly line–like organization of DNA replication and methylation, which might explain the high fidelity of the

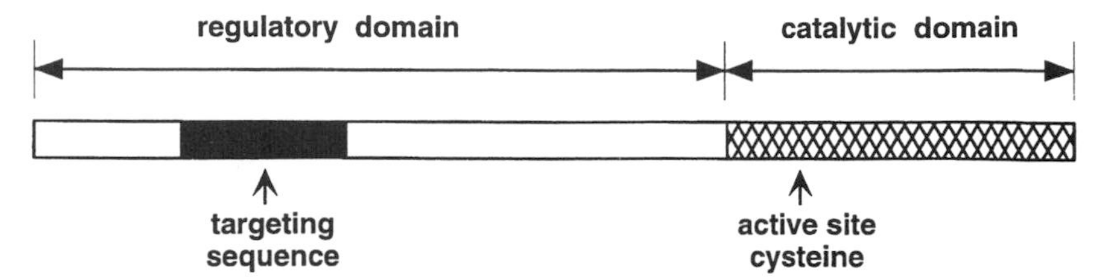

Figure 7.6. Structure of the mammalian DNA MTase. The known DNA MTase has a catalytic domain of about 500 amino acids that is very similar to the group of prokaryotic C5 DNA MTase and in fact has all the conserved peptide motifs described for this group (Leonhardt and Bestor, 1993). The regulatory domain contains the targeting sequence that is necessary and sufficient to direct DNA MTase to nuclear sites of DNA replication (Leonhardt et al., 1992). At this moment the exact size and structure of the regulatory domain is unclear, because new upstream sequences were discovered that could potentially extend the open reading frame by up to 171 amino acids and include also a testis-specific exon (Tucker et al., 1996b). The role of these tissue-specific forms in the maintenance and change of DNA methylation patterns remains to be elucidated.

maintenance of DNA methylation in mammalian cells (Leonhardt et al., 1992; Leonhardt and Cardoso, 1995).

The role of this maintenance MTase was clearly demonstrated by site-directed gene disruption. Homozygous mice ($Dnmt^{-/-}$) carrying a deletion in both alleles die at midgestation (E. Li et al., 1992). Interestingly, $Dnmt^{-/-}$ ES cells contain residual amounts of 5-methylcytosine and retain some de novo methyltransferase activity (Lei et al., 1996). These results suggest that, in addition to the known maintenance MTase (see Fig. 7.6) there are also one or more de novo MTases, which might introduce changes in the methylation pattern at specific stages, as was postulated more than 20 years ago (Holliday and Pugh, 1975). The postulated de novo MTase remained undetected despite intensive searches, and only recently were ESTs (expressed sequence tags) published that are homologous but not identical to the already known maintenance MTase, suggesting that another MTase exists in mammalian cells. The intriguingly complex structure of the known MTase suggests that it may be more than "just" a maintenance enzyme and that it might also participate in the change of methylation pattern. A first hint in support of this hypothesis comes from the recent identification of tissue-specific MTase mRNA forms (Tucker et al., 1996b). It was furthermore shown that partial in vitro proteolysis can convert this DNA MTase into an enzyme with de novo specificity (Bestor, 1992).

Another way of changing the methylation pattern is a loss of methylation. As outlined in Figure 7.5, there are two possible pathways, an active demethylation and a passive loss. The latter requires two rounds of DNA replication, in which a particular site is not methylated. This loss could be caused by transcription factors either blocking the site or preventing DNA MTase from accessing the site, e.g., by preventing DNA MTase from binding to the replication foci or by retaining it in the cytoplasm.

Interestingly, DNA MTase was found in the cytoplasm during preimplantation development, when the total concentration of 5-methylcytosine in the genome is decreasing (Carlson et al., 1992).

Direct evidence exists for active demethylation, i.e., the loss of methylation at specific sites in the absence of DNA replication. Well-documented examples of active demethylation include the chicken vitellogenin gene (Wilks et al., 1982, 1984), the rat α-actin gene (Paroush et al., 1990), and the chicken δ-crystallin (Sullivan and Grainger, 1987). Thus far, two different mechanisms of demethylation have been described. The vitellogenin gene becomes demethylated by an excision repair-like process (Jost, 1993). Demethylation of the α-actin gene was shown in vitro to be protease resistant and to require a catalytic RNA (Weiss et al., 1996). In summary, great progress has been made over the past years, but we are still far from understanding the regulation of DNA methylation. From the first results it is likely to be as complex as the regulation of gene expression itself.

DNA Methylation and the Control of Gene Expression

DNA methylation sites (CpG dinucleotides) are not evenly distributed throughout the genome, and their methylation status is not random. CpG dinucleotides are clearly underrepresented in the genome and occur only at about one-fifth of the statistically expected frequency (Russell et al., 1976). In the course of evolution, CpG sites were most likely destroyed by C to T transitions (see below and Fig. 7.4). The few remaining CpG sites within the coding part of genes are highly methylated, while unmethylated CpG sites are clustered at the 5′ end of all housekeeping genes and about 40% of all tissue-specific genes (Bird, 1986; Larsen et al., 1992). Due to the high density of CpG sites in the promoter region, these clusters are generally referred to as *CpG islands.*

There are literally hundreds of studies showing an inverse relationship between DNA methylation in the promoter region and transcriptional activity and the respective changes throughout development and differentiation. Just to mention one out of many studies in the early 1980s, it was shown that in the process of lymphocyte maturation the transcribed immunoglobin genes were always hypomethylated and that the methylation in these loci decreased as maturation progressed (Storb and Arp, 1983). This type of observation raised the question of whether it was just a correlation or whether DNA methylation was indeed the cause of gene inactivation or maybe only a consequence of gene inactivation that was brought about by other means. The effect of DNA methylation on gene expression was directly analyzed by introducing unmethylated and in vitro methylated reporter constructs into mammalian cells. Using, for example, the γ-globin gene, it was shown that methylation in the promoter region, but not in the coding part of the gene, prevents transcription (Busslinger et al., 1983).

In principle, two different models involving a direct or an indirect mechanism can be envisaged to explain the effect of DNA methylation on gene transcription. Studies on the expression control of the tyrosine aminotransferase gene showed that all promoter-binding factors were present in both expressing and nonexpressing cells. However, the promoter region was methylated in nonexpressing cells, and that prevented binding of these factors (Becker et al., 1987). Similar results were obtained with an adenovirus major late promoter (Watt and Molloy, 1988) and the cAMP-responsive element (Iguchi-Ariga and Schaffner, 1989). Interestingly, methylation of the SP1-binding site, which is typical for housekeeping promoters, did not prevent transcription factor binding (Höller et al., 1988). These experiments clearly show that DNA methylation can specifically interfere with protein–DNA interactions and thus influence gene expression.

Even though the methyl group is strategically well positioned in the major groove of the DNA double helix, where it can directly interfere with the binding of regulatory factors, in most cases the effects on transcription control are probably exerted by nonspecific means. In that regard, it was shown that DNA methylation prevents the formation of active chromatin and renders integrated DNA sequences inactive and DNase I insensitive (Keshet et al., 1986). The formation of inactive chromatin is mediated by nuclear factor(s) that specifically bind to methylated CpG sites (Boyes and Bird, 1991). Thus far, two different methyl-CpG–binding proteins (MeCP1 and MeCP2) have been identified (Meehan et al., 1989; Lewis et al., 1992). MeCP2 is, like DNA MTase itself, dispensable in embryonic stem cells but is required for embryonic development (Tate et al., 1996).

In summary, by now there is no question that DNA methylation can and does control gene expression. It is also clear that DNA methylation preserves the inactive state of genes, but it remains to be elucidated how the methylation of CpG islands is regulated. Whatever the final picture may look like, DNA methylation is undoubtedly an essential element in the control of gene expression during mammalian development. Most of the last paragraphs dealt with its effect on the expression of endogenous genes, but also the expression of foreign DNA is affected. Mammalian cells seem to be able to recognize foreign DNA and inactivate it by DNA methylation. Integrated viral sequences, e.g., are highly methylated, and infectivity as well as viral gene expression are inversely correlated with methylation (Stuhlmann et al., 1981; Jähner et al., 1982). In fact, viral sequences are scattered throughout our genome, but—fortunately for us—they are highly methylated and transcriptionally inactive. What appears like a blessing in the case of viral infection constitutes a serious problem for gene therapy trials, where transgene expression decreases with time.

DNA Methylation During Differentiation, Development, and Aging

As discussed throughout this section, DNA methylation often correlates with gene activity and changes as genes are being turned on or off during development and differentiation. The regulation of these developmental changes in the methylation pattern is still far from being understood, but DNA methylation is clearly more than an accompanying molecular ornamentation. DNA methylation by itself can indeed induce entire differentiation programs. This is best illustrated by some early experiments treating undifferentiated fibroblasts (10T1/2 and 3T3 cells) with the demethylating agent 5-azacytidine. Within several days after addition of the drug clear phenotypical conversions took place, and several types of differentiated cells were observed, including mostly contractile striated muscle cells, adipocytes, and chondrocytes (Taylor and Jones, 1979). Also other differentiation programs like smooth muscle cell differentiation (Iehara et al., 1996) and neuronal differentiation (Bartolucci et al., 1989) can be induced by forced hypomethylation. By now, much progress has been made studying the induction of myogenic differentiation. One factor that is directly and/or indirectly activated by hypomethylation and can convert fibroblasts to myoblasts is the myogenic transcription factor MyoD (Davis et al., 1987). The MyoD promoter is partially methylated in most tissues, but it is unmethylated in skeletal muscle (Zingg et al., 1994). A distal enhancer element is also methylated in nonmuscle tissues, and demethylation precedes transcription of MyoD (Brunk et al., 1996). The role of DNA methylation is furthermore supported by experiments showing myogenic differentiation upon overexpression of a DNA MTase antisense mRNA (Szyf et al., 1992). Curiously, overexpression of the DNA MTase itself in myoblasts was also reported to accelerate myogenic differentiation and myotube formation (Takagi et al., 1995). Altogether, these results clearly show a regulatory role of DNA methylation during myogenic differentiation; however, the mechanisms causing these changes in the methylation pattern during differentiation remain to be elucidated.

In the first part of this chapter, it is discussed that transformation/proliferation and differentiation are mutually exclusive, which requires that differentiation factors be inactivated in the process of cellular transformation. In support of this hypothesis, hypermethylation of the MyoD promoter region was observed during immortalization and transformation of fibroblast cell lines (Jones et al., 1990), as well as in the process of oncogenic transformation (Rideout et al., 1994). The reverse also seems possible and, for example, forced hypomethylation induces neuronal differentiation in neuroblastoma cells (Bartolucci et al., 1989).

Among these developmental changes in the methylation pattern, two phenomena, chromosomal imprinting and X-inactivation, stand out in view of their particular regulation and significance in biology. Early pronuclei transplantation experiments on one-cell-stage embryos indicated that the genomes inherited from the father and the mother are not equivalent. Embryos with either two maternally or two paternally derived genomes could not complete embryogenesis (McGrath and Solter, 1984). Because both genomes should be identical at the DNA sequence level, there had to be something else that distinguished both genomes and rendered them nonequivalent. This distinctive mark also had to be of a transient nature because, after passage through the germline, an initially maternally derived genome might then be inherited from a father. In brief, this distinctive mark is a difference in the methylation pattern between maternal and paternal chromosomes. As a consequence of these different patterns, different sets of genes are expressed from the paternal and the maternal genomes. Genes exhibiting this parent-specific expression pattern are called *imprinted genes* (for recent reviews on this rapidly progressing area, see Barlow, 1995; Latham et al., 1995). There are altogether 17 imprinted genes known, including insulin-like growth factor-II (IGF-II) which is discussed in more detail below. The most recent entry to this list of imprinted genes is the human potassium channel gene KVLQT1 (Lee et al., 1997; Neyroud et al., 1997), and more imprinted genes are likely to be discovered soon.

Whatever the intricate regulatory mechanisms may be that lead to the transcriptional silencing of imprinted genes, the disruption of the mouse DNA MTase gene (Dnmt) leading to genome-wide hypomethylation clearly demonstrated that DNA methylation is required to maintain this inactive status (Li et al., 1993). The parent-specific imprint, i.e., methylation pattern, is acquired during oogenesis and spermatogenesis. Once this imprint is erased by genetic manipulations it can only be restored by renewed passage through the germline (Tucker et al., 1996a).

A special case of developmentally controlled monoallelic expression is the X-chromosome inactivation. In this case, an entire chromosome (one of the two X chromosomes in female somatic cells) is transcriptionally silenced and methylated. Methylation does not appear to be the cause, however, but rather a secondary phenomenon because, in the case of the X-linked Hprt gene, methylation clearly occurs after inactivation and transcriptional silencing (Lock et al., 1987). Instead, inactivation seems to be initiated by the cis-acting regulatory RNA Xist, which is expressed before X-inactivation (Kay et al., 1993). The Xist gene itself is one of the above-mentioned imprinted genes and is expressed from the inactive X-chromosome. DNA hypomethylation causes the ectopic expression of the Xist gene on the active X chromosome, which then causes the inactivation of this chromosome (Panning and Jaenisch, 1996).

Finally, the DNA methylation pattern also changes in the process of aging. In some protooncogenes, e.g., c-fos and c-myc, age-dependent changes in the DNA methylation pattern were observed (Ono et al., 1989; Uehara et al., 1989), suggesting that altered gene expression caused by DNA methylation may affect cellular vitality in the senescent phase. In some cases, similar changes were observed in the process of aging and tumor formation. In the aging colorectal mucosa, for example, increasing methylation of the estrogen receptor was observed, which might constitute an age-dependent predisposition to sporadic colon tumors (Issa et al., 1994). Moreover, the age-dependent changes in the methylation pattern of the c-fos gene can also be found in the early stages of hepatocarcinogenesis (Choi et al., 1996). Although it is still too early for general conclusions, the changes in the methylation pattern occurring in the aging organism might affect the overall fitness of cells and also represent a predisposition for neoplastic transformations.

DNA Methylation and Cancer

DNA methylation can affect cell cycle and growth control by two different mechanisms. First, ectopic changes in the methylation pattern can cause either transcriptional silencing of growth suppressing factors or activation of growth promoting factors. Second, 5-methylcytosine is also a hot spot for mutations, which can cause inactivation of growth suppressors. Both mechanisms can unbalance the regulation of growth control and thus cause tumor formation (Fig. 7.7). For a better understanding of these different roles of DNA methylation during neoplastic transformation and tumor outgrowth, one has to take into account the distribution of methylated and unmethylated CpG sites in the genome (Fig. 7.8). Methylation of the CpG island in the promoter region causes transcriptional silencing, and deamination of methylated sites within the coding sequence causes mutations.

It was early observed that in the process of tumor formation the overall level of 5-methylcytosine decreases (Feinberg and Vogelstein,

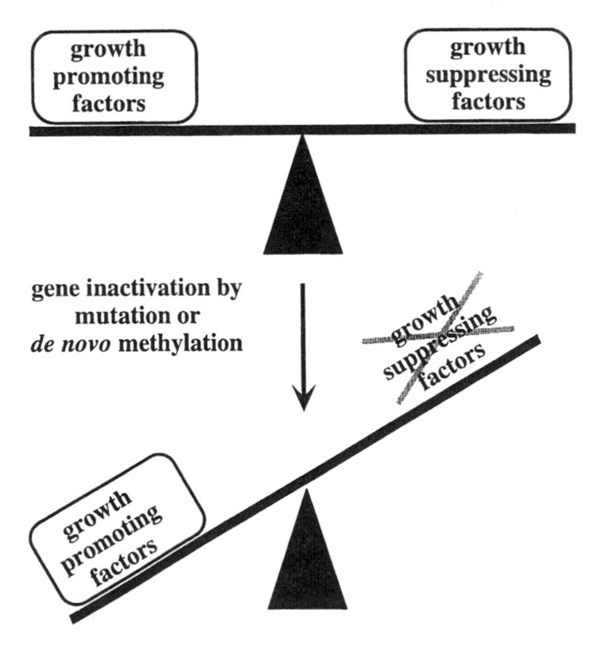

Figure 7.7. Balance of growth regulators. The successful control of cell proliferation in normal tissues depends on a balance of growth promoting and growth suppressing factors, as well as apoptosis. This balance can be disturbed by a variety of different mechanisms potentially leading to uncontrolled cell proliferation and neoplasia. Two of these mechanisms, transcriptional silencing and C to T transition mutations, involve DNA methylation (see Figs. 7.4 and 7.8).

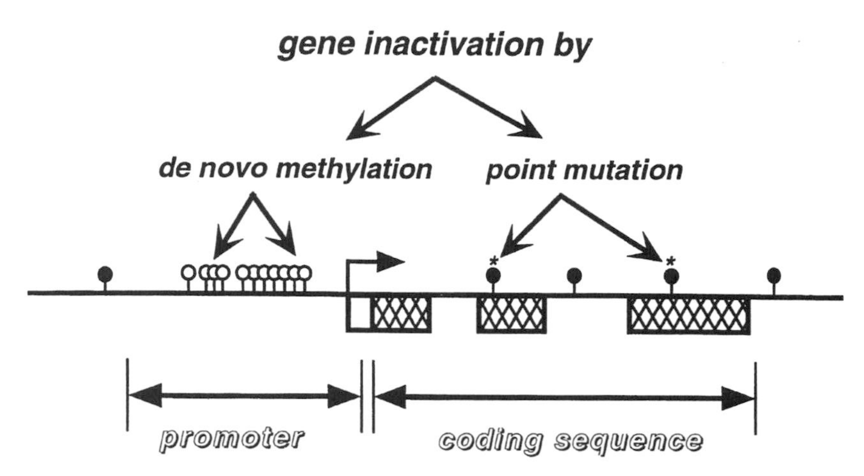

Figure 7.8. Inactivation of genes by two methylation-dependent mechanisms. Unlike all other dinucleotides, CpG sites are not evenly distributed throughout the genome. They tend to be clustered in the promoter region, where they form the so-called CpG island (see text), and they are clearly underrepresented within the coding sequence of genes. Moreover, cytosine residues in the promoter region are usually unmethylated, whereas cytosine residues within the coding sequence are mostly methylated. Genes can be inactivated by either de novo methylation in the promoter region or by C to T transition mutations within the coding sequence. Transcriptional silencing by de novo methylation is a normal and integral mechanism of transcription control during development, but errors may cause accidental inactivation of genes with dramatic consequences for the organism.

1983; Gama-Sosa et al., 1983; Feinberg et al., 1988), whereas selected CpG islands are methylated (Baylin et al., 1987). Practically nothing is known about how these changes in the methylation pattern are caused. Several studies showed an elevated level of DNA methyltransferase activity in tumor cells, which seems to increase during the progression of neoplasms (Kautiainen and Jones, 1986; El Deiry et al., 1991). However, it remains obscure how this increased activity can be compatible with the observed general hypomethylation and the specific methylation of CpG islands. Moreover, tumors are characterized by a higher proliferation, which is accompanied by a general upregulation of proliferation-associated genes, and higher DNA methyltransferase levels may simply be a reflection of this overall change (Lee et al., 1996).

The pRb was the first identified tumor suppressor protein and served as a paradigm for tumor formation. In most cases, inactivation of the Rb gene is caused by a mutation (deletion and point mutations), but in about 10%–20% of unilateral retinoblastomas hypermethylation was found in the 5′ region of the Rb gene (Greger et al., 1989; 1994; Sakai et al., 1991). In some sporadic retinoblastomas (16% of all unilateral tumors), no mutations were detected, and it was found that inactivation was caused by hypermethylation, which prevented binding of transcription factors (Ohtani-Fujita et al., 1993). This type of analysis was later extended to the newly discovered family of cyclin-dependent kinase inhibitors, and hypermethylation with transcriptional silencing was found for p15 (Herman et al., 1996) and p16 (Gonzalez-Zulueta et al., 1995; Herman et al., 1995; Merlo et al., 1995; Otterson et al., 1995).

In more than half of human breast cancers, hypermethylation and absence of mRNA of the mammary-derived growth inhibitor were observed (Huynh et al., 1996). In addition to these suppressors, other growth regulators like the estrogen receptor (Issa et al., 1994) and even a DNA repair enzyme (Harris et al., 1996) were found to be hypermethylated and transcriptionally silenced during transformation. Finally, differentiation factors like the MyoD transcription factor were also found to become hypermethylated during oncogenic transformation (Rideout et al., 1994).

Many of these studies show a correlation between tumor formation, transcriptional silencing, and hypermethylation, but in some cases a direct causal relationship could be established. In some renal carcinoma cell lines, re-expression of the von Hippel-Lindau gene was achieved by treatment with the demethylating agent 5-azacytosine (Herman et al., 1994). The promoter of the cell adhesion molecule E-cadherin was also found to be hypermethylated in breast and prostate carcinomas. These cells were shown to contain the necessary transcription factors, and 5-azacytosine treatment indeed caused re-expression of E-cadherin as well as concomitant reversal to epithelial cell morphology and adhesive properties (Graff et al., 1995; Yoshiura et al., 1995).

This growing evidence that hypermethylation of tumor suppressor genes may play an important role during carcinogenesis also inspired the reverse approach, i.e., to screen for hypermethylated sites in tumors. Using this approach, a new potential tumor suppressor gene (HIC-1, *h*ypermethylated *i*n *c*ancer) was identified (Wales et al., 1995). In light of these results, the action of known carcinogens was also revisited and, e.g., nickel seems to cause chromatin condensation and DNA methylation (Lee et al., 1995).

The above discussion concerns the inactivation of tumor suppressors in a broad sense. However, the activation of growth factors can also be caused by changes in the DNA methylation pattern and lead to cancer. As discussed above, most genes are expressed from both alleles, but some genes are expressed exclusively from either the paternal or the maternal chromosome, a process termed *imprinting*. Transcription of the imprinted genes is blocked by DNA methylation. Imprinted genes generally include factors involved in cell cycle and growth control, and normal development depends on the balanced expression of these factors. Relaxation of imprinted genes was first observed in Wilms' tumor patients. Normally, IGF-II is expressed from the paternal chromosome, but about two-thirds of Wilms' tumors showed biallelic IGF-II expression (Ogawa et al., 1993; Rainier et al., 1993). Loss of imprinting of the IGF-II gene was also discovered in a number of other human tumors.

DNA methylation is also a hot spot for mutations, and it can affect cell cycle and growth control. From early studies in *E. coli,* it is clear that methylated cytosines are hot spots for C to T transitions. Analyse of mutations causing human genetic diseases have shown that this is also true in humans, and it was estimated that about 35% of all point mutations occur at CpG sites (Cooper and Youssoufian, 1988). This mutational hot spot is also involved in the inactivation of cell cycle regulators. In the case of the p53 tumor suppressor gene, it was calculated that on average about 25% of all mutations reported in all human tumors occur at the CpG dinucleotide (Greenblatt et al., 1994). These data clearly indicate that "normal" DNA methylation by itself can pose a risk to human health. In general, it seems that less methylation is better, and in fact forced hypomethylation appears to decrease the incidence of intestinal neoplasia in some tumor-prone mice strains (Laird et al., 1995).

The molecular basis for this high mutation rate at CpG sites lies in the similarity between 5-methylcytosine and thymine (note that both have a methyl group at the C-5 position; see Fig. 7.4). Hydrolytic deamination of 5-methylcytosine, but not cytosine, causes a direct conversion from C to T. It was shown that this conversion can be enhanced by DNA MTase itself under conditions of limiting S-adenosylmethionine (SAM) (Shen et al., 1992). These results concur with an earlier observation that a methyl-deficient diet lowers the intracellular concentration of SAM and enhances liver tumor formation in rats (Mikol et al., 1983). Admittedly, the conditions used for these experiments were rather extreme and are highly unlikely to occur in human nutrition, but less dramatic changes in the intracellular SAM levels might be effective in the long run. The latter idea was indeed supported by two large health studies in the United States showing that a low dietary intake of methionine and folate and a high consumption of alcohol, which seems to influence the methyl group availability, increases the risk for colorectal neoplasia (Giovannucci et al., 1993). Thus, the old adage to "eat fresh fruit and vegetables and not drink too much alcohol" makes sense from the point of view of DNA methylation and cancer. It is still too early for definite conclusions, but these results may link DNA methylation, dietary habits, and carcinogenesis.

PERSPECTIVES

Most of the above section deals with all of the problems that can be caused by DNA methylation, raising the question of what is it good for. The results with DNA MTase–deficient mice, which die at midgestation, indicate that we cannot live without it. After millions of years with DNA methylation, it became an indispensable regulatory factor in the coordinated and developmentally controlled expression of genes.

At first, the demethylating agents like 5-azacytidine and 5-aza-2′-deoxycytidine appeared to be miracle drugs. They reactivate silenced tumor suppressors, reverse the process of oncogenic transformation, lower the risks of mutation, and so forth, but—as so often in life—there are side effects involved, and more specific approaches are needed. For a long time little was known about the regulation of DNA methylation. Now, since the cloning of mammalian DNA MTase and the new possibilities of genetic manipulations in mice, rapid progress is being made and will lead to a better understanding of the regulation and, most importantly, the deregulation of DNA methylation in mammals. Future studies should elucidate genetic, dietary, and environmental factors affecting DNA methylation. With this knowledge, we might some day be able to positively influence DNA methylation in the prevention of cancer and in the process of aging. We may also be able to use it for antiviral defenses and bypass it in gene therapy approaches.

REFERENCES

Adachi M, Suematsu S, Kondo T, Ogasawara J, Tanaka T, Yoshida N, Nagata S (1995): Targeted mutation in the Fas gene causes hyperplasia in peripheral lymphoid organs and liver. Nature Genet 11:294–300.

Alema S, Tato F (1994): Oncogenes and muscle differentiation: Multiple mechanisms of interference. Semin Cancer Biol 5:147–156.

Alnemri ES, Livingston DJ, Nicholson DW, Salvesen G, Thornberry NA, Wong WW, Yuan J (1996): Human ICE/CED-3 protease nomenclature. Cell 87:171.

Barlow DP (1995): Gametic imprinting in mammals. Science 270:1610–1613.

Bartolucci S, Estenoz M, Longo A, Santoro B, Momparler RL, Rossi M, Augusti-Tocco G (1989): 5-Aza-2′-deoxycytidine as inducer of differentiation and growth inhibition in mouse neuroblastoma cells. Cell Differ Dev 27:47–55.

Baylin SB, Fearon ER, Vogelstein B, de Bustros A, Sharkis SJ, Burke PJ, Staal SP, Nelkin BD (1987): Hypermethylation of the 5′ region of the calcitonin gene is a property of human lymphoid and acute myeloid malignancies. Blood 70:412–417.

Becker PB, Ruppert S, Schutz G (1987): Genomic footprinting reveals cell type–specific DNA binding of ubiquitous factors. Cell 51:435–443.

Benezra R, Davis RL, Lassar A, Tapscott S, Thayer M, Lockshon D, Weintraub H (1990): Id: A negative regulator of helix-loop-helix DNA binding proteins. Control of terminal myogenic differentiation. Ann NY Acad Sci 599:1–11.

Bengal E, Ransone L, Scharfmann R, Dwarki VJ, Tapscott SJ, Weintraub H, Verma IM (1992): Functional antagonism between c-Jun and MyoD proteins: A direct physical association. Cell 68:507–519.

Bestor TH (1992): Activation of mammalian DNA methyltransferase by cleavage of a Zn binding regulatory domain. EMBO J 11:2611–2617.

Bestor T, Laudano A, Mattaliano R, Ingram V (1988): Cloning and sequencing of a cDNA encoding DNA methyltransferase of mouse cells. The carboxyl-terminal domain of the mammalian enzymes is related to bacterial restriction methyltransferases. J Mol Biol 203:971–983.

Bird AP (1986): CpG-rich islands and the function of DNA methylation. Nature 321:209–213.

Blau HM (1992): Differentiation requires continuous active control. Annu Rev Biochem 61:1213–1230.

Blau HM, Pavlath GK, Hardeman EC, Chiu CP, Silberstein L, Webster SG,

Miller SC, Webster C (1985): Plasticity of the differentiated state. Science 230:758–766.

Boyes J, Bird A (1991): DNA methylation inhibits transcription indirectly via a methyl-CpG binding protein. Cell 64:1123–1134.

Braun T, Bober E, Winter B, Rosenthal N, Arnold HH (1990): Myf-6, a new member of the human gene family of myogenic determination factors: Evidence for a gene cluster on chromosome 12. EMBO J 9:821–831.

Braun T, Buschhausen-Denker G, Bober E, Tannich E, Arnold HH (1989): A novel human muscle factor related to but distinct from MyoD1 induces myogenic conversion in 10T1/2 fibroblasts. EMBO J 8:701–709.

Bravo R, Macdonald-Bravo H (1985): Changes in the nuclear distribution of cyclin (PCNA) but not its synthesis depend on DNA replication. EMBO J 4:655–661.

Breitbart RE, Liang CS, Smoot LB, Laheru DA, Mahdavi V, Nadal-Ginard B (1993): A fourth human MEF2 transcription factor, hMEF2D, is an early marker of the myogenic lineage. Development 118:1095–1106.

Brunk BP, Goldhamer DJ, Emerson CP Jr (1996): Regulated demethylation of the myoD distal enhancer during skeletal myogenesis. Dev Biol 177:490–503.

Busslinger M, Hurst J, Flavell RA (1983): DNA methylation and the regulation of globin gene expression. Cell 34:197–206.

Cardoso MC, Leonhardt H, Nadal-Ginard B (1993): Reversal of terminal differentiation and control of DNA replication: Cyclin A and Cdk2 specifically localize at subnuclear sites of DNA replication. Cell 74:979–992.

Carlson LL, Page AW, Bestor TH (1992): Properties and localization of DNA methyltransferase in preimplantation mouse embryos: Implications for genomic imprinting. Genes Dev 6:2536–2541.

Caruso M, Martelli F, Giordano A, Felsani A (1993): Regulation of MyoD gene transcription and protein function by the transforming domains of the adenovirus E1A oncoprotein. Oncogene 8:267–278.

Chen PL, Scully P, Shew JY, Wang JY, Lee WH (1989): Phosphorylation of the retinoblastoma gene product is modulated during the cell cycle and cellular differentiation. Cell 58:1193–1198.

Cherington V, Brown M, Paucha E, ST Louis J, Spiegelman BM, Roberts TM (1988): Separation of simian virus 40 large-T-antigen–transforming and origin-binding functions from the ability to block differentiation. Mol Cell Biol 8:1380–1384.

Choi EK, Uyeno S, Nishida N, Okumoto T, Fujimura S, Aoki Y, Nata M, Sagisaka K, Fukuda Y, Nakao K, Yoshimoto T, Kim YS, Ono T (1996): Alterations of c-fos gene methylation in the processes of aging and tumorigenesis in human liver. Mutat Res 354:123–128.

Clarke AR, Maandag ER, van Roon M, van der Lugt NM, van der Valk M, Hooper ML, Berns A, te Riele H (1992): Requirement for a functional Rb-1 gene in murine development. Nature 359:328–330.

Clarke PGH, Clarke S (1996): Nineteenth century research on naturally occurring cell death and related phenomena. Anat Embryol 193:81–99.

Clegg CH, Linkhart TA, Olwin BB, Hauschka SD (1987): Growth factor control of skeletal muscle differentiation: Commitment to terminal differentiation occurs in G_1 phase and is repressed by fibroblast growth factor. J Cell Biol 105:949–956.

Colombel M, Olsson CA, Ng PY, Buttyan R (1992): Hormone-regulated apoptosis results from reentry of differentiated prostate cells onto a defective cell cycle. Cancer Res 52:4313–4319.

Cooper DN, Youssoufian H (1988): The CpG dinucleotide and human genetic disease. Hum Genet 78:151–155.

Corbeil HB, Whyte P, Branton PE (1995): Characterization of transcription factor E2F complexes during muscle and neuronal differentiation. Oncogene 11:909–920.

Cosulich S, Clarke P (1996): Apoptosis: Does stress kill? Curr Biol 6:1586–1588.

Cowan WM, Fawcett JW, O'Leary DD, Stanfield BB (1984): Regressive events in neurogenesis. Science 225:1258–1265.

Crescenzi M, Fleming TP, Lassar AB, Weintraub H, Aaronson SA (1990): MyoD induces growth arrest independent of differentiation in normal and transformed cells. Proc Natl Acad Sci USA 87:8442–8446.

Crescenzi M, Soddu S, Tato F (1995): Mitotic cycle reactivation in terminally differentiated cells by adenovirus infection. J Cell Physiol 162:26–35.

Davis RL, Weintraub H, Lassar AB (1987): Expression of a single transfected cDNA converts fibroblasts to myoblasts. Cell 51:987–1000.

Debbas M, White E (1993): Wild-type p53 mediates apoptosis by E1A, which is inhibited by E1B. Genes Dev 7:546–554.

DeCaprio JA, Ludlow JW, Lynch D, Furukawa Y, Griffin J, Piwnica-Worms H, Huang CM, Livingston DM (1989): The product of the retinoblastoma susceptibility gene has properties of a cell cycle regulatory element. Cell 58:1085–1095.

Deckwerth TL, Elliott JL, Knudson CM, Johnson EM Jr, Snider WD, Korsmeyer SJ (1996): BAX is required for neuronal death after trophic factor deprivation and during development. Neuron 17:401–411.

Dulbecco R, Hartwell LH, Vogt M (1965): Induction of cellular DNA synthesis by polyoma virus. Proc Natl Acad Sci USA 53:403–410.

Edmondson DG, Olson EN (1989): A gene with homology to the myc similarity region of MyoD1 is expressed during myogenesis and is sufficient to activate the muscle differentiation program. Genes Dev 3:628–640.

El Deiry WS, Nelkin BD, Celano P, Yen RW, Falco JP, Hamilton SR, Baylin SB (1991): High expression of the DNA methyltransferase gene characterizes human neoplastic cells and progression stages of colon cancer. Proc Natl Acad Sci USA 88:3470–3474.

El Deiry WS, Tokino T, Velculescu VE, Levy DB, Parsons R, Trent JM, Lin D, Mercer WE, Kinzler KW, Vogelstein B (1993): WAF1, a potential mediator of p53 tumor suppression. Cell 75:817–825.

Ellis HM, Horvitz HR (1986): Genetic control of programmed cell death in the nematode *C. elegans.* Cell 44:817–829.

Endo T, Nadal-Ginard B (1986): Transcriptional and posttranscriptional control of c-myc during myogenesis: Its mRNA remains inducible in differentiated cells and does not suppress the differentiated phenotype. Mol Cell Biol 6:1412–1421.

Endo T, Nadal-Ginard B (1989): SV40 large T-antigen induces reentry of terminally differentiated myotubes into the cell cycle. In Kedes LH, Stockdale FE (eds): The Cellular and Molecular Biology of Muscle Development. New York: Alan R. Liss, Inc., pp 95–104.

Evan GI, Wyllie AH, Gilbert CS, Littlewood TD, Land H, Brooks M, Waters CM, Penn LZ, Hancock DC (1992): Induction of apoptosis in fibroblasts by c-myc protein. Cell 69:119–128.

Ewton DZ, Florini JR (1981): Effects of the somatomedins and insulin on myoblast differentiation in vitro. Dev Biol 86:31–39.

Falcone G, Alema S, Tato F (1991): Transcription of muscle-specific genes is repressed by reactivation of pp60v–src in postmitotic quail myotubes. Mol Cell Biol 11:3331–3338.

Falcone G, Boettiger D, Alema S, Tato F (1984): Role of cell division in differentiation of myoblasts infected with a temperature-sensitive mutant of Rous sarcoma virus. EMBO J 3:1327–1331.

Feddersen RM, Ehlenfeldt R, Yunis WS, Clark HB, Orr HT (1992): Disrupted cerebellar cortical development and progressive degeneration of Purkinje cells in SV40 T antigen transgenic mice. Neuron 9:955–966.

Feinberg AP, Gehrke CW, Kuo KC, Ehrlich M (1988): Reduced genomic 5-methylcytosine content in human colonic neoplasia. Cancer Res 48:1159–1161.

Feinberg AP, Vogelstein B (1983): Hypomethylation distinguishes genes of some human cancers from their normal counterparts. Nature 301:89–92.

Fogel M, Defendi V (1967): Infection of muscle cultures from various species with oncogenic DNA viruses (SV40 and polyoma). Proc Natl Acad Sci USA 58:967–973.

Franklin DS, Xiong Y (1996): Induction of p18INK4c and its predominant association with CDK4 and CDK6 during myogenic differentiation. Mol Biol Cell 7:1587–1599.

Freeman RS, Estus S, Johnson EM Jr (1994): Analysis of cell cycle–related gene expression in postmitotic neurons: Selective induction of cyclin D1 during programmed cell death. Neuron 12:343–355.

Frommer M, McDonald LE, Millar DS, Collis CM, Watt F, Grigg GW, Molloy PL, Paul CL (1992): A genomic sequencing protocol that yields a positive display of 5-methylcytosine residues in individual DNA strands. Proc Natl Acad Sci USA 89:1827–1831.

Gama-Sosa MA, Slagel VA, Trewyn RW, Oxenhandler R, Kuo KC, Gehrke CW, Ehrlich M (1983): The 5-methylcytosine content of DNA from human tumors. Nucleic Acids Res 11:6883–6894.

Giovannucci E, Stampfer MJ, Colditz GA, Rimm EB, Trichopoulos D, Rosner BA, Speizer FE, Willett WC (1993): Folate, methionine, and alcohol intake and risk of colorectal adenoma. J Natl Cancer Inst 85:875–884.

Gonzalez-Zulueta M, Bender CM, Yang AS, Nguyen T, Beart RW, Van Tornout JM, Jones PA (1995): Methylation of the 5′ CpG island of the p16/CDKN2 tumor suppressor gene in normal and transformed human tissues correlates with gene silencing. Cancer Res 55:4531–4535.

Goss RJ (1966): Hypertrophy versus hyperplasia. Science 153:1615–1620.

Graessmann A, Graessmann M, Fogel M (1973): The relationship of polyoma virus-induced tumor (T) antigen to activation of DNA synthesis in rat myotubes. Dev Biol 35:180–186.

Graff JR, Herman JG, Lapidus RG, Chopra H, Xu R, Jarrard DF, Isaacs WB, Pitha PM, Davidson NE, Baylin SB (1995): E-cadherin expression is silenced by DNA hypermethylation in human breast and prostate carcinomas. Cancer Res 55:5195–5199.

Greenblatt MS, Bennett WP, Hollstein M, Harris CC (1994): Mutations in the p53 tumor suppressor gene: Clues to cancer etiology and molecular pathogenesis. Cancer Res 54:4855–4878.

Greger V, Debus N, Lohmann D, Hopping W, Passarge E, Horsthemke B (1994): Frequency and parental origin of hypermethylated RB1 alleles in retinoblastoma. Hum Genet 94:491–496.

Greger V, Passarge E, Hopping W, Messmer E, Horsthemke B (1989): Epigenetic changes may contribute to the formation and spontaneous regression of retinoblastoma. Hum Genet 83:155–158.

Grigg G, Clark S (1994): Sequencing 5-methylcytosine residues in genomic DNA. BioEssays 16:431–436.

Gu W, Schneider JW, Condorelli G, Kaushal S, Mahdavi V, Nadal-Ginard B (1993): Interaction of myogenic factors and the retinoblastoma protein mediates muscle cell commitment and differentiation. Cell 72:309–324.

Guo K, Walsh K (1997): Inhibition of myogenesis by multiple cyclin–Cdk complexes. Coordinate regulation of myogenesis and cell cycle activity at the level of E2F. J Biol Chem 272:791–797.

Haas-Kogan DA, Kogan SC, Levi D, Dazin P, T'Ang A, Fung YK, Israel MA (1995): Inhibition of apoptosis by the retinoblastoma gene product. EMBO J 14:461–472.

Halevy O, Novitch BG, Spicer DB, Skapek SX, Rhee J, Hannon GJ, Beach D, Lassar AB (1995): Correlation of terminal cell cycle arrest of skeletal muscle with induction of p21 by MyoD. Science 267:1018–1021.

Hardy S, Kong Y, Konieczny SF (1993): Fibroblast growth factor inhibits MRF4 activity independently of the phosphorylation status of a conserved threonine residue within the DNA-binding domain. Mol Cell Biol 13:5943–5956.

Harris H (1990): The role of differentiation in the suppression of malignancy. J Cell Sci 97:5–10.

Harris LC, von Wronski MA, Venable CC, Remack JS, Howell SR, Brent TP (1996): Changes in O6-methylguanine–DNA methyltransferase expression during immortalization of cloned human fibroblasts. Carcinogenesis 17:219–224.

Hengartner MO, Ellis RE, Horvitz HR (1992): *Caenorhabditis elegans* gene ced-9 protects cells from programmed cell death. Nature 356:494–499.

Hengartner MO, Horvitz HR (1994): *C. elegans* cell survival gene ced-9 encodes a functional homolog of the mammalian proto-oncogene bcl-2. Cell 76:665–676.

Herman JG, Jen J, Merlo A, Baylin SB (1996): Hypermethylation-associated inactivation indicates a tumor suppressor role for p15INK4B. Cancer Res 56:722–727.

Herman JG, Latif F, Weng Y, Lerman MI, Zbar B, Liu S, Samid D, Duan DS, Gnarra JR, Linehan WM, et al. (1994): Silencing of the VHL tumor-suppressor gene by DNA methylation in renal carcinoma. Proc Natl Acad Sci USA 91:9700–9704.

Herman JG, Merlo A, Mao L, Lapidus RG, Issa JP, Davidson NE, Sidransky D, Baylin SB (1995): Inactivation of the CDKN2/p16/MTS1 gene is frequently associated with aberrant DNA methylation in all common human cancers. Cancer Res 55:4525–4530.

Hiebert SW, Chellappan SP, Horowitz JM, Nevins JR (1992): The interaction of RB with E2F coincides with an inhibition of the transcriptional activity of E2F. Genes Dev 6:177–85.

Höller M, Westin G, Jiricny J, Schaffner W (1988): Sp1 transcription factor binds DNA and activates transcription even when the binding site is CpG methylated. Genes Dev 2:1127–1135.

Holliday R, Pugh JE (1975): DNA modification mechanisms and gene activity during development. Science 187:226–232.

Holtzer H, Biehl J, Yeoh G, Meganathan R, Kaji A (1975): Effect of oncogenic virus on muscle differentiation. Proc Natl Acad Sci USA 72:4051–4055.

Hozak P, Hassan AB, Jackson DA, Cook PR (1993): Visualization of replication factories attached to nucleoskeleton. Cell 73:361–373.

Hu JS, Olson EN (1990): Functional receptors for transforming growth factor-beta are retained by biochemically differentiated C2 myocytes in growth factor–deficient medium containing EGTA but down-regulated during terminal differentiation. J Biol Chem 265:7914–7919.

Huynh H, Alpert L, Pollak M (1996): Silencing of the mammary-derived growth inhibitor (MDGI) gene in breast neoplasms is associated with epigenetic changes. Cancer Res 56:4865–4870.

Iehara N, Takeoka H, Tsuji H, Imabayashi T, Foster DN, Strauch AR, Yamada Y, Kita T, Doi T (1996): Differentiation of smooth muscle phenotypes in mouse mesangial cells. Kidney Int 49:1330–1341.

Iguchi-Ariga SM, Schaffner W (1989): CpG methylation of the cAMP-responsive enhancer/promoter sequence TGACGTCA abolishes specific factor binding as well as transcriptional activation. Genes Dev 3:612–619.

Issa JP, Ottaviano YL, Celano P, Hamilton SR, Davidson NE, Baylin SB (1994): Methylation of the oestrogen receptor CpG island links ageing and neoplasia in human colon. Nature Genet 7:536–540.

Jacks T, Fazeli A, Schmitt EM, Bronson RT, Goodell MA, Weinberg RA (1992): Effects of an Rb mutation in the mouse. Nature 359:295–300.

Jähner D, Stuhlmann H, Stewart CL, Harbers K, Löhler J, Simon I, Jaenisch R (1982): De novo methylation and expression of retroviral genomes during mouse embryogenesis. Nature 298:623–628.

Jen Y, Weintraub H, Benezra R (1992): Overexpression of Id protein inhibits the muscle differentiation program: In vivo association of Id with E2A proteins. Genes Dev 6:1466–1479.

Jones PA, Wolkowicz MJ, Rideout WMd, Gonzales FA, Marziasz CM, Coetzee GA, Tapscott SJ (1990): De novo methylation of the MyoD1 CpG island during the establishment of immortal cell lines. Proc Natl Acad Sci USA 87:6117–6121.

Jost JP (1993): Nuclear extracts of chicken embryos promote an active demethylation of DNA by excision repair of 5-methyldeoxycytidine. Proc Natl Acad Sci USA 90:4684–4688.

Jurgens G (1995): Axis formation in plant embryogenesis: Cues and clues. Cell 81:467–470.

Kautiainen TL, Jones PA (1986): DNA methyltransferase levels in tumorigenic and nontumorigenic cells in culture. J Biol Chem 261:1594–1598.

Kay GF, Penny GD, Patel D, Ashworth A, Brockdorff N, Rastan S (1993): Expression of Xist during mouse development suggests a role in the initiation of X chromosome inactivation. Cell 72:171–182.

Kerr JF, Wyllie AH, Currie AR (1972): Apoptosis: A basic biological phenomenon with wide-ranging implications in tissue kinetics. Br J Cancer 26:239–257.

Keshet I, Lieman-Hurwitz J, Cedar H (1986): DNA methylation affects the formation of active chromatin. Cell 44:535–543.

Kiess M, Gill RM, Hamel PA (1995): Expression and activity of the retinoblastoma protein (pRB)–family proteins, p107 and p130, during L6 myoblast differentiation. Cell Growth Differ 6:1287–1298.

Kirshenbaum LA, Abdellatif M, Chakraborty S, Schneider MD (1996): Human E2F-1 reactivates cell cycle progression in ventricular myocytes and represses cardiac gene transcription. Dev Biol 179:402–411.

Kirshenbaum LA, Schneider MD (1995): Adenovirus E1A represses cardiac gene transcription and reactivates DNA synthesis in ventricular myocytes, via alternative pocket protein– and p300-binding domains. J Biol Chem 270:7791–7794.

Konieczny SF, Drobes BL, Menke SL, Taparowsky EJ (1989): Inhibition of myogenic differentiation by the H-ras oncogene is associated with the down-regulation of the MyoD1 gene. Oncogene 4:473–81.

Konigsberg IR, McElvain N, Tootle M, Herrmann H (1960): The dissociability of deoxyribonucleic acid synthesis from the development of multinuclearity of muscle cells in culture. J Biophys Biochem Cytol 8:333–343.

Korsmeyer SJ (1995): Regulators of cell death. Trends Genet 11:101–105.

Kuida K, Zheng TS, Na S, Kuan C, Yang D, Karasuyama H, Rakic P, Flavell RA (1996): Decreased apoptosis in the brain and premature lethality in CPP32-deficient mice. Nature 384:368–372.

Laird PW, Jackson-Grusby L, Fazeli A, Dickinson SL, Jung WE, Li E, Weinberg RA, Jaenisch R (1995): Suppression of intestinal neoplasia by DNA hypomethylation. Cell 81:197–205.

Larsen F, Gundersen G, Lopez R, Prydz H (1992): CpG islands as gene markers in the human genome. Genomics 13:1095–1107.

Lassar AB, Thayer MJ, Overell RW, Weintraub H (1989): Transformation by activated ras or fos prevents myogenesis by inhibiting expression of MyoD1. Cell 58:659–667.

Latham KE, McGrath J, Solter D (1995): Mechanistic and developmental aspects of genetic imprinting in mammals. Int Rev Cytol 160:53–98.

Lee EY, Chang CY, Hu N, Wang YC, Lai CC, Herrup K, Lee WH, Bradley A (1992): Mice deficient for Rb are nonviable and show defects in neurogenesis and haematopoiesis. Nature 359:288–294.

Lee HH, Kaighn ME, Ebert JD (1968): Induction of thymidine-3H incorporation in multinucleated myotubes by Rous sarcoma virus. Int J Cancer 3:126–136.

Lee MP, Hu RJ, Johnson LA, Feinberg AP (1997): Human KVLQT1 gene shows tissue-specific imprinting and encompasses Beckwith-Wiedemann syndrome chromosomal rearrangements. Nature Genet 15:181–185.

Lee PJ, Washer LL, Law DJ, Boland CR, Horon IL, Feinberg AP (1996): Limited upregulation of DNA methyltransferase in human colon cancer reflecting increased cell proliferation. Proc Natl Acad Sci USA 93:10366–10370.

Lee YW, Klein CB, Kargacin B, Salnikow K, Kitahara J, Dowjat K, Zhitkovich A, Christie NT, Costa M (1995): Carcinogenic nickel silences gene expression by chromatin condensation and DNA methylation: A new model for epigenetic carcinogens. Mol Cell Biol 15:2547–2557.

Lei H, Oh SP, Okano M, Juttermann R, Goss KA, Jaenisch R, Li E (1996): De novo DNA cytosine methyltransferase activities in mouse embryonic stem cells. Development 122:3195–3205.

Leonhardt H, Bestor TH (1993): Structure, function and regulation of mammalian DNA methyltransferase. In Jost JP, Saluz HP (eds): DNA Methylation: Molecular Biology and Biological Significance. Basel: Birkhäuser Verlag, pp 109–119.

Leonhardt H, Cardoso MC (1995): Targeting and association of proteins with functional domains in the nucleus: The insoluble solution. Int Rev Cytol 162B:303–335.

Leonhardt H, Page AW, Weier HU, Bestor TH (1992): A targeting sequence directs DNA methyltransferase to sites of DNA replication in mammalian nuclei. Cell 71:865–873.

Levine AJ (1997): p53, the cellular gatekeeper for growth and division. Cell 88:323–331.

Lewis JD, Meehan RR, Henzel WJ, Maurer-Fogy I, Jeppesen P, Klein F, Bird A (1992): Purification, sequence, and cellular localization of a novel chromosomal protein that binds to methylated DNA. Cell 69:905–914.

Li E, Beard C, Jaenisch R (1993): Role for DNA methylation in genomic imprinting. Nature 366:362–365.

Li E, Bestor TH, Jaenisch R (1992): Targeted mutation of the DNA methyltransferase gene results in embryonic lethality. Cell 69:915–926.

Li L, Zhou J, James G, Heller-Harrison R, Czech MP, Olson EN (1992): FGF inactivates myogenic helix-loop-helix proteins through phosphorylation of a conserved protein kinase C site in their DNA-binding domains. Cell 71:1181–1194.

Liu Y, Kitsis RN (1996): Induction of DNA synthesis and apoptosis in cardiac myocytes by E1A oncoprotein. J Cell Biol 133:325–334.

Lo DC, Allen F, Brockes JP (1993): Reversal of muscle differentiation during urodele limb regeneration. Proc Natl Acad Sci USA 90:7230–7234.

Lock LF, Takagi N, Martin GR (1987): Methylation of the Hprt gene on the inactive X occurs after chromosome inactivation. Cell 48:39–46.

Ludlow JW, DeCaprio JA, Huang CM, Lee WH, Paucha E, Livingston DM (1989): SV40 large T antigen binds preferentially to an underphosphorylated member of the retinoblastoma susceptibility gene product family. Cell 56:57–65.

Martelli F, Cenciarelli C, Santarelli G, Polikar B, Felsani A, Caruso M (1994): MyoD induces retinoblastoma gene expression during myogenic differentiation. Oncogene 9:3579–3590.

Martin JF, Miano JM, Hustad CM, Copeland NG, Jenkins NA, Olson EN (1994): A Mef2 gene that generates a muscle-specific isoform via alternative mRNA splicing. Mol Cell Biol 14:1647–1656.

Martin JF, Schwarz JJ, Olson EN (1993): Myocyte enhancer factor (MEF) 2C: A tissue-restricted member of the MEF-2 family of transcription factors. Proc Natl Acad Sci USA 90:5282–5286.

Martin SJ, Reutelingsperger CP, McGahon AJ, Rader JA, van Schie RC, LaFace DM, Green DR (1995): Early redistribution of plasma membrane phosphatidylserine is a general feature of apoptosis regardless of the initiating stimulus: Inhibition by overexpression of Bcl-2 and Abl. J Exp Med 182:1545–1556.

Massague J, Cheifetz S, Endo T, Nadal-Ginard B (1986): Type beta transforming growth factor is an inhibitor of myogenic differentiation. Proc Natl Acad Sci USA 83:8206–8210.

McDermott JC, Cardoso MC, Yu YT, Andres V, Leifer D, Krainc D, Lipton SA, Nadal-Ginard B (1993): hMEF2C gene encodes skeletal muscle- and brain-specific transcription factors. Mol Cell Biol 13:2564–2577.

McGrath J, Solter D (1984): Completion of mouse embryogenesis requires both the maternal and paternal genomes. Cell 37:179–183.

Meehan RR, Lewis JD, McKay S, Kleiner EL, Bird AP (1989): Identification of a mammalian protein that binds specifically to DNA containing methylated CpGs. Cell 58:499–507.

Merlo A, Herman JG, Mao L, Lee DJ, Gabrielson E, Burger PC, Baylin SB, Sidransky D (1995): 5′ CpG island methylation is associated with transcriptional silencing of the tumour suppressor p16/CDKN2/MTS1 in human cancers. Nature Med 1:686–692.

Meyerowitz EM (1997): Genetic control of cell division patterns in developing plants. Cell 88:299–308.

Mikol YB, Hoover KL, Creasia D, Poirier LA (1983): Hepatocarcinogenesis in rats fed methyl-deficient, amino acid–defined diets. Carcinogenesis 4:1619–1629.

Miner JH, Wold B (1990): Herculin, a fourth member of the MyoD family of myogenic regulatory genes. Proc Natl Acad Sci USA 87:1089–1093.

Miyashita T, Reed JC (1995): Tumor suppressor p53 is a direct transcriptional activator of the human bax gene. Cell 80:293–299.

Motoyama N, Wang F, Roth KA, Sawa H, Nakayama K, Nakayama K, Negishi I, Senju S, Zhang Q, Fujii S, et al. (1995): Massive cell death of immature hematopoietic cells and neurons in Bcl-x–deficient mice. Science 267:1506–1510.

Muchmore SW, Sattler M, Liang H, Meadows RP, Harlan JE, Yoon HS, Nettesheim D, Chang BS, Thompson CB, Wong SL, Ng SL, Fesik SW (1996): X-ray and NMR structure of human Bcl-xL, an inhibitor of programmed cell death. Nature 381:335–341.

Muller R (1995): Transcriptional regulation during the mammalian cell cycle. Trends Genet 11:173–178.

Mymryk JS, Lee RW, Bayley ST (1992): Ability of adenovirus 5 E1A proteins to suppress differentiation of BC3H1 myoblasts correlates with their binding to a 300 kDa cellular protein. Mol Biol Cell 3:1107–1115.

Nadal-Ginard B (1978): Commitment, fusion and biochemical differentiation of a myogenic cell line in the absence of DNA synthesis. Cell 15:855–864.

Nagata S (1997): Apoptosis by death factor. Cell 88:355–365.

Nevins JR (1992): E2F: A link between the Rb tumor suppressor protein and viral oncoproteins. Science 258:424–429.

Neyroud N, Tesson F, Denjoy I, Leibovici M, Donger C, Barhanin J, Faure S, Gary F, Coumel P, Petit C, Schwartz K, Guicheney P (1997): A novel mutation in the potassium channel gene KVLQT1 causes the Jervell and Lange-Nielsen cardioauditory syndrome. Nature Genet 15:186–189.

Nishikawa A, Hayashi H (1995): Spatial, temporal and hormonal regulation of programmed muscle cell death during metamorphosis of the frog *Xenopus laevis*. Differentiation 59:207–214.

Novitch BG, Mulligan GJ, Jacks T, Lassar AB (1996): Skeletal muscle cells lacking the retinoblastoma protein display defects in muscle gene expression and accumulate in S and G2 phases of the cell cycle. J Cell Biol 135:441–456.

Ogawa O, Eccles MR, Szeto J, McNoe LA, Yun K, Maw MA, Smith PJ, Reeve AE (1993): Relaxation of insulin-like growth factor II gene imprinting implicated in Wilms' tumour. Nature 362:749–751.

Ohtani-Fujita N, Fujita T, Aoike A, Osifchin NE, Robbins PD, Sakai T (1993): CpG methylation inactivates the promoter activity of the human retinoblastoma tumor-suppressor gene. Oncogene 8:1063–1067.

Olson EN (1992): Interplay between proliferation and differentiation within the myogenic lineage. Dev Biol 154:261–272.

Olson EN, Klein WH (1994): bHLH factors in muscle development: Dead lines and commitments, what to leave in and what to leave out. Genes Dev 8:1–8.

Olson EN, Perry M, Schulz RA (1995): Regulation of muscle differentiation by the MEF2 family of MADS box transcription factors. Dev Biol 172:2–14.

Olson EN, Sternberg E, Hu JS, Spizz G, Wilcox C (1986): Regulation of myogenic differentiation by type beta transforming growth factor. J Cell Biol 103:1799–1805.

Olwin BB, Hauschka SD (1988): Cell surface fibroblast growth factor and epidermal growth factor receptors are permanently lost during skeletal muscle terminal differentiation in culture. J Cell Biol 107:761–769.

Ono T, Takahashi N, Okada S (1989): Age-associated changes in DNA methylation and mRNA level of the c-myc gene in spleen and liver of mice. Mutat Res 219:39–50.

Otterson GA, Khleif SN, Chen W, Coxon AB, Kaye FJ (1995): CDKN2 gene silencing in lung cancer by DNA hypermethylation and kinetics of p16INK4 protein induction by 5-aza 2′deoxycytidine. Oncogene 11:1211–1216.

Pan H, Griep AE (1994): Altered cell cycle regulation in the lens of HPV-16 E6 or E7 transgenic mice: Implications for tumor suppressor gene function in development. Genes Dev 8:1285–1299.

Panning B, Jaenisch R (1996): DNA hypomethylation can activate Xist expression and silence X-linked genes. Genes Dev 10:1991–2002.

Parker SB, Eichele G, Zhang P, Rawls A, Sands AT, Bradley A, Olson EN, Harper JW, Elledge SJ (1995): p53-independent expression of p21Cip1 in muscle and other terminally differentiating cells. Science 267:1024–1027.

Paroush Z, Keshet I, Yisraeli J, Cedar H (1990): Dynamics of demethylation and activation of the alpha-actin gene in myoblasts. Cell 63:1229–1237.

Paul CL, Clark SJ (1996): Cytosine methylation: Quantitation by automated genomic sequencing and GENESCAN analysis. BioTechniques 21:126–133.

Pexieder T (1975): Cell death in the morphogenesis and teratogenesis of the heart. Adv Anat Embryol Cell Biol 51:3–99.

Pollock R, Treisman R (1991): Human SRF-related proteins: DNA binding properties and potential regulatory targets. Genes Dev 5:2327–2341.

Puri PL, Avantaggiati ML, Balsano C, Sang N, Graessmann A, Giordano A, Levrero M (1997): p300 is required for MyoD-dependent cell cycle arrest and muscle-specific gene transcription. EMBO J 16:369–383.

Rainier S, Johnson LA, Dobry CJ, Ping AJ, Grundy PE, Feinberg AP (1993): Relaxation of imprinted genes in human cancer. Nature 362:747–749.

Rao L, Perez D, White E (1996): Lamin proteolysis facilitates nuclear events during apoptosis. J Cell Biol 135:1441–1455.

Rao SS, Chu C, Kohtz DS (1994): Ectopic expression of cyclin D1 prevents activation of gene transcription by myogenic basic helix-loop-helix regulators. Mol Cell Biol 14:5259–5267.

Rao SS, Kohtz DS (1995): Positive and negative regulation of D-type cyclin expression in skeletal myoblasts by basic fibroblast growth factor and transforming growth factor beta. A role for cyclin D1 in control of myoblast differentiation. J Biol Chem 270:4093–4100.

Rhodes SJ, Konieczny SF (1989): Identification of MRF4: A new member of the muscle regulatory factor gene family. Genes Dev 3:2050–2061.

Rideout WM, Eversole-Cire P, Spruck CH, Hustad CM, Coetzee GA, Gonzales FA, Jones PA (1994): Progressive increases in the methylation status and heterochromatinization of the myoD CpG island during oncogenic transformation. Mol Cell Biol 14:6143–6152.

Russell GJ, Walker PM, Elton RA, Subak-Sharpe JH (1976): Doublet frequency analysis of fractionated vertebrate nuclear DNA. J Mol Biol 108:1–23.

Sakai T, Toguchida J, Ohtani N, Yandell DW, Rapaport JM, Dryja TP (1991): Allele-specific hypermethylation of the retinoblastoma tumor-suppressor gene. Am J Hum Genet 48:880–888.

Sanders EJ, Wride MA (1995): Programmed cell death in development. Int Rev Cytol 163:105–173.

Schneider JW, Gu W, Zhu L, Mahdavi V, Nadal-Ginard B (1994): Reversal of terminal differentiation mediated by p107 in $Rb^{-/-}$ muscle cells. Science 264:1467–1471.

Shen JC, Rideout WMd, Jones PA (1992): High frequency mutagenesis by a DNA methyltransferase. Cell 71:1073–1080.

Shin EK, Shin A, Paulding C, Schaffhausen B, Yee AS (1995): Multiple change in E2F function and regulation occur upon muscle differentiation. Mol Cell Biol 15:2252–2262.

Sicinski P, Donaher JL, Geng Y, Parker SB, Gardner H, Park MY, Robker RL, Richards JS, McGinnis LK, Biggers JD, Eppig JJ, Bronson RT, Elledge SJ, Weinberg RA (1996): Cyclin D2 is an FSH-responsive gene involved in gonadal cell proliferation and oncogenesis. Nature 384:470–474.

Sicinski P, Donaher JL, Parker SB, Li T, Fazeli A, Gardner H, Haslam SZ, Bronson RT, Elledge SJ, Weinberg RA (1995): Cyclin D1 provides a link between development and oncogenesis in the retina and breast. Cell 82:621–630.

Skapek SX, Rhee J, Kim PS, Novitch BG, Lassar AB (1996): Cyclin-mediated inhibition of muscle gene expression via a mechanism that is independent of pRB hyperphosphorylation. Mol Cell Biol 16:7043–7053.

Skapek SX, Rhee J, Spicer DB, Lassar AB (1995): Inhibition of myogenic differentiation in proliferating myoblasts by cyclin D1–dependent kinase. Science 267:1022–1024.

Song K, Wang Y, Sassoon D (1992): Expression of Hox-7.1 in myoblasts inhibits terminal differentiation and induces cell transformation. Nature 360:477–481.

Sorrentino V, Pepperkok R, Davis RL, Ansorge W, Philipson L (1990): Cell proliferation inhibited by MyoD1 independently of myogenic differentiation. Nature 345:813–815.

Stockdale FE, Holtzer H (1961): DNA synthesis and myogenesis. Exp Cell Res 24:508–520.

Storb U, Arp B (1983): Methylation patterns of immunoglobulin genes in lymphoid cells: Correlation of expression and differentiation with undermethylation. Proc Natl Acad Sci USA 80:6642–6646.

Stuhlmann H, Jahner D, Jaenisch R (1981): Infectivity and methylation of retroviral genomes is correlated with expression in the animal. Cell 26:221–232.

Sullivan CH, Grainger RM (1987): Delta-crystallin genes become hypomethylated in postmitotic lens cells during chicken development. Proc Natl Acad Sci USA 84:329–333.

Szyf M, Rouleau J, Theberge J, Bozovic V (1992): Induction of myogenic differentiation by an expression vector encoding the DNA methyltransferase cDNA sequence in the antisense orientation. J Biol Chem 267:12831–12836.

Takagi H, Tajima S, Asano A (1995): Overexpression of DNA methyltransferase in myoblast cells accelerates myotube formation. Eur J Biochem 231:282–291.

Takahashi T, Tanaka M, Brannan CI, Jenkins NA, Copeland NG, Suda T, Nagata S (1994): Generalized lymphoproliferative disease in mice, caused by a point mutation in the Fas ligand. Cell 76:969–976.

Tanaka EM, Gann AA, Gates PB, Brockes JP (1997): Newt myotubes reenter the cell cycle by phosphorylation of the retinoblastoma protein. J Cell Biol 136:155–165.

Tate P, Skarnes W, Bird A (1996): The methyl-CpG binding protein MeCP2 is essential for embryonic development in the mouse. Nature Genet 12:205–208.

Taylor SM, Jones PA (1979): Multiple new phenotypes induced in 10T1/2 and 3T3 cells treated with 5-azacytidine. Cell 17:771–779.

Thorburn AM, Walton PA, Feramisco JR (1993): MyoD induced cell cycle arrest is associated with increased nuclear affinity of the Rb protein. Mol Biol Cell 4:705–713.

Tiainen M, Pajalunga D, Ferrantelli F, Soddu S, Salvatori G, Sacchi A, Crescenzi M (1996a): Terminally differentiated skeletal myotubes are not confined to G_0 but can enter G_1 upon growth factor stimulation. Cell Growth Differ 7:1039–1050.

Tiainen M, Spitkovsky D, Jansen-Durr P, Sacchi A, Crescenzi M (1996b): Expression of E1A in terminally differentiated muscle cells reactivates the cell cycle and suppresses tissue-specific genes by separable mechanisms. Mol Cell Biol 16:5302–5312.

Tucker KL, Beard C, Dausmann J, Jackson-Grusby L, Laird PW, Lei H, Li E, Jaenisch R (1996a): Germ-line passage is required for establishment of methylation and expression patterns of imprinted but not of nonimprinted genes. Genes Dev 10:1008–1020.

Tucker KL, Talbot D, Lee MA, Leonhardt H, Jaenisch R (1996b): Complementation of methylation deficiency in embryonic stem cells by a DNA methyltransferase minigene. Proc Natl Acad Sci USA 93:12920–12925.

Uehara Y, Ono T, Kurishita A, Kokuryu H, Okada S (1989): Age-dependent and tissue-specific changes of DNA methylation within and around the c-fos gene in mice. Oncogene 4:1023–1028.

Vaux DL, Weissman IL, Kim SK (1992): Prevention of programmed cell death in *Caenorhabditis elegans* by human bcl-2. Science 258:1955–1957.

Veis DJ, Sorenson CM, Shutter JR, Korsmeyer SJ (1993): Bcl-2–deficient mice demonstrate fulminant lymphoid apoptosis, polycystic kidneys, and hypopigmented hair. Cell 75:229–240.

Wales MM, Biel MA, El Deiry W, Nelkin BD, Issa JP, Cavenee WK, Kuerbitz SJ, Baylin SB (1995): p53 activates expression of HIC-1, a new candidate tumour suppressor gene on 17p13.3. Nature Med 1:570–577.

Watanabe-Fukunaga R, Brannan CI, Copeland NG, Jenkins NA, Nagata S (1992): Lymphoproliferation disorder in mice explained by defects in Fas antigen that mediates apoptosis. Nature 356:314–317.

Watt F, Molloy PL (1988): Cytosine methylation prevents binding to DNA of a HeLa cell transcription factor required for optimal expression of the adenovirus major late promoter. Genes Dev 2:1136–1143.

Weintraub H (1993): The MyoD family and myogenesis: Redundancy, networks, and thresholds. Cell 75:1241–1244.

Weintraub SJ, Prater CA, Dean DC (1992): Retinoblastoma protein switches the E2F site from positive to negative element. Nature 358:259–261.

Weiss A, Keshet I, Razin A, Cedar H (1996): DNA demethylation in vitro: Involvement of RNA. Cell 86:709–718.

Wigler M, Levy D, Perucho M (1981): The somatic replication of DNA methylation. Cell 24:33–40.

Wilks AF, Cozens PJ, Mattaj IW, Jost JP (1982): Estrogen induces a demethylation at the 5′ end region of the chicken vitellogenin gene. Proc Natl Acad Sci USA 79:4252–4255.

Wilks A, Seldran M, Jost JP (1984): An estrogen-dependent demethylation at the 5′ end of the chicken vitellogenin gene is independent of DNA synthesis. Nucleic Acids Res 12:1163–1177.

Woloshin P, Song K, Degnin C, Killary AM, Goldhamer DJ, Sassoon D, Thayer MJ (1995): MSX1 inhibits myoD expression in fibroblast × 10T1/2 cell hybrids. Cell 82:611–620.

Wright WE, Sassoon DA, Lin VK (1989): Myogenin, a factor regulating myogenesis, has a domain homologous to myoD. Cell 56:607–617.

Yaffe D, Gershon D (1967): Multinucleated muscle fibres: Induction of DNA synthesis and mitosis by polyoma virus infection. Nature 215:421–424.

Yaffe D, Saxel O (1977): Serial passaging and differentiation of myogenic cells isolated from dystrophic mouse muscle. Nature 270:725–727.

Yanai N, Obinata M (1994): Apoptosis is induced at nonpermissive temperature by a transient increase in p53 in cell lines immortalized with temperature-sensitive SV40 large T-antigen gene. Exp Cell Res 211:296–300.

Yen RW, Vertino PM, Nelkin BD, Yu JJ, el-Deiry W, Cumaraswamy A, Lennon GG, Trask BJ, Celano P, Baylin SB (1992): Isolation and characterization of the cDNA encoding human DNA methyltransferase. Nucleic Acids Res 20:2287–2291.

Yoshiura K, Kanai Y, Ochiai A, Shimoyama Y, Sugimura T, Hirohashi S (1995): Silencing of the E-cadherin invasion-suppressor gene by CpG methylation in human carcinomas. Proc Natl Acad Sci USA 92:7416–7419.

Yu YT, Breitbart RE, Smoot LB, Lee Y, Mahdavi V, Nadal-Ginard B (1992): Human myocyte-specific enhancer factor 2 comprises a group of tissue-restricted MADS box transcription factors. Genes Dev 6:1783–1798.

Yuan J, Shaham S, Ledoux S, Ellis HM, Horvitz HR (1993): The *C. elegans* cell death gene ced-3 encodes a protein similar to mammalian interleukin-1 beta-converting enzyme. Cell 75:641–652.

Zacksenhaus E, Jiang Z, Chung D, Marth JD, Phillips RA, Gallie BL (1996): pRb controls proliferation, differentiation, and death of skeletal muscle cells and other lineages during embryogenesis. Genes Dev 10:3051–3064.

Zingg JM, Pedraza-Alva G, Jost JP (1994): MyoD1 promoter autoregulation is mediated by two proximal E-boxes. Nucleic Acids Res 22:2234–2241.

CHAPTER 8

REPLICATIVE SENESCENCE AND IMMORTALIZATION

JUDITH CAMPISI
Department of Cell and Molecular Biology, Berkeley National Laboratory, Berkeley, CA 94720

INTRODUCTION

Normal somatic cells do not divide indefinitely. This property is termed the *finite replicative life span* of cells, and it restricts cell division by a process known as *replicative* or *cellular senescence*. Senescent cells irreversibly arrest proliferation with a G_1 DNA content. Senescent cells also become resistant to apoptotic death and show selective changes in differentiated cell functions. Several lines of evidence suggest that replicative senescence is a powerful, albeit imperfect, tumor suppressive mechanism and that it also contributes to organismic aging. Replicative senescence appears to be controlled by multiple, dominant-acting genetic loci. Cells escape senescence and acquire an infinite replicative life span, or an immortal phenotype, due to a loss of gene function or genetic loci, some of which have been mapped to specific human chromosomes.

What determines replicative life span? What prevents cell division and alters differentiation once replicative capacity is exhausted? Cell division may be limited by progressive telomere shortening. A critically short telomere may induce the expression of dominant growth inhibitors, which institute and maintain the growth arrest. Presenescent and senescent cells express many genes in common. However, some genes whose activities are needed for cell cycle progression are repressed in senescent cells, whereas at least two inhibitors of cell cycle progression are overexpressed by senescent cells. Whether and how these changes in the expression of cell cycle regulatory genes participate in the senescence-associated changes in differentiation is not yet known. Nonetheless, both the changes in cell proliferation and the changes in differentiation that are the hallmark of replicative senescence are likely to have physiological consequences, particularly for mammalian organisms.

The Molecular Basis of Cell Cycle and Growth Control, Edited by G.S. Stein, R. Baserga, A. Giordano, and D.T. Denhardt
ISBN 0-471-15706-6, pages 348–373.

REPLICATIVE SENESCENCE—WHAT IS IT?

Cell Division Potential Is Limited

Normal somatic cells of higher animals do not divide indefinitely. This property, the finite replicative life span of cells, limits cell division by a process of replicative or cellular senescence. Although noted earlier, the finite replicative life span of cells was first systematically described nearly 40 years ago in cultures of human fibroblasts (Hayflick and Moorhead, 1961; Hayflick, 1965). Since then, many cell types from several animal species have been shown to have a limited division potential (for review, see Stanulis-Praeger, 1987; Cristofalo and Pignolo, 1993; Campisi et al., 1996). Replicative senescence may be a very primitive phenotype because it occurs in the simple single-cell organism *Saccharomyces cerevisiae* (Jazwinski, 1996) In mammals, it has been proposed that replicative senescence is a tumor suppressive mechanism that may also contribute to the aging of mitotically competent tissues (Campisi, 1996).

Cellular senescence is generally studied in culture. In culture, cell *proliferation* (used here interchangeably with *growth*) can be easily monitored for the many doublings that are needed for most or all cells in a population to senesce. In contrast, cell proliferation is not easily monitored or manipulated in vivo. There are, nevertheless, a limited number of studies on the proliferative capacity of some tissues and cell types in animals (Krohn, 1962, 1966; Daniel et al., 1968; Daniel, 1972). Although still controversial (Rubin, 1997), the results from these studies, together with evidence discussed below, strongly suggest that replicative senescence is not an artifact of cell culture.

Replicative senescence depends on cell division—not time. Thus, maintaining proliferative cells such as fibroblasts as quiescent cultures does not reduce their replicative potential upon subsequent stimulation (Hayflick, 1965; Dell'Orco et al., 1973; Goldstein and Singal, 1974). Likewise, the cell cycle duration does not affect the number of divisions through which cells can proceed. On the other hand, replicative potential depends very much on the cell type and the species and age of the donor (Stanulis-Praeger, 1987; Cristofalo and Pignolo, 1993; Campisi, 1996).

Only two, perhaps three, types of cells may have an infinite or immortal replicative life span. Certainly the germline is capable of unlimited replication. In addition, many tumors contain cells that are immortal. Finally, some primitive stem cells may divide indefinitely, but this has yet to be proven. Replicative senescence is exceedingly stringent in human cells. In sharp contrast to many rodent cells, human cells rarely spontaneously fail to senesce (for review, see Ponten, 1976; Sager, 1984; McCormick and Maher, 1988; Shay and Wright, 1991).

The Senescent Phenotype

The fraction of senescent cells (cells that have reached the end of their replicative life span) rises more or less exponentially with each doubling of the cell population (Smith and Hayflick, 1974; Smith and Whitney, 1980). Eventually, the culture consists entirely of senescent cells. Senescent cells remain viable for long periods of time. They metabolize RNA and protein, respond to environmental signals, and retain many presenescent characteristics. Moreover, presenescent and senescent cells express many genes in common, several of which remain inducible by external stimuli throughout the replicative life span (Hornsby et al., 1986; Rittling et al., 1986; Seshadri and Campisi, 1990; Choi et al., 1995). Three features distinguish senescent cells from their presenescent counterparts. First, senescent cells are incapable of proliferation in response to mitogenic stimuli. This growth arrest is essentially irreversible. Second, senescent cells are resistant to apoptosis. Third, senescent cells show selected changes in differentiated functions. Thus, replicative senescence results in a complex phenotype, one that entails much more than a simple terminal arrest of cell proliferation.

Irreversible Growth Arrest. One of the defining characteristics of replicative senescence is a stable, essentially irreversible arrest of cell proliferation. This feature of the senescent phenotype has been much more extensively studied than the resistance to apoptosis or altered functions.

Senescent cells arrest growth with a G_1 DNA content. Once this growth arrest occurs, physiological mitogens cannot stimulate them to enter the S phase of the cell cycle (for reviews, see Goldstein, 1990; McCormick and Campisi, 1991; Shay and Wright, 1991; Cristofalo and Pignolo, 1993; Campisi et al., 1996). Senescent cells do not fail to initiate DNA replication in response to mitogens because of a *general* breakdown in growth factor signal transduction. This has been most clearly established for human fibroblast cultures in which replicative senescence and the mitogen responsiveness of many genes have been well characterized (Goldstein, 1990; Campisi et al., 1996). Many genes, including at least three protooncogenes (c-jun, c-myc, and c-ras-Ha), remain fully mitogen-inducible in senescent human fibroblasts (Rittling et al., 1986; Chang and Chen, 1988; Seshadri and Campisi, 1990; Phillips et al., 1992; Hara et al., 1994). However, a small number of mitogen-inducible, growth regulatory genes remain repressed in senescent cells, and at least two growth inhibitors are overexpressed by senescent cells. The details of the growth arrest associated with senescence and its implications for tumor suppression are discussed later.

Resistance to Apoptosis. A second feature that distinguishes senescent from presenescent cells is their susceptibility to programmed cell death, or apoptosis. Although some cells may die during passage in culture (Pignolo et al., 1994), senescent cells are substantially more resistant than presenescent cells to dying in response to apoptotic stimuli (Wang et al., 1994). The mechanism by which senescent cells resist apoptotic death is not well understood. However, resistance to apoptosis may explain why senescent cells accumulate in vivo, as discussed further below. That senescent cells are resistant to apoptosis underscores the distinction, often blurred in the literature, between cell death and cell senescence (Linskens et al., 1995).

Altered Differentiation. In addition to an arrest of cell proliferation, replicative senescence entails changes in cell function (for reviews, see Campisi et al., 1996; Campisi, 1997a). The altered function of senescent cells is frequently overlooked. Unlike the growth arrest, and possibly the resistance to apoptosis, the functional changes that occur when cells senesce depend on the cell type. By the criteria of a stable irreversible growth arrest and alterations in function, senescent cells resemble terminally differentiated cells. Unlike terminal differentiation, however, replicative senescence is the consequence of a finite number of cell divisions.

Most cell types show some general phenotypic changes when they senesce. These changes include an enlarged cell size, increased lysosome biogenesis, and decreased rates of protein synthesis and degradation (reviewed in Stanulis-Praeger, 1987). In addition, there are changes in the expression or regulation of cell type–specific genes. For example, there is a marked increase in the expression of interleukin-1α and the cell-specific adhesion molecule I-CAM in senescent human endothelial cells (Maier et al., 1990, 1993). In human adrenocortical epithelial cells, senescence causes a selective loss in the ability to induce 17α-hydroxylase, a key enzyme in cortisol biosynthesis (Hornsby et al., 1987; Cheng et al., 1989). Senescent dermal fibroblasts show a marked increase in expression of collagenase and stromelysin (proteases that degrade extracellular matrix proteins) and reduced expression of the protease inhibitors TIMP-1 and -3 (tissue inhibitor of metalloproteinase-1 and -3) (West et al., 1989; Millis et al., 1992; Wick et al., 1994). Thus, replicative senescence entails changes in differentiation that can have rather substantial consequences for cell—and, at least in principle, tissue—function and integrity. It is likely that these functional changes may be more important for the role of replicative senescence in aging than the failure of senescent cells to proliferate.

REPLICATIVE SENESCENCE—SIGNIFICANCE

There are two ideas, which are not mutually exclusive, regarding the physiological significance of replicative senescence. One holds that replicative senescence is a tumor suppressive mechanism; the other holds that senescent cells contribute to the age-related decline in tissue function and integrity. The evidence for each of these ideas is discussed below. First, however, let us consider the basis for the concept that cellular senescence may play a role in both tumor suppression and aging.

At first glance, the idea that replicative senescence is partly responsible for protection from cancer as well as some of the manifestations of aging may seem contradictory. The evolutionary theory of aging holds that the phenotypes characteristic of advanced age are a consequence of the decline in the force of natural selection with increasing age. This idea suggests that some traits that were selected to maintain health during early reproductive fitness may have unselected and deleterious effects later in life (see Finch, 1990; Rose, 1991). Thus, at least in higher eukaryotes, replicative senescence may contribute to the relative freedom from cancer that is seen early in life. Later in life, however, replicative senescence may be deleterious because dysfunctional senescent cells accumulate. The altered functions of senescent cells, which can have far-ranging effects due to the overexpression of secreted molecules, may be particularly important in this regard (Campisi, 1997a).

At least among mammals, there is a very close relationship between the rate of aging and the rate at which cancers arise. Cancer rates generally rise with sharply exponential kinetics after about the first 50% of a mammalian organism's life span. Thus, in humans, with a life span of 100–120 years, cancer incidence rises sharply after about 50 years of age, whereas in mice, with a life span of about 3 years, cancer incidence rises after about 1.5 years of age. In fact, it has been proposed that the rates of cancer incidence and aging may derive from a shared mechanism (Miller, 1991). One possibility is that the finite replicative life span of cells links the rate of aging with that of tumor incidence (Campisi, 1997a).

Replicative Senescence and Aging

The idea that replicative senescence is related to organismic aging derives from several lines of evidence, much of it correlative. First, cells cultures from old donors tend to senesce after fewer doublings (PD) than cultures from young donors (Martin et al., 1970; Le Guilly et al., 1973; Rheinwald and Green, 1975; Schneider and Mitsui, 1976; Bierman, 1978; Bruce et al., 1986). There is considerable scatter in the data, particularly among humans, some of which may be due to genetic and life history differences. Consequently, there is very little predictive value in the relationship between the replicative life span of a cell culture and the age of the donor. In general, however, human fetal fibroblasts typically senesce after 50–80 PD, fibroblasts from young to middle-aged adults may do so after 20–50 PD, and cells from old adults may senesce after 10–30 PD. The most optimistic interpretation of these findings is that the replicative life span of cells in renewable tissues may be progressively exhausted during the chronological life span of organisms.

A second line of correlative evidence derives from limited interspecies comparisons. In general, cells from short-lived species tend to senesce after fewer PD than comparable cells from long-lived species (Goldstein, 1974; Hayflick, 1977; Rohme, 1981). Thus, mouse fetal fibroblasts senesce after 10–15 PD, whereas fibroblasts from the Galapagos tortoise proliferate for more than 100 PD. Although other interpretations are possible, these results raise the possibility that the genes that control the chronological life span of organisms may overlap, at least in part, with genes that control the replicative life span of cells.

Third, cells from humans with hereditary premature aging syndromes senesce much more rapidly than cells from age-matched controls

(Goldstein, 1978; Martin, 1978; Salk et al., 1985; Brown, 1990). This has been best studied in cells from donors with the autosomal recessive Werner's syndrome, a segmental, adult-onset, premature aging syndrome. Patients with Werner's syndrome develop several, but not all, of the pathologies associated with aging, including atherosclerosis, cancer, type II diabetes, cataracts, and osteoporosis, about 20 years prematurely. The gene that is defective in this disorder (WRN) was recently cloned (Yu et al., 1996). WRN appears to encode a DNA helicase and may be involved in one or more aspect of DNA metabolism. Cells from young adult donors with Werner's syndrome generally senesce within 10–20 PD, well ahead of cells from age-matched controls. These findings further support the idea that organismic life span and replicative life span are related and may be controlled by overlapping genes.

Finally, the regulation of a few genes is altered similarly by replicative senescence in culture and aging in vivo. For example, the c-fos protooncogene becomes refractory to induction by mitogens (Seshadri and Campisi, 1990) (but not ultraviolet radiation; Choi et al., 1995) in human fibroblasts that have undergone senescence in culture. c-fos also becomes refractory to induction by a pressure overload to the heart (Takahashi et al., 1992) or an electroconvulsive shock to the brain (D'Costa et al., 1991) in aged rats or mice. Likewise, induction of the heat shock protein 70 (hsp70) by stress is attenuated by replicative senescence in cultured human fibroblasts (Liu et al., 1989) and by chronological age in rat tissues (Fargnoli et al., 1990; Heydari et al., 1993). In addition, the telomeres of human fibroblasts and T lymphocytes have been shown to shorten progressively with cell division in culture and with age in vivo.

Despite mounting correlative evidence, direct evidence for the idea that senescent cells accumulate with age has been lacking. Recently, however, an enzymatic marker for senescent human cells was described that provided the first in situ evidence that senescent cells may accumulate with age in human tissue. This marker, the senescence-associated β-galactosidase (SA-β-gal), was expressed by several human cell types upon senescence in culture (Dimri et al., 1995). The activity is detectable in individual cells by histochemical staining at pH 6 using the artificial substrate X-gal. SA-β-gal differs in pH optimum from the lysosomal β-gal (pH 4) and bacterial β-gal (pH 7.5) commonly used as a reporter gene.

In culture, SA-β-gal is readily detectable in senescent human fibroblasts, keratinocytes, endothelial cells, and mammary epithelial cells. However, the activity is very low or undetectable in quiescent fibroblasts or terminally differentiated keratinocytes, as well as in several immortal human cells, including HeLa. However, HeLa cells expressed SA-β-gal after introduction (by microcell fusion) of a normal human chromosome 4, which induces senescence in these cells. Thus, SA-β-gal expression is linked to replicative senescence, but not to quiescence or terminal differentiation, at least in human fibroblasts and keratinocytes (Dimri et al., 1995). The origin and function of SA-β-gal is not known. Nonetheless, it provided a means with which to distinguish senescent cells from quiescent or terminally differentiated cells in tissue. In frozen sections of skin samples from young and old donors, dermal fibroblasts and epidermal keratinocytes expressing SA-β-gal activity were clearly more abundant in skin from older donors (Dimri et al., 1995).

These findings support the idea that replicative senescence occurs during organismic aging and that senescent cells accumulate in aged tissue. Whether this accumulation contributes to age-associated pathology has yet to be determined. Nonetheless, it is possible that the age-associated decline in regenerative capacity and function may at least partly derive from the accumulation of senescent cells that cannot proliferate, are resistant to apoptotic death, and have an altered differentiated phenotype.

Replicative Senescence and Tumor Suppression

The idea that replicative senescence is a powerful, albeit imperfect, tumor suppressive mechanism derives from several lines of evidence.

First, many tumors contain cells that have bypassed the limit to cell division imposed by replicative senescence (Ponten, 1976; Sager, 1991). That is, tumors very often contain immortal cells or cells that have an extended replicative life span. These findings do not argue that immortality is required for tumorigenesis (although it may be for metastasis). Rather, they suggest that there may be strong selection for cells with an increased proliferative potential during the development of malignant tumors. Immortality, or even an extended replicative life span, greatly increases the ability of cells to progress toward malignancy because it permits the extensive cell division that is needed to acquire the multiple, successive mutations that allow inappropriate growth, invade the surrounding tissue, and metastasize.

Second, certain oncogenes act, at least in part, by immortalizing or extending the replicative life span of cells (Ruley, 1983; Weinberg, 1985; Galloway and McDougall, 1989; Hinds et al., 1989; Shay and Wright, 1991). These oncogenes include mutated or unregulated cellular genes (such as p53, which readily immortalizes rodent cells and extends the proliferative potential of human cells). They also include viral oncogenes—foreign genes carried by certain oncogenic DNA viruses—that are thought to play a role in the etiology of some human cancers (e.g., the Epstein-Barr virus or certain serotypes of the human papillomaviruses). Thus, mutations in cellular genes that lead to tumorigenesis, or the strategies of some oncogenic viruses, can permit cells to escape the constraints on cell proliferation that are imposed by replicative senescence.

Third, among the genes that are essential for establishing and maintaining replicative senescence are two well-recognized tumor suppressor genes: The p53 gene and the retinoblastoma susceptibility gene (Rb) (Hara et al., 1991; Shay and Wright, 1991; Dimri and Campisi, 1994). The evidence that p53 and Rb are critical regulators of replicative senescence is discussed later. For the moment, suffice it to say that the activities of p53 and Rb are together perhaps the most commonly lost gene functions in all of human cancers (Bookstein and Lee, 1991; Hollstein et al., 1991; Weinberg, 1991), and the activities of both of these genes are necessary for replicative senescence.

Finally, there are at least two examples of targeted gene disruptions in mice that give rise to multiple, early onset cancers in the animal and a loss or substantial reduction in the ability of cells to undergo replicative senescence. These examples are mice with homozygous germline disruptions in either the p53 or p16 tumor suppressor genes. These mice show normal embryonic and neonatal development, but develop tumors in a variety of tissues very early in life (within a few months). Cells cultured from these animals do not appear to senesce; at the very least, loss of p53 or p16 function may very dramatically increase the rate of immortalization (Donehower et al., 1992; Harvey et al., 1993; Jacks et al., 1994; Serrano et al., 1996).

Taken together, these data from molecular, cell, and organismic biology strongly suggest that replicative senescence is a powerful tumor suppressive mechanism. As such, cell senescence may have evolved to help maintain the relative freedom from cancer that is characteristic of the first half of the life span of mammalian organisms.

REPLICATIVE SENESCENCE—GENETICS

The idea that replicative senescence is controlled by specific genetic loci is now well substantiated. This is not to say that extrinsic factors do not influence replicative senescence. For example, the PD at which a cell culture completely senesces depends on the oxygen tension in which they are grown (low oxygen tension extends replicative life span) (Packer and Fuehr, 1977; Balin et al., 1977; Chen et al., 1995). Thus, oxidative damage may influence the replicative life span of cells. As is the case for the growth and senescence of organisms, the replicative senescence of cells is very likely governed by genetic programs that are influenced by nongenetic and environmental factors.

Somatic Cell Genetics

Dominance. The growth arrest associated with cell senescence and the finite replicative life span of cells are dominant traits in somatic cell fusion experiments (Norwood et al., 1974; Muggleton-Harris and Hayflick, 1976; Bunn and Tarrant, 1980; Muggleton-Harris and DeSimone, 1980; Stein et al., 1982, 1985; Pereira-Smith and Smith, 1983).

When proliferating presenescent cells are fused to senescent cells, they fail to initiate DNA replication in response to mitogens. Moreover, when most immortal tumor cells are fused to senescent cells, DNA replication is inhibited. These findings suggest that the growth stimulatory genes expressed by proliferating cells, as well as the cellular oncogenes expressed by tumor cells, do not overcome the growth arrest associated with senescence. Moreover, they suggest that senescent cells do not fail to proliferate due to a deficiency in positive growth signals or to defects in critical cellular components. Rather, senescent cells must express one or more factor that acts in a trans-dominant fashion to inhibit progression into the S phase of the cell cycle.

Similarly, when presenescent cells are fused to immortal tumor cells, the hybrid cells acquire a finite replicative capacity. Immortal cells that emerge from these hybrid cell populations frequently show a loss of normal chromosomes. These findings suggest that normal cells express dominant genes that limit replicative life span and are lost or inactivated in immortal cells. This idea is strengthened by the finding that different immortal cell lines often (but not always, as discussed below) produce hybrids that have a limited cell division potential. Thus, a finite replicative life span is a dominant trait, whereas immortality is a recessive trait.

Complementation Groups. Most often, when two immortal cell lines are fused, the hybrid cells senesce. However, some immortal cell pairs form hybrids that are immortal, even though each member of that pair, upon fusion to a third cell line, may produce hybrids that senesce. These findings suggest that inactivation of more than one gene can lead to immortality. Thus, immortal cells with lesions in different genes can complement one another, producing hybrids with a finite replicative life span; immortal cells with lesions in the same gene cannot complement, and the hybrids remain immortal. At least four complementation groups for immortality (designated A, B, C, and D) have now been described (Pereira-Smith and Smith, 1988). It is particularly interesting that cells of diverse tissue origins may belong to the same complementation group. This finding suggests that the genes that control replicative senescence do not act in a cell type– or tissue-specific manner. Taken together, the evidence strongly suggests that the finite replicative life span of cells is controlled by multiple, dominant genetic loci.

Human Chromosomes. Several human chromosomes have been reported to induced a finite replicative life span and senescence when introduced into specific immortal cell lines. These chromosomes include 1, 3, 4, 6, 7, 18, and X, among others (Sugawara et al., 1990; Yamada et al., 1990; Klein et al., 1991; Ning et al., 1991; Wang et al., 1992; Ogata et al., 1993; Barbanti-Brodano and Croce, 1994; Gaulandi et al., 1994; Hensler et al., 1994; Sandhu et al., 1994; Uejima et al., 1995; Ohmura et al., 1995; Oshimura and Barrett, 1997). Of particular importance, the abilities of chromosomes 1, 4, and 7 to limit replicative potential is restricted to cells assigned to complementation groups C, B, and D, respectively. Thus, the chromosomes on which three of the four complementation group genes reside are known, although the genes of interest have yet to be identified.

The number of human chromosomes reported to induce senescence exceeds the number of complementation groups (Campisi et al., 1996; Sasaki et al., 1994; Oshimura and Barrett, 1997). Moreover, some immortal cells can be assigned to more than one complementation group (Duncan et al., 1993; Berry et al., 1994). It is possible that some immortal cells have lost more than one senescence-regulatory gene (and thus may assign to more than one complementation group). It is also possible that some chromosomes may arrest cell division by mechanisms that are reminiscent but independent of the growth arrest associated with senescence. Until the genes that induce senescence are identified, the basis for the complementation groups and the mechanisms by which normal genetic loci induce immortal cells to senesce remain speculative.

Viral Genes that Overcome Replicative Senescence

The somatic cell fusion experiments suggest that growth stimulatory cellular genes, including cellular oncogenes, cannot immortalize normal cells or induce DNA synthesis in senescent cells. However, the oncogenes of certain DNA tumor viruses can extend the replicative life span of human cells and reactivate DNA synthesis in senescent cells. The best studied of these is the large T antigen of simian virus 40 (SV40) (Fanning, 1992). T antigen is the only single gene that can stimulate most senescent human cells to synthesize DNA (Ide et al., 1983, 1984; Gorman and Cristofalo, 1985; Shay and Wright, 1991). Certain viruses other than T antigen encode two genes that together provide the activities of T antigen. These include the E1A and E1B genes of oncogenic adenoviruses (Boulanger and Blair, 1991) and the E6 and E7 genes of oncogenic papillomaviruses (Galloway and McDougall, 1989).

Immortalization. When introduced into presenescent cells, T antigen (or the combinations E1A and E1B, or E6 and E7) extends their replicative life span and promotes immortalization. The efficiency of immortalization is highly species dependent. Immortalization by viral oncogenes is fairly efficient in rodent cells (Ponten, 1976; Ruley, 1983; Shay and Wright, 1991), but much rarer in human cells (discussed below).

The extension of replicative life span caused by viral oncogenes such as T antigen is about 20 PD for human fetal fibroblasts (Ide et al., 1984; Shay and Wright, 1989; Hara et al., 1991), but can be more than 60 PD for some cell types (Rinehart et al., 1991; Ryan et al., 1992). Cells expressing T antigen do not eventually senesce. Rather, they enter a state termed *crisis* in which cell growth is nearly balanced by cell death and senescence. Human cell cultures may spend weeks or months in crisis before either the culture dies or rare immortal cells emerge (Girardi et al., 1965; Ide et al., 1984; Neufeld et al., 1987; Shay and Wright, 1989, 1991; Hawley-Nelson et al., 1989; Munger et al., 1989; Boulanger and Blair, 1991; Hara et al., 1991; Shay et al., 1993). Cells expressing either E6 or E7 also have an extended replicative life span. However, the majority of these cells eventually do senesce (Hawley-Nelson, et al., 1989; Munger et al., 1989; Band et al., 1991; Halbert et al., 1991; Reznikoff et al., 1994).

T antigen immortalizes about 3 in 10^{-7} human fetal lung fibroblasts (Shay and Wright, 1989). This low frequency suggests that immortality requires at least one mutational event in addition to the actions of T antigen. Surprisingly, T antigen appears to immortalize human epithelial cells (from adult mammary gland) about 100-fold more readily (about 1 in 10^{-5}) than human fetal lung fibroblasts (Shay et al., 1993). T antigen–expressing fibroblasts become pseudotetraploid as they approached crisis, whereas T antigen–expressing epithelial cells remain diploid or pseudodiploid. If one allele of a dominant senescence-inducing gene spontaneously mutates, the genomic instability induced by T antigen may favor conversion to homozygosity. Thus, fewer nondysjunction events may be needed to immortalize some epithelial cells (Shay et al., 1993). This may explain why some epithelial cells, but not fibroblasts, are immortalized by E6 or E7 alone (Band et al., 1991; Halbert et al., 1991; Reznikoff et al., 1994). An intrinsic difference in genomic stability between mesenchyme versus epithelial-derived cells may also contribute to the fact that most adult human cancers arise from epithelial cells.

The above findings suggest that a finite replicative life span entails at least two mechanisms, termed M1 and M2 (mortality mechanisms 1 and 2) (Wright et al., 1989; Shay and Wright, 1991). T antigen inactivates one mechanism (M1), most likely by inactivating the p53 and pRb tumor suppressor proteins (discussed below). However, both the M1 and M2 mechanisms must be inactivated before human cells completely escape senescence and acquire an immortal replicative life span.

Reactivation of DNA Synthesis in Senescent Cells. T antigen not only extends the replicative life span of presenescent cells, but also stimulates DNA replication in cells that have already reached senescence (Tsuji et al., 1983; Ide et al., 1983; Gorman and Cristofalo, 1985; Lumpkin et al., 1986; Dimri and Campisi, 1994). Senescent cells progress through and complete DNA replication under the influence of T antigen; however, T antigen–expressing senescent cells do not undergo mitosis (Ide et al., 1983; Gorman and Cristofalo, 1985). Furthermore, human fibroblasts immortalized by a conditional T antigen reversibly arrest growth in G_1 when T antigen is conditionally inactivated (Radna et al., 1989; Wright et al., 1989). Thus, the growth arrest associated with replicative senescence appears to entail two blocks to cell cycle progression. One block prevents cells from entering the S phase. This G_1 block is overcome by T antigen, and thus is very likely controlled by p53 and pRb (discussed below). A second, T antigen–insensitive block prevents cells from undergoing mitosis.

Roles for the p53 and Rb Tumor Suppressor Genes. T antigen (or the combinations E1A and E1B or E6 and E7) have two functions in common: they bind and/or inactivate the p53 and pRb tumor suppressor proteins (Galloway and McDougall, 1989; Boulanger and Blair, 1991; Fanning, 1992). T antigen binds and thereby inactivates both p53 and pRb, as well as the pRb-related proteins p107 and p130. E1A and E7, on the other hand, bind only pRb and its related proteins, whereas E1B binds, and E6 inactivates, only p53. Thus, the p53 and pRb tumor suppressor proteins are prime candidates for cellular genes that control the T antigen–sensitive mechanisms that establish and maintain the replicative senescence of normal cells.

It appears that both p53 and pRb must be inactivated to maximally extend replicative life span and induce crisis in presenescent human cells and reactivate DNA synthesis in senescent cells. This conclusion derives from several lines of evidence. First, Rb and p53 antisense oligomers, which substantially reduced pRb and p53 expression, extended the replicative life span of human fibroblasts to the same extent as T antigen (Hara et al., 1991). Similarly, E6 and E7 were both required for maximal life span extension of human fibroblasts; alone, each extended to a lesser extent than the two together (White et al., 1994). Second, when human fibroblasts immortalized by a conditional T antigen arrested growth when T antigen was inactivated, E6 and E7 together, but not either alone, rescued the growth arrest (Shay et al., 1991). Third, T antigen mutants that are defective either in pRb binding (Sakamoto et al., 1993; Hara et al., 1996b) or p53 binding were very ineffective at reactivating DNA synthesis in senescent human fibroblasts (Dimri and Campisi, 1994).

Taken together, the above findings suggest that at least two cellular tumor suppressor genes—p53 and Rb—are responsible for establishing and maintaining the G_1 growth arrest characteristic of senescent cells. Inactivation of both Rb and p53 is necessary for any mechanism, including T antigen and other viral oncogenes, to overcome this M1 block to unlimited cell proliferation.

MOLECULAR CONTROL OF REPLICATIVE SENESCENCE

Molecular insights into three critical questions are just emerging. What is the "counting" mechanism? Why do senescent cells fail to initiate DNA synthesis? What dominant inhibitors are expressed by senescent cells?

The Counting Mechanism and Telomere Shortening Hypothesis

How do cells sense the number of divisions they have completed? This number can be more than 60–80 for human fetal fibroblasts and fairly reproducible for a given culture. The telomere shortening hypothesis is perhaps the most viable explanation for a cell division "counting" mechanism (Levy et al., 1992).

Telomeres, the ends of linear chromosomes, consist of a repetitive DNA sequence (TTAGGG in human and other vertebrates) and specialized proteins. Because DNA replication is bidirectional, and DNA polymerases move only $5' \rightarrow 3'$ and require a labile primer to initiate replication, each round of DNA replication leaves some $3'$ bases at the telomere unreplicated. Germ cells express telomerase, which adds telomeric repeats to chromosome ends de novo. However, most somatic cells do not express this enzyme.

The telomere shortening hypothesis predicts that telomeres should progressively shorten throughout the replicative life span of cells (Harley et al., 1990; Allsopp et al., 1992; Levy et al., 1992). The terminal restriction fragment (TRF), which contains the telomeric repeat and some subtelomeric DNA, is 12–15 kb in germ cells and 8–10 kb in many somatic cells (depending on the cell type and donor age). Indeed, mean telomere length decreases about 50 bp per doubling for cultured human fibroblasts (and about 15 bp per year in lymphoid cells from different aged donors). At senescence, the average TRF is about 4–6 kb. Because homologous chromosomes segregate randomly during mitosis, there may be considerable variability in telomere length from chromosome to chromosome and from cell to cell. In fact, senescent cells may have one or more chromosome with a TRF considerably shorter than 4–6 kb. These findings are consistent with the idea that a critically short telomere may induce cells to enter an irreversible senescent state.

Telomeres continue to shorten in human cells having an extended replicative life span (e.g., cells expressing T antigen). Human telomeres may shorten to an average TRF of about 2 kb, at which point the culture enters crisis. Telomere length then stabilizes, and sometimes increases, in most immortal human cells. This is usually due to expression of telomerase (Counter et al., 1992; Kim et al., 1994). Although most normal somatic cells do not express this enzyme, germ cells and most immortal cells are telomerase positive. Thus, reactivation of telomerase activity may be one mechanism by which cells can acquire an unlimited cell division potential. However, telomerase is not needed for extension of replicative life span. Human cells expressing T antigen, or E1A and E1B, remain devoid of telomerase activity throughout their extended life span and during crisis. In contrast, immortal variants arising from these cultures expressed the activity (Counter et al., 1992). Thus, the T antigen–insensitive, or M2, mechanism that ensures a finite replicative life span may entail, at least in part, repression of telomerase activity (Shay and Wright, 1991; Holt et al., 1997; Oshimura and Barrett, 1997).

Despite its frequent association with replicative immortality, telomerase is neither sufficient nor necessary for unlimited cell division. For example, following mitogenic stimulation, normal human T cells express telomerase, but the telomeres nonetheless shorten and the cells eventually senesce (Vaziri et al., 1993; Buchkovich and Greider, 1996; Effros, 1996; Weng et al., 1996). Similarly, telomerase is expressed by human basal epidermal keratinocytes, which nonetheless senesce (Harlebachor and Boukamp, 1996). Thus, telomerase activity per se can be insufficient to prevent telomere shortening and cell senescence. The level of telomerase expressed by human T cells and keratinocytes may be inadequate to completely prevent telomere shortening. Alternatively, telomerase may have only limited accessibility to the telomeres in these cells. In addition, there are now several examples of immortal human tumor cells that do not express any detectable telomerase (Bryan et al., 1995; Bryan and Reddel, 1997). Thus, there are clearly telomerase-independent mechanisms for maintaining telomere length.

Most studies on the relationship between telomere length, telomerase activity, and proliferative capacity have been confined to human cells, which have relatively short (<20 kb) telomeres. The common mouse species, for example, has an average telomere length of 100 kb or more. Whether and to what extent telomere length controls of the replicative life span of nonhuman cells is, at present, not at all clear.

If telomere length is a critical determinant of the senescence of human cells, how might telomere shortening alter growth and differentiation? Very little is known about this in higher eukaryotic cells. However, telomeres and telomerase can be easily manipulated by genetic means in lower eukaryotes such as yeast. Three models for how telomere shortening may alter cell phenotype have emerged from studies in yeast (reviewed in Campisi, 1997b). These models are not mutually exclusive and propose that a short telomere may (1) induce an irreversible DNA damage response; (2) alter the availability of transcription factors and thus alter gene expression; or (3) decrease heterochromatin, thereby activating the expression of genes near the telomeres. The recent cloning of human telomerase components (Feng et al., 1995; Blasco et al., 1995; Meyerson et al., 1997; Nakamura et al., 1997) should greatly facilitate critical testing of the telomere shortening hypothesis in human cells.

Growth Arrest Mechanisms

Several growth-related genes have now been identified whose expression or regulation is altered by replicative senescence, principally in human fibroblasts. Although there is little doubt that the growth arrest associated with senescence is controlled by multiple mechanisms, it is unlikely that each change in gene expression is independently controlled. Thus, the challenge at present is sorting out how the multiple changes in gene expression are related to each other and how they relate to the dominance of the growth arrest, the inability to enter S phase and mitosis, and the functions of p53 and pRb.

There is now at least a superficial understanding of some of the immediate cause(s) for the failure of senescent human fibroblasts to proliferate. These prime causes fall into two categories. First, several essential and positive regulators of cell cycle progression appear to be repressed or refractory to induction by mitogens in senescent cells. In addition, certain growth inhibitory genes are overexpressed or constitutively active.

Repression of Early Response Genes. Early response genes are induced within about an hour of mitogenic stimulation, independent of prior gene expression. Many early response genes are induced to approximately normal levels and with approximately normal kinetics in senescent human fibroblasts (Rittling et al., 1986; Chang and Chen, 1988; Seshadri and Campisi, 1990; Phillips et al., 1992). These include at least three protooncogenes (c-jun, c-myc, and c-ras-Ha). These findings suggest that senescent cells do not fail to synthesize DNA in response to mitogenic stimuli due to an *overall* breakdown in growth factor signal transduction mechanisms. Rather, replicative senescence causes the selective repression of a few positive-acting, growth regulatory genes.

At least two early response genes whose expression appears to be essential for fibroblasts to initiate DNA synthesis do fail to respond to mitogens. These are the c-fos protooncogene (Seshadri and Campisi, 1990), which encodes a component of the AP1 transcription factor, and the Id1 and Id2 genes (Hara et al., 1994), which encode negative regulators of basic helix-loop-helix (bHLH) transcription factors. Because expression of both c-fos and the Id genes is required for entry into S phase, the repression of either gene can explain why senescent cells fail to proliferate.

The mechanism(s) responsible for the repression of Id1 and Id2 are unknown. c-fos, on the other hand, may remain repressed in senescent cells because the serum response factor (SRF) fails to bind the serum response element (SRE) in the c-fos promoter. SRF is hyperphosphorylated in senescent cells, and this may prevent its binding to the SRE and thus prevent mitogen-inducible c-fos expression (Atadia et al., 1994). Although mitogen-dependent c-fos induction is lost in senescent cells, c-fos remains inducible by DNA-damaging agents (Choi et al., 1995). It is not known whether or how the repression of c-fos and Id are related, nor is it known how the repression of these early response genes relates to the repression of genes normally induced at the G_1/S boundary (discussed below).

Repression of G_1/S Genes. Several genes that are normally induced in late G_1 or at the G_1/S boundary are not expressed by senescent cells. The replication-dependent histone genes are not expressed (Seshadri and Campisi, 1990; Zambetti et al., 1987), nor are the genes encoding cyclins A and B and the cyclin-dependent kinase cdc2 (Stein et al., 1991). In addition, senescent cells do not express genes encoding a variety of enzymes needed for DNA metabolism, including thymidine kinase (TK), thymidylate synthetase, DNA polymerase α, and dihydrofolate reductase (DHFR) (Pang and Chen, 1993, 1994).

The mechanism responsible for the repression of at least some of the G_1/S genes is now clear: Senescent human fibroblasts are markedly deficient in the activity of E2F (Dimri et al., 1994; Good et al., 1996). E2F is a heterodimeric transcription factor that is essential for the induction of many G_1/S genes and is negatively regulated by the unphosphorylated form of pRb (Nevins, 1992). The deficiency in E2F activity is very likely due at least in part to repression of the E2F1 (Dimri et al., 1994) and E2F5 (Good et al., 1996) components of E2F. The E2F5 and E2F1 mRNAs are normally induced 1–2 hours after mitogenic stimulation and about 4 hours before S phase, respectively. Neither gene is expressed in senescent cells, and the mechanism(s) responsible for this repression are not yet known. Senescent human fibroblasts do express other E2F components (DP1, DP2, and E2F3) (Dimri et al., 1994; Good et al., 1996). These components may be under continuous negative regulation by pRb, which remains in its unphosphorylated form in senescent cells (Stein et al., 1990) and may form novel, senescence-specific E2F complexes containing pRb-related proteins and the cyclin-dependent kinase inhibitor p21 (Afshari et al., 1996).

The repression of cyclin A and cdc2 may also be a downstream consequence of the deficiency in E2F activity in senescent cells. Both cyclin A and cdc2 require E2F activity for their expression (Yamamoto et al., 1994; Dalton, 1992). In addition, because cyclin B and cdc2 are important for regulating progression through G_2, the repression of cyclin B and cdc2 may be among the immediate causes of the G_2/M block in senescent cells. If so, the repression of E2F1 and deficiency in E2F activity may be the link that connects the G_1 block to the G_2/M block in senescent cells.

Expression of Growth Inhibitors. Senescent cells overexpress two inhibitors of cyclin-dependent protein kinases (CDKS), which very likely also constitutes an immediate cause for their failure to proliferate. The first such inhibitor, p21, was independently cloned as an overexpressed gene in senescent human fibroblasts (Noda et al., 1994), a p53-inducible gene (El Deiry et al., 1993), and a CDK-interacting protein (Gu et al., 1993; Harper et al., 1993; Xiong et al., 1993). p21 inhibits the kinase activity of many cyclin/CDK complexes, including the G_1 CDK complexes such as cyclin E- and cyclin D/CDKs 2 and 4. The second such inhibitor, p16, is overexpressed in senescent cells and frequently deleted in tumor cells; it is a specific inhibitor of cyclin D/CDK complexes (Hara et al., 1996a; Serrano et al., 1996).

Senescent cells are markedly deficient in cyclin E and D–dependent CDK activity, despite adequate levels of the proteins that comprise these complexes (Dulic et al., 1993). Clearly, the overexpression of p21 and p16 are prime candidates for causing this deficiency. Because pRb is an important substrate for G_1 CDKs, p21 and p16 overexpression is also very likely responsible for the constitutive underphosphorylation of pRb in senescent cells (Stein et al., 1990) and subsequent inhibition of E2F activity (Dimri et al., 1996). Thus, several immediate causes for the G_1 block of senescent cells (inhibition of CDK activity, pRb underphosphorylation, and E2F inhibition) can be traced to the overexpression of p21 and p16. In support of this idea, targeted inactivation of the p21 gene in normal human fibroblasts was recently shown to extend the replicative life span, similar to that which occurs upon pRb inactivation (Brown et al., 1997). The mechanism responsible for the overexpression of the p21 and p16 CDK inhibitors in senescent cells is not yet known.

In summary, the immediate causes for the failure of senescent cells to initiate DNA replication appears to be the repression of a few key positive growth regulators and the overexpression of two key growth inhibitors. Minimally, these include the repression of three transcriptional regulators (c-fos, Id, and E2F) and the overexpression of two CDK inhibitors (p21 and p16). Together, these changes in gene expression may account for many of the other changes in senescence-associated changes in gene expression. In addition, the p53 and pRb tumor suppressors are rendered constitutive growth inhibitors in senescent cells.

IMPLICATIONS FOR TUMOR SUPPRESSION AND AGING

There is now substantial evidence that replicative senescence constitutes a powerful, albeit imperfect, tumor suppressive mechanism. There is also correlative evidence, and some direct evidence, that replicative senescence occurs in vivo and that senescent cells accumulate in tissues with age. The key features of senescent cells are their inability to proliferate, resistance to apoptosis, and altered differentiated functions. The failure to proliferate would clearly be the most important feature vis à vis the role of this process in tumor suppression. The failure to proliferate might also retard the ability of tissues that accumulate senescent cells to regenerate or repair. However, substantial numbers of proliferative cells can often be recovered from even very old tissues, and wounds do heal even in very old mammals. In contrast, the altered differentiation of senescent cells might have a substantial impact on tissue function and integrity with age. Moreover, the functional senescence-associated changes would be expected to affect neighboring (nonsenescent) cells, or even other tissues, and the fraction of senescent cells needed to have adverse local or systemic effects need not be large. Thus, replicative senescence may be a double-edged biological sword. It may have evolved to retard tumorigenesis early in life, but may have unselected deleterious effects late in life.

REFERENCES

Allsopp RC, Vaziri H, Patterson C, Goldstein S, Younglai EV, Futcher AB, Greider CW, Harley CB (1992): Telomere length predicts replicative capacity of human fibroblasts. Proc Natl Acad Sci USA 89:10114–10118.

Afshari CA, Nichols MA, Xiong Y, Mudryj M (1996): A role for a p21-E2F interaction during senescence arrest of normal human fibroblasts. Cell Growth Differ 7:979–988.

Atadia PW, Stringer KF, Riabowol KT (1994): Loss of serum response element binding activity and hyperphosphorylation of serum response factor during cellular aging. Mol Cell Biol 14:4991–4999.

Balin AK, Goodman DBP, Rasmussen H, Cristofalo VJ (1977): The effect of oxygen and vitamin E on the lifespan of human diploid cells in vitro. J Cell Biol 74:58–67.

Band V, DeCaprio JA, Delmolino L, Kulesa V, Sager R (1991): Loss of p53 protein in human papillomavirus type 16 E6-immortalized human mammary epithelial cells. J Virol 65:6671–6676.

Barbanti-Brodano D, Croce CM (1994): Suppression of tumorigenicity of breast cancer cells by microcell-mediated chromosome transfer: Studies on chromosomes 6 and 11. Cancer Res 54:1331–1336.

Berry IJ, Burns JE, Parkinson EK (1994): Assignment of two human epidermal squamous cell carcinoma cell lines to more than one complementation group for the immortal phenotype. Mol Carcinog 9:134–142.

Bierman EL (1978): The effect of donor age on the in vitro lifespan of cultured human arterial smooth-muscle cells. In Vitro 14:951–955.

Blasco MA, Funk W, Villeponteau B, Greider CW (1995): Functional characterization and developmental regulation of mouse telomerase RNA. Science 269:1267–1270.

Bookstein R, Lee WH (1991): Molecular genetics of the retinoblastoma tumor suppressor gene. Crit Rev Oncol 2:211–227.

Boulanger PA, Blair GE (1991): Expression and interactions of human adenovirus oncoproteins. Biochem J 275:281–299.

Brown JP, Wei W, Sedivy JM (1997): Bypass of senescence after disruption of p21CIP1/WAF1 gene in normal diploid human fibroblasts. Science 277:831–834.

Brown WT (1990): Genetic diseases of premature aging as models of senescence. Annu Rev Gerontol Geriatr 10:23–42.

Bruce SA, Deamond SF, T'so POP (1986): In vitro senescence of Syrian hamster mesenchymal cells of fetal to aged adult origin. Inverse relationship between in vivo donor age and in vitro proliferative capacity. Mech Aging Dev 34:151–173.

Bryan TM, Englezou A, Gupta J, Bacchetti S, Reddel RR (1995): Telomere elongation in immortal human cells without detectable telomerase activity. EMBO J 14:4240–4248.

Bryan TM, Reddel RR (1997): Telomere dynamics and telomerase activity in vitro immortalized human cells. Eur J Cancer 33:767–773.

Buchkovich KJ, Greider CW (1996): Telomerase regulation during entry into the cell cycle in normal human T cells. Mol Cell Biol 7:1443–1454.

Bunn CL, Tarrant GM (1980): Limited lifespan in somatic cell hybrids and cybrids. Exp Cell Res 127:385–396.

Campisi J (1996): Replicative senescence: An old lives tale? Cell 84:497–500.

Campisi J (1997a): Aging and cancer: The double-edged sword of replicative senescence. J Am Geriatr Soc 45:1–6.

Campisi J (1997b): The biology of replicative senescence. Eur J Cancer 33:703–709.

Campisi J, Dimri GP, Hara E (1996): In Control of replicative senescence. Schneider E, Rowe J (eds): Handbook of the Biology of Aging. New York: Academic Press, pp 121–149.

Chang ZF, Chen KY (1988): Regulation of ornithine decarboxylase and other cell-cycle dependent genes during senescence of IMR90 human fibroblasts. J Biol Chem 263:11431–11435.

Chen Q, Fischer A, Reagan JD, Yan LJ, Ames BN (1995): Oxidative DNA damage and senescence of human diploid fibroblast cells. Proc Natl Acad Sci USA 92:4337–4341.

Cheng CY, Ryan RF, Vo TP, Hornsby PJ (1989): Cellular senescence involves stochastic processes causing loss of expression of differentiated functions: SV40 as a means for dissociating effects of senescence on growth and on differentiated function gene expression. Exp Cell Res 180:49–62.

Choi AMK, Pignolo RJ, ap Rhys CMI, Cristofalo VJ, Holbrook NJ (1995): Alterations in the molecular response to DNA damage during cellular aging of cultured fibroblasts: Reduced AP1 activation and collagenase gene expression. J Cell Physiol 164:65–73.

Counter CM, Avillon AA, LeFeuvre CE, Stewart NG, Greider CW, Harley CB, Bacchetti S (1992): Telomere shortening associated with chromosome instability is arrested in immortal cells which express telomerase activity. EMBO J 11:1921–1929.

Cristofalo VJ, Pignolo RJ (1993): Replicative senescence of human fibroblast-like cells in culture. Physiol Rev 73:617–638.

Dalton S (1992): Cell cycle regulation of the human cdc2 gene. EMBO J 11:1979–1984.

Daniel CW, De Orne KB, Young JT, Blair PB, Faulkin LJ (1968): The in vivo life span of normal and preneoplastic mouse mammary glands: A serial transplantation study. Proc Natl Acad Sci USA 61:53–60.

Daniel CW (1972): Aging of cells during serial propagation in vivo. Adv Gerontol Res 4:167–199.

D'Costa A, Breese CR, Boyd RL, Booze RM, Sonntag WE (1991): Attenuation of fos-like immunoreactivity induced by a single electroconvulsive shock in brains of aged mice. Brain Res 567:204–211.

Dell'Orco RT, Mertens JG, Kruse PF (1973): Doubling potential, calendar time and senescence of human diploid cells in culture. Exp Cell Res 77:356–360.

Dimri GP, Lee X, Basile G, Acosta M, Scott G, Roskelley C, Medrano EE, Linskens M, Rubelj I, Pereira-Smith O, Peacocke M, Campisi J (1995): A novel biomarker identifies senescent human cells in culture and aging skin in vivo. Proc Natl Acad Sci USA 92:9363–9367.

Dimri GP, Campisi J (1994): Molecular and cell biology of replicative senescence. Cold Spring Harbor Symp Quant Biol Mol Gen Cancer 54:67–73.

Dimri GP, Hara E, Campisi J (1994): Regulation of two E2F related genes in presenescent and senescent human fibroblasts. J Biol Chem 269:16180–16186.

Dimri GP, Nakanishi M, Desprez PY, Smith JR, Campisi J (1996): Inhibition of E2F activity by the p21 inhibitor of cyclin-dependent protein kinases in cells expressing or lacking a functional retinoblastoma protein. Mol Cell Biol 16:2987–2997.

Donehower LA, Harvey M, Slagke BL, McArthur MJ, Montgomery CA Jr, Butel JS, Bradley A (1992): Mice deficient for p53 are developmentally normal but susceptible to spontaneous tumors. Nature 356:215–221.

Dulic V, Drullinger LF, Lees E, Reed SI, Stein GH (1993): Altered regulation of G1 cyclins in senescent human diploid fibroblasts: Accumulation of inactive cyclin E–cdk and cyclin D–cdk complexes. Proc Natl Acad Sci USA 90:11043–11038.

Duncan EL, Whitaker NJ, Moy EL, Reddel RR (1993): Assignment of SV40-immortalized cells to more than one complementation group for immortalization. Exp Cell Res 205:337–344.

El Deiry WS, Tokino T, Velculescu VE, Levy DB, Parsons R, Trent JM, Lin D, Mercer WE, Kinzler KW, Vogelstein B (1993): WAF1, a potential mediator of p53 tumor suppression. Cell 75:817–825.

Effros RB (1996): Insights on immunological aging derived from the T lymphocyte cellular senescence model. Exp Gerontol 31:21–27.

Fanning E (1992): Structure and function of simian virus 40 large tumor antigen. Annu Rev Biochem 61:55–85.

Fargnoli J, Kunisada T, Fornace AJ, Schneider EL, Holbrook NJ (1990): Decreased expression of heat shock protein 70 mRNA and protein after heat treatment in cells of aged rats. Proc Natl Acad Sci USA 87:846–850.

Feng JL, Funk WD, Wang SS, Weinrich S, Avilion AA, Chiu CP, Adams BR, Chang E, Allsopp, RC, Yu J, Le S, West MD, Harley CB, Andrews WH, Greider CW, Villeponteau B (1995): The RNA component of human telomerase. Science 269:1236–1241.

Finch CR (1990): Longevity, Senescence, and the Genome. Chicago: University of Chicago Press.

Galloway DA, McDougall JK (1989): Human papillomaviruses and carcinomas. Adv Virus Res 37:125–171.

Girardi AJ, Fensen FC, Koprowski H (1965): SV40-induced transformation of human diploid cells: Crisis and recovery. J Cell Comp Physiol 65:69–84.

Goldstein S (1974): Aging in vitro: Growth of cultured cells from the Galapagos tortoise. Exp Cell Res 83:297–302.

Goldstein S (1978): Human genetic disorders that feature premature onset and accelerated progression of biological aging. In Schneider E (ed): The Genetics of Aging. New York: Plenum Press, pp 171–224.

Goldstein S (1990): Replicative senescence: The human fibroblast comes of age. Science 249:1129–1133.

Goldstein S, Singal DP (1974): Senescence of cultured human fibroblasts: Mitotic versus metabolic time. Exp Cell Res 88:359–364.

Good LF, Dimri GP, Campisi J, Chen KY (1996): Regulation of dihydrofolate reductase gene expression and E2F components in human diploid fibroblasts during growth and senescence. J Cell Physiol 168:580–588.

Gorman SD, Cristofalo VJ (1985): Reinitiation of cellular DNA synthesis in BrdU-selected nondividing senescent WI38 cells by simian virus 40 infection. J Cell Physiol 125:122–126.

Gu Y, Turek CW, Morgan DO (1993): Inhibition of cdk2 activity in vivo by an associated 20 K regulatory subunit. Nature 366:707–710.

Gualandi F, Morelli C, Pavan JV, Rimessi P, Sensi A, Bonfatti A, Gruppioni R, Possati L, Stanbridge EJ, Barbanti-Brodano G (1994): Induction of senescence and control of tumorigenicity in BK virus transformed mouse cells by human chromosome 6. Genes, Chromosomes Cancer 10:77–84.

Halbert CL, Demers GW, Galloway DA (1991): The E7 gene of human papillomavirus type 16 is sufficient for immortalization of human epithelial cells. J Virol 65:473–478.

Hara E, Smith R, Parry D, Tahara H, Peters G (1996a): Regulation of p16 (CdkN2) expression and its implications for cell immortalization and senescence. Mol Cell Biol 16:859–867.

Hara E, Tsuri H, Shinozaki S, Oda K (1991): Cooperative effect of antisense-Rb and antisense-p53 oligomers on the extension of lifespan in human diploid fibroblasts, TIG-1. Biochem Biophys Res Commun 179:528–534.

Hara E, Uzman JA, Dimri GP, Nehlin JO, Testori A, Campisi J (1996b): The helix-loop-helix protein Id-1 and a retinoblastoma protein binding mutant of SV40 T antigen synergize to reactivate DNA synthesis in senescent human fibroblasts. Dev Genet 18:161–172.

Hara E, Yamaguchi T, Nojima H, Ide T, Campisi J, Okayama H, Oda K (1994): Id related genes encoding helix loop helix proteins are required for G_1 progression and are repressed in senescent human fibroblasts. J Biol Chem 269:2139–2145.

Harlebachor C, Boukamp P (1996): Telomerase activity in the regenerative basal layer of the epidermis in human skin and immortal and carcinoma-derived skin keratinocytes. Proc Natl Acad Sci USA 93:6476–6481.

Harley CB, Futcher AB, Greider CW (1990): Telomeres shorten during ageing of human fibroblasts. Nature 345:458–460.

Harper JW, Adami GR, Wei N, Keyomarsi K, Elledge SJ (1993): The p21-cdk–interacting protein Cip1 is a potent inhibitor of G_1 cyclin–dependent kinases. Cell 75:805–816.

Harvey M, Sands AT, Weiss RS, Hegi ME, Wiseman RW, Pantazis P, Giovanella BC, Tainsky MA, Bradley A, Donehower LA (1993): In vitro growth characteristics of embryo fibroblasts isolated from p53-deficient mice. Oncogene 8:2457–2467.

Hawley-Nelson P, Vousden KH, Hubbert NL, Lowy DR, Schiller JT (1989): HPV16 E6 and E7 proteins cooperate to immortalize human foreskin keratinocytes. EMBO J 8:3905–3910.

Hayflick L (1965): The limited in vitro lifetime of human diploid cell strains. Exp Cell Res 37:614–636.

Hayflick L (1977): The cellular basis for biological aging. In Finch CE, Hayflick L (eds): Handbook of the Biology of Aging. New York: Van Nostrand Reinhold, pp 159–186.

Hayflick L, Moorhead PS (1961): The serial cultivation of human diploid cell strains. Exp Cell Res 25:585–621.

Hensler P, Annab LA, Barrett JC, Pereira-Smith OM (1994): A gene involved in control of human cellular senescence on human chromosome 1q. Mol Cell Biol 14:2292–2297.

Heydari AR, Wu B, Takahashi R, Strong R, Richardson A (1993): Expression of heat shock protein 70 is altered by age and diet at the level of transcription. Mol Cell Biol 13:2909–2918.

Hinds P, Finlay C, Levine AJ (1989): Mutation is required to activate the p53 gene for cooperation with the ras oncogene and transformation. J Virol 63:739–746.

Hollstein M, Sidransky D, Vogelstein B, Harris CC (1991): p53 mutation in human cancer. Science 253:49–53.

Holt SE, Wright WE, Shay JW (1997): Multiple pathways for the regulation of telomerase activity. Eur J Cancer 33:761–766.

Hornsby PJ, Aldern KA Harris SE (1986): Clonal variation in response to adrenocorticotropin in cultured bovine adrenocortical cells: Relationship to senescence. J Cell Physiol 129:395–402.

Hornsby PJ, Hancock JP, Vo TP, Nason LM, Ryan RF, McAllister JM (1987): Loss of expression of a differentiated function gene, steroid 17α-hydroxylase, as adrenocortical cells senesce in culture. Proc Natl Acad Sci USA 84:1580–1584.

Ide T, Tsuji Y, Ishibashi S, Mitsui Y (1983): Reinitiation of host DNA synthesis in senescent human diploid cells by infection with simian virus 40. Exp Cell Res 143:343–349.

Ide T, Tsuji Y, Nakashima T, Ishibashi S (1984): Progress of aging in human diploid cells transformed with a tsA mutant of simian virus 40. Exp Cell Res 150:321–328.

Jacks T, Remington L, Williams BO, Schmitt EM, Halachmi S, Bronson RT, Weinberg RA (1994): Tumor spectrum analysis in p53-mutant mice. Curr Biol 4:1–7.

Jazwinski SM (1996): Longevity, genes and aging. Science 273:54–59.

Kim NW, Platyszek M, Prowse KR, Harley CB, West MD, Ho PL, Coviello GM, Wright WE, Weinrich SL, Shay JW (1994): Specific association of human telomerase activity with immortal cells and cancer. Science 266:2011–2015.

Klein CB, Conway K, Wang XW, Bhamra L, Cohen MD, Annab L, Barrett JC, Costa M (1991): Senescence of nickel-transformed cells by an X chromosome: Possible epigenetic control. Science 251:796–799.

Krohn PL (1962): Review lectures on senescence. II. Heterochronic transplantation in the study of aging. Proc R Soc Lond 157:128–147.

Krohn PL (1966): Transplantation and aging. In PL Krohn (ed): Topics of the Biology of Aging. New York: John Wiley, pp 125–138.

Le Guilly Y, Simon M, Lenoir P, Bourel M (1973): Long-term culture of human adult liver cells: Morphological changes related to in vitro senescence and effect of donor's age on growth potential. Gerontology 19:303–313.

Levy MZ, Allsopp RC, Futcher AB, Greider CW, Harley CB (1992): Telomere end-replication problem and cell aging. J Mol Biol 225:951–960.

Linskens M, Harley CB, West MD, Campisi J, Hayflick L (1995): Replicative senescence and cell death. Science 267:17.

Liu AY, Lin Z, Choi HS, Sorhage F, Boshan L (1989): Attenuated induction of heat shock gene expression in aging diploid fibroblasts. J Biol Chem 264:12037–12045.

Lumpkin CK, Knepper JE, Butel JS, Smith JR, Pereira-Smith OM (1986): Mitogenic effects of the protooncogene and oncogene forms of c-H-ras DNA in human diploid fibroblasts. Mol Cell Biol 6:2990–2993.

Maier JAM, Statuto M, Ragnoti G (1993): Senescence stimulates U037-endothelial cell interactions. Exp Cell Res 208:270–274.

Maier JAM, Voulalas P, Roeder D, Maciag T (1990): Extension of the lifespan of human endothelial cells by an interleukin-1α antisense oligomer. Science 249:1570–1574.

Martin GM (1978): In Genetic syndromes in man with potential relevance to the pathobiology of aging. Bergsma D, Harrison DE (eds): Genetic Effects on Aging and Birth Defects. New York: Alan Liss, pp 5–39.

Martin GM, Sprague CA, Epstein CJ (1970): Replicative life-span of cultivated human cells. Effect of donor's age, tissue and genotype. Lab Invest 23:86–92.

McCormick A, Campisi J (1990): Cellular aging and senescence. Curr Opin Cell Biol 3:230–234.

McCormick JJ, Maher VM (1988): Towards an understanding of the malignant transformation of diploid human fibroblasts. Mutat Res 199:273–291.

Meyerson M, Counter CM, Eaton EN, Ellisen LW, Steiner P, Caddie SD, Ziaugra L, Beijersbergen RL, Davidoff MJ, Liu Q, Bacchetti S, Haber DA, Weinberg RA (1997): hEST2, the putative human telomerase catalytic subunit gene is upregulated in tumor cells and during immortalization. Cell 90:785–796.

Miller RA (1991): Gerontology as oncology: Research on aging as a key to the understading of cancer. Cancer 68:2496–2501.

Millis AJ, Hoyle M, McCue HM, Martini H (1992): Differential expression of metalloproteinase and tissue inhibitor of metalloproteinase genes in aged human fibroblasts. Exp Cell Res 201:373–379.

Muggleton-Harris AL, DeSimone DW (1980): Replicative potentials of fusion products between WI38 and SV40-transformed WI38 cells and their components. Somat Cell Genet 6:689–698.

Muggleton-Harris AL, Hayflick L (1976): Cellular aging studied by the reconstruction of replicating cells from nuclei and cytoplasms isolated from normal human diploid cells. Exp Cell Res 103:321–330.

Munger K, Phelps WC, Bubb V, Howley PM, Schlegel R (1989): The E6 and E7 genes of human papillomavirus type 16 together are necessary and sufficient for transformation of primary human keratinocytes. J Virol 63:4417–4421.

Nakamura TM, Morin GB, Chapman KB, Weinrich SL, Andrews WH, Linger J, Harley CB, Cech TR (1997): Telomerase catalytic subunit homologs from fission yeast and humans. Science 277:955–959.

Neufeld DS, Ripley S, Henderson A, Ozer HL (1987): Immortalization of human fibroblasts transformed by origin-defective simian virus 40. Mol Cell Biol 7:2794–2802.

Nevins J (1992): E2F: A link between the Rb tumor suppressor protein and viral oncoproteins. Science 258:1300–1303.

Ning Y, Weber JL, Killary AM, Ledbetter DH, Smith JR, Pereira-Smith OM (1991): Genetic analysis of indefinite division in human cells: Evidence for a senescence-related gene(s) on human chromosome 4. Proc Natl Acad Sci USA 88:5635–5639.

Noda A, Ning Y, Venable SF, Pereira-Smith OM, Smith JR (1994): Cloning of senescent cell derived inhibitors of DNA synthesis using an expression screen. Exp Cell Res 211:90–98.

Norwood TH, Pendergrass WR, Sprague CA, Martin GM (1974): Dominance of the senescent phenotype in heterokaryons between replicative and postreplicative human fibroblast-like cells. Proc Natl Acad Sci USA 71:2231–2235.

Ogata T, Ayusawa D, Namba M, Takahashi E, Oshimura M, Oishi M (1993): Chromosome 7 suppresses indefinite division of nontumorigenic immortalized human fibroblast cell lines KMST-6 and SUSM-1. Mol Cell Biol 13:6036–6043.

Ohmura H, Tahara H, Suzuki M, Ide T, Shimizu M, Yoshida MA, Tahara E, Shay JW, Barrett JC, Oshimura M (1995): Restoration of the cellular senescence program and repression of telomerase by human chromosomes. Jpn J Cancer Res 86:899–904.

Oshimura M, Barrett JC (1997): Multiple pathways to cellular senescence: Role of telomerase repressors. Eur J Cancer 33:710–715.

Packer L, Fuehr K (1977): Low oxygen concentration extends the lifespan of cultured human diploid cells. Nature 267:423–425.

Pang JH, Chen KY (1993): A specific CCAAT binding protein CBP/tk may be involved in the regulation of thymidine kinase gene expression in human IMR-90 diploid fibroblasts during senescence. J Biol Chem 268:2909–2916.

Pang JH, Chen KY (1994): Global change of gene expression at late G_1/S boundary may occur in human IMR-90 diploid fibroblasts during senescence. J Cell Physiol 160:531–536.

Pereira-Smith OM, Smith JR (1983): Evidence for the recessive nature of cellular immortality. Science 221:964–967.

Pereira-Smith OM, Smith JR (1988): Genetic analysis of indefinite division in human cells: Identification of four complementation groups. Proc Natl Acad Sci USA 85:6042–6046.

Phillips PD, Pignolo RJ, Nishikura K, Cristofalo VJ (1992): Renewed DNA synthesis in senescent WI38 cells by expression of an inducible chimeric c-fos construct. J Cell Physiol 151:206–212.

Pignolo RJ, Rotenberg MO, Cristofalo VJ (1994): Alterations in contact and density-dependent arrest state in senescent WI-38 cells. In Vitro Cell Dev Biol 30A:471–476.

Ponten J (1976): The relationship between in vitro transformation and tumor formation in vivo. Biochim Biophys Acta 458:397–422.

Radna RL, Caton Y, Jha KK, Kaplan P, Li G, Traganos F, Ozer HL (1989): Growth of immortal simian virus 40 ts-A transformed human fibroblasts is temperature dependent. Mol Cell Biol 9:3093–3096.

Reznikoff CA, Belair C, Savelieva E, Zhai Y, Pfeifer K, Yeager T, Thompson KJ, De Vries S, Bindley C, Newton MA, Kekhon G, Waldman F (1994): Long-term genome stability and minimal genotypic and phenotypic alterations in HPV16 E7-, but not E6-, immortalized human uroepithelial cells. Genes Dev 8:2227–2240.

Rheinwald JR, Green H (1975): Serial cultivation of strains of human epidermal keratinocytes: The formation of keratinizing colonies from single cells. Cell 6:331–344.

Rinehart CA, Haskill JS, Morris JS, Butler TD, Kaufman DG (1991): Extended life span of human endometrial stromal cells transfected with cloned origin-defective temperature sensitive simian virus 40. J Virol 65:1458–1465.

Rittling SR, Brooks KM, Cristofalo VJ, Baserga R (1986): Expression of cell cycle dependent genes in young and senescent WI38 fibroblasts. Proc Natl Acad Sci USA 83:3316–3320.

Rohme D (1981): Evidence for a relationship between longevity of mammalian species and lifespans of normal fibroblasts in vitro and erythrocytes in vivo. Proc Natl Acad Sci USA 78:5009–5013.

Rose MR (1991): The Evolutionary Biology of Aging. Oxford: Oxford University Press.

Rubin H (1997): Cell aging in vivo and in vitro. Mech Aging Dev 98:1–25.

Ruley HE (1983): Adenovirus early region 1A enables viral and cellular transforming genes to transform primary cells in culture. Nature 304:602–606.

Ryan QC, Goonewardene IM, Murasko DM (1992): Extension of lifespan of human T lymphocytes by transfection with SV40 large T antigen. Exp Cell Res 199:387–391.

Sager R (1984): Resistance of human cells to oncogenic transformation. Cancer Cells 2:487–493.

Sager R (1991): Senescence as a mode of tumor suppression. Env Health Perspect 93:59–62.

Sakamoto K, Howard T, Ogryzko V, Xu NZ, Corsico CC, Jones DH, Howard B: (1993): Relative mitogenic activities of wild-type and retinoblastoma binding defective SV40 T antigens in serum deprived and senescent human fibroblasts. Oncogene 8:1887–1893.

Salk D, Bryant E, Hoehn H, Johnston P, Martin GM (1985): Growth characteristics of Werner syndrome cells in vitro. Adv Exp Med Biol 190:305–311.

Sandhu AK, Hubbard K, Kaur GP, Jha KK, Ozer HL, Athwal RS (1994): Senescence of immortal human fibroblasts by introduction of normal human chromosome 6. Proc Natl Acad Sci USA 91:5498–5502.

Sasaki M, Honda T, Yamada H, Wake N, Barrett JC, Oshimura M (1994): Evidence for multiple pathways to cellular senescence. Cancer Res 54:6090–6093.

Schneider EL, Mitsui Y (1976): The relationship between in vitro cellular aging and in vivo human aging. Proc Natl Acad Sci USA 73:3584–3588.

Serrano M, Lee H, Chin L, Cordon-Cardo C, Beach D, DePinho RA (1996): Role of the INK4A locus in tumor suppression and cell mortality. Cell 85:27–37.

Seshadri T, Campisi J (1990): c-fos repression and an altered genetic program in senescent human fibroblasts. Science 247:205–209.

Shay JW, Pereira-Smith OM, Wright WE (1991): A role for both Rb and p53 in the regulation of human cellular senescence. Exp Cell Res 196:33–39.

Shay JW, Van Der Haegen BA, Ying Y, Wright WE (1993): The frequency of immortalization of human fibroblasts and mammary epithelial cells transfected with SV40 large T antigen. Exp Cell Res 209:45–52.

Shay JW, Wright WE (1989): Quantitation of the frequency of immortalization of normal human diploid fibroblasts by SV40 large T antigen. Exp Cell Res 184:109–118.

Shay JW, Wright WR (1991): Defining the molecular mechanisms of human cell immortalization. Biochim Biophys Acta 1071:1–7.

Smith JR, Hayflick L (1974): Variation in the lifespan of clones derived from human diploid strains. J Cell Biol 62:48–53.

Smith JR, Whitney RG (1980): Intraclonal variation in proliferative potential of human diploid fibroblasts: Stochastic mechanism for cellular aging. Science 207:82–84.

Stanulis-Praeger B (1987): Cellular senescence revisited: A review. Mech Aging Dev 38:1–48.

Stein GH, Beeson M, Gordon L (1990): Failure to phosphorylate the retinoblastoma gene product in senescent human fibroblasts. Science 249:666–669.

Stein GH, Drullinger LF, Robetorye RS, Pereira-Smith OM, Smith JR (1991): Senescent cells fail to express cdc2, cycA and cycB in response to mitogen stimulation. Proc Natl Acad Sci USA 88:11012–11016.

Stein GH, Namba M, Corsaro CM (1985): Relationship of finite proliferative lifespan, senescence, and quiescence in human cells. J Cell Physiol 122:343–349.

Stein GH, Yanishevsky RM, Gordon L, Beeson M (1982): Carcinogen-transformed human cells are inhibited from entry into S phase by fusion to senescent cells but cells transformed by DNA tumor viruses overcome the inhibition. Proc Natl Acad Sci USA 79:5287–5291.

Sugawara O, Oshimura M, Koi M, Annab LA, Barrett JC (1990): Induction of cellular senescence in immortalized cells by human chromosome 1. Science 247:707–710.

Takahashi T, Schunkert H, Isoyama S, Wei JY, Nadal-Ginard B, Grossman W, Izumo S (1992): Age-related differences in the expression of protooncogene and contractile protein genes in response to pressure overload in the rat myocardium. J Clin Invest 89:939–946.

Tsuji Y, Ide T, Ishibani S (1983): Correlation between the presence of T antigen and the reinitiation of host DNA synthesis in senescent human diploid fibroblasts after SV40 infection. Exp Cell Res 144:165–169.

Uejima H, Mitsuya K, Kugoh H, Horikawa I, Oshimura M (1995): Normal human chromosome 2 induces cellular senescence in the human cervical carcinoma cell line SiHa. Genes Chromosomes Cancer 14:120–127.

Vaziri H, Schacter F, Uchida I (1993): Loss of telomeric DNA during aging of normal and trisomy 21 human lymphocytes. Am J Hum Genet 52:661–667.

Wang E, Lee MJ, Pandey S (1994): Control of fibroblast senescence and activation of programmed cell death. J Cell Biochem 54:432–439.

Wang XW, Lin X, Klein CB, Bhamra RK, Lee YW, Costa M (1992): A conserved region in human and Chinese hamster X chromosomes can induce cellular senescence of nickel-transformed Chinese hamster cell lines. Carcinogenesis 13:555–561.

Weinberg RA (1985): The action of oncogenes in the cytoplasm and nucleus. Science 230:770–774.

Weinberg RA (1991): Tumor suppressor genes. Science 254:1138–1146.

Weng NP, Levine BL, June CH, Hodes RJ (1996): Regulated expression of telomerase activity in human T lymphocyte development and activation. J Exp Med 183:2471–2479.

West MD, Pereira-Smith OM, Smith JR (1989): Replicative senescence of human skin fibroblasts correlates with a loss of regulation and overexpression of collagenase activity. Exp Cell Res 184:138–137.

White AE, Livanos EM, Tlsty TD (1994): Differential disruption of genomic integrity and cell cycle regulation in normal human fibroblasts by the HPV oncoproteins. Genes Dev 8:666–677.

Wick M, Burger C, Brusselbach S, Lucibello FC, Muller R (1994): A novel member of human tissue inhibitor of metalloproteinases (TIMP) gene family is regulated during G_1 progression, mitogenic stimulation, differentiation and senescence. J Cell Biochem 269:18953–18960.

Wright WE, Pereira-Smith OM, Shay JW (1989): Reversible cellular senescence: Implications for immortalization of normal human diploid fibroblasts. Mol Cell Biol 9:3088–3092.

Xiong Y, Hannon G, Zhang H, Casso D, Beach D (1993): p21 is a universal inhibitor of cyclin kinases. Nature 366:701–704.

Yamada H, Wake N, Fujumoto S, Barrett JC, Oshimura M (1990): Multiple chromosomes carrying tumor suppressor activity for a uterine endometrial carcinoma cell line identified by microcell-mediated chromosome transfer. Oncogene 5:1141–1147.

Yamamoto M, Yoshida M, Ono K, Fujita T, Ohtani-Fujita N, Sakai T, Nikaido T (1994): Effect of tumor suppressors on cell cycle-regulatory genes: RB suppresses p34/cdc2 expression and normal p53 suppresses cyclin A expression. Exp Cell Res 210:94–101.

Yu CE, Oshima J, Fu YH, Wijsman EM, Hisama F, Alisch R, Matthews S, Nakura J, Miki T, Ouais S, Martin GM, Mulligan J, Schellenberg GD (1996): Positional cloning of the Werner's syndrome gene. Science 272:258–262.

Zambetti G, Dell'Orco R, Stein G, Stein J (1987): Histone gene expression remains coupled to DNA synthesis during in vitro cellular senescence. Exp Cell Res 172:397–403.

CHAPTER 9

ANTISENSE STRATEGIES IN THE TREATMENT OF LEUKEMIA

BRUNO CALABRETTA and TOMASZ SKORSKI
Thomas Jefferson University, Department of Microbiology and Immunology, Kimmel Cancer Center, Philadelphia, PA 19107

INTRODUCTION

In the last 10–15 years, an extraordinary amount of information has accumulated linking various steps in the process of neoplastic transformation with very specific genetic alterations. This has generated novel conceptual approaches to the design of therapeutic strategies, and it is now widely recognized that effective cancer treatment can only be devised if the genetic and pathogenetic events underlying neoplastic transformation are fully understood.

To date, the best-characterized example of a malignant disease in which precise genetic alterations correlate with distinct histopathological and clinical stages is colorectal cancer (Vogelstein et al., 1988; Kinzler and Vogelstein, 1996). In such cancers, the frequency and order of appearance of genetic abnormalities are well defined, and there has been significant progress in understanding the pathogenetic role of the various genetic abnormalities. In human leukemias, the most common genetic abnormalities have also been dissected, and the pathogenetic significance of such abnormalities is under intense investigation.

Furthermore, whereas reliable animal models of solid tumors are scarce, existing animal models of human leukemia closely reproduce the disease process in humans. For example, in severe combined immunodeficiency (SCID) mice, injected leukemia cells first infiltrate hematopoietic tissues and then spread to nonhematopoietic organs (Kamel-Reid et al., 1989). This provides a relevant disease model in which to assess the effects of novel therapeutic strategies, including those involving oncogene targeting.

The Molecular Basis of Cell Cycle and Growth Control, Edited by G.S. Stein, R. Baserga, A. Giordano, and D.T. Denhardt
ISBN 0-471-15706-6, pages 374–384.

CHRONIC MYELOGENOUS LEUKEMIA: A PARADIGM FOR GENE-TARGETED THERAPIES

One of the best understood and widely investigated hematological malignancies, both at the genetic level and in its clinical manifestation, is chronic myelogenous leukemia (CML). This clonal disorder, which arises from neoplastic transformation of the hematopoietic stem cell, typically involves progression from a protracted chronic phase to an accelerated phase followed by a rapidly fatal blastic phase (Kantarjan et al., 1993).

In the chronic phase, the disease process is characterized by increased numbers of immature myeloid precursors in bone marrow, peripheral blood, spleen, and liver. Such precursors retain the ability to terminally differentiate so that the hallmark of the disease is a marked granulocytosis. Spleen and liver are also sites of extramedullary hematopoiesis, which contributes to the prominent hepatosplenomegaly typically found in CML patients. The transition from chronic phase to blast crisis is characterized by a dramatic increase in the number of blast cells in hematopoietic tissues. Such cells exhibit a proliferative advantage over normal cells and loss of differentiation potential. In blastic crisis, the therapeutic options are very limited and patients usually succumb within a few months.

Thus, therapies of CML are primarily directed to delaying as much as possible the onset of blastic transformation. Cytogenetically, CML is characterized by the t(9;22) translocation (Rowley, 1980) in which sequences of the c-Abl and the breakpoint cluster region (Bcr) genes on chromosomes 9 and 22, respectively, join to form the Bcr/Abl fusion genes of the Philadelphia chromosome (de Klein et al., 1982). Such genes encode the Bcr/Abl oncogenic tyrosine kinase (Konopka et al., 1984), which plays an essential role in the pathogenesis of CML (Daley et al., 1990; Heisterkamp et al., 1990).

The crucial role of the oncogenic Bcr/Abl protein in the pathogenesis of CML has been well demonstrated in several in vitro and in vivo studies showing that (1) normal marrow progenitor cells infected with retroviruses carrying Bcr/Abl chimeric constructs became growth factor–independent and less susceptible to undergo apoptosis (Gishizky and Witte 1992; Skorski et al., 1996b); (2) upon injection in syngeneic or immunodeficient mice, Bcr/Abl-infected cells induce a disease process reminiscent of that occurring in humans (Daley et al., 1990; Kelliher et al., 1990); and (3) Bcr/Abl transgenic mice develop hematological malignancies that resemble acute lymphocytic leukemia (Heisterkamp et al., 1990).

The Bcr/Abl oncogene belongs to the Src family of nonreceptor tyrosine kinases. The tyrosine kinase activity of the Bcr/Abl oncoprotein is essential in the process of transformation, because tyrosine kinase–deficient Bcr/Abl mutants do not transform hematopoietic cells (Lugo et al., 1990). This inability to transform hematopoietic cells correlates with lack of activation of secondary effectors such as ras, PI-3 kinase, jun kinase, and STATs (Cortez et al., 1995; Skorski et al., 1995b; Raitano et al., 1995; Carlesso et al., 1995). Moreover, wild-type Bcr/Abl, but not its tyrosine kinase–deficient mutant, protects hematopoietic cells from apoptosis induced by growth factor deprivation, in part by inducing the expression of the antiapoptotic gene bcl-2 (Sanchez-Garcia and Grutz, 1995), although the contribution of other mechanisms such as post-translational modification of Bcl-2 and/or of other members of the Bcl-2 family cannot be excluded. Thus, inhibition of Bcr/Abl expression and/or tyrosine kinase activity is expected to exert antileukemia effects in vitro and in vivo.

Tyrosine kinase inhibitors with apparent specificity for the Bcr/Abl kinase have recently been developed (Levitzki and Gazit, 1995; Amati et al., 1993). One such compound, CGP57148, induced a 92%–98% decrease in the ability of leukemic cells carrying the Bcr/Abl oncoprotein to form colonies in semisolid media (Druker et al., 1996). Of equal importance, such compounds had no apparent effect on colony formation from normal hematopoietic cells (Druker et al., 1996). Futhermore, CGP57148 demonstrated in vivo activity against Bcr/Abl–induced tumors (Druker et al., 1996).

The Bcr/Abl chimeric genes of the Philadelphia chromosome translocation generate Bcr/Abl chimeric transcripts that are uniquely suited for RNA-directed gene inhibition strategies involving ribozymes or "antisense" oligonucleotides (ODNs). The latter are short DNA sequences synthesized as exact reverse complements of the desired mRNA target's nucleotide sequences.

In theory, antisense DNA molecules can only hybridize with their mRNA targets via Watson-Crick base-pairing, a process that has an affinity and specificity many orders of magnitude higher than that achieved using traditional pharmacological approaches. Once the RNA–DNA duplex forms, translation of the message is prevented, and RNA degradation by RNase H is promoted. By directly targeting the gene product(s) pathogenetically involved in the disease process, the antisense strategy promises a degree of specificity in its mechanism of action that is generally not attainable with conventional anticancer chemotherapeutic agents or even with small molecular protein inhibitors.

Successful antisense ODN strategies in vivo require that the ODNs be (1) metabolically stable in a physiological environment; (2) able to bind the target RNA with high affinity and specificity; (3) efficiently taken up by target cells in tissues; and (4) pharmacokinetically and pharmacodynamically appropriate. Whereas no available ODN meets all of these criteria, the significant anticancer activity in animals treated with modified ODNs (especially phosphorothioate derivatives) has generated confidence that newly developed compounds will have more favorable properties.

TARGETING BCR/ABL mRNA IN A SCID MOUSE MODEL OF PHILADELPHIA[1] LEUKEMIA

When transplanted into SCID mice, CML blast crisis cells obtained directly from the bone marrow of patients show a pattern of leukemic

TABLE 9.1. Survival of Leukemic SCID mice Treated With Antisense ODN–Based Therapies

Treatment	Leukemia	Survival (Weeks)
Control oligonucleotides (ODNs)	BV173[1]	7–9
bcr/abl antisense ODNs	BV173	16–23
Control ODNs	CML-BC[2]	17–20
bcr/abl antisense ODNs	CML-BC	24–28
bcr/abl + c-myc antisense ODNs	BV173	26–38
bcr/abl + c-myc antisense ODNs	CML-BC	32–50
bcr/abl antisense ODNs and cyclophosphamide	CML-BC	38–60 (50%); >60 (50%)

[1]BV173 is a Philadelphia[1] cell line derived from a patient with lymphoid CML blast crisis.
[2]CML-BC are primary cells obtained from a patient with myeloid CML blast crisis.

spread reminiscent of that observed during the natural course of the disease in humans. Systemic treatment with nuclease-resistant 16- or 26-mer antisense phosphorothioate ODNs (1 mg/day for 9–12 days) targeting the Bcr-Abl chimeric transcript at the junction between the second exon of Bcr and the second exon of c-abl (b2/a2 antisense ODNs) led to a marked suppression of the disease process as indicated by analyses of different measures of leukemia burden. Specifically, the number of CALLA-positive cells, number of clonogenic leukemic cells in hematopoietic tissues, and levels of bcr/abl transcripts in hematopoietic and nonhematopoietic organs were all reduced by treatment with ODNs directed to the junction of the Bcr/Abl transcripts (Skorski et al., 1994, 1995a,b).

SCID mice injected with Philadelphia[1] cells and then treated with Bcr/Abl antisense ODNs evidenced only residual disease at 42 and 56 days after injection of leukemic cells. In contrast, untreated mice and mice treated with control ODNs had microscopic, macroscopic, and molecular evidence of active leukemia (Skorski et al., 1994, 1995a,b).

Consistent with the results of leukemia cell burden assays, survival of the antisense-treated mice was significantly prolonged compared with that of control mice (19.4 ± 1.4 weeks vs. 9.7 ± 0.9 weeks, $p < 0.001$, using BV173 cells; 23.3 ± 1.4 weeks vs. 15.5 ± 2.1, $p < 0.001$, using CML blast crisis primary cells) (Table 9.1).

TARGETING BCR/ABL mRNA AND A SECONDARY BCR/ABL DOWNSTREAM EFFECTOR IN LEUKEMIC SCID MICE

Although the disease process was temporarily suppressed and survival was prolonged in leukemic mice treated with Bcr/Abl antisense ODNs, mice invariably succumbed to their disease. This outcome may have reflected the inability to maintain a sufficient tissue concentration of ODNs after systemic injection to induce a sustained suppression of oncogenic expression sufficient for leukemic cell death and/or an intrinsic biological limitation of targeting a single oncogene.

In this regard, it is now becoming clear that the ability of the Bcr/Abl oncoprotein to transform hematopoietic cells rests in the activation of downstream effectors that transmit the oncogenic signal from the cytoplasm to the nucleus (Sawyers et al., 1995; Goga et al., 1995; Skorski et al., 1995b). One such effector is the protooncogene c-myc, which is required for Bcr/Abl or v-Abl transformation of hematopoietic cells (Sawyers et al., 1992). Thus, use of a combination of ODNs targeting two different components of the signal transduction pathway required for proliferation of CML cells might yield more profound cell killing in vitro and in vivo. In fact, results of analytical assays (immunofluorescence detection of leukemic cells in tissues, colony formation from mouse tissue cell suspensions, and detection of Bcr/Abl transcripts in mouse tissues) showed that the leukemic cell load in SCID mice systemically treated with the combination of Bcr/Abl and c-myc ODNs was reduced compared with that of mice treated with the individual ODNs (Skorski et al., 1995a).

The more potent effects of the ODN combination therapy were reflected in the different survival rates of the antisense ODN-treated mice. Median survival of the control mice was 15.5 ± 2.1 weeks; median survival of the mice treated with Bcr/Abl or c-myc antisense ODN was 23.0 ± 1.1 and 23.3 ± 1.4 weeks, respectively; mice treated with an equal dose of both antisense [S]ODNs survived significantly longer (32.5 ± 6.9 weeks; $p < 0.001$ compared with mice receiving a single ODN) (Table 9.1).

The reasons underlying the more potent antileukemia effects of the ODN combination therapy remain unclear. One possibility is that ODNs targeting a second oncogene involved in the disease process arrest the growth of cells that escape the proliferation inhibitory effect associated with individual gene targeting. It is also possible that downregulation of gene expression by a single antisense ODN at the relatively low concentration reached in vivo (Agrawal et al., 1991) might be insufficient to inhibit cell proliferation, whereas "partial" inhibition of two cooperating oncogenes might induce a longer lasting suppression of cell proliferation. The more potent antileukemia effects obtained by targeting two cooperating oncogenes is reminiscent of the effects induced by combination chemotherapy strategies in which different drugs are designed to target tumor cells as they progress through distinct stages of the cell cycle; however, the potential advantage of antioncogene therapy rests in its selectivity for disease-inducing agents and, in turn, its sparing of normal cells.

The therapeutic potential of the combination therapy approach can be further improved by optimal combinations of agents that inhibit gene expression at different stages of the disease process once the underlying pathogenic mechanisms are better known.

COMBINATION THERAPY OF LEUKEMIC SCID MICE USING CYCLOPHOSPHAMIDE AND BCR/ABL ANTISENSE ODN

Another strategy that might enhance the therapeutic potential of oncogene-targeted ODNs involves the use of these compounds in conjunction with conventional antineoplastic drugs. It is becoming clear that the therapeutic effects of most of these drugs depend not only on their interference with distinct aspects of tumor cell metabolism but also on their ability to modulate the levels of proteins involved in regulating the process of apoptosis. This raises the possibility of adjusting drug concentrations to limit the effects on metabolic processes (which are also toxic for normal cells) while preserving the apoptosis-inducing function.

CML cells are effectively killed in vitro when exposed to suboptimal concentrations of the cyclophosphamide in vitro–active derivative mafosfamide and Bcr/Abl antisense ODNs (Skorski et al., 1993). Such improvement over the effects of either mafosfamide or antisense ODNs alone correlated with an enhanced susceptibility of leukemic cells to undergo apoptosis when exposed to increasing concentrations of mafosfamide followed by treatment with a constant amount of Bcr/Abl ODNs (Skorski, et al., 1997). At the highest concentration of mafosfamide (100 μg/ml), almost all cells underwent apoptosis; however, in cultures exposed to lower concentrations of the drug, only subsequent treatment with b2/a2 antisense ODNs dramatically increased the proportion of apoptotic cells. Treatment with mafosfamide induced an increase in the levels of the apoptosis-associated p53 protein and of Bax, a member of the Bcl-2 family, which is positively regulated by p53 (Oltvai et al., 1993; Miyashita and Reed, 1995). In cultures treated with b2/a2 antisense ODNs, the most marked effect was the suppression of Bcl-2 expression, whereas p53 and Bax levels were essentially unchanged. Overall, cultures treated with mafosfamide and b2/a2 antisense ODNs exhibited increased expression of positive effectors (p53 and Bax) and decreased expression of a negative regulator (Bcl-2) in the pathway leading to apoptotic cell death (Skorski et al., 1997).

In leukemic SCID mice treated with a single injection of cyclophosphamide (25 mg/kg) followed by multiple injections of antisense ODNs (1 mg/day for 12 consecutive days), the disease process was slowed (as revealed by histopathological analysis and by detection of leukemic cells in peripheral blood) to a much greater extent than that induced by treatment with either agent alone. Survival of the mice in the various treatment groups was in good agreement with the effects on the disease process.

Mice treated with bcr/abl antisense ODNs or with a single injection of cyclophosphamide had a longer survival of control mice (median survival of 22 ± 2.5 weeks vs. 14 ± 1 week) (Table 9.1). However, mice treated with bcr/abl antisense ODNs plus cyclophosphamide showed the most favorable therapeutic response; 50% were healthy more than 60 weeks after the injection of leukemic cells, whereas the remaining 50% died of leukemia after 36–48 weeks (median survival 39 ± weeks; $p <$

0.001 compared with survival of mice treated with cyclosphosphamide or Bcr/Abl antisense ODNs alone) (Table 9.1). Consistent with the increased susceptibility of leukemic cells to apoptosis after in vitro treatment with mafosfamide and bcr/abl antisense ODNs, the improved therapeutic response of mice to treatment with cyclophosphamide and bcr/abl ODNs likely reflects an enhanced propensity of leukemic cells for apoptosis in vivo.

CONCLUSIONS

CML can serve as a paradigm of the types of gene-directed strategies that can be developed based on an understanding of the pathogenetic mechanisms involved in the disease.

By virtue of its transforming properties when introduced into hematopoietic cells, the Bcr/Abl oncoprotein plays a causative role in the deregulated growth of hematopoietic progenitor cells that characterizes CML. Moreover, the Philadelphia chromosome, which generates Bcr/Abl chimeric genes, is the only genetic abnormality detected during the protracted chronic phase of the disease, providing further evidence of the importance of the Bcr/Abl oncoprotein in sustaining the initial phase of the disease.

The Bcr/Abl oncoprotein also plays an essential role during the more aggressive blast crisis, as indicated by the inhibition of proliferation and semisolid media colony formation of CML blast crisis cells using antisense strategies (antisense ODNs or plasmids expressing Bcr/Abl sequences in the antisense orientation). Furthermore, a second Philadelphia chromosome can be the sole additional genetic abnormality in a cohort of CML blast crisis patients. Together, the data indicate that the use of ODNs directed to the chimeric leukemia-specific Bcr/Abl transcripts or the use of inhibitors of the Bcr/Abl-encoded tyrosine kinase activity to perturb Bcr/Abl function is a rational strategy for the treatment of CML. Indeed, such approaches have induced suppression of Philadelphia[1] proliferation both in vitro and in vivo (Druker et al., 1996; Skorski et al., 1994, 1995a, 1996a).

In the case of ODN targeting, the antileukemia effects were only temporary (Skorski et al., 1994, 1995a, 1996a). ODN combinations or combination therapies with conventional anticancer drugs improved the results achieved using individual ODNs directed to Bcr/Abl transcripts. In the case of ODN combination therapies, the rationale for using two or more oncogene-targeted ODNs rests in the recognition that transformation of hematopoietic cells by the Bcr/Abl oncoprotein is accompanied by activation of several downstream effectors that transduce Bcr/Abl oncogenic signals. Thus, a more effective therapeutic strategy consists in targeting different components of the signal transduction apparatus activated by Bcr/Abl. In SCID mouse experiments, the targeting of c-myc as a Bcr/Abl secondary downstream effector has proved to be a therapeutically appropriate choice; however, additional targets can be selected on the basis of the current understanding of the role of Bcr/Abl-activated pathways in Bcr/Abl leukemogenesis. For example, it has been suggested that cyclin D1 cooperates with Bcr/Abl in leukemogenesis, based on its ability to rescue the transformation-deficient phenotype of the SH2 domain Bcr/Abl mutants (Afar, et al., 1995). Other downstream effectors such as c-jun are apparently required for Bcr/Abl transformation of rodent fibroblasts (Raitano et al., 1995) and may also be involved in transformation of hematopoietic cells. If jun expression proves necessary for the proliferation of Philadelphia[1] cells, it might be fruitful to simultaneously target c-myc and c-jun in addition to Bcr/Abl. Like c-myc, c-jun has a short half-life, thus making its transcript an ideal target for antisense-based therapies.

An attractive possibility in antisense-based therapeutic strategies is the use of oncogene-targeted ODNs together with conventional anticancer chemotherapeutic agents. It is becoming clear that chemotherapeutic agents with DNA-damaging properties exert their effects not only via inhibition of specific enzymatic pathways necessary for RNA or DNA synthesis but also by their ability to induce apoptosis.

A feature of CML is the reduced susceptibility of myeloid precursor cells to apoptotic processes upon growth factor deprivation (Bedi et al., 1994). This is consistent with the ability of Bcr/Abl to protect growth factor–dependent myeloid precursor cell lines from apoptosis induced by growth factor deprivation and is apparently mediated by Bcr/Abl regulation of Bcl-2 levels (Sanchez-Garcia and Grutz, 1995).

The combined use of agents such as cyclophosphamide and Bcr/Abl antisense ODNs that alter the ratio of the apoptotic regulators so that levels of the death-inducing Bax exceed those of the apoptosis-inhibitor Bcl-2 provides the opportunity to modulate the genetic program that regulates cell survival and death. This quasiphysiological modulation of apoptosis by low doses of an anticancer agent and ODNs directed to a leukemia-specific gene is designed to spare a higher number of normal cells relative to leukemic cells, thus attaining the degree of selectivity expected from gene-targeted therapies of human malignancies.

The susceptibility of CML cells to undergo apoptosis after treatment with mafosfamide is, in part, due to the effects on p53. Cells engineered to express an oncogene or cells naturally carrying one (such as the Bcr/Abl oncogene of CML blast crisis cells) are more prone to undergo apoptosis in vitro and in vivo if treated with DNA-damaging agents and if they carry a wild-type p53 gene (Lowe et al., 1993, 1994).

However, structural alterations (deletions, rearrangements, or point mutations) of p53 are detected in 25%–30% of CML blast crisis cells (Feinstein et al., 1991) and in more than 50% of all cancers (Harris, 1996). Thus, to exploit the apoptosis-inducing function of conventional chemotherapeutic agents, strategies must be devised to convert mutant p53 into a protein that mimicks the effects of the wild-type protein.

Most of the p53 mutations in human neoplasias involve the segment from amino acid 102 to 292, which is required for sequence-specific DNA binding. The most common mutation in this region such as those involving R248 and R273 result in defective contact with the DNA and loss of the ability of p53 to act as a transcription factor (Harris, 1996). The specific DNA-binding function of p53 appears to be allosterically regulated, at least in part, by the C-terminal region located between amino acid residues 363 and 393 (Hupp et al., 1992). Interestingly, the monoclonal antibody Pab421, whose epitope maps to amino acids 370–378, was shown to restore DNA-binding activity to a subset of p53 mutants (Hupp et al., 1995) and, most importantly, to restore the transcription activation function of mutant p53 in cancer cells (Abarzua et al., 1995).

On the basis of the results of these experiments, a chemically modified peptide corresponding to residues 369–382 of p53 was shown to restore the transcription activation function of the endogenous p53-273^{his} mutant present in SW480 colon cancer cells (Abarzua et al., 1996). Thus, the development of small molecules that allosterically alter DNA binding by mutant p53 might restore drug-sensitivity to drug-resistant tumor cells carrying p53 mutations.

Alternatively, chemotherapeutic agents to be used in anticancer therapies could be selected based on their ability to activate p53 downstream effectors. For example, Bax is a p53-regulated gene with apoptosis-inducing functions; thus, drugs that induce Bax with a p53-independent mechanism might prove therapeutically useful in the treatment of cancers with mutant p53.

Our studies of combination therapies with bcr/abl antisense ODNs and cyclophosphamide showed that suppression of Bcr/Abl expression induces Bcl-2 downregulation because of the functional link between these two genes (Sanchez-Garcia and Grutz, 1995). Thus, strategies directed to inhibiting Bcl-2 expression and/or function might decrease the threshold at which anticancer chemotherapeutic agents induce cell killing. In this regard, antisense ODNs directed to Bcl-2 mRNA have been shown to increase the susceptibility of tumor cell lines to killing by chemotherapeutic agents (Kitada et al., 1994; Keith et al., 1995). Furthermore, clinical trials are now being conducted involving the use of bcl-2 antisense ODNs to treat malignant lymphomas.

In summary, understanding the genetic mechanisms underlying the growth advantage and susceptibility to apoptosis of tumor cells raises the possibility of developing rational therapies that will ultimately improve the outcome of anticancer treatment.

REFERENCES

Abarzua et al. (1995): Microinjection of monoclonal antibody PAb421 into human SW480 colorectal carcinoma cells restores the transcription activation function to mutant p53. Cancer Res 55:3490–3494.

Abarzua et al. (1996): Restoration of the transcription activation function to mutant p53 in human cancer cells. Oncogene 13:2477–2482.

Afar et al. (1995): Signaling by ABL oncogenes through cyclin D1. Proc Natl Acad Sci USA 12:9540–9544.

Agrawal et al. (1991): Pharmacokinetics, biodistribution, and stability of oligodeoxynucleotide phosphorothioates in mice. Proc Natl Acad Sci USA 88:7595–7599.

Amati et al. (1993): Tyrphostin-induced inhibition of $p210^{brc/abl}$ tyrosine kinase activity induces K562 to differentiate. Blood 82:3524–3528.

Bedi et al. (1994): Inhibition of apoptosis by BCR/ABL in chronic myeloid leukemia. Blood 83:1179–1187.

Carlesso et al. (1995): Tyrosyl phosphorylation and DNA binding activity of signal transducers and activators of transcription (STAT) proteins in hematopoietic cells transformed by bcr/abl. J Exp Med 183:811–820.

Cortez et al. (1995): Structural and signaling requirements for bcr/abl-mediated transformation and inhibition of apoptosis. Mol Cell Biol 15:5531–5541.

Daley et al. (1990): Induction of chronic myelogenous leukemia in mice by the $p210^{bcr/abl}$ gene of the Philadelphia chromosome. Science 247:824–830.

de Klein et al. (1982): A cellular oncogene is translocated to the Philadelphia chromosome in chronic myelocytic leukemia. Nature 300:765–767.

Druker et al. (1996): Effects of a selective inhibitor of the abl tyrosine kinase on the growth of BCR/ABL positive cells. Nature Med 2:561–566.

Feinstein et al. (1991): p53 in chronic myelogenous leukemia in acute phase. Proc Natl Acad Sci USA 88:6293–6297.

Gishizky ML, Witte ON (1992): Initiation of deregulated growth of multipotent cells by bcr-abl in vitro. Science 256:836–839.

Goga et al. (1995): Alternative signals to RAS for hematopoietic transformation by the BCR/ABL oncogene. Cell 82:981–984.

Harris CC (1996): Structure and function of the p53 tumor suppressor gene: Clues for rational cancer therapeutic strategies. J Natl Cancer Inst 88:1442–1455.

Heisterkamp et al. (1990): Acute leukemia in bcr-abl transgenic mice. Nature 344:251–253.

Hupp et al. (1992): Regulation of the specific DNA binding function of p53. Cell 71:875–886.

Hupp et al. (1995): Small peptides activate the latent sequence-specific DNA binding function of p53. Cell 83:237–245.

Kamel-Reid et al. (1989): A model of human acute lymphoblastic leukemia in immunodeficient SCID mice. Science 246:1597–1600.

Kantarjan et al. (1993): Chronic myelogenous leukemia: A concise update. Blood 82:691–703.

Keith et al. (1995): Inhibition of bcl-2 with antisense oligonucleotides induces apoptosis and increases the sensitivity of AML blasts to Ara-C. Leukemia 9:131–138.

Kelliher et al. (1990): Induction of a chronic myelogenous leukemia-like syndrome in mice with v-abl and bcr/abl. Proc Natl Acad Sci USA 87:6649–6693.

Kinzler KW, Vogelstein B (1996): Lessons from hereditary colorectal cancer. 87:159–170.

Kitada et al. (1994): Reversal of chemoresistance of lymphoma cells by antisense-mediated reduction of bcl-2 gene expression. Antisense Res Dev 4:71–78.

Konopka et al. (1984): An alteration of the human c-abl protein in K562 leukemia cells unmasks associated tyrosine kinase activity. Cell 37:1035–1042.

Levitzki A, Gazit A (1995): Tyrosine kinase inhibition: An approach to drug development. Science 267:1782–1788.

Lowe et al. (1993): p53-dependent apoptosis modulates the cytotoxicity of anticancer agents. Cell 74:959–967.

Lowe et al. (1994): p53 status and the efficacy of cancer therapy in vivo. Science 266:807–810.

Lugo et al. (1990): Tyrosine kinase activity and transformation potency of bcr/abl oncogene products. Science 2207:1079–1084.

Miyashita T, Reed JC (1995): Tumor suppressor p53 is a direct transcriptional activator of the human bax gene. Cell 80:293–299.

Oltvai et al. (1993): Bcl-2 heterodimerizes in vivo with a conserved homolog, bax, that accelerates programmed cell death. Cell 74:609–619.

Raitano et al. (1995): The bcr/abl leukemic oncogene activates JUN kinase and requires JUN for transformation. Proc Natl Acad Sci USA 92:11746–11750.

Rowley JD (1980): Chromosome abnormalities in human leukemia. Annu Rev Genet 14:17–48.

Sanchez-Garcia I, Grutz G (1995): Tumorigenic activity of the BCR/ABL oncogenes is mediated by bcl-2. Proc Natl Acad Sci USA 92:5287–5291.

Sawyers et al. (1992): Dominant negative myc blocks transformation of ABL oncogenes. Cell 70:901–910.

Sawyers et al. (1995): Genetic requirement for RAS in the transformation of fibroblasts and hematopoietic cells by the BCR/ABL oncogene. J Exp Med 181:307–313.

Skorski et al. (1993): Highly efficient eliminators of Philadelphia[1] leukemic cells by exposure to bcr/abl antisense oligodeoxynucleotides combined with mafosfamide. J Clin Invest 92:194–202.

Skorski et al. (1994): Suppression of Philadelphia[1] leukemia cell growth in mice by BCR/ABL antisense oligodeoxynucleotides. Proc Natl Acad Sci USA 91:4504–4508.

Skorski et al. (1995a): Leukemia treatment in severe combined immunodeficiency mice by antisense oligodeoxynucleotides targeting cooperating oncogenes. J Exp Med 182:1645–1653.

Skorski et al. (1995b): Phosphatidylinositol-3 kinase activity is regulated by BCR/ABL and is required for the growth of Philadelphia chromosome positive cells. Blood 86:726–736.

Skorski et al. (1996a): Antisense oligodeoxynucleotide combination therapy of primary chronic myelogenous leukemic blast crisis in SCID mice. Blood 88:1005–1012.

Skorski et al. (1996b): Blastic transformation of p53-deficient bone marrow cells by $p210^{bcr/abl}$ tyrosine kinase. Proc Natl Acad Sci USA 93:13137–13142.

Skorski et al. (1997a): Treatment of Philadelphia[1] leukemia in severe combined immunodependent mice by combination of cyclophosphamide and bcr/abl antisense oligodeoxynucleotides. J Natl Cancer Inst 89:124–133.

Vogelstein et al. (1988): Genetic alterations during colorectal tumor development. N Engl J Med 319:525–532.

INDEX